The Flagellates

The Systematics Association Special Volume Series

Series Editor

Alan Warren
Department of Zoology, The Natural History Museum,
Cromwell Road, London, SW7 5BD, UK.

The Systematics Association provides a forum for discussing systematic problems and integrating new information from genetics, ecology and other specific fields into taxonomic concepts and activities. It has achieved great success since the Association was founded in 1937 by promoting major meetings covering all areas of biology and palaeontology, supporting systematic research and training courses through the award of grants, production of a membership newsletter and publication of review volumes by its publishers Taylor & Francis. Its membership is open to both amateurs and professional scientists in all branches of biology who are entitled to purchase its volumes at a discounted price.

The first of the Systematics Association's publications, *The New Systematics*, edited by its then president Sir Julian Huxley, was a classic work. Over fifty volumes have now been published in the Association's 'Special Volume' series often in rapidly expanding areas of science where a modern synthesis is required. Its *modus operandi* is to encourage leading exponents to organize a symposium with a view to publishing a multi-authored volume in its series based upon the meeting. The Association also publishes volumes that are not linked to meetings in its 'Volume' series.

Anyone wishing to know more about the Systematics Association and its volume series is invited to contact the series editor.

Forthcoming titles in the series:

The Changing Wildlife of Great Britain and Ireland
Edited by D. Hawksworth

Other Systematics Association publications are listed after the index for this volume.

The Systematics Association Special Volume Series 59

The Flagellates

Unity, diversity and evolution

Edited by Barry S. C. Leadbeater
and J. C. Green

LONDON AND NEW YORK

First published 2000 by Routledge

2 Park Square, Milton Park, Abingdon, Oxfordshire OX14 4RN
52 Vanderbilt Avenue, New York, NY 10017

Routledge is an imprint of the Taylor & Francis Group, an informa business

First issued in paperback 2019

Typeset in Sabon by
Curran Publishing Services Ltd.

British Library Cataloguing in Publication Data
A catalogue record for this book is available from the British Library

Library of Congress Cataloging-in-Publication Data
A catalog record for this book has been requested

ISBN 978-0-7484-0914-3 (hbk)
ISBN 978-0-367-39850-7 (pbk)

Contents

Contributors

Arndt, H., Zoologisches Institut (Ökologie), University of Cologne, Weyertal 119, D-50923 Köln (Cologne), Germany.

Auer, B., Zoologisches Institut, University of Cologne, Weyertal 119, D-50923, Köln (Cologne), Germany.

Becker, B., Botanisches Institut, University of Cologne, Gyrhofstrasse 15, D-50931 Köln (Cologne), Germany.

Brugerolle, G., Laboratoire de biologie des Protistes, UPRES A 6203, Université Blaise Pascal de Clermond-Ferrand, 63177 Aubière, France.

Cavalier-Smith, T., Department of Zoology, University of Oxford, South Parks Road, Oxford, OX1 3PS, UK.

Cleven, E-J., Zoological Institute, University of Cologne, D-50923, Köln (Cologne), Germany.

Darley, W. Marshall, Department of Botany, University of Georgia, Athens, GA 30602, USA.

Dietrich, D., Zoological Institute, University of Cologne, D-50923, Köln (Cologne), Germany.

Edvardsen, B., University of Oslo, Department of Biology, PO Box 1069, Blindern, N-0316, Oslo, Norway.

Farmer, M. A., Department of Cellular Biology, Biological Sciences, University of Georgia, Athens, GA 30602, USA.

Gräfenhan, T., Zoological Institute, University of Cologne, D-50923, Köln (Cologne), Germany.

Green, J. C., Electron Microscope Unit, University of Plymouth, Plymouth, PL4 8AA, UK.

Hart, P. E., Tumor Biology Program, Mayo Clinic/Foundation, Rochester, MN 55905, USA.

Holwill, M. E. J., Department of Physics, King's College, University of London, The Strand, London WC2R 2LS, UK.

Karpov, S. A., Biological Research Institute, State University of St Petersburg, Oranienbaumskoye Sch. 2, Stary Peterhof, 198904 St Petersburg, Russia.

Kawai, H., Department of Biology, Faculty of Science, Kobe University, Rokkodai, Nadaku, Kobe 657, Japan.

Kreimer, G., Botanisches Institut, University of Cologne, Gyrhofstr. 15, D-50931, Köln (Cologne), Germany.

Lange, M., Alfred-Wegener-Institute, Am Handelshafen 12, D-27515 Bremerhaven, Germany.

Larsen, A., University of Bergen, Department of Fisheries and Marine Biology, Bergen High Technology Center, N-5020, Bergen, Norway.

Laybourn-Parry, J., Department of Physiology, University of Nottingham, Sutton Bonington, Loughborough, LE12 5RD, UK.

Leadbeater, B. S. C., School of Biosciences, University of Birmingham, Edgbaston, Birmingham, B15 2TT, UK.

Lee, W. J., School of Biological Sciences, University of Sydney, New South Wales 2006, Australia.

McCready, S. M. M., School of Biosciences, University of Birmingham, Edgbaston, Birmingham, B15 2TT, UK.

Medlin, L. K., Alfred-Wegener-Institute, Am Handelshafen 12, D-27515 Bremerhaven, Germany.

Moestrup, Ø., Department of Phycology, Botanical Institute, University of Copenhagen, Øster Farimagsgade 2D, DK-1353 Copenhagen, Denmark.

Müller, M., Rockefeller University, 1230 York Avenue, New York, NY 10021, USA.

Mylnikov, A. P., Institute for the Biology of Inland Waters, Russian Academy of Sciences, Borok, Yaroslavskaya obl., 152 742, Russia.

Parry, J., Institute of Environmental and Natural Sciences, University of Lancaster, Lancaster, LA1 4YQ, UK.

Patterson, D. J., School of Biological Sciences, Zoology Building A08, University of Sydney, New South Wales 2006, Australia.

Raven, J. A., Department of Biological Sciences, University of Dundee, Dundee, DD1 4HN, Scotland.

Salisbury, J. L., Tumor Biology Program, Department of Biochemistry and Molecular Biology, Mayo Clinic/Foundation,Rochester, MN 55905, USA.

Sleigh, M. A., Department of Biology, Medical and Biological Sciences Building, University of Southampton, Bassett Crescent East, Southampton, SO16 7PX, UK.

Taylor, H. C., Physics Department, King's College London, Strand, London, WC2R 2LS, UK.

Vickerman, K., Division of Environmental and Evolutionary Biology, Institute of Biomedical and Life Sciences, University of Glasgow, Glasgow, G12 8QQ, Scotland.

Weitere, M., Zoologisches Institut (Ökologie), University of Cologne, Weyertal 119, D-50923 Köln (Cologne),Germany.

Preface

Ever since the early observations of Antony van Leeuwenhoek, biologists have been fascinated by the *protozoa*, including the flagellates. For many years, the protozoa were treated for systematic and taxonomic purposes as a unified major animal group, but with increasing knowledge, the introduction of new techniques in microscopy, and recently the development of molecular genetics, the artificiality of this treatment has been fully appreciated, and the flagellates are now regarded as a paraphyletic group.

The justification for a volume such as this on *The Flagellates* may therefore be called into question, but we would argue that in spite of their diverse origins all flagellates have much in common, and the study of flagellate cells of various kinds has contributed much to our understanding of cell structure and function in general. Thus, the evolution of the flagellum (or cilium) has been inextricably linked with, for example, the development of the eukaryotic cell and its requirement for movement. This is true whether it is an individual monadoid species, a reproductive or dispersal stage within a lifecycle, or a specialized cell of a multicellular organism. Furthermore, flagellar basal and root systems have been intimately associated with the evolution of mitosis in the eukaryotic cell.

As well as having diverse evolutionary histories, flagellates are ubiquitous and occupy an infinite variety of habitats. They may, for example, be free-living and autotrophic, or heterotrophic; symbiotic or parasitic. However, in spite of their great diversity, and whatever their particular habit or niche, they are united by the 'flagellate condition'.

This book sets out to examine flagellates from a multidisciplinary standpoint. Of primary concern are the unifying structures, mechanisms and processes involved in flagellate biology. We begin with a review of the complex history of flagellate studies from the first use of microscopes in the mid-seventeenth century to the present, and this is followed by a series of chapters on common aspects of flagellates. These include a discussion of the problems inherent in being a flagellate, and reviews of the structure and function of the flagellum itself, the cytoskeleton, surface structures and sensory mechanisms.

The diversity of flagellates is recognized in the next series of chapters, which include reviews of trophic strategies of both free-living and parasitic groups, and contributions on ecology, biogeography and population genetics. The final chapters of the book are concerned with the occurrence and loss of organelles, and other aspects of flagellate evolution and phylogeny.

It should be noted that the emphasis of this volume is on flagellates rather than on flagellates and ciliates. Flagellates and ciliates do have much in common, and many of the features associated with the flagellate condition are also attributable to ciliates. However, the latter have other properties peculiar to themselves, and the editors felt that it was not possible within the scope of the present work to do justice to both groups of protists.

While it is not a symposial volume, this book was a joint venture with a symposium on 'The Flagellates' held at the University of Birmingham (UK) in September 1998. We are very pleased, therefore, to acknowledge the invaluable support received from the University of Birmingham, the Systematics Association, the British Section of the Society of Protozoologists, the British Phycological Society and the Annals of Botany Company. We are also very grateful to the referees for their time and expertise in reviewing the various contributions.

Barry S. C. Leadbeater
Birmingham
January 2000

J. C. Green
Plymouth
January 2000

Chapter 1

The flagellates

Historical perspectives

Barry S. C. Leadbeater and Sharon M. M. McCready

ABSTRACT

Although references to flagellates *en masse* are apparently recorded in biblical and other early texts, the earliest authenticated observations of individual flagellates using a microscope are attributed to Antony van Leeuwenhoek during the second half of the seventeenth century. Leeuwenhoek is credited with seeing, for the first time, an impressive range of free-living and parasitic species. Otto Friderich Müller, in two major works dated 1773 and 1786, is one of the first authors to record flagellates with formal (latinized) descriptions using the system of nomenclature devised by his Scandinavian contemporary Carl von Linnaeus. Many of the taxa that Müller described are still recognized today. The nineteenth century was a time of great expansion in science and witnessed the publication of major works on the protozoa in general, and the flagellates in particular. The monographs of Christian Gottfried Ehrenberg, Félix Dujardin, Fredrich Ritter von Stein, William Saville Kent and Otto Bütschli are of particular note. By the end of the nineteenth century an ordered systematics of the protozoa was established and this remained in place, with a few modifications, for the better part of a century. However, from the earliest attempts to devise such a system for the protozoa, our understanding of the evolutionary relationships between the many free-living and parasitic flagellates has remained problematical because of the exceptional degree of paraphyly displayed by these organisms.

The advent of electron microscopy, particularly from the 1960s onwards, has provided valuable information which has helped to clarify systematic relationships among the protists. However, it has been the contribution of molecular biology and, in particular, the information gathered from sequencing of genes or gene products, that has led to a major reappraisal of flagellate systematics. At the same time as these advances have been made, and partly as a result of them, evidence has accumulated which lends support to the hypothesis that organelles, such as chloroplasts and mitochondria, have been acquired by endosymbiosis. At the time of writing, flagellate systematics and our understanding of evolutionary relationships remain in a state of flux. However, there is now a general feeling of confidence that the combination of molecular studies supplemented by electron microscopy will continue to provide further new and valuable insights into flagellate systematics for the future.

1.1 Early history of flagellate studies

Excluding biblical and other early literary references to what are probably flagellates *en masse*, the history of the study of flagellates spans a period of about 350 years. It is important to remember that these years have seen vast changes in the technology of science, in our understanding of life around us and, in particular, in evolutionary theory. In the mid-seventeenth century, spontaneous generation was still in fashion; it was just under eighty years until Carl von Linneaus (1735) developed a standard system of nomenclature and nearly 200 years before Charles Darwin (1859) published the *The Origin of Species*. The funding of science has changed dramatically during this period of time. It is only in the last century that science has been funded from public expenditure; from the mid-seventeenth to mid-nineteenth centuries scientific study was the pursuit of the aristocracy, philanthropic entrepreneurs, or amateurs who had access to instruments such as microscopes and, perhaps more important, the time available for such work.

The history of flagellate studies, as for the protists in general, has been intimately linked to the development of microscopy and, without exception, major developments in microscopes have led to important advances in our understanding of the protists. By general consent, Antony van Leeuwenhoek (1632–1723) is considered to be the father of protozoology (Figure 1.2). In some ways it is surprising that a merchant draper from Delft, Holland, who as a pastime developed a hand-held microscope, an unusual instrument even by mid-seventeenth century standards, should have been so perceptive as to have observed so many firsts in the fields of protistology and microbiology. His keen powers of observation and his commitment to detail, so carefully recorded in his letters to the Royal Society of London and elsewhere, make his contribution of the first rank. Dobell (1932), who has faithfully pieced together the life and works of van Leeuwenhoek, has attempted taxonomic identification of many of his illustrations and descriptions for which there were no formal (latinized) names.

Leeuwenhoek's observations on the free-living protozoa probably began with his discovery of certain *'very little animalcules'* seen in freshwater in 1674. In a letter to the Secretary of the Royal Society of London he referred to:

> These animalcules had divers colours . . . others again were green in the middle, and before and behind white. . . . And the motion of most of these animalcules was so swift and so various, upwards, downwards, and round about, that 'twas wonderful to see: I judge that some of these little creatures were above a thousand times smaller than the smallest ones I have ever yet seen.
>
> (Leeuwenhoek, 1674)

Dobell (1932) interprets these green and white animalcules as being *Euglena viridis*.

In a subsequent epistle to the Royal Society dated 9 October 1676, famously known as the *Letter on the Protozoa*, Leeuwenhoek (1677) wrote at length about animalcules that he observed in standing rainwater, sea water, pepper infusions and vinegar. Among the flagellates, Dobell (1932) considers that Leeuwenhoek observed *Monas* sp., probably *M. vulgata*, and *Bodo* sp. Later, Leeuwenhoek (1700) recorded in detail a description of *Volvox* colonies (Figure 1.1) and their reproduction. In

1702 Leeuwenhoek provided descriptions of what Dobell (1932) considers to be *Chlamydomonas* and *Haematococcus* with probably the first description of flagella:

> their bodies seemed to be composed of particles that presented an oval figure; therewithal they had short thin instruments which stuck out a little way from the round contour, and wherewith they performed the motions of rolling round and going forward.
>
> (Leeuwenhoek, 1702)

Other seventeenth century microscopists who probably observed flagellates include Christiaan Huygens, who in 1678 described *Astasia,* and John Harris, who in 1696 rediscovered *Euglena* (see Dobell, 1932).

By the beginning of the eighteenth century, minute animalcules were closely associated with infusions of one sort or another which led to the introduction of the popular term *infusion animals* by Ledermuller (1763). This was soon formalized to the *Infusoria* by Wrisberg (1765). The first separate treatise on microscopic life was by Joblot (1718). In the mid-eighteenth century, systematics and protozoology were dominated respectively by two Scandinavian contemporaries, Carl von Linnaeus (1707–1778) and Otto Fredrich Müller (1730–1784). Linnaeus, a Swedish botanist, devoted most of his efforts to the systematics of plants and higher animals and it was not until the tenth edition of *Systema Naturae* (Linnaeus, 1758), the starting date for zoological nomenclature, that the flagellate *Volvox* was recorded. The contribution of Otto Friderich Müller (Figure 1.3), a Danish marine invertebrate zoologist, to the taxonomy of protozoa is seminal. He observed and described many Infusoria giving them distinctive generic and specific titles in line with Linnaeus's then newly introduced binomial system of nomenclature. In *Vermium Terrestrium et Fluviatilium* (Müller, 1773), he was the first to recognize dinoflagellates (*Bursaria* (= *Ceratium*) *hirundinella* and *Vorticella* (= *Peridinium*) *cincta*) which he included in the *Animalium Infusonium.* In his later, posthumous work, *Animalcula Infusoria Fluviatilia et Marina* (Müller, 1786), he included within the Order *Infusoria*

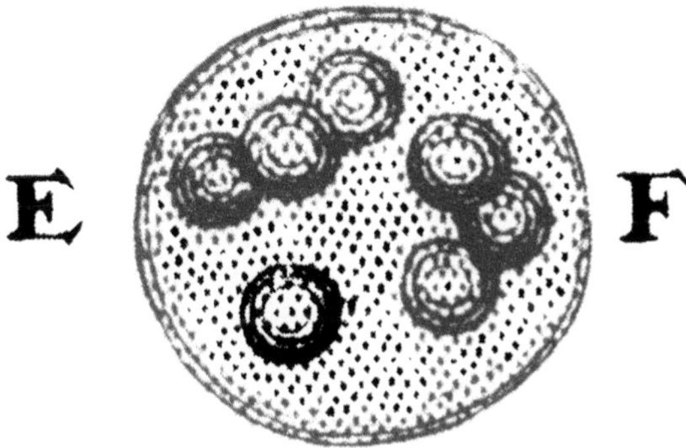

Figure 1.1 A *Volvox* colony containing seven daughter colonies
Source: Reproduced from Fig. 2 of Leeuwenhoek's letter 122 to Sir Hans Sloane, 2 January 1700. *Philosophical Transactions of the Royal Society*, **22** (261), 509

2
3
4
5
6
7

descriptions of many ciliates, flagellates and heliozoa that are still recognized today. In particular, *Gonium* and *Monas* are flagellate genera originally named by Müller (1786). In addition to Protozoa as now construed, Muller's *Infusoria* also included bacteria, diatoms, vinegar worms, planarian worms, cercaria larvae, rotifers and other members of the subsequently instituted Metazoa.

The nineteenth century was to see enormous advances in all branches of science. It was also a time when many basic theories were established: in biology two of the most important were the cell theory (Schleiden, 1838; Schwann, 1839) and evolutionary theory, with the publication in 1859 of Darwin's *The Origin of Species*. The contribution of these theories was to bring together disparate observations in a unifying and secure framework. However, at the time, matters were not plain sailing and throughout the century major new insights were often bedevilled by countervailing ideas based on flawed or inaccurate observations. Nevertheless, by the end of the century the foundations of protozoology, that were set to dominate much of the twentieth century, were securely established.

At the beginning of the nineteenth century the Infusoria comprised a vast array of organisms spanning in size and complexity a range from bacteria to small invertebrates including worms and crustacea. In 1817 Goldfuss (1817) (Figure 1.4) introduced the term *Protozoa*, without definition or explanation which might suggest that the term had an earlier origin. This grouping (Goldfuss, 1820) included polyps, medusae, infusoria and phytozoa (plant-animals) (Figure 1.14). By 1839 Schleiden (1838) and Schwann (1839) had developed what became known as the 'cell theory' in which it was acknowledged that the cell was basic to all organisms and therefore organisms consisted of one or more cells. Following this discovery, matters moved fast with respect to the infusoria. Meyen (1839) and Dujardin (1841) independently recognized that many protozoa were single cells with a high level of internal organization comparable to plant cells. Barry (1843), in a publication entitled *The Infusoria Compared with Cells*, argued that *Monas* and other flagellates comprised single cells that had a nucleus which corresponded to the cell nucleus of higher forms. Von Siebold (1845) (Figure 1.5) formulated the *doctrine of the unicellular nature of Protozoa*. He divided the Protozoa into two classes, the *Infusoria* and *Rhizopoda* (see Dujardin discussed later), the former being subdivided into two groups, the ciliates (*Stomatoda*) and the flagellates (*Astomata*). The rotifers he removed to the *Vermes*, and the bacteria and *Volvox* were placed in the plant kingdom.

In 1852 Perty instituted the *Archezoa*, equivalent to von Siebold's (1845) Protozoa, with the Stomatoda renamed the *Ciliata* and the Astomata (flagellates) renamed the *Phytozoida* (plant-animals) (Perty, 1852). Cohn (1853) (Figure 1.6) is recognized as the originator of the term *Flagellata*. He confirmed that the green colouration of some flagellates was due to chlorophyll and that pigmented flagellates

(left)
Figure 1.2 Antony van Leeuwenhoek (1632–1723)
Figure 1.3 Otto Friderich Müller (1730–1784)
Figure 1.4 Georg August Goldfuss (1782–1848)
Figure 1.5 Karl Theodor von Siebold (1804–1855)
Figure 1.6 Ferdinand Cohn (1828–1898)
Figure 1.7 Karl Moritz Diesing (1800–1867)

8
9
10
11
12
13

were photosynthetic plants. This led to a conundrum that has taken many years to resolve, namely that both plants and animals were contained within the Protozoa, and yet they were also classified separately in the plant and animal kingdoms. Diesing (1866) (Figure 1.7) instituted the name *Mastigophora* (meaning whip bearer) for flagellates, and many subsequent classificatory systems have used this terminology.

Just as the studies on the Infusoria during the eighteenth century had been dominated by the monographs of Otto Friderich Müller, so in the nineteenth they were dominated by the monographs of Ehrenberg, Dujardin, Stein and Kent. Christian Gottfried Ehrenberg's (1795–1876) (Figure 1.8) monumental monograph, entitled *Die Infusionthierchen als Vollkommene Organismen* (The infusion animals as complete organisms), published in 1838, established the Infusoria, including protozoa, bacteria, diatoms, desmids and rotifers, as a major separate group of animals. Ehrenberg, a native of Leipzig and trained in medicine, had great observational ability which is manifest in the quality of his illustrations in this and other works. He was also interested in micropalaeontology and described many fossils. However, his classification of the single-celled Infusoria, including the flagellates, into the *Polygastrica* on account of their many stomachs, and his belief that these organisms were organized in the same way as higher animals, were to deflect his attention from a more fundamental understanding of cell structure. Unfortunately, to everyone's great loss, Ehrenberg ignored evolution and cell theory but he opposed spontaneous generation. Nevertheless, in spite of his ill-founded beliefs, he added a large number of new genera, including *Euglena*, *Bodo*, and many diatom species to the literature, most of which are still in valid use today (Schlegel and Hausmann, 1996; Williams and Huxley, 1998).

Félix Dujardin (1801–1860) (Figure 1.9), a Frenchman and contemporary of Ehrenberg, was a physiologist, morphologist and taxonomist who concentrated his attention on living foraminifera, amoebae and certain ciliates. He introduced the term *sarcode* to describe the streaming cytoplasm of amoeboid cells. His observations on cell structure, including feeding experiments involving pigmented granules, convinced him of the error of Ehrenberg's polygastrica theory and rapidly lead to the theory's demise. Dujardin's (1841) treatise entitled *Histoire Naturelle des Zoophytes*, like Müller's and Ehrenberg's before, is noted for the number of new genera and species described and also for the quality of the illustrations. Dujardin was the first to describe three types of free-living forms, *les flagélles, les rhizopods* and *les ciliés*. He discovered the flagellate cells of sponges, although he did not observe their collars, and held that they were nothing more than colonies of Infusoria. Dujardin's (1841) classification of the Infusoria into four orders according to the means of locomotion established the system for classifying protozoans that was ultimately used throughout the nineteenth and twentieth centuries.

(left)
Figure 1.8 Christian Gottfried Ehrenberg (1795–1876)
Figure 1.9 Félix Dujardin (1801–1860)
Figure 1.10 Friedrich Ritter von Stein (1818–1885)
Figure 1.11 Ernst Haeckel (1834–1919)
Figure 1.12 William Saville Kent (1845–1908)
Figure 1.13 Otto Bütschli (1848–1920)

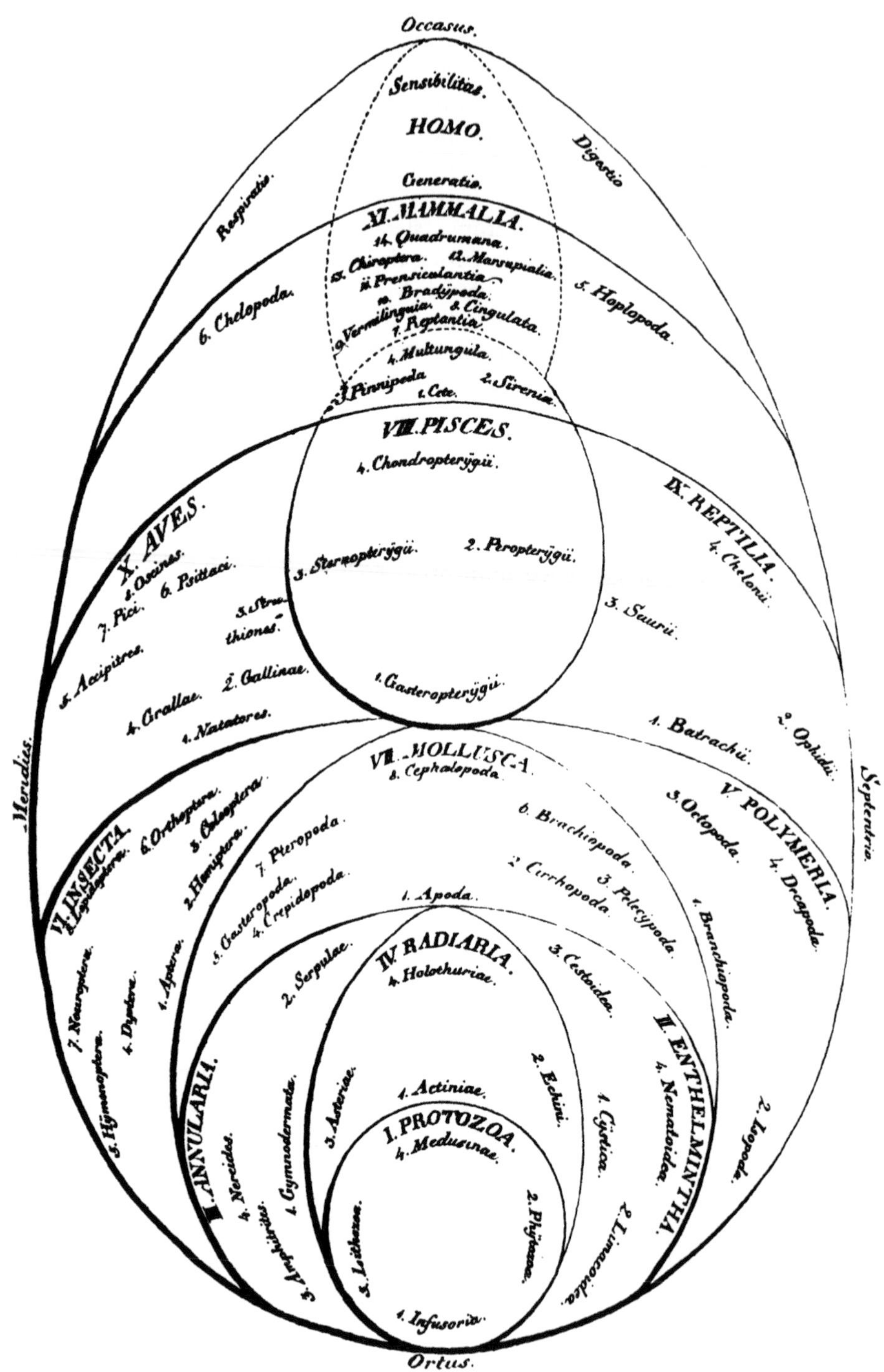

Figure 1.14 Georg August Goldfuss's phylogenetic tree published in 1820
Source: Reproduced from Goldfuss, 1820

Friedrich Ritter von Stein (1818–1885) (Figure 1.10), another in the spirit of Müller and Ehrenberg, working at the German-speaking University of Prague, produced four beautifully illustrated volumes on free-living protozoa entitled *Der Organismus der Infusionthiere* (Stein, 1859/83). Again he was the originator of many names still in use today. He also became involved in controversy over the so-called 'Acineta theory', the details of which now seem arcane. What is of lasting importance is that, in spite of controversies at the time, so much of Stein's work is still of relevance today.

As we have seen, Dujardin (1841) laid the foundations for a classification of the Protozoa. Dujardin considered the sponges to be colonies of Infusoria. However, Haeckel (1869, 1878) (Figure 1.11) argued that sponges should be excluded from the Protozoa, and in doing so identified himself with von Siebold's (1845) view that the Protozoa were unicellular animals. In 1878 Haeckel published his most famous systematic generalization, the division of the animal kingdom into the Protozoa and Metazoa. He gave the sponges independent status within the Metazoa since, in his opinion, they more nearly approximated to the Coelenterata (Haeckel, 1878). A long, and at times acrimonious, dispute raged within the literature between those who considered that the Sponges were a separate metazoan group (Haeckel, 1869, 1870, 1873) and those, including Carter (1871), James-Clark (1868) and Kent (1880/2) (Figure 1.12), who considered the sponges to be closely related to the Infusoria, and in particular the recently recognized choanoflagellates. The heart of this argument is now somewhat esoteric, although it is still of importance and has not fully been resolved a century later. Haeckel's defense of the argument involved his Gastraea theory, whereby he postulated that all Metazoa have a common ancestor involving a *gastrula* stage. James-Clark (1878) and Kent (1880/2) argued against the sponges having a gastrula phase in the sense in which Haeckel interpreted it, and laid great emphasis on the morphological similarity between the choanoflagellates and the choanocytes of sponges. The latter, by this time, were fully recognized as having conical-shaped *hyaline* collars (James-Clark 1866, 1868, 1878). The discovery by Kent (1880/2) of *Proterospongia*, a colonial choanoflagellate, appeared to settle the proximity of the relationship between choanoflagellates and sponges, but in the light of recent information the closeness of this relationship is still in doubt.

Kent (1880/2) considered that amoeboid forms were the progenitors of all other groups of protozoa (Figure 1.15). The choanoflagellates were separated from the remainder of the flagellates and were associated with the sponges. The ciliates were separated from the Suctoria (Tentaculifera) and the opalinids. Kent omitted the recently instituted Sporozoa (Leuckart, 1879) although he included some gregarines in his overall scheme (Figure 1.15).

Otto Bütschli (1848–1920) (Figure 1.13), the great architect of systematic protozoology (Dobell, 1951; Corliss, 1978), also published a classification and evolutionary arrangement of the Protozoa. The basis of his system at the class level was still primarily the means of locomotion. Bütschli's (1880/9) system instituted five classes, namely the *Sarcodina* (amoebae), *Mastigophora* (flagellates), *Infusoria* (ciliates), *Sporozoa* and *Radiolaria*. The flagellates occupied four orders (*Flagellata*, *Choanoflagellata*, *Dinoflagellata* and *Cystoflagellata*) within the class

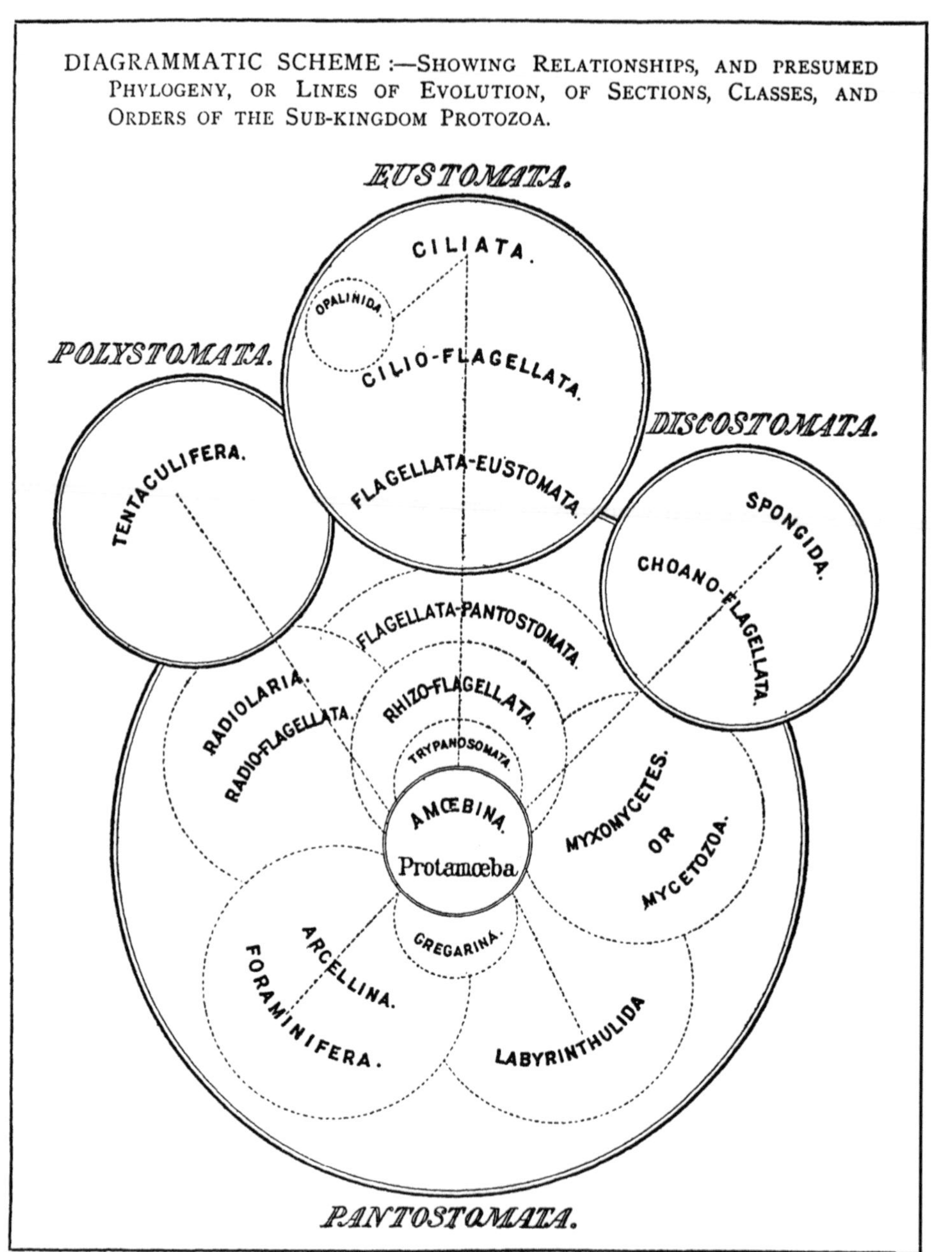

Figure 1.15 William Saville Kent's phylogenetic tree published in 1880/2
Source: Reproduced from Kent, 1880/2

Mastigophora. He grouped the ciliates and Suctoria within the class Infusoria and from this time onwards the term *infusoria* has been restricted to the ciliates.

Bütschli's (1880/9) system had the advantages of order and relative simplicity, and probably for this reason has formed the basis of the systematics of Protozoa for the better part of a century. However, from the outset there were uneasy compromises. In the case of the Mastigophora, to try to justify the monophyletic status of flagellates using such a fundamental (plesiomorphic) character as the flagellum proved to be problematic even during Bütschli's era. Bütschli tried, until his death in 1920, to prove an evolutionary relationship between rhizopods and flagellates. He emphasized the homology of flagella and slowly swinging pseudopods of some amoebae. In 1910 he devised the term *undulipodium* to emphasize the evolutionary origin of flagella and pseudopodia from a common ancestral structure (see Hausmann and Hülsmann, 1996). However, with limited exceptions (Margulis *et al.*, 1990) this term has fallen into disuse (Hülsmann, 1992).

Following Bütschli, many general texts on protozoa have been published, the majority of which have used Bütschli's system as the basis of their particular systematic treatments. In many cases the terms *Flagellata* and *Mastigophora* have been used interchangeably. Hartog (1906) acknowledged the artificiality of the grouping within the Flagellata. He distinguished flagellates from some sarcodines and sporozoans, which have flagella-bearing stages in their life-cycles, by the relative permanence of flagella on true flagellate cells. In Hartog's (1906) scheme the Flagellata were subdivided into two orders and ten classes, eight of which were autotrophic, equivalent to already established algal classes, and two heterotrophic.

Doflein (1916) divided the flagellates into two subclasses, *Phytomastigina* and *Zoomastigina*, based on their autotrophic and heterotrophic modes of nutrition respectively. Similar systems were adopted by Wenyon (1926) and Kudo (1966). However, on the basis that many amoebae had flagella-bearing stages in their life-cycles and that some flagellates could become amoeboid, Grassé (1952) proposed a union of the Flagellata and Rhizopoda to form a single subphylum, *Rhizoflagellata*, with two superclasses namely, *Flagellata* and *Rhizopoda*. The Flagellata *sensu* Grassé included eleven classes, ten of which comprised autotrophs. The eleventh contained the zooflagellates which Grassé acknowledged at the time as being polyphyletic.

The union of the Sarcodina with the Flagellata (as Mastigophora) was also adopted in the classification proposed by the Society of Protozoologists (Honigberg *et al.*, 1964). In this system the subphylum *Sarcomastigophora* is divided into two classes, namely the *Phytomastigophorea*, for autotrophic flagellates, and *Zoomastigophorea,* for heterotrophic flagellates. In the Levine *et al.* (1980) revision of the Honigberg *et al.* (1964) classification, Sarcomastigophora is raised to the rank of phylum with attendant upgrading to subphyla of the Mastigophora and Sarcodina, the former retaining the two classes noted in the classification of Honigberg *et al.* (1964). The classification of Levine *et al.* (1980) took note of information available at that time from electron microscopy, but it was more a system of convenience and the authors freely acknowledged that it did not 'necessarily indicate evolutionary relationships'. In fact, in spite of hoping that the scheme 'would last for many years', this attempt to bring order to the Protozoa essentially marked the end of the traditional (classical) system of classification based on Bütschli's (1880/9) scheme, and was outmoded even as it was being constructed. Other forces, including advances in electron microscopy and molecular biology, were to revolutionize conventional thinking on protozoan classification and systematics.

1.2 Parasitic flagellates

According to Dobell (1932), Antony van Leeuwenhoek is credited with the first descriptions of parasitic flagellates. Bearing in mind that Leeuwenhoek did not attempt to name the organisms he observed, and we only have his descriptions occasionally accompanied by sketches or drawings, it is difficult to be certain what he did, in fact, see. However, he is generally credited with being the first to observe *Opalina* (really an opalinid species of the genus *Cepedea* (according to Corliss, 1975)), *Chilomastix*, *Giardia* and *Trichomonas sensu lato*. In 1683 Leeuwenhoek reported:

> In the month of June, I met with some frogs whose excrement was full of a innumerable company of living creatures, of different sorts and sizes. . . . The whole excrement was so full of living things that it seemed to move.
>
> (Leeuwenhoek, 1683)

Otto Friderich Müller was the first to recognize an oral trichomonad (under his genus *Cercaria*) from humans. Donné (1836), a physician, described the parasite *Trichomonas vaginalis* in the genital secretions of humans.

It was not until late in the nineteenth century that the involvement of protozoa in diseases of animals and humans was established. With respect to protozoan parasites of the blood, the first mention of the disease known as sleeping sickness was recorded in 1841. Valentin (1841) found flagellates (*Trypanoplasma*) in the blood of salmon. Soon after, Gluge (1842) and Gruby (1843) discovered trypanosomes in the blood of frogs, for which the latter author instituted the genus *Trypanosoma* (*T. sanguinis*) (Gruby, 1843). In 1880 Evans, a veterinarian working in India, showed that Surra, a disease of horses and camels, was associated with infection of the blood by a trypanosome that subsequently came to bear his name, *T. evansi* (Evans, 1880). However, it was not until Bruce's (1895) work on the trypanosmatid disease, nagana (cattle trypanosomiasis), was published, that the links between a flagellate blood parasite *T. brucei*, its transmission by tsetse flies and the abilty of game animals to act as a reservoir of infection for domestic animals was fully appreciated (Vickerman, 1997). The end of the nineteenth century saw the unravelling of numerous multiphasic life-cycles of human protozoan parasites, including sleeping sickness, Chaga's disease and malaria.

1.3 Protoctistology, protistology, protozoology

The kingdom system of living organisms – the Plantae (Vegetabilia) and Animalia – dates back to Linnaeus (1735). This system, with its implied rigid division of living organisms into plants and animals (botany and zoology), has had a major effect on subsequent ideas on classification and systematics throughout the nineteenth and twentieth centuries. Although, as described earlier, eighteenth and nineteenth century microscopists grouped diverse minute organisms together within the *Infusoria*, the direction of systematics in the nineteenth century was to delineate and separate groups more precisely. Nevertheless, for a brief interlude in the mid-nineteenth century attention focused on the possible evolutionary relationships between autotrophic and heterotrophic microorganisms. In 1860, Hogg instituted the kingdom *Primigenum* as a fourth kingdom (the others being plants, animals and minerals) comprising 'all lower beings – Protoctista (first created beings) – to include both the Protophyta and Protozoa' (Hogg, 1860). Haeckel (1866), from a slightly different standpoint, instituted the kingdom *Protista* as a third kingdom between the animals and plants (Figure 1.16) comprising bacteria (Moneres), cyanobacteria (Myxocystoda), diatoms, flagellates, rhizopods, myxomycetes and sponges (prior to separating the latter into the Metazoa (Haeckel, 1869, 1873)). The terms *Protoctista* and *Protista* were given little recognition at the time and fell into disuse. The convention for the rest of the nineteenth century was to separate autotrophs from heterotrophs. Occasional reference to the protists (for example Copeland, 1956; Whittaker, 1959; Corliss, 1984) was made during the first three quarters of the twentieth century only to be quickly forgotten (see fuller historical accounts in Rothschild, 1989; Lipscomb, 1991; Ragan, 1997; Corliss 1998a, b).

A number of major unrelated developments occurred during the latter part of the twentieth century that were to culminate in a dramatic change to conventional thinking about protistan systematics and evolution, and these heralded a revival of the 'protist concept' (Corliss, 1986). These developments included, first, the systematic separation of prokaryotes from eukaryotes, second, the advent of electron microscopy to descriptive protistology, third, the concept that certain organelles evolved by symbiosis (symbiogenesis) and fourth, the application of molecular analytical techniques to protistan systematics.

1.3.1 Separation of prokaryotes from eukaryotes

The separation of prokaryotes was first suggested by Chatton (1925) and later developed and formalized by Stanier and colleagues (Stanier and van Niel, 1962; Stanier *et al.*, 1963). Although this separation is now fully acknowledged, it took some time for Chatton's work to be generally accepted. Corliss (1986) attributes this, in part, to botanists and phycologists resisting the removal of the cyanobacteria from algal classificatory systems.

Although the recognition of the Prokaryota was not directly related to the subsequent systematic fate of the flagellates, nevertheless the formal recognition of the Prokaryota focused attention on the need to re-examine the evolutionary relationships among all forms of life. Furthermore, it focused attention on the significance of the discontinuity between prokaryotes and eukaryotes, and the apparent absence of intermediates.

1.3.2 Advent of electron microscopy

The use of electron microscopy started as early as the 1930s. Use of electron microscopes for biological studies in the 1940s was limited to a few laboratories, and the first whole mount preparations date from this period (for example, Schmitt *et al.*,

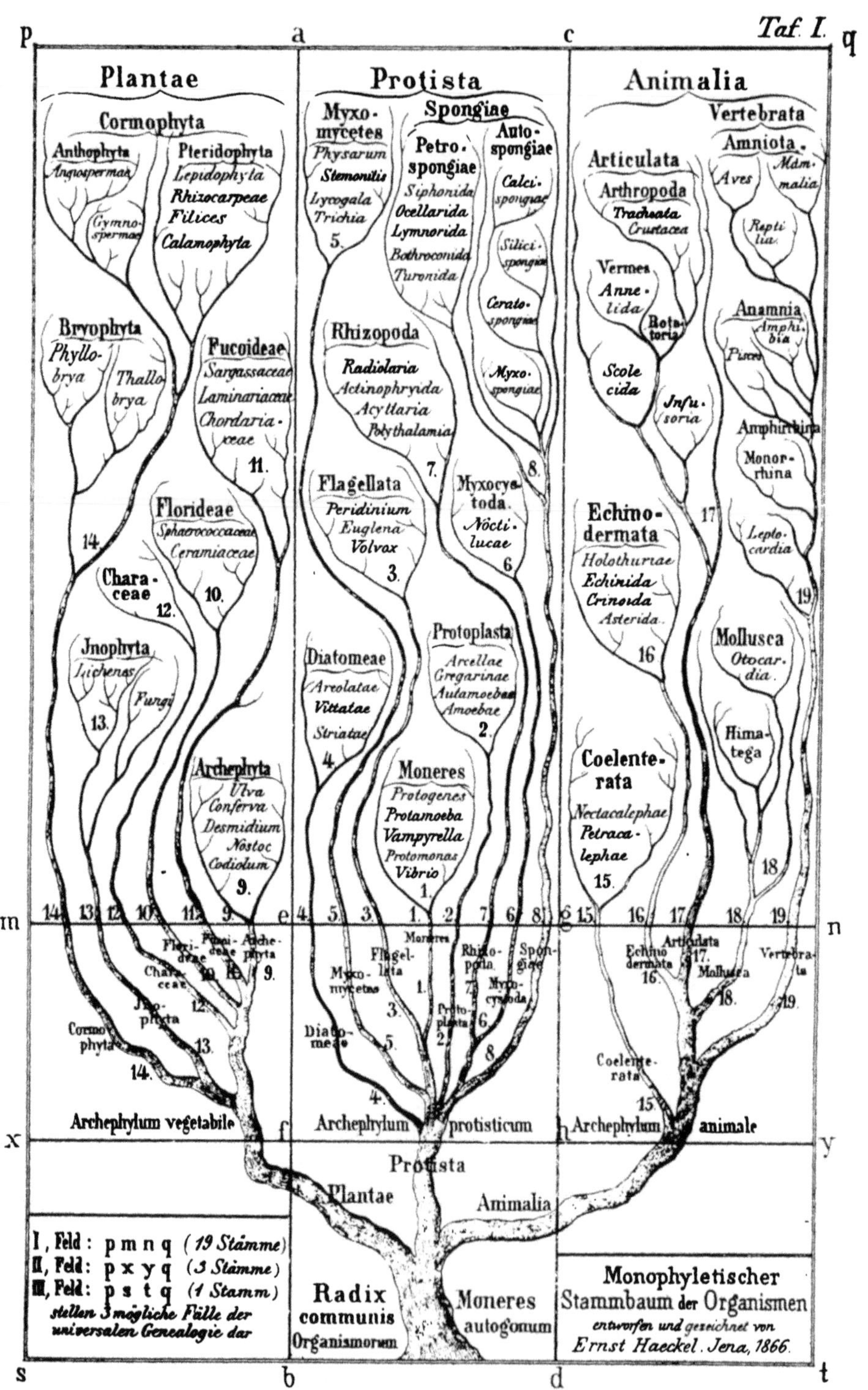

Figure 1.16 Ernst Haeckel's phylogenetic tree published in 1866

Source: Reproduced from Haeckel, 1866

1943). Whole mount reconstructions of the flagellum date from the late 1940s and early 1950s (see later in this chapter). The first thin sections were cut and viewed in 1954 (Fawcett and Porter, 1954). However, it was not until the early 1960s that the great refinements in electron micrsocopes and fixation procedures, involving the use of aldehydes, made possible the widespread use of electron microscopy.

The literature on electron microscopy of protists is so enormous that it is impossible to summarize it here. From the point of view of protistan classification and systematics details relating to flagellar systems, cytoskeletal structures, nuclear division patterns, appearance of mitochondria, chloroplasts and endomembrane systems have been of key importance (Patterson, 1994). Initially the impact of these studies on evolutionary relationships was greater on autotrophic flagellates, but more recently attention has focused on heterotrophic flagellates (Patterson and Larsen, 1991). Among other matters, these studies have led to the acceptance that the flagellates, including both autotrophs and heterotrophs, are an extreme example of paraphyly.

1.3.3 Symbiotic origin of organelles

By the end of the nineteenth century, symbiosis was an acknowledged phenomenon and had been well described for a number of organisms including lichens (de Bary, 1879). However, it was the Russian school of scientists, and in particular Merezhkovsky and Famintsyn, who first suggested that some cell organelles might be derived from other free-living organisms in a symbiotic union. Merezhkovsky (1905, 1906), while studying diatoms, came to the conclusion that chloroplasts were capable of division and did not arise *de novo* when the cell was exposed to light. He considered that the great self-sufficiency and functional independence of chromatophores from the nucleus of the cell supported the hypothesis that they were of symbiotic origin. Merezhkovsky (1905) attempted to isolate chromatophores, rather like zoochlorellae, but was unable to grow them outside the host cell. Nevertheless, he considered that they were derived from symbiotic cyanobacteria. Comparison of volume, size, inner structure and method of reproduction showed the great resemblance between chloroplasts and cyanobacteria. Merezhkovsky (1909) introduced the term *symbiogenesis* which he defined as 'the origin of organisms through the combination and unification of two or many beings in symbiosis'.

With regard to mitochondria, it was the American biologist Wallin (1925a, 1925b) who first suggested that mitochondria were bacterial symbionts. This idea was based on the similar staining properties of bacteria and mitochondria. Again, all attempts to isolate mitochondria and grow them like bacteria outside the host cell were unsuccessful.

In the 1970s, a reappraisal of the origin and evolution of chloroplasts and mitochondria arose out of electron microscopical, biochemical and molecular studies. The similarities in the genome structure and ribosomes between chloroplasts and mitochondria on the one hand, and prokaroytes on the other, gave fresh impetus to the 'symbiosis theory'. Suddenly it seemed that symbiosis might explain the sudden change from the prokaryote to eukaryote pattern of organization without any obvious intermediates. Subsequently, biochemical and molecular biological studies have demonstrated further the similarity of chloroplasts and mitochondria to prokaryotes, and now it is accepted wisdom that these organelles were probably acquired as a result of symbiotic events. According to Corliss (1986), 1975 marks the watershed when the concept of protistology was revived. The debate has now moved on from whether symbiosis occurred, to how frequently the organelles evolved and by which mechanism (Margulis, 1970, 1993; Taylor, 1974; Margulis *et al.,* 1990; Patterson, 1994; Cavalier-Smith, Chapter 17 this volume).

1.4 Flagella

Antony van Leeuwenhoek is generally credited as the first person to have seen and recorded the presence of cilia on minute animalcules. In a letter to the Royal Society Leeuwenhoek (1677) described what Dobell (1932) considers to be a ciliate as having 'little paws . . . which were many in number, in proportion to the animalcule; and also a much bigger animalcule which was furnished with little legs'. It would appear that Otto Friderich Müller (1786) was the first to use the word *cilium*, meaning a hair, which probably reflects the fact that rows of cilia on ciliates resembled eyelashes. The term *flagellum* (a whip) was introduced later and is generally attributed to Dujardin (1841).

By 1835 cilia and flagella had been observed widely and the first reviews on these structures were written by Purkinje and Valentin (1835) and Sharpey (1835).

1.4.1 Substructure of flagella

The essential elements of the flagellar apparatus, namely the axoneme of longitudinal fibrils, a basal body and the flagellar root, were recognized by the end of the nineteenth century, although the combination of these parts was not recorded together for one organism until much later.

As early as 1881, Englemann (1881) reported the existence of fibrils within *cilia*. However, it is Ballowitz (1888) who is generally regarded as the first to illustrate indisputably the longitudinal fibrils that comprise the flagellar axoneme. His

(right)
Figure 1.17 Flagellum of *Euglena viridis* treated with bacterial stain showing a single array of long flagellar hairs. Magnification x c. 5000
Source: Reproduced from Fischer, 1894
Figures 1.18 and 1.19 Two *Monas* cells stained with bacterial stain showing a bilateral array on the flagellum. Magnification x 3000
Source: Reproduced from Hoeffler, 1889
Figure 1.20 Dismembered sperm tail of fowl showing eleven longitudinal fibrils
Source: Redrawn after Ballowitz, 1888
Figure 1.21 Dismembered sperm tails of chicken showing eleven longitudinal fibrils; nine of the fibrils (L-fibrils) are thicker and encircle the two central fibrils (M-fibrils) which are thinner and more delicate. Magnification x c. 20,000
Source: Reproduced from Grigg and Hodge, 1949
Figure 1.22 Dismembered flagellum of *Dryopteris villarsii* showing periodic projections on fibrils which probably represent linkages. Magnification x c. 30,000.
Source: Reproduced from Manton and Clarke, 1952
Figure 1.23 Reproduction of Manton and Clarke's (1952) diagrammatic reconstruction of the arrangment of fibrils (microtubules) within the axoneme of the flagellum of *Sphagnum*

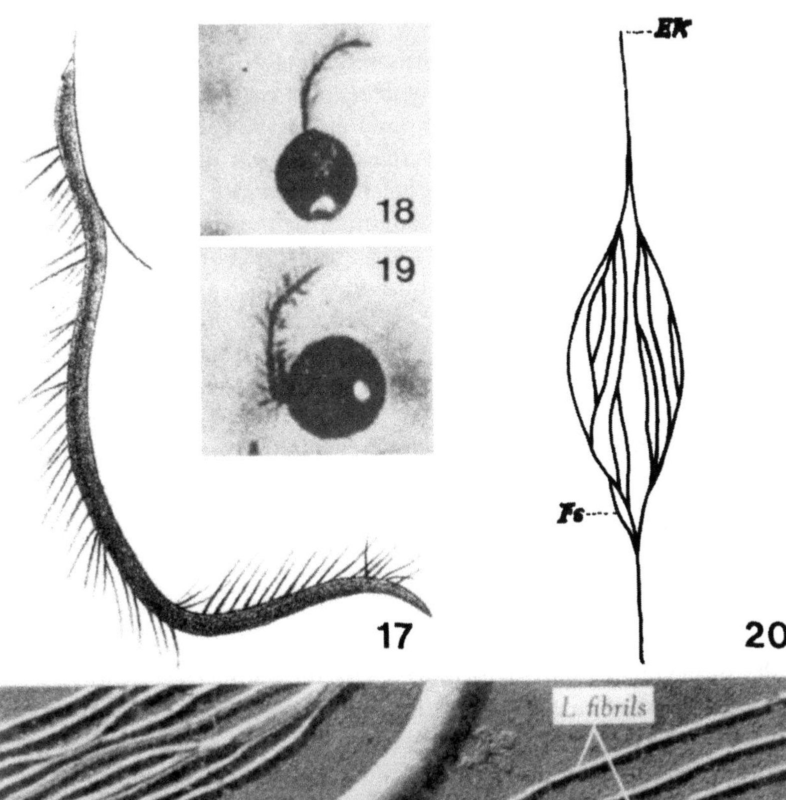
18
19
EK
Fs
17
20

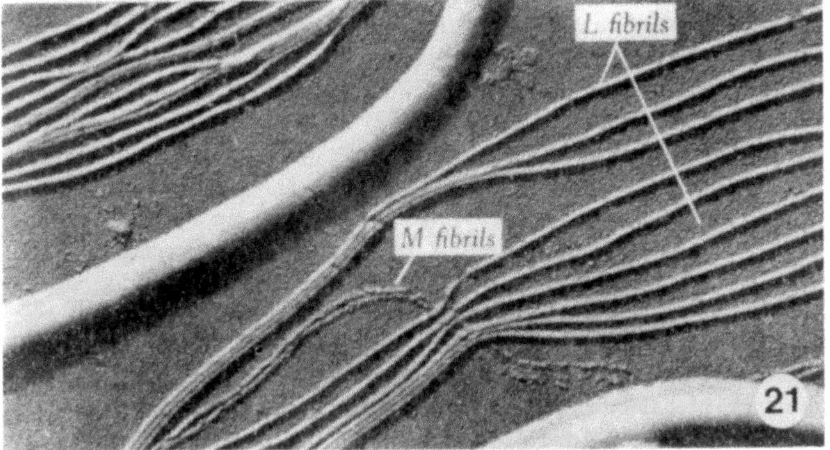
L. fibrils
M fibrils
21

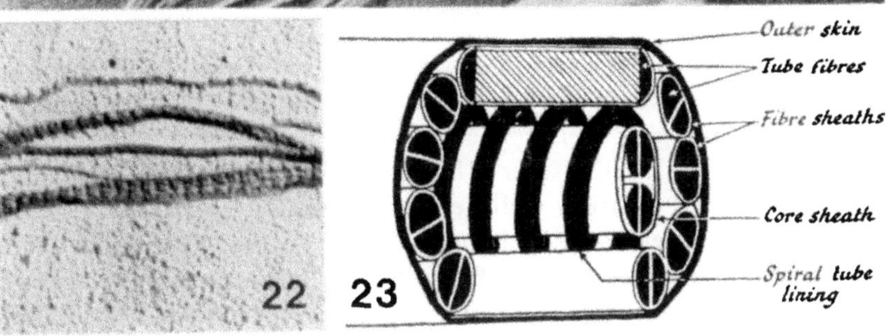
22
23
Outer skin
Tube fibres
Fibre sheaths
Core sheath
Spiral tube lining

drawings of dismembered sperm tails of fowl are exquisite in their detail, and in one illustration he actually records eleven fibrils (Figure 1.20), with some variation in width. Sperm tails appear to dismember with relative ease when dried on to slides, subjected to osmotic shock or treated with mild digestive agents. With the advent of electron microscopy and its early use for the observation of dried whole mounts, animal and plant spermatozoids made ideal subjects for investigation. Frayed axonemes of sperm tails allowed precise counting of the eleven component fibrils (Schmitt *et al.,* 1943; Pitelka, 1949). Grigg and Hodge (1949) recognized the difference between nine thicker, more robust fibrils which they termed *L (lateral) fibrils* and the two thinner, more delicate, *M (middle) fibrils* (Figure 1.21). The L fibrils appeared to form a lamellar sheet that was rolled around the two thinner fibrils which formed the axis of the axoneme. Transverse striations were present on the fibrils, which Grigg and Hodge (1949) interpreted as being similar to the banding seen on structural proteins such as collagen, but which probably represent linkages between the various components of the axoneme.

At almost the same time as Grigg and Hodge (1949) were observing dismembered animal flagella, Manton and her co-workers were studying the frayed flagella of plant spermatozoids. In 1950 Manton (1950) illustrated spermatozoids of *Dryopteris villarsii* with dismembered flagella viewed with ultraviolet microscopy. Subsequently, from 1950 to 1952, Manton and co-workers published electron micrographs of the dismembered spermatozoid flagella of many crypytogams (Manton, 1952). In these publications they also noted the difference in width between the outer nine and inner two fibrils and observed various types of periodic striations and projections (Figure 1.22). The culminating achievement of this burst of activity was the publication by Manton and Clarke (1952) of a reconstruction of the 9+2 axoneme (Figure 1.23), which in most respects anticipated the details that were to be revealed later by sectioning (Fawcett and Porter, 1954). The subsequent development of negative staining techniques and improvements in fixation for electron microscopy in the 1960s and onwards made possible observation of great structural detail in flagella and cilia. This, together with advances in biochemistry and molecular biology, has led to a burgeoning of information on flagella and flagellar movement, summarized in early review articles by Dryl and Zurzycki (1972) and Sleigh (1974) and later ones by Warner *et al.* (1989), Bloodgood (1990) and others.

The similarity of ciliary bases to centrosomes was recognized by Henneguy (1898) and von Lenhossek (1898), and ciliary roots were observed by Schuberg (1891) and Maier (1903).

Flagellar appendages were first reported by Loeffler (1889), who observed *Monas* in a mordant stained preparation of bacteria. The stain accumulated around the flagellar hairs, thereby enlarging their size and making them visible with the light microscope (Figures 1.18, 1.19). Fischer (1894), using a modification of Loeffler's staining technique, viewed a variety of flagellates, including *Monas*. He confirmed Loeffler's (1889) earlier observations and found that the flagellum of *Euglena* bore a single row of hairs (Figure 1.17). Fischer (1894) introduced the terms *Flimmergeisseln* (ciliated flagellum) and *Peitschengeisseln* (whip flagellum) for hairy and smooth flagella respectively. In general, these terms have fallen into disuse, although the word *flimmer* is still used by some authors today.

Of the many workers who have observed flagellar appendages subsequently, only four require mention. Petersen (1929) reported the presence of hairs on the flagella of many flagellates. He observed combinations of smooth and hairy flagella on single cells and in the case of the choanoflagellate *Salpingoeca* a single flagellum that bore hairs in the proximal region and was smooth distally. Deflandre (1934) introduced the name *mastigoneme* (whip (flagellar) thread) and a suite of other terms to describe the location and organization of mastigonemes on the flagellum. Terms such as *isokont, anisokont* and *heterokont*, the latter introduced by Adolph Luther (1899), came into common usage, and subsequently the term *heterokont* has taken on a specific meaning in referring to a combination of a forwardly directed flagellum with two rows of tripartite hairs and a second flagellum without these appendages.

1.4.2 Flagellar movement

Historically, views about flagellar and ciliary movement have been greatly influenced by concurrent studies on other contractile structures such as muscle cells. Thus from an early time the association of fibrils with contractile structures has been established for both flagella and muscle cells (Engelmann, 1881; Ballowitz, 1888). Much time and space were devoted to trying to establish whether the undulation of flagella resulted from a force initiated at the base of the structure, or whether movement was generated along the length of the flagellum (Fauré-Fremiet, 1961). Gray (1928), in his book on ciliary movement, argued on a number of counts that the flagellum is an active organelle with the ability to generate a mechanical force along its length. Theories concerning the mechanism of flagellar movement ranged from *hyaloplasm* flowing into and out of a curved hollow outgrowth (Grant, 1835) to rhythmic contraction of the flagellar membrane (Heidenhain, 1907). The suggestion of Sharpey (1835) that cilia possess internal contractility was a precursor of what we understand today. Studies of isolated, permeabilized flagella demonstrated that flagellar movement was brought about by a process within the axoneme (Brokaw, 1961; Bishop, 1962). That this movement might result from an ATP-induced sliding of non-contractile filaments in the axoneme, analogous to the sliding mechanism of muscle contraction, was recognized by Afzelius (1959). However, it was the analysis of microtubules within the ciliary tips of the mussel *Eliptio* by Satir (1965, 1968) that ultimately revealed that the outer doublets do slide with respect to one another during ciliary bending, and that they do not undergo contraction. Furthermore, ATP treatment of glycerinated models resulted in *in vitro* movement.

1.5 End of the old order and the beginning of the new

The 1970s saw the eclipse of the old order which had served for the better part of a century. While it had been recognized that the traditional systematics of Protozoa, based on the system of Bütschli (1880/9), was a scheme of convenience rather than reflecting in any real sense evolutionary relationships, nevertheless it had brought a much appreciated degree of order and security. However, with the daily acquistion of new EM information together with novel molecular data, the system became untenable except as a convenience for teaching purposes. The domination of microscopy, and thus morphology, in influencing our views on protist evolution had, after more than 300 years, yielded to other technologies. As Corliss so aptly summarized the situation that still exists today:

> at the present time we are in a state of flux – we are frustatingly trapped between existing classifications of protists that are recognized to be faulty and some future schemes not yet available. The latter, hopefully closer to the ideal natural system long awaited, probably will not be ready for at least several years, perhaps not until the turn of the century.
>
> (Corliss, 1994)

The major area of expansion in the past fifteen years has been the contribution of molecular biology to our understanding of protistan systematics. The advantage of molecular information over morphological characters is that it overcomes the problems associated with the comparison of phenotypic traits with their often complex genetic backgrounds (Schlegel, 1991). Instead, the genes themselves, or secondary gene products, the proteins, are the characters of importance. Molecular markers have the advantage of being either present or absent without gradual transitions. This is particularly true of DNA sequences, which can usually be studied independently of inner (physiological) or outer (environmental) influences. They are referable to a defined gene locus and their investigation is not complicated by recessive or dominant modes of inheritance which may complicate comparative studies using other characters (Schlegel, 1991). More important, molecular characters evolve, in the vast majority of instances, independently of other characters and under different circumstances.

Molecular markers are not without their disadvantages, such as the occurrence of back mutations and the possibility of lateral gene transfer. Phylogenetic relationships can be inferred from sequence data only if two conditions are met, namely that the sequences compared must be homologous and orthologous (Schlegel, 1991). Also while the sequences of bases obtained can be absolutely correct, the models of relatedness generated from them do not carry the same degree of certainty. Alignment of molecules prior to comparisons has a subjective component, homology cannot unambiguously be differentiated from analogy, and algorithms which seek to identify the best tree are unable to evaluate every option (Patterson, 1994). Nonetheless, there has been broad agreement between morphological and molecular approaches.

The first molecular markers used in the study of protistan phylogeny were amino acid sequences of proteins such as cytochrome *c*, ferredoxins and superoxiode dismutases (Schlegel, 1991). All are widely distributed, highly conserved, polymeric molecules. In the late 1970s and early 1980s emphasis shifted to studies of universal, conservative and manageable nucleic acids such as the RNA molecules of ribosomes. Initially, information was gained for 5s rRNAs with about 125 nucleotide bases, but as technology improved the small subunit rRNA, 16s for prokaryotes and 18s for eukaryotes, became the molecule of choice. The latter, because of its large size, ranging from 1,200–2,300 nucleotide bases, and the occurrence of both highly conserved and partially conserved sequence elements, has permitted the study of both close and distant relationships. Many phylogenetic comparisons have now been made on the basis of 18s rRNA data. The result of these studies has been to revolutionize completely our understanding of protist relationships and, in particular, those of the flagellates (Cavalier-Smith, 1993, 1998).

What now emerges is that the ancestor (cenancestor in Cavalier-Smith's terminology, Chapter 17 this volume) of all extant eukaryotes was almost certainly a flagellate, even if it was an amoeboflagellate. There is some agreement between the different protagonists on the sequence in which various cell components appeared (Patterson, 1994; Sleigh, 1995; Cavalier-Smith, Chapter 17 this volume). Corliss (1998a) rightly sums up the current situation as being 'in a state of flux rather than in a state of chaos'. The problem is how the protists can be conveniently treated. Corliss (1998a, 1998b) makes five specific suggestions: one is to consider the protists as representing a separate kingdom, as Margulis (for example Margulis *et al.*, 1990; Margulis and Schwartz, 1998) and a few others have persistently done. But, as Patterson points out, 'protists are still paraphyletic – there are no traits that unify all protists to the exclusion of all other eukaryotes. The protists cannot be defined and their composition remains open to dispute' (Patterson, 1994).

Another possibility is to disperse the protists among diverse eukaryotic kingdoms, a view especially supported by Cavalier-Smith (1993, 1998) and Corliss (1994, 1998a, 1998b). This has attracted certain favour but implies the recognition of several to many eukaryotic kingdoms. A third possibility is to recognize the protists as merely an evolutionary grade or level of structural organization on the way to the evolution of other eukaryotes. A fourth possible solution would recognize a number of independent evolutionary lines or phylogenetic clades without expressing explicitly their possible taxonomic interrelationships. Finally, the protists might be amenable to classification purely on ecological grounds, again, however, irrespective of their high-level conventional taxonomy.

This is a story without an ending, and it is sure to extend well into the twenty-first century. However, in contemplating the 350 years that have passed since Antony van Leeuwenhoek (1674) first observed his 'very little animalcules', we may reflect that it has been necessary to undergo the trials and tribulations experienced here in order to delineate and separate protist groupings so that in the end, knowing the building blocks, we can construct the evolutionary schemes of the future.

ACKNOWLEDGEMENTS

We are very grateful to Professor John Corliss for reading and commenting on this manuscript and for giving much helpful advice. We are grateful also to Professor Corliss for generously making available the following illustrations: Figures 1.2, 1.3, 1.8, 1.9, 1.10, 1.11, 1.12 and 1.13, and giving us permission to reproduce them. We are grateful to the following publishers and journals for permissions to reproduce the following illustrations: Swets and Zeitlinger, Amsterdam for Figure 1.1; Urban and Fischer Verlag, Stuttgart for Figure 1.4; Universitätsbibliothek Erlangen-Nürnberg for Figure 1.7; *Handbuch der Zoologie*, Nurnberg for Figure 1.14; David Bogue, London for Figure 1.15; *Jahrbücher für wissenschaftliche Botanik* for Figure 1.17; *Zentralblatt für Bakteriologie und Parasitenkunde* for Figures 1.18 and 1.19; *Australian Journal of Scientific Research* for Figure 1.21; *Journal of Experimental Botany* for Figures 1.22 and 1.23.

REFERENCES

Afzelius, B. (1959) Electron microscopy of the sperm tail. Results obtained with a new fixative. *Journal of Biophysical and Biochemical Cytology*, 5, 269–278.

Ballowitz, E. (1888) Untersuchungen über die Struktur der Spermatozoën, zugleich ein Beitrag zur Lehr vom feineren Bau der contraktilen Elemente. *Archiv für mikroskopische Anatomie*, **32**, 401–473.

Barry, M. (1843) On fissiparous generation. *Edinburgh New Philosophical Journal*, **35**, 205–221.

Bary, A. de. (1879) *Die Erscheinung der Symbiose*. Strasburg.

Bishop, D. W. (1962) Sperm motility. *Physiological Review*, **42**, 1–59.

Bloodgood, R. A. ed. (1990) *Ciliary and Flagellar Membranes*. New York: Plenum Press.

Brokaw, C. J. (1961) Movement and nucleoside polyphosphatase activity of isolated flagella from *Polytoma uvella*. *Experimental Cell Research*, **22**, 151–162.

Bruce, D. (1895) *Preliminary Report on the Tsetse Fly Disease or Nagana in Zululand*. Durban: Bennett and David.

Bütschli, O. (1880/9) Protozoa. In *Klassen ünd Ordungen des Thier-reichs* (ed. H. G. Bronn), vol. 1. Leipzig: C. F. Winter.

Carter, H. J. (1871) A description of two new Calciospongiae, to which is added confirmation of Prof. James-Clark's discovery of the true form of the sponge cell (animal), and an account of the polype–like pore–area of *Cliona corallinoides* contrasted with Prof. E. Häckel's view on the relationship of the sponges to the corals. *Annals and Magazine of Natural History*, Ser 4, **8**, (43), 1–27.

Cavalier-Smith, T. (1993) Kingdom Protozoa and its 18 phyla. *Microbiological Reviews*, 57, 953–994.

Cavalier-Smith, T. (1998) A revised six-kingdom system of life. *Biological Reviews*, **73**, 203–266.

Chatton, E. (1925) *Pansporella perplexa*, amoebien a spores protégées parasite des daphnies. Réflexions sur la biologie et la phylogénie des protozoaires. *Annales Science Naturelle Zoologie*, **8**, 5–84.

Cohn, F. J. (1853) Flagellata. *Zeitschrift für wissenschaftliche Zoologie*, **6**.

Copeland, H. F. (1956) *The Classification of Lower Organisms*. Palo Alto, California: Pacific Books.

Corliss, J. O. (1975) Three centuries of protozoology: a brief tribute to its founding father, A. von Leeuwenhoek of Delft. *Journal of Protozoology*, **22**, 3–7.

Corliss, J. O. (1978) A salute to fifty-four great microscopists of the past: a pictorial footnote to the history of Protozoology. Part 1. *Transactions of the American Microscopical Society*, **97**, 419–458.

Corliss, J. O. (1984) The kingdom Protista and its 45 phyla. *Biosystems*, **17**, 87–126.

Corliss, J. O. (1986) Progress in protistology during the first decade following the re-emergence of the field as a respectable interdisciplinary area in modern biological research. *Progress in Protistology*, **1**, 11–63.

Corliss, J. O. (1994) An interim utilitarian ('user–friendly') hierarchical classification and characterization of the protists. *Acta Protozoologica*, **33**, 1–51.

Corliss, J. O. (1998a) Classification of protozoa and protists: the current status. In *Evolutionary Relationships Among Protozoa*, (eds G. H. Coombs, K. Vickerman, M. A. Sleigh and A. Warren). Systematics Association Special Volume 56, Dordrecht: Kluwer, pp. 409–447.

Corliss, J. O. (1998b) Haeckel's kingdom Protista and current concepts in systematic protozoology. *Stapfia*, **56**, 85–104.

Darwin, C. (1859) *On the origin of species by means of natural selection, or the preservation of favoured races in the struggle for life*. London: John Murray.

Deflandre, G. (1934) Existence sur les flagélles de filaments lateraux ou terminaux (mastigonemes). *Comptes Rendus de l'Academie des Sciences (Paris)*, **198**, 497–499.
Diesing, K. M. (1866) Revision der Prothelminthen. Abtheilung: Mastigophoren. *Sitzungsberichte – Akademie der Wissenschaften in Wien, Mathematisch–naturwissenschaftliche*, **52**, 287–401.
Dobell, C. (1932) *Antony van Leeuwenhoek and his 'Little Animals'*. Amsterdam: Swets and Zeitlinger.
Dobell, C. (1951) In Memoriam. Otto Bütschli (1848–1920), "Architect of Protozoology." *Isis*, **42**, 20–22.
Doflein, F. J. T. (1916) *Lehrbuch der Protozoenkunde*. 5th edn. Jena: Fischer.
Donné, A. (1836) Animalcules obsérves dans les matiéres purulentes et le produit des sécrétions des organes génitaux de l' homme et de la femme. *Comptes Rendus de l'Academie des Sciences* **3**, 385–386.
Dryl, S. and Zurzycki, J. eds (1972) International symposium on motile systems of cells. *Acta Protozoologica*, **11**.
Dujardin, F. (1841) *Histoire Naturelle des Zoophytes. Infusoires*. Paris: Roret.
Ehrenberg, C. G. (1838) *Die Infusionthierchen als Vollkommene Organismen*. Leipzig: L. Voss.
Engelmann, T. W. (1881) Über den faserigen Bau der contractilen Substanzen u.s.w. *Pflügers Archiv für die gesamte Physiologie des Menschen und der Tiereu*, **25**, 538–65.
Evans, G. (1880) Report of Surra disease in the Dera Ismael Khan district. *Punjab Goverment Military Department*, **493**, 446.
Fauré-Fremiet, E. (1961) Cils vibratiles et flagelles. *Biological Reviews*, **36**, 464–536.
Fawcett, D. W. and Porter, K. R. (1954) A study on the fine structure of ciliated epithelia. *Journal of Morphology*, **94**, 221–282.
Fischer, A. (1894) Über die Geisseln einiger Flagellaten Princs. *Jahrbücher für wissenschaftliche Botanik*, **26**, 187–235.
Gluge, G. (1842) Über ein eigenthümliches Entozoon im Blute des Frosches. *Archiv für Anatomie, Physiologie und wissenschaftliche Medizin*, p. 148.
Goldfuss, G. A. (1817) *Über die Entwicklungsstufen des Thieres*. Nurenberg: Leonard Schrag.
Goldfuss, G. A. (1820) *Handbuch der Zoologie*, vol. 1. Nuremberg.
Grant, R. E. (1835) On the nervous system of *Beroé pileus* Lam. and on the structure of its cilia, (Communicated 1833). *Transactions of the Zoological Society of London*, **1**, 9–12.
Grassé, P.-P. (1952) *Traité de Zoologie. Anatomie, Systématique, Biologie*. 2 vols. Paris: Masson.
Gray, J. (1928) *Ciliary Movement*. Cambridge: Cambridge University Press.
Grigg, G. W. and Hodge, A. J. (1949) Electron microscopic studies of spermatazoa. *Australian Journal of Scientific Research*, **2**, 271–286.
Gruby, D. (1843) Recherches et observations sur une nouvelle espéce d'haematozoaire, *Trypanosoma sanguinis*. *Comptes Rendus de l'Academie des Sciences*, **17**, 1134–1136.
Haeckel, E. H. P. A. (1866) *Generelle Morphologie der Organismen*. Berlin: G. Reimer.
Haeckel, E. H. P. A. (1869) Über den Organismus der Schwämme und ihre Verwandtschaft mit der Corallen. *Jenaische Zeitschrift*, **5**, 207–254.
Haeckel, E. H. P. A. (1870) On the organisation of the sponges and their relationship to the corals. *Annals and Magazine of Natural History*, Ser. 4, Vol. 5, 1–13, 107–120.
Haeckel, E. H. P. A. (1873) On the Calciospongiae, their position in the animal kingdom and their relation to the theory of descendence. *Annals and Magazine of Natural History*, Ser. 4, 241–262, 421–431.
Haeckel, E. H. P. A. (1878) *Die Protistenkreich*. Berlin.
Hartog, M. (1906) Protozoa. In *The Cambridge Natural History – Protozoa, Porifera, Coelenterata and Echinodermata*, (eds S. F. Harmer and A. E. Shipley). London: Macmillan.

Hausmann, K. and Hülsmann, N. (1996) *Protozoology*. Stuttgart: Georg Thieme Verlag.

Heidenhain, M. (1907) *Plasma und Zelle*. Jena: Gustav Fischer.

Henneguy, L.F. (1898) Sur les rapports des cils vibratiles avec les centrosomes. *Archiv fur microskopische Anatomie*, **1**, 482–496.

Hogg, J. (1860) On the distinctions of a plant and an animal, and on a fourth kingdom of nature. *Edinburgh New Philosophical Journal*, **12**, 216–225.

Honigberg, B. M. (Chairman), Balamuth, W., Bovee, E. C., Corliss, J. O., Gojdics, M., Hall, R. P., Kudo, R. R., Levine, N. D., Loeblich, A. R. Jr., Weiser, J. and Wenrich, D. H. (1964) A revised classification of the Phylum Protozoa. *Journal of Protozoology*, **11**, 7–20.

Hülsmann, N. (1992) Undulipodium: end of a useless discussion. *European Journal of Protistology*, **28**, 253–257.

James-Clark, H. (1866) Conclusive proofs on the animality of the ciliate sponges, and their affinities with the Infusoria Flagellata. *Annals and Magazine of Natural History*, Ser. 4, vol. 19, 13–19.

James-Clark, H. (1868) On the Spongiae Ciliatae as Infusoria Flagellata: or observations on the structure, animality and relationship of *Leucosolenia botryoides* Bowerbank. *Annals and Magazine of Natural History*, Ser. 4, vol. 1, 133–142, 188–215, 250–264.

James-Clark, H. (1878) On the nature of sponges. *Proceedings of Boston Natural History Society*, **11**, 16–17.

Joblot, L. (1718) *Descriptions et Usages de Plusiers Nouveaux Microscopes tout simples que composez avec de Nouvelles Obsevations*. Paris: J. Collombat.

Kent, W. S. (1880/2) *A Manual of the Infusoria*. Vols 1–3. London: David Bogue.

Kudo, R. R. (1966) *Protozoology*, 5th edn. Springfield, Illinois: Thomas.

Ledermuller, M. F. (1763) *Mikroskopische Gemüthsergötzungen und Augenergötzungen*. Nuremberg.

Leeuwenhoek, A. van (1674) More observations from Mr Leeuwenhoek in a letter of September 7, 1674, sent to the publisher. *Philosophical Transactions of the Royal Society*, **9**, 178–182.

Leeuwenhoek, A. van (1677) Observations communicated to the publisher by Mr Antony van Leewenhoeck in a Dutch letter of the 9th of Octob. 1676 here English'd: Concerning little animals by him observed in rain-, well-, sea- and snow-water, as also in water wherein pepper had lain infused. *Philosophical Transactions of the Royal Society*, **12**, 821–831.

Leeuwenhoek, A. van (1683) Letter 38 – To Christopher Wren, 16 July 1683. Short English translation. *Philosophical Transactions of the Royal Society*, **13**, (152), 347.

Leeuwenhoek, A. van (1700) Letter 122 – To Sir Hans Sloane, 2 January 1700. *Philosophical Transactions of the Royal Society*, **22**, (261), 509.

Leeuwenhoek, A. van (1702) Letter 144 – To Hendril van Bleyswyk, 9 February 1702. *Brieven* (7de Vervolg), III, 400; *Op. Omnium (Epist. Soc. Reg.)*, III, 380.

Lenhossek, M. von (1898) Über Flimmerzellen. *Verhandlungen der Anatomischen Gesellschaft*, **12**, 106–128.

Leuckart, K. (1879) *Allgemeine Naturgeschichte der Parasiten*. Leipzig: C. F. Winter.

Levine, N. D. (Chairman), Corliss, J. O., Cox, F. E. G., Deroux, G., Grain, J., Honiberg, B. M., Leedale, G. F., Loeblich, A. R., Lom, J., Lynn, D., Merinfeld, E. G., Page, F. C., Poljansky, G., Sprague, V., Vavra, J. and Wallace, F. G. (1980) A newly revised classification of the Protozoa. *Journal of Protozoology*, **27**, 37–58.

Linnaeus, C. (1735) *Systema naturae, sive regna tria naturae systematice proposita per classes, ordines, genera et species*. Lugduni Batavorum.

Linnaeus, C. (1758) *Systema Naturae*. Vol. 1. 10th edn. Holmii: Impensis Laurentii Salvii.

Lipscomb (1991) Broad classification: the kingdoms and the protozoa. In *Parasitic Protozoa*, (eds J. P. Kreier and J. R. Baker), vol. 1, 2nd edn, New York: Academic Press, pp. 81–136.

Loeffler, F. (1889) Eine neue Methode zum Färbern der Mikroorganismen, im besonderen ihrer Wimperhaare und Geisseln. *Zentralblatt für Bakteriologie und Parasitenkunde*, **6**, 209–224.

Luther, A. (1899) Über *Chlorosaccus*, eine neue Gattung der Süsswasseralgen, nebst einigen Bemerkungen zur Systematik verwandter Algen. *Bihang till kungliga Svenska vetenskapsakademiens handlingar*, 24, Afd. III, 1–22.

Maier, H. N. (1903) Über den feineren Bau der Wimperapparate der Infusorien. *Archiv für Protistenkunde*, **2**, 73–179.

Manton, I. (1950) Demonstration of compound cilia by means of the ultraviolet microscope. *Journal of Experimental Botany*, **1**, 69–70.

Manton, I. (1952) The fine structure of plant cilia. Structural aspects of cell physiology. *Symposia of the Society for Experimental Biology*, **6**, 306–319.

Manton, I. and Clarke, B. (1952) An electron microscope study of the spermatozoid of *Sphagnum*. *Journal of Experimental Botany*, **3**, 265–275.

Margulis, L. (1970) *Origin of eukaryotic cells*. New Haven: Yale University Press.

Margulis, L. (1993) *Symbiosis in cell evolution*. New York: W. H. Freeman.

Margulis, L. and Schwartz, K. V. (1998) *Five kingdoms. An illustrated guide to the phyla of life on earth*. 3rd edn. San Francisco and New York: W. H. Freeman.

Margulis, L., Chapman D. J., Corliss J. O. and Melkonian, M., eds (1990) *The Handbook of Protoctists*. Boston: Jones and Bartlett.

Merezhkovsky, K. S. (1905) Über Natur und Ursprung der Chromatophoren in Pflanzenreiche. *Biologisches Zentralblatt*, **25** (18), 593–604.

Merezhkovsky, K. S. (1906) Principles of endochromes. In *Memoirs of the University of Kazan*, vol. 73, bks 2, 3, 5–6, pp.1–176, 177–288, 289–402. (In Russian.)

Merezhkovsky, K. S. (1909) The theory of two plasms as the basis of symbiogenesis, a new study on the origins of organisms. *Proceedings of Studies of the Imperial Kazan University*, Publishing Office of the Imperial University. (In Russian)

Meyen, F. J. F. (1839) Einige Bemerkungen über den Verdau–ungs–Apparat der Infusorian. *Archiv für Anatomie, Physiologie und wissenschaftliche Medizin*, pp. 74–79.

Müller, O. F. (1773) *Vermium Terrestrium et Fluviatilium, seu animalium infusorium, helminthicorum et testaceorum, non marinorum, succincta historia*. Havniae [Copenhagen] et Lipsiae [Leipzig]: Heineck and Faber.

Müller, O. F. (1786) *Animalcula Infusoria Fluviatilia et Marina, quae detexit, systematice descripsit et ad vivum delineari craviti*. Havniae [Copenhagen] et Lipsiae [Leipzig]: Mölleri.

Patterson, D. J. (1994) Protozoa: evolution and systematics. In *Progress in Protozoology*, (eds K. Hausman and N. Hülsmann), Proceedings of the IX Congress of Protozoology, Berlin 1993. Stuttgart: Gustav Fischer Verlag, pp. 1–14.

Patterson, D. J. and Larsen, J. eds (1991) *The biology of free-living heterotrophic flagellates*. Oxford: Clarendon Press.

Perty, J. A. M. (1852) *Zur Kentniss kleinster Lebensformen nach Bau, Funktionen, Systematik, mit Specialverzeichniss der in der Schweiz, beobachteten*, Berlin: Jent und Reinert.

Petersen, J. B. (1929) Beitrage zur Kenntnis der Flagellatengeisseln. *Botanisk Tidsskrift*, **40**, 373–389.

Pitelka, D. R. (1949) Observations on flagellum structure in Flagellata. *University of California Publications in Zoology*, **53**, 377–403.

Purkinje, J. E. and Valentin, G. G. (1835) De phaenomeno generali et fundamentali motus vibratorii continui in membranis cum externis tum internis animalium plurimorum et superiorum et inferiorum ordinum obvii. In *Opera Omnia*, J. E. Purkinje, vol. 1, Vratislaviae (Bratislava), pp. 277–371.

Ragan, M. A. (1997) A third kingdom of eukaryotic life: history of an idea. *Archiv für Protistenkunde*, **148**, 225–243.

Rothschild, L. J. (1989) Protozoa, Protista, Protoctista: what's in a name? *Journal of Historical Biology*, **22**, 277–305.

Satir, P. (1965) Studies on cilia. II. Examination of the distal region of the ciliary shaft and the role of filaments in motility. *Journal of Cell Biology*, **26**, 805–834.

Satir, P. (1968) Studies on cilia. III. Further studies on the cilium tip and a sliding filament model of ciliary motility. *Journal of Cell Biology*, 39, 77–94.

Schlegel, M. (1991) Protist evolution and phylogeny as discerned from small subunit ribosomal RNA sequence comparisons. *European Journal of Protistology*, **27**, 207–19.

Schlegel, M. and Hausmann, K. eds (1996) *Christian Gottfried Ehrenberg-Festschrift*. Leipzig: Leipziger Universitätsverlag.

Schleiden, M. J. (1838) Beiträge zur Phytogenesis. *Archiv für Anatomie, Physiologie und wissenschaftliche Medizin*, Leipzig, pp. 137–176.

Schmitt, F. O., Hall, C. E. and Jakus, M. A. (1943) The ultrastructure of protoplasmic fibrils. *Biological Symposia*, Lancaster, Pa.: Jacques Cattell Press, **10**, 261–276.

Schuberg, A. (1891) Zur Kenntnis des *Stentor coeruleus. Zoologische Jahrbücher*, **4**, 197–238.

Schwann (1839) *Mikroskopische Untersuchungen über die Übereinstimmung in der Struktur und dem Wachsthum der Thiere und Pflanzen*. Berlin: G. E. Reimer.

Sharpey, W. (1835) Cilia. In *Cyclopoedia of Anatomy and Physiology*, (ed. R. B. Todd). London: Longman, Brown, Green, Longman and Roberts, pp 606–638.

Siebold, C. T. E. von (1845) Lehrbuch der vergleichenden Anatomie der Wirbellossen Thiere. In *Lehrbuch der Vergleichenden Anatomie*, (eds C. T. E. von Siebold and H. Stannius), vol. 1. Berlin: von Veit.

Sleigh, M. A. (1974) *The Biology of Cilia and Flagella*. London: Pergamon.

Sleigh, M. A. (1995) Progress in understanding the phlogeny of flagellates. In *The Biology of Free-Living Heterotrophic Flagellates*, (ed. S. A. Karpov), *Cytology*, **37**, 985–1009.

Stanier, R. Y. and Niel, C. B. van (1962) The concept of bacterium. *Archiv für Mikrobiologie*, **42**, 17–35.

Stanier, R. Y., Doudoroff, M. and Adelberg, E. A. (1963) *The Microbial World*. 2nd edn. Englewood Cliffs, N.J.: Prentice-Hall.

Stein, F. (1859/83) *Der Organismus der Infusionsthiere*. Leipzig: W. Englemann.

Taylor, F. J. R. (1974) Implications and extensions of the serial endosymbiosis theory of the origin of eukaryotes. *Taxon*, **23**, 229–258.

Vickerman, K. (1997) Landmarks in trypanosome research. In *Trypanosomiasis and Leishmaniasis* (eds J. C. Mottram, G. H. Coombs and P. H. Holmes), London: CAB International, pp. 1–37.

Valentin, G. G. (1841) Über ein Entozoon im Blute von *Salmo fario. Archiv für Anatomie, Physiologie und wissenschaftliche Medizin*, 435–436.

Wallin, I. E. (1925a) On the nature of mitochondria, viii. Further experiments in the cultivation of mitochondria. *American Journal of Anatomy*, **35**, 403–415.

Wallin, I. E. (1925b) On the nature of mitochondria, ix. Demonstration of the bacterial nature of mitochondria. *American Journal of Anatomy*, **36**, 131–146.

Warner, F. D., Satir, P. and Gibbons, I. R. eds (1989) *Cell movement*, vol. 1. New York: Liss.

Wenyon, C. M. (1926) *Protozoology. A Manual for Medical Men, Veteranarians, and Zoologists*. 2 vols. London: Balliére, Tyndall and Cox.

Whittaker, R. H. (1959) On the broad classification of organisms. *Quarterly Review of Biology*, **34**, 210–226.

Williams, D. M. and Huxley, R. (1998) Christian Gottfried Ehrenberg (1795–1876): the man and his legacy. *The Linnean*, Special Issue 1, 1–88.

Wrisberg, H. A. (1765) *Observationum de Animalculis Infusoriis Satura*. Goettingae (Gottingen): Vendenhoek.

Chapter 2

The flagellate condition

J. A. Raven

ABSTRACT

Flagellates comprise a taxonomic diversity of unicellular and colonial organisms (or flagellate stages in life-cycles) in the size range 2–2000 μm which are predominantly free-swimming in a very wide range of free-living and symbiotic situations. They variously exhibit photolithotrophy and the phagotrophic and saprotrophic manifestations of chemoorganotrophy, as well as combinations of these modes, in addition to associations with other organisms in symbiosis *sensu lato*. Flagellates compete with organisms having the same range of trophic modes, but with alternative life forms, such as non-motile or gliding unicellular or filamentous (including mycelial) organisms in sediments, and non-flagellate organisms in the plankton. In comparison with these other life forms occupying similar habitats we can identify costs of the flagellate condition and benefits which presumably outweigh the costs in extant flagellates. Identifiable costs of flagella are those of construction of flagella (energy, C, N, P), of flagellar operation (energy, with an upper limit on energy cost imposed by the maximum catalytic capacity for energy transformation by the measured content of dynein ATPases, and a lower limit imposed by the minimum energy dissipated against drag forces) and of flagellar maintenance. There does not seem to be a measurable cost of the flagellate condition in reducing specific growth rate relative to comparable (in size and phylogeny) organisms under either resource-saturated or resource-limited conditions.

The benefits of flagellar activity are not always easy to express in the same units as those of the costs; this is especially the case for reproductive benefits, such as dispersal and sex. The benefits in terms of resource acquisition and retention are more readily related to costs. Benefits in resource acquisition terms include the capacity to allow photolithotrophs to reach the 'optimal' position in the light and in the dark in opposing vertical gradients of light and nutrients in unmixed water-bodies, and to allow phagotrophs to maximize particle extraction from the medium for both swimming and attached flagellates as well as allowing attached phototrophs to maximize inorganic C uptake. In terms of resource retention flagellar mobility can help to avoid predation and avoid damage caused by high fluxes of photosynthetically active radiation and of UV-B. At the risk of being Panglossian, it appears to be possible to explain some at least of the occurrences of flagellates in various habitats, either as the dominant phase in the life-cycle or a temporally restricted phase in the life-cycle, in terms of cost-benefit analyses. However, agreement between cost-benefit prediction and what happens in the real world needs further investigation in terms of causal evolutionary (inclusive fitness) terms.

2.1 Introduction

The presence of flagella gives ecological and hence evolutionary opportunities to organisms, but also imposes certain constraints. This contribution aims to describe, and where possible to quantify, the costs and benefits of the flagellate condition in the hope of providing some evolutionary rationale for the phylogenetic and ecological occurrence of flagella. In pursuing this end there will be brief consideration of alternative means of achieving some flagellar functions.

The occurrence of flagella appears to be ancestral in the eukaryotes, and flagella (sometimes in their ciliary manifestations) are widespread among the eighteen phyla of the kingdom Protozoa (Cavalier-Smith, 1993; cf. Margulis *et al.,* 1994). Flagella are also common in the kingdom Viridiplantae, but are absent from the Biliphyta (red algae) and all extant seed plants except *Ginkgo* and the cycads (Cavalier-Smith, 1993, 1996). Flagella are commonly found in metazoa as ciliated epithelia and as sperm propellants, and also in the 'lower' fungi *sensu stricto* as well as in the oomycetes which are heterokont Chromista (Cavalier-Smith, 1993). The evolution, structure and mechanism of energizing bacterial and archaeal flagella are very different from the eukaryotic flagella, and they will only be mentioned in connection with constraints in certain organisms which are imposed by the presence of eukaryotic flagella which do not apply to Bacteria or Archaea (e.g. volume regulation in fresh water).

This contribution deals mainly with flagellates in the sense of flagellate cells or colonies in the size range 2 μm – 2 mm as the main phase in the life-cycle, but also considers the occurrence of flagella in the vegetative or trophic part of the life-cycle of larger organisms, and in reproduction and dispersal of such organisms (Fagerström *et al.*, 1998; Grosberg and Strathmann, 1998).

2.2 The occurrence and putative roles of flagella in relation to nutrition, life-cycle and habitat of organisms

2.2.1 *Nutritional range of flagellates*

The range of flagellate organisms depends on photolithotrophy and both the saprotrophic and the phagotrophic variants of chemoorganotrophy. There are also mixotrophs which combine various of these modes of nutrition (see for example Raven, 1995, 1997a; Sleigh, Chapter 8 this volume), and a number of chemoorganotrophic flagellates are obligate or facultative anaerobes (Fenchel and Finlay, 1995). These different modes of energy metabolism are very largely dependent on the endosymbiotic incorporation of other organisms into the flagellate. Thus, the capacity for aerobic chemoorganotrophy depends on an (endo)symbiosis involving an ancestral flagellate and a ß-proteobacterium capable of tricarboxylic acid cycle and oxidative phosphorylation. This symbiont subsequently became genetically integrated by loss of some of the bacterial genes needed for independent existence, and transfer of most of the genes, required for functioning of the endosymbiont as a mitochondrion, to the flagellate nucleus. The hydrogenosomes (where, in all but one reported case, all symbiont genes have been lost or transferred to the host nucleus: Akhananova *et al.*, 1998) of some obligately anaerobic flagellates probably shared a common origin with the mitochondrion, and the perceived likelihood that any extant flagellates are primitively amitochondrial has diminished over the last few years (see Martin and Müller, 1998; Brugerolle, Chapter 9 this volume).

The plastids which give many flagellates their capacity to photosynthesize are also genetically integrated products of (endo)symbiosis, in this case of a primary endosymbiosis involving a cyanobacterium (Chlorophyta and the non-flagellate Rhodophyta) or a secondary endosymbiosis involving a eukaryotic photosynthetic unicell resulting from a primary symbiotic event in the case of the Euglenophyta, the Chlorarachniophyta and (the not now phototrophic) apicomplexians (endosymbiosis of a unicellular member of the Chlorophyta) and the Chromista (Heterokontophyta, Cryptophyta and Haptophyta) and (?) Dinophyta (endosymbiosis of a unicellular member of the Rhodophyta) (Martin *et al.*, 1998; Roberts *et al.*, 1998).

In addition to genetically integrated photosynthetic organelles, some flagellates (and ciliates) are photosynthetic by virtue of a photosynthetic apparatus which has only recently (within one or a few generations) been acquired (Raven, 1997a). Thus, the occurrence of both mitochondria and plastids as genetically integrated organelles in eukaryotes may have resulted from phagotrophy by flagellate ancestors (Martin and Müller, 1998). In the case of genetically chemoorganotrophic flagellates these are generally kleptoplastids, which are derived from phagotrophically ingested food items; since many of the genes needed for their synthesis and maintenance are in the food item genome rather than host flagellate genome, these kleptoplastids cannot function for more than days or weeks (which is a long time in terms of the minimum generation time of most flagellates).

There is no reason why the ingested food item yielding the kleptoplastids cannot itself be a flagellate. Ingestion of a flagellate is the rule in those marine invertebrate (Cnidaria, tridacnid bivalves) and protistan (Foramenifera, Radiolaria) photosynthetic endosymbioses involving the dinoflagellate *Symbiodinium*. This flagellate is ingested anew in each host cell generation except in those cnidarians in which the symbiont is transferred between generations in the eggs of the host. *Symbiodinium* is capable of independent existence.

2.2.2 Flagella in life-cycles

The 'mission statement' of this article (as stated at 2.1) is to deal mainly with the 2 µm – 2 mm (equivalent spherical diameter) cells and organisms which are flagellate for all or most of their life-cycle, although there may be encysted, non-flagellate stages which permit the organism to survive circumstances inimical to survival of the flagellate phases (Raven, 1982). There are flagellate cells that are stages in the life-cycle of organisms which are non-flagellate during the major phases of nutrient acquisition but produce flagellate cells involved in assexual and/or sexual reproduction and dispersal. This is the situation in, for example, centric diatoms (male gametes), brown algae and oomycetes (assexual spores and gametes), male gametes of many metazoa, and zoospores and male gametes of many green algae as well as male gametes of bryophytes, pteridophytes, cycads and *Ginkgo*.

2.2.3 *Functioning of flagella in relation to habitat*

Here (Table 2.1) we deal with flagellar functions in an ecological context. Flagellar motility of unattached cells or organisms can increase resource acquisition by photolithotrophs, saprotrophs and phagotrophs in a relatively homogenous (over the scale of the organism) environment. Photolithotrophic cells and organisms can rotate by flagellar activity, allowing a more equal distribution of photons to chromophores in an organism in a vector light environment. Many photolithotrophic flagellates show such rotation as they swim with frequencies which could permit constructive variations in the photon supply to different parts of the photosynthetic apparatus, that is, give photosynthetic rates in excess of that found if the cell was not rotating while swimming in the same radiation field (see Raven, 1994b). Furthermore, Harz *et al.* (1992) propose that rotation during swimming by chlamydomonad green algae integrates the vectorial light signal, providing information on the light direction for phototactic responses. The cells orient themselves such that modulation of the light signal detected with the eyespot during rotation is minimal, ensuring that swimming occurs either away from or toward the light.

For acquisition of chemical resources there are also advantages in movement relative to a homogenous (but growth rate-limiting) resource supply, including being stationery (benthic) and causing a flow of resource-containing water over the organism. These stimulatory effects are generally more important for particle capture in phagotrophy than for the uptake of dissolved material in photolithotrophy and saprotrophy in small (< 200 μm) organisms (Fenchel, 1987; Pahlow *et al.*, 1997; Sand-Jensen *et al.*, 1997). However, in larger organisms (for example, symbiotically photosynthetic sponges) the water currents induced by flagellar activity may be important in supplying inorganic C for photosynthesis (Section 2.3.3a). Flagella-stimulated particle ingestion by the host cells also provides them with those photobionts which are capable of independent life and which are acquired anew in each host generation, and also kleptoplastids with an even shorter potential life-span in the host (Section 2.2.1). Phagotrophy combined with phototrophy (mixotrophy) can help to provide an organism with a more nearly optimal ratio of resources (energy, C, N, P) in a given environment than would chemoorganotrophic phagotrophy, or photolithotrophy, alone (Raven, 1981, 1997a; Haustrom and Riemann, 1996; cf. Barbeau *et al.*, 1997).

A further potential benefit of the flagellate condition is found in heterogenous environments in which the resources are not equally available in all parts of the habitat. Here flagellar motility can, in theory, permit the organism to be positioned so that the organism can obtain as near as possible to an optimal (for its growth and maintenance) ratio of resources. In the case of planktonic photolithotrophs there is usually a gradient of photosynthetically active radiation (highest near the surface) in the opposite direction to the gradient for (inorganic) nutrients (highest at depth, where regeneration by chemoorganotrophic activities exceeds photolithotrophic consumption). Here the optimal ratio of resource availabilities over a twenty-four-hour

Table 2.1 Benefits of flagella

Benefits of flagella and restrictions on their applicability	*Alternatives to flagella in providing the benefits supplied by flagella and restrictions on their applicability*	*References*
(1) Rotation of cells during mobility can help to constructively alternate the (limiting) photon flux density incident on different parts of the photosynthetic apparatus in phototrophs. Limited to vector radiation fields, organisms large enough to have a significant self-shading effect.	Rotation occurs naturally (thermal motion or microturbulence) at 'constructive' frequencies in cells less than a few μm in diameter, although such small cells may not have enough self-shading to benefit from this rotation.	Raven (1994b) Section 2.3.3
(2) Movement through a 'homogenous' environment increases nutrient availability by decreasing diffusion boundary layer thickness (only significant for larger organisms) and hence increasing uptake of dissolved nutrients from growth-rate-limiting concentrations for phototrophs and saprotrophs. More quantitatively important for phagotrophy by flagellates. Particles can contribute information (genes) as well as matter and energy (endosymbiosis).	Larger free-swimming animals rely on muscular activity.	Pahlow, Riebesell and Wolf-Gladrow (1997) Section 2.2.3
(3) Movement of water over, or through, benthic organisms (e.g. sponges) aided by flagella activity aids uptake of solutes, particles.	'Passive' flow due to currents, waves, relative to attached organisms with no flow-generating flagella or muscles; actomyosin driven pumping in wall-less organisms.	Vogel (1981) Section 2.3.3
(4) Flagellates can maximize acquisition of resources from a spatially heterogeneous environment, e.g. where resource gradients are in opposite directions in waters with little vertical water movement, photosynthetic flagellates can position themselves with near-optimal light availability in daylight but deeper, in high nutrient concentrations, at night. Can work in free water bodies and in sediments/soils, but limited to where vertical water movement is slower than speed of swimming.	Muscle-based movements can also provide positioning in the water column which, for large organisms, can occur despite large vertical water movements. Muscle-based and pseudopodial (also actomyosin-based) movements can occur in sediments, as can gliding motility, all at velocities adequate to move through the opposing gradients to appropriate day and night positions. 'Passive' movements related to buoyancy variations occur in cephalopods, teleosts, cyanobacteria and diatoms.	Raven and Richardson (1984) Raven (1997c) Walsby (1994) Section 2.3.3

/continued overleaf

Table 2.1 (continued)

Benefits of flagella and restrictions on their applicability	*Alternatives to flagella in providing the benefits supplied by flagella and restrictions on their applicability*	*References*
(5) Flagellar mobility may help to remove motile organisms from physiochemically or biotically damaging environments, although the flagellar apparatus is more sensitive to UV-B than many cell functions, and movement from a damaging radiation environment using flagella depends on appropriate water movement regimes.	Other (e.g. actomyosin-based, 'passive' buoyancy-related, gliding) can be involved in avoiding damaging environments. Chrysophyte trichocysts (as offensive/defensive 'ballistic missiles') are only effective over a few μm even with a muzzle velocity of 260 m s^{-1}; 'cruise missiles' (flagella-powered 'torpedoes') would have a longer range.	Gordon (1987) Häder and Häder (1988) Raven (1988) Raven (1991) Section 2.3.3
(6) Dispersal and sexual reproduction involves flagella in organisms which are flagellate throughout their life as well as many of those which spend most of their time as non-flagellate unicells or as multicellular organisms with few or no flagellate cells. Limitations as for resource acquisition and avoidance of damaging environments; only works in water or water films on terrestrial organisms, or in internal fertilization.	Passive (borne on water currents) fertilization and dispersal is an important adjunct to flagellar motility of zoospores, flagellate sperm. Aquatic organisms with male gametes lacking flagella (red algae; pollen grains of seagrasses) can achieve submerged external fertilization. Ballistic' dispersal (fertilization) procedures much more effective in aerial than aquatic habitats, especially on a scale which is competitive with flagellar activity.	Van der Haage (1996) Serrão *et al.* (1996) Raven (1998a) Section 2.3.3

diel cycle can be achieved by diel vertical migrations such that the organisms are closer to the surface in the daytime but deeper in the water body where nutrient concentrations are higher at night (Raven and Richardson, 1984; cf. Rivkin *et al.*, 1984). This will be considered later (2.3.3b), but for the moment we note two further significant points. First, the observed diel vertical migrations are sometimes in the 'wrong' direction (according to the analysis earlier). Second, the execution of these diel migrations, and indeed of any positioning of a flagellate relative to the surface, depends on the vertical movement of water being slower than the speed at which the flagellate can swim.

The gradients of resources are much steeper in another major flagellate habitat, that is, the surface of submerged and (moist) soils (Fenchel, 1987; Fenchel and Finlay, 1995). Here photosynthetically active radiation gradients found over metres or tens of metres in the water body are found over millimetres in sediments. Flagellates in these circumstances can also show diel migrations, modified by responses to tidal cycles in marine and estuarine sediments (Round, 1981).

In addition to positive tactic (or phobic) responses to resources present in quantities which are insufficient to saturate growth, there are also negative tactic or phobic responses permitting flagellate organisms to move from environments with an excessive level of a resource, for example, photosynthetically active radiation (Häder and Griebenow, 1988; Raven, 1989). In many cases the environmentally correlated, and damaging rather than nutritional, UV-B radiation acts to inhibit flagella activity, thus preventing movement of the organism from the damaging UV-B and photosynthetically active radiations (Häder and Häder, 1988; Raven, 1991).

Flagellar function is especially prone to damage by UV-B since it is impossible to screen UV-B by sacrificial UV-B-absorbing materials on the medium side of the flagellar plasmalemma (Raven 1991). The flagellar movements described so far in this section are nutritional, although with genetic overtones in the acquisition of photobionts. Motility using flagella also has a very obvious genetic role in those organisms which have flagellate gametes and thus involve flagella motility in syngamy in essentially full-time flagellates (for example *Chlamydomonas* spp.) as well as macroscopic organisms with gametes (sometimes with assexual zoospores) as the only flagellate stage in the life-cycle (see 2.2.2).

As with the specifically nutritional responses, the functioning of flagella in sexual reproduction and in the location of physically and chemically appropriate places for settlement of zoospores of benthic (attached) organisms are dependent on a hydrodynamic environment which does not involve water movements at speeds greater than those which can be countered by flagellar motility. Serrão *et al.* (1996) report an important behavioural response in the brown alga *Fucus vesiculosus*; gamete (motile flagellate spermatozoa; non-motile eggs) release is delayed until water movement is not so fast as to eliminate pheromone-directed flagellar movement of sperms to eggs.

A concomitant of these different roles of the flagella in nutrition, dispersal and sexual reproduction is that the flagellate organism or stage in the life-cycle must have some sensor for environmental cues which permits it to perceive aspects of the physics and chemistry of its environment, and in particular to perceive gradients in these physicochemical properties, and to have a signal transduction pathway which yields appropriate tactic or phobic responses. These responses involve cues such as electromagnetic radiation (Häder, 1979; Häder and Griebenow, 1988), gravity (Häder and Griebenow, 1988), (geo)magnetism (Torres de Araujo *et al.*, 1986) and dissolved chemicals including nutrient solutes (Amsler and Neushul, 1989) and (sex) pheromones (Maier, 1995).

2.2.4 *Alternatives to flagella*

So far in section 2.2 the roles of flagella have, as is appropriate in this volume, been described as a prelude to discussing (section 2.3) the costs and benefits of the flagellate condition. However, there are alternative means of achieving many of the outcomes of flagellar activity; these can also be subject to cost-benefit analysis.

First, the role of flagellar motility in rotating photolithotrophs as they swim, with the consequent possibility of photosynthesis-enhancing fluctuations in photon supply from a vector radiation field to different parts of the photosynthetic apparatus can be (indeed, must be) mimicked for very small (picoplankton) cells by small-scale turbulence causing rotation at the appropriate frequencies (Raven, 1994b). However, in these cases the small size of the cells means a very low absorptance even with the maximum possible pigment content so that the gradient of light absorption across the cell diameter in a vector light field is small, so the potential gain in terms of intermittency effects enhancing photosynthesis are also small (Raven, 1994b).

Second, enhancement of chemical resource uptake by movement of the organism relative to its environment can be also achieved for planktonic organisms by sinking (or rising) relative to the surrounding water via density differences between the organism and its environment (Raven, 1986, 1997c; Walsby, 1994; Pahlow *et al.*, 1997) without the need for flagella.

Third, movement of water over the nutrient-absorptive surface of a benthic organism as a means of preventing depletion of the local part of the bulk phase and/or disrupting boundary layers which limit the rate of nutrient uptake (including respiratory and photosynthetic gases, and particles) occurs by means of natural water movements over attached organisms, and can even lead to 'internal' ventilation of macroscopic multicellular organisms such as sponges (Vogel, 1981; Raven, 1993). Metazoa also have the possibility of muscular (acto-myosin) activity in bringing about such movements (see Raven, 1993).

Fourth, positioning organisms vertically in an aqueous environment with limited vertical fluid flow, including the case of diel vertical migration, can also be accomplished by muscular activity in larger metazoa (larger than a few mm) where the limitations on the size of the organism which can be thus positioned and the vertical water movement regime in which a given position can be maintained are both greater than for the flagellar motility case (Section 2.3). Regulation of density relative to that of the medium via gas vesicles (in prokaryotes), low-density solutions in aqueous vacuoles or extracellular spaces (in marine eukaryotes), or gas bladders in teleosts, as a means of reducing overall density, and solid or dissolved ballast as a means of increasing overall density is a further means of achieving a given vertical position in a water body (Raven, 1997c; Walsby, 1994).

Fifth, for wet soils and sediments, vertical positioning can be achieved by gliding motility (on the solid surface by some bacteria, including cyanobacteria, and eukaryotes, including many desmids and pennate diatoms) in the same size range as those with flagellar motility. Although gliding motility typically gives movement rates which are one to three orders of magnitude lower than flagellar motility, the distances through which movement occurs in achieving a given fractional change in resource availability are correspondingly smaller in the much steeper gradients in particulate habitats (Fenchel, 1987; Fenchel and Finlay, 1995; Raven, 1993, 1997); non-motile (filamentous) organisms in or on soil or sediment can compete with motile cells (Carlile, 1994; Raven, 1981).

Sixth, reproductive (sex, dispersal) roles of flagella are replaced by a number of other mechanisms in organisms which lack flagella. Amoeboid gametes (those motile via an actomyosin-based mechanochemical transducer) characterize several taxa with gamete transfer via a conjugation channel between walled cells (for example oomycetes, conjugalean green algae) and in some organisms where gametes are not constrained by walls at the time of syngamy (for example, pennate diatoms). Here motility is constrained relative to that of flagellate gametes by the need for a solid surface (provided by the inside surface of cell walls in oomycetes and conjugalean algae) and by the slow speed of amoeboid movement which, like gliding motility (see point five), is at least an order of magnitude lower than that of typical flagellar motility.

The Rhodophyta (like the conjugalean charophyceans) lack flagella, and their male gametes and spores (tetraspores and carpospores) are dispersed by water currents as are the pollen grains of seagrasses (van der Haage, 1996; Raven, 1998a) The limited data on fertilization success in these organisms suggest that it is not lower than for macroalgae with flagellate iso- or aniso-gametes, or male gametes, in similar habitats (Table 2.1).

2.3 Cost-benefit analyses of the flagellate condition

2.3.1 *Background to cost-benefit analyses*

Measurements, or failing that quantitative estimates, of the benefits accruing from the production, maintenance and operation of a given structural or functional trait can help to evaluate the possible evolutionary significance of that trait in cost-benefit analyses. Such analyses ultimately depend on assumptions related to optimal allocation of resources in the organism, that is, that any allocation of resources other than the one observed in a given environment means a lower inclusive fitness for the organism (Dennett, 1996). More immediately, the cost-benefit analysis depends on the cost(s) and the benefits(s) being expressible in the same units. This is particularly problematic when the costs are in energy or chemical resource units while the benefit is in terms of dispersal or genetic recombination. The analysis is complicated (or even vitiated) if some costs are neglected and/or some benefits are not identified. In the case of multiple benefits the question of weighting the various benefits is also a significant problem. Undeterred by these considerations, Section 2.3.2 considers costs of flagella while 2.3.3 aims to quantify some benefits.

2.3.2 *Costs of flagella*

Consideration of the costs of flagella includes the resource costs of producing, maintaining and operating flagella, the resource costs of necessary corollaries of flagella such as the contractile vacuoles of freshwater flagellate cells, and the growth and maintenance costs of flagellate as opposed to non-flagellate cells (Table 2.2).

Capital costs

The resource costs of producing flagella are discussed by Raven and Richardson (1984) in the context of the overall resource costs of cell synthesis discussed by Raven (1982; cf. Raven and Beardall, 1981). For a dinophyte cell of 25 µm effective spherical radius and with a total flagella length of 250 µm, the energy, C, N and P needed to synthesise the flagella is not more than 10^{-3} that of synthesis of the rest of the cell (Raven and Richardson, 1984). These calculations take into account the genetic costs of flagella in the sense of the cost of replicating and maintaining, as well as of transcribing and translating, the genes associated with the flagella.

Table 2.2 Costs of flagella

Costs of flagella	*Costs of alternatives to flagella*	*References*
(1) Capital costs Flagellar apparatus (including basal bodies) for a 25 μm equivalent spherical radius cell costs not more than 10^{-3} of synthesis of rest of cell in carbon, nitrogen, phosphorus, or energy terms. For freshwater organisms, a contractile vacuole mechanism of volume regulation (active water efflux) is needed; this again costs less than 10^{-3} of the rest of the cell in C, N, P or energy terms for its synthesis.	Costs of amoeboid or eukaryotic 'gliding' motility on a surface are less readily quantified than are costs of the flagellar apparatus, but costs are unlikely to be significantly less than thoseof flagella. The same applies to muscular mechanisms of motility in a bulk water phase, or a solid surface or in flying. 'Passive' means of adjusting buoyancy (e.g. gas vesicles/ ballast of prokaryotes; aqueous vacuoles of marine phyto-plankton) not less than costs of flagella. For freshwater organisms amoeboid movement at least also involves contractile vacuoles. Capital costs of all walls as volume-regulating organelles cost two orders of magnitude more than contractile vacuoles in energy and carbon terms.	Raven (1982) Raven and Beardall (1981) Raven and Richardson (1984) Walsby (1994) Section 2.4.1
(2) Running costs Minimum work done against viscous drag for a 25 μm equivalent spherical radius cell moving at 500 μm s^{-1} uses less than 10^{-4} of energy consumed in growth. Maximum mechanistic cost based on dynein content and a specific reaction rate of 100 s^{-1} is still less than 10^{-3} of energy consumed in growth at a specific growth rate of 8.10^{-6} s^{-1}.	Minimum work done against viscous drag for an organism of a given size moving at a given speed is the same regardless of the means of mechanochemical conversion. Mechanistic cost depends on the effectiveness of coupling of the molecular water to macroscopic mechanism of movement.	Raven (1982) Raven and Beardall (1981) Raven and Richardson (1984) Section 2.4.1
(3) Maintenance costs Replacement of damaged proteins in flagella for a 25 μm equivalent spherical radius cell with 250 μm of flagella costs less than 10^{-3} that of protein 'repair' of the rest of the cell; a similar argument can be made for the costs of recouping solute leakage through membranes. The cost of operating the contractile vacuole apparatus in freshwater flagellates may be as	Replacement of damaged components of other material proteins (mainly actomyosin) and associated skeletal elements in amoeboid and muscular movement is likely to be at least as costly as in the case of flagella. The cost of operating the volume-regulating contractile vacuole apparatus in amoeboid freshwater organisms is similar	Raven (1982) Raven and Beardall (1981) Raven and Richardson (1984) Section 2.4.1

Table 2.2 continued

Costs of flagella	*Costs of alternatives to flagella*
little as 10^{-4} of the energy used in growth of a specific growth rate of 6.10^{-8} s^{-1} when considered on a minimum thermodynamic cost basis, while a mechanistic estimate is 2.10^{-3} of the energy used in growth at the maximum specific growth rate, using a very low estimate 10^{-15} m s^{-1} Pa^{-1} for the plasmalemma. Even with a higher 10^{-14} m s^{-1} Pa^{-1} estimate of hydraulic conductivity the minimum thermodynamic and mechanistic estimates are only 10^{-3} and 2.10^{-2} respectively of the energy used in growth at the maximum specific growth rate.	to that of flagellates, as is that of operating the volume-regulating system of freshwater metazoa. Cell walls as volume regulators in non-flagellate organisms (and, in flagellates, cells, as structures without the possibility of volume regulation) have low maintenance costs relative to those of contractile vacuole function.

Running costs

The resource costs of operating the flagellar apparatus were approached by Raven and Richardson (1984) by computing both the minimum energy input needed to overcome viscous forces to yield movement at the observed velocity, and the maximum energy which can be dissipated if the mechanochemical conversion apparatus (tubulin-dynein) is operating at its most likely maximal rate. For the model dinophyte considered in the capital costs section, moving at the relatively high rate of 500 μm s^{-1}, the work done against viscous forces is $1.19.10^{-13}$ W per cell, which is less than 0.1 of the computed power output as ATP in maintenance respiration for an otherwise similar non-motile cell, and less than 10^{-4} of the maximum mechanistic power consumption in growth at a specific growth rate of 8.10^{-6} s^{-1} (Raven and Richardson, 1984). For the maximum mechanistic power requirement for motility based on the content, and specific reaction rate, of the dynein ATPase, Raven and Richardson (1984) computed a power consumption in flagellar activity, with a dynein specific reaction rate of 100 s^{-1}, of $1.38.10^{-12}$ W per cell for the model dinophyte cell discussed in the section on capital costs. This value is equal to the energy output of maintenance respiration rate of a non-flagellate cell, but is still less than 10^{-3} of the maximum mechanistic power consumption in growth at a specific growth rate of 8.10^{-6} s^{-1} (Raven and Richardson, 1984). Motility has, of course, no direct running costs in terms of consumption of resources such as N and P, but uses C in the sense of producing carbon dioxide in respiratory processes energizing maintenance.

Maintenance

The maintenance of flagella (even in the absence of flagellar operation) also has energy costs, in that flagella become damaged and must be replaced. Furthermore, the flagellar plasmalemma leaks solutes and water when these are maintained out of energetic equilibrium across the membrane, with energetic costs of maintaining these gradients via energized fluxes countering the leakage fluxes, as well as costs of protein turnover.

The model dinophyte considered in the section on capital costs has less than 10^{-3} of its total protein in the flagella, so that if the protein damage (and hence resynthesis) rate is the same for flagellar proteins as for the mean of total cell proteins, then the maintenance energy costs of flagellar protein 'repair' are less than 10^{-3} of the total protein resynthesis costs for an equivalent non-flagellate cell. A similar argument can be made for solute leakage, since the flagellar plasmalemma area of the model dinophyte is only 2.10^{-2} of the total plasmalemma area. However, the occurrence of functional flagella in freshwater organisms dictates that the necessarily higher internal than external osmolarity and the attendant tendency for water to enter cannot be constrained by a rigid cell wall which permits turgor generation and volume regulation (Guillard, 1960; Raven, 1976, 1982, 1993); the need for movement of part of the plasmalemma relative to another part in flagellar motility means that a turgor-resistant wall is not compatible with motility. These considerations also apply to actomyosin-based motility, whether this involves amoeboid movement of single cells or muscular activity in metazoa, and to centrin (spasmoneme)-based movement in certain ciliates. Here the whole plasmalemma area of the cells is involved, and not just (in the case of flagellates) the plasmalemma of the flagella. We note that the frequent flagellar correlate of phagotrophy is not obviously compatible with the occurrence of a cell wall covering the whole cell with the exception of the flagellar axes, although Raven (1993) points out that multiple entry symbionts into walled higher plant cells occurs many times in the ontogeny of individual plants with intracellular N_2-fixing symbionts.

Raven (1982) has computed the minimum and mechanistic energy costs of operation of the contractile vacuole mechanism of water expulsion in freshwater flagellates. Raven (1982) deals with spherical flagellate cells of 5 μm and 10 μm radius. Applying these computations to the model dinophyte considered earlier and in Raven and Richardson (1984) shows that with L_p (hydraulic conductivity of the plasmalemma) of 10^{-14} m s^{-1} Pa^{-1} and a driving force for water entry of 2.10^5 Pa (Raven 1982) the water influx is $1.57.10^{-17}$ m^3 per second in a 50 μm equivalent spherical radius cell. The minimum rate of energy consumption in water expulsion is then the rate of water influx (= rate of water efflux in the steady state) of 1.57. 10^{-17} m^3 per second per cell times the driving force for water entry (2.10^5 Pa = 2.10^5 N m^{-2} = 2.10^5 J m^{-3}) or $3.14.10^{-12}$ W per cell. Since the power consumption in growth of $4.89.10^{-9}$ W per cell for a growth rate of 8.10^{-6} s^{-1} (generation time of twenty-four hours), contractile vacuole operation uses a *minimum* of $6.42.10^{-4}$ of the energy used in growth.

Turning to a mechanistic estimate of the energy costs of water expulsion, the estimates of Raven (1982) must be considered in relation to recent findings that the contractile vacuole membrane is energized by a V-type H^+ ATPase (Heuser *et al.*, 1993; Robinson *et al.*, 1998) with a presumed H^+: ATP of 2. This H^+: ATP stoichiometry would permit the KCl: ATP stoichiometry of 1 proposed by Raven (1982), so that the ATP cost of water expulsion by a mechanism of KCl secretion (with water) into the contractile vacuole with subsequent KCl resorption (without water) and water expulsion is 100 mol ATP per m^3 water expelled. With 55 kJ per mol ATP hydrolyzed this involves an energy consumption of 5.5 MJ per m^3 water expelled; with $1.57.10^{-17}$ m^3 water expelled per second per cell (see above) the power required is $8.66.10^{-11}$ W per cell, or 0.018 of the mechanistic energy requirement for growth at a specific growth rate of 8.10^{-6} s^{-1}, and significantly more than the energy available from maintenance respiration in a comparable non-flagellate cell. Raven (1982) points out that the L_p value of the plasmalemma of freshwater flagellate and amoeboid cells may be as low as 10^{-15} m s^{-1} Pa^{-1}, presumably involving the absence of aquaporins (Maurel, 1997), reducing the energy costs for water expulsion by an order of magnitude to $3.14.10^{-13}$ W per cell as the minimum energetic cost and $8.64.10^{-12}$ W per cell as the mechanistic cost, respectively $6.42.10^{-5}$ and $1.8.10^{-3}$ of the energy used in growth at 8.10^{-6} s^{-1}.

The mechanistic considerations by Heuser *et al.* (1993) of contractile vacuole operation, suggesting that water may be excreted as a NH_4^+ HCO_3^- solution iso-osmotic with the cytosol using excretory products of chemoorganotrophic metabolism (respiratory CO_2, and NH_3) is not readily related to the metabolism of photolithotrophs (Robinson *et al.*, 1998; cf. Raven, 1997b). At all events the energy cost of the mechanism of water excretion (iso-osmotic with cytosol) of Heuser *et al.* (1993) might, at its most efficient, be half that suggested by Raven (1982). However, the operation of the contractile vacuole system would still be significantly more energetically expensive than the operation of the flagella. Such considerations would also apply to actomyosin- or centrin-based motility systems in freshwater protists.

A final consideration here is the occurrence of cell walls. Raven (1976, 1982) pointed out that cell walls could not act in turgor resistance in flagellates, since there must be (flagellar) plasmalemma not covered by cell wall if the flagella are to function in motility. It is also clear that cell walls (and the looser-fitting loricas) of flagellates cannot grow by plastic extension driven by turgor (see Harold *et al.*, 1995, 1996; Pickett-Heaps, 1998). Computations in Raven (1982) show that the capital resource cost of volume regulation by cell walls (which have no running costs) exceeds the capital and running resource costs of water excretion in unicells growing at high specific growth rates, but that similar cells with slow, or no, growth have lower energy costs of volume regulation by cell wall synthesis than by water excretion. This resource costs argument could help explain the occurrence of walled cysts as resting stages in some flagellates (chrysophytes *sensu lato*; dinophytes) (Raven, 1982).

The multiple functions of cell walls could help to explain in selective terms their occurrence in some flagellates, for example *Chlamydomonas* spp. (Chlorophyceae) where protection could be a cell wall role, and in *Dinobryon* spp. (Chrysophyceae) where the lorica not only acts in protection but also alters the hydrodynamics of swimming cells and increases the chance of phagotrophy, albeit at the expense of increasing drag during swimming. In concluding this consideration of cell walls and flagella motility, we return to a point presaged in Section 2.1. This concerns the prokaryotic flagella of Bacteria and Archaea. These function in motility of walled cells using proton or sodium free energy gradients across the plasmalemma by acting as ion channels; since these channels do not conduct water to a significant extent they can and do function in walled, turgid cells. Similar considerations probably apply to the case of gliding motility in prokaryotes.

Summation of costs in terms of growth and survival

The measured and (more usually) estimated costs discussed in this section should be detected in terms of growth rates or survival times measured under conditions in which the benefits of flagella (sections 2.2, 2.3.3) are not expressed. For growth the conditions which eliminate the benefits of flagella are clearest when all resources are present in optimal supply, so that any role of flagella in stimulating resource acquisition is discounted. Here the diversion of resources to producing, operating and maintaining flagella might be expected, using optimal allocation arguments, to reduce the maximum specific growth rate to that of an otherwise comparable non-flagellate organism (Raven, 1986, 1994a). For resource-limited growth the resource whose supply is limiting growth must not be one whose supply is enhanced by flagella activity; an example is light limitation with fluctuations in photon flux density at the frequency of rotation during flagella motion (section 2.2.3). In the case of survival under conditions in which growth cannot occur, there is little possibility that flagella activity could enhance resource supply to an extent that would increase the potential for energy-requiring maintenance processes.

The available data on growth rates of flagellate and the most directly (phylogenetically) comparable non-flagellate organisms do not yield any obvious differences in maximum specific growth rate between flagellates and non-flagellates (data in Raven, 1986, 1994a, 1994c, 1998b, 1999; Strom and Morello, 1998). However, no studies seem to have been conducted which set out to compare the maximum specific growth rate of flagellates and non-flagellates. There seem to be no data available which fulfil the criteria set out above for comparisons of resource-limited growth rate and survival of non-growing conditions of flagellates and of closely comparable non-flagellates. Further work is clearly needed here, bearing in mind constraints on the resolution of differences in growth rate between otherwise similar flagellate and non-flagellate cells, as a result of genetic variation in the specific growth rates among clones of the same morphological species isolated from the same water body (Brand, 1981). The ratio of minimum/maximum growth rates varies among species from 0.3–0.8 with a coefficient of variation ranging from 3–13 per cent (Brand, 1981).

2.3.3 Benefits of flagella

This attempt to quantify the benefits of flagella does not attempt to cover all of the benefits of flagella mentioned in a qualitative context in Section 2.2. The emphasis here is on the benefits which can be most readily expressed in terms of the costs of flagella production, maintenance and activity, that is, energy, C, N, and P. Accordingly, the first cases considered concern the benefits of flagellar activity in acquisition of resources in, respectively, media which are (at the spatial scale of a benthic organism) homogenous with respect to the resource under consideration (i), and media which are (at the spatial scale of the planktonic organism's capacity to move in space) heterogeneous with respect to the resources under consideration (ii). After this there is a brief consideration (iii) of the benefits which are less readily related quantitatively to the costs of the flagellate condition.

(i) This consideration of the quantitative benefits of flagella in resource acquisition by benthic organisms in what is, at the scale of the organism, a homogenous resource supply environment takes as an example benthic photosynthetic flagellates (Sand-Jensen *et al.*, 1997) and sponges. For marine sponges with cyanobacterial photobionts (Cheshire and Wilkinson, 1991; Wilkinson *et al.*, 1988; Wilkinson, 1992) it is likely that the flagella-driven flow of seawater through the colony is required to supply inorganic carbon from seawater to the cells containing cyanobacteria at the rate required to account for the observed rate of photosynthesis (Raven, 1993; Raven and Cheshire, unpublished). Furthermore, the energy costs of flagellar synthesis, maintenance and operation are less than the benefits from increased rates of light energy acquisition in photosynthesis as a result of the flagella-enhanced inorganic carbon supply in natural light environments (Raven and Cheshire, unpublished). The nutritional benefits of flagella activity for benthic organisms in a homogenous environment can also be obtained, probably for similar resource costs, by actomyosin-based amoeboid and muscular movements (see Raven, 1981, 1993) for phototrophs, saprotrophs and phagotrophs.

(ii) The quantitative benefits of flagellar motility on resource acquisition from a spatially heterogeneous resource supply environment in relation to the costs of flagella activity have been considered by Raven and Richardson (1984; see Lieberman *et al.*, 1994) for planktonic flagellates. As was pointed out earlier (Section 2.2.2) the diel vertical migrations of photosynthetic flagellates could only yield resource acquisition benefits in excess of costs when vertical water movements are slower than the swimming speed of the motile organisms.

This analysis (Raven and Richardson, 1984) shows that the resource (energy, nitrogen, phosphorus) acquisition benefits of diel vertical migrations can exceed the resource costs (energy, nitrogen, phosphorus) of producing, maintaining and operating flagella (Table 2.1). This conclusion is based on the observed rates and extents of diel vertical migrations in natural populations, the attenuation of photosynthetically active radiation down the water column, and the concentration gradient of available nitrogen and phosphorus sources from the deeper, high-nutrient, predominantly chemoorganotrophic deeper waters to the low nutrient, predominantly phototrophic surface waters (Raven and Richardson, 1984). The conclusions from this analysis apply to freshwater as well as to marine flagellates despite the necessary resource costs of contractile vacuoles and their operation in freshwater flagellates, and also to flagellates in the steeper light and nutrient gradients in soil and aquatic sediments.

As with the discussion under (i), it is desirable, and to some extent possible, to compare the benefits and costs of motility induced by non-flagellar mechanisms with those of motility involving other mechanisms. For organisms in open water, the alternatives for diel (or longer-periodicity) vertical migrations are gas vesicles (in cyanobacteria) and aqueous vacuoles containing a low-density solution which occupy a large fraction of the volume of the protoplasts (in large-celled eukaryotes, a mechanism which can only function in higher-density salt waters). Walsby (1994) has shown that the resource costs of the production of gas vesicles in cyanobacteria exceed that of flagella in eukaryotes, a conclusion which stands even when resource costs of flagellar operation are considered. Nevertheless, the resource acquisition benefits of vertical migrations related to gas vesicle operation can exceed the resource costs of their production (and of carbohydrate ballast production) (Walsby, 1994). The resource-acquisition benefits of vertical migrations related to (aqueous) vacuolar density changes can also exceed the costs (Raven, 1997c).

In soil or sediments the alternatives to flagella motility in maximizing resource acquisition on a twenty-four-hour basis by diel vertical migration are gliding motility and amoeboid movement over a solid (sediment) surface. Here the steep gradients of light and of nutrients means that the slow speeds possible with gliding and amoeboid movements are adequate to permit diel vertical excursions of a few mm, which spans the whole of the depth sediment with sufficient light for net photosynthesis, and a substantial nitrogen and phosphorus gradient while the minimum energy cost of movement deduced from minimum energy expended in overcoming friction is similar to that of flagellar movement. Thus, the resource acquisition benefits of motility exceed the resource costs in certain environments for gliding as for flagellar motility (Raven, 1983; Raven and Richardson, 1984; Table 2.1).

(iii) For the potential benefits of flagellar activity which do not involve the acquisition of resources which can be expressed in the same units as the costs of flagellar motility, quantitative cost-benefit analyses are more difficult. No such quantitative analysis will be attempted for the reproductive (sex; dispersal) advantages of flagellar motility. However, such analyses can be performed for the avoidance of damage to the photosynthetic apparatus by high flux densities of photosynthetically active or UV-B radiation by flagellar (or other) motility. In the case of UV-B, the use of motility to move the organism to a lower flux density of UV-B is, as was shown in 2.2.2, reduced to the extent that motility is very sensitive to UV-B damage. For photoinhibition by photosynthetically active radiation, the sort of cost-benefit analysis performed by Raven (1989, 1994d) for the leaf-folding mechanism of the terrestrial flowering plant *Oxalis* spp. can be applied to motility by flagellates or to gliding motility. Such an analysis yields similar results in terms of a net energy gain from motility as from the avoidance reaction, that is, the cost of producing and operating the motility mechanisms can be less than the benefit of reducing the amount of foregone photosynthesis due to photodamage or thylakoid-level avoidance mechanisms. The requirements for a net energy gain by this motility-dependent mechanism of limiting photoinhibitory energy costs relative to the absence of motility includes the speed at which the organisms can move relative to the photosynthetically active radiation attenuation coefficient of the environment and the rate of increase of surface photon flux density, and the frequency with which such rapid increases of photon flux density occur. In view of the much greater light attenuation relative to the speed of swimming in sediments, the motility option for minimizing the effects of photoinhibition may show a higher benefit:cost ratio than it would in open water.

This consideration of the cost-benefit analysis of avoidance of photoinhibitory damage returns us to the problem (section 2.3.1) of multiple benefits of a given cost. In this case the cost is that of flagella production while the benefits are the optimization of resource (photons, nitrogen and phosphorus) acquisition over a twenty-four-hour period including, in this case, the benefit of maximizing net energy acquisition by motility-related avoidance of photoinhibition. Here the additional benefit of photoinhibition avoidance can be regarded as an extension (in the same resource units) of the original cost–benefit analysis.

2.4 Flagellate opportunities and limitations

2.4.1 *Opportunities*

The wide range of functions of flagella discussed in Section 2.2, and the quantitation of some of the benefits relative to costs in 2.3, show that flagella have contributed to a great diversity of functions in eukaryote evolution, and that in some cases the evolutionary fitness of the functions can be, at least in part, rationalized in quantitative cost–benefit analyses.

2.4.2 *Limitations*

There are constraints on the range of sizes of flagellate organisms which can live as self-motile organisms. Thus, there is an upper limit on the size of free-swimming flagellates (including ciliates) in the form of ctenophores which are much smaller than the largest muscle-driven motile organisms (blue whales). For photosynthetic organisms, regardless of their mechanisms of mechanochemical transduction, there is an upper limit on size related to the area for photon absorption for motile photolithotrophs or 'solarmobiles' (Grassmann, 1988).

There is also a minimum size of flagellate related to the fixed diameter of the axoneme and the minimum length of axonemes which is functional in generating motility as a function of the size of the rest of the cell (Raven, 1986). This imposes a lower size limit not much smaller than the smallest known flagellates (Raven, 1998b, 1999; Sieburth and Johnston, 1989). Phagotrophic flagellates may have a lower size limit imposed by predator:prey size ratio considerations, and the smallest size cells which could be prey (Hansen *et al.*, 1997; Raven, 1998b 1999); other phagotrophs (for example, amoebae) would be similarly constrained.

2.4.3 *Limitations and opportunities: the case of* Volvox

Kirk (1998) has produced a scholarly account of the ecology, evolution, development and molecular cell biology of the polyphyletic genus *Volvox*, the largest motile phytoplankton organism, distinguished by predominantly motile spheroids of a thousand or more chlamydomonad cells at least 1 mm in diameter swimming at 1 mm per second or more and executing diel vertical migrations of up to 18 m. The distinction between 'mortal' somatic cells and 'immortal' germ cells in the asexual reproductive cycle of *Volvox* spp. has important evolutionary and developmental implications. Kirk's (1998) account puts these various aspects of the biology, ecology and phylogeny of Volvox into the context of its flagellate condition, drawing particularly on the work of Bell (1985), Koufopanou (1994) and Koufopanou and Bell (1993; see also Bell and Mooers, 1997): see Raven (1998c).

It is clear that the volvocines are dependent on their flagella in their natural environment, and there is reasonable evidence that the possession of flagella whose basal bodies act as centrioles in cell division in a walled cell whose ecology is based on continuous flagellar function favoured the evolution of multiple fission (a way to multicellular organisms), and also favours a germ/soma distinction, keeping the organism motile during cell division (Kirk, 1998). Clearly colonial motile photosynthetic flagellates are common not only among the Chlorophyceae but also in the Haptophyceae (*Corymbellus*) and the Chrysophyceae *sensu lato* (that is, including Synurophyceae), where *Uroglena* (Chrysophyceae *sensu stricto*, in gelatinous colonies up to 150 μm diameter) and *Synura* (Synurophyceae, lacking gelatinous material between cells, colonies up to 100 μm diameter), although there are larger planktonic non-photosynthetic choanoflagellate colonies (Patterson, 1996; van den Hoek *et al.*, 1995). However, none of these colonies achieve the size of *Volvox*, nor are they differentiated into soma and germ. Furthermore, their cell division involves binary fission, flagellar basal bodies do not function as centrioles, so even when there is a rigid cell covering (for example the loose-fitting lorica of *Dinobryon*), swimming is not necessarily prejudiced by cell division.

2.5 The flagellate condition: conclusions

The flagellates are major primary producers and phagotrophs (and significant saprotrophs) in the 2 μm – 2mm size range in many habitats with a continuous aqueous phase. Their key feature of motility away from a solid surface (and of causing water motion over cells and over or through multicellular organisms) has important resource retention, resource acquisition and reproductive benefits (Section 2.2) which, in some cases, can be shown to exceed very significantly the direct and indirect costs of flagellar motility as shown by quantitative cost-benefit analyses (Section 2.3). Such cost-benefit analyses are more readily conducted for benefits (energy, C, N, P) which can be easily expressed in the same units as the direct and indirect costs of flagella synthesis, maintenance and operation.

Flagellates cannot have a cell wall which is functional in turgor generation or in protecting the whole of the plasmalemma from physical or biotic damage. The absence of a cell wall is prerequisite for phagotrophy, although a lorica (*Dinobryon*) or collar (choanoflagellates) can aid phagotrophy by modulating flagella-induced feeding currents. Mixotrophy (photosynthesis by phagotrophs) is common among free swimming flagellates (and ciliates), where it can be a result of kleptoplasty or maintenance of microalgal food items as endosymbionts by essentially chemo-organotrophic phagotrophic flagellates, as well as by phagotrophy by photosynthetic flagellates with (genetically integrated) plastids. Mixotrophy can help to achieve a balanced resource input (energy, C, N, P) in habitats where phagotrophic chemoorganotrophy or photolithotrophy alone cannot.

Some of the benefits of the flagellate condition can be achieved in other ways. Examples are motility, buoyancy regulation and muscular activity in causing motility in open water, and of gliding and amoeboid movement over surfaces and particulate environments, and phagotrophy by amoeboid cells. Where quantitative comparisons are possible of costs and benefits of flagellar operation with that of the alternative means of attaining similar ends, it appears that the flagellates have no greater costs per unit benefit, and sometimes lower costs, than in the case of the alternative mechanism of motility. To some extent the superiority may be a result of comparing the performance of flagellates with other organisms under flagellate-optimized conditions (for example of organism size).

ACKNOWLEDGEMENTS

Work on resource acquisition and retention in algae in the author's laboratory is funded by the Natural Environment Research Council. Dr Anthony Cheshire was very helpful in reawakening my interest in sponges and in correcting misapprehensions that I harboured about them. I am very grateful to Dr Richard Geider and another, anonymous, referee for their perceptive comments and for drawing my attention to some references.

REFERENCES

Akhananova, A., Voncken, F., van Alen, T., van Hoek, A., Boxma, B., Vogels, G., Veenhuis, M. and Hackstein, J. H. P. (1998) A hydrogenosome with a genome. *Nature*, **396**, 527–528.

Amsler, C. D. and Neushul, M. (1989) Chemotactic effects of nutrients on spores of the kelps *Macrocystis pyrifera* and *Pterygophora californica*. *Marine Biology*, **102**, 557–564.

Barbeau, K., Moffet, J. W., Caren, D. A., Groot, P. L. and Erdner, D. L. (1997) Role of protozoan grazing in relieving iron limitation of phytoplankton. *Nature*, **380**, 61–64.

Bell, G. (1985) The origin and early evolution of germ cells as illustrated by the Volvocales. In *The Origin and Evolution of Sex* (eds H. O. Halverson and A. Monroy), New York: Alan and Liss, pp. 221–256.

Bell, G. and Mooers, A. O. (1997) Size and complexity in multicellular organisms. *Biological Journal of the Linnean Society*, **60**, 345–363.

Brand, L. E. (1981) Genetic variability in reproductive rates in marine phytoplankton populations. *Evolution*, **35**, 1117–1127.

Carlile, M. J. (1994) The success of the hyphae and mycelium. Unitary *vs* modular organisms. In *The Growing Fungus* (eds N. A. R. Gow and G. M. Gadd), London: Chapman and Hall, pp. 1–19.

Cavalier-Smith, T. (1993) Kingdom protozoa and its 18 Phyla. *Microbiological Reviews*, **57**, 953–994.

Cavalier-Smith, T. (1996) The kingdom Chromista: origin and systematics. *Progress in Phycological Research*, **4**, 309–347.

Cheshire, A. C. and Wilkinson, C. R. (1991) Modelling the photosynthetic production by sponges on Davies Reef, Great Barrier Reef. *Marine Biology*, **109**, 13–18.

Dennett, D. C. (1996) *Darwin's Dangerous Idea*. London: Penguin.

Fagerström, T., Briscoe, D. A. and Sunnucks, P. (1998) Evolution of mitotic cell-lineages in multicellular organisms. *Trends in Ecology and Evolution*, **13**, 117–120.

Fenchel, T. (1987) *The Ecology of Protozoa*. Madison: Science Tech Publishers; Berlin: Springer Verlag.

Fenchel, T. and Finlay, B. J. (1995) *Ecology and Evolution in Anoxic Worlds*. Oxford: Oxford University Press.

Gordon, R. (1987) A retaliatory role for algal projectiles, with implications for the mechanochemistry of diatom gliding motility. *Journal of Theoretical Biology*, **126**, 419–436.

Grassmann, P. (1988) The separation of plants and animals and solarmobiles. *Die Naturwissenschaften*, **75**, 43–44.
Grosberg, P. K. and Strathmann, R. R. (1998) One cell, two cell, red cell, blue cell: the persistence of a unicellular stage in multicellular life histories. *Trends in Ecology and Evolution*, **13**, 112–116.
Guillard, R. R. L. (1960) A mutant of *Chlamydomonas moewusii* lacking contractile vacuoles. *Journal of Protozoology*, **7**, 262–268.
Häder, D-P. (1979) Control of locomotion: photomovement. In *Encyclopaedia of Plant Physiology*, New Series, vol. 7 (eds W. Haupt and M. E. Feinbeck), Berlin: Springer-Verlag, pp. 268–309.
Häder, D-P. and Griebenow, K. (1988) Orientation of the green flagellate *Euglena gracilis* in a vertical column of water. *FEMS Microbiology Ecology*, **53**, 159–167.
Häder, D-P and Häder, M. A. (1988) Inhibition of motility and phototaxis in the green flagellate, *Euglena gracilis*, by UV–B radiation. *Archives of Microbiology*, **150**, 20–25.
Hansen, P .J., Bjørnsen, P. K and Hansen, B. W. (1997) Zooplankton grazing and growth: scaling within the 2–2000 µm body size range. *Limnology and Oceanography*, **42**, 687–704.
Harold, F. M., Harold, R. L. and Money, N. P. (1995) What drives cell wall expansion? *Canadian Journal of Botany*, **73** (Supplement 1), S379–S383.
Harold, R. L., Money, N. P. and Harold, F. M. (1996) Growth and morphogenesis in *Saprolegnia ferax*: is turgor required? *Protoplasma*, **191**, 105–114.
Harz, H., Nonnengässer, C. and Hegemann, P. (1992) The photoreceptor current of the green alga *Chlamydomonas*. *Philosophical Transactions of the Royal Society of London*, Series B, **338**, 39–52.
Haustrom, H. and Riemann, B. (1996) Ecological importance of bacterivorous, pigmented flagellates (mixotrophs) in the Bay of Aarhus, Denmark. *Marine Ecology Progress Series*, **137**, 251–263.
Heuser, J., Zhu, Q. and Clarke, M. (1993) Proton pumps populate the contractile vacuoles of *Dictyostelium* amoebae. *Journal of Cell Biology*, **121**, 1311–1327.
House, C. R. (1974) *Water Transport in Cells and Tissues*. London: Edward Arnold.
Kirk, D. L. (1998) *Volvox. Molecular-Genetic Origins of Multicellularity and Cellular Differentiation*. Cambridge: Cambridge University Press.
Koufopanou, V. (1994) The evolution of soma in the Volvocales. *American Naturalist*, **143**, 907–931.
Koufopanou, V. and Bell, G. (1993) Soma and germ: an experimental approach using *Volvox*. *Proceedings of the Royal Society of London*, Series B, **254**, 107–113.
Lieberman, O. S., Shilo, M. and van Rijn, J. (1994) The physiological ecology of a freshwater dinoflagellate bloom population: vertical migration, nitrogen limitation and nutrient uptake kinetics. *Journal of Phycology*, **30**, 964–971.
Maier, I. (1995) Brown algal pheromones. *Progress in Phycological Research*, **11**, 51–62.
Margulis, L., Schwartz, K. V. and Dolan, M. (1994) *The Illustrated Five Kingdoms. A Guide to the Diversity of Life on Earth*. New York: Harper Collins.
Martin, W. and Müller, M. (1998) The hydrogen hypothesis for the first eukaryote. *Nature*, **392**, 37–49.
Martin, W., Stoebe, B., Goremykin, V., Hansmann, S., Hasegawa, M. and Kowallik, K. V. (1998) Gene transfer to the nucleus and the evolution of chloroplasts. *Nature*, **393**, 162–165.
Maurel, C. (1997) Aquaporins and water permeability of plant membranes. *Annual Review of Plant Physiology and Plant Molecular Biology*, **48**, 399–429.
Pahlow, M., Riebesell, U. and Wolf-Gladrow, D. A. (1997) Impact of cell shape and chain formation on nutrient acquisition by marine diatoms. *Limnology and Oceanography*, **42**, 1660–1672.

Patterson, D. J. (1996) *Free-Living Freshwater Protozoa. A Colour Guide*. London: Manson.
Pickett-Heaps, J. D. (1998) Cell division and morphogenesis of the centric diatom *Chaetoceros decipiens* (Bacillariophyceae). II. Electron microscopy and a new paradigm for tip growth. *Journal of Phycology*, **34**, 995–1004.
Prosser, C. L. ed. (1973) *Comparative Animal Physiology*. Philadelphia: W. B. Saunders.
Raven, J. A. (1976) Transport in algal cells. In *Encyclopedia of Plant Physiology*, New Series, vol. 2A (eds U. Lüttge and M. G. Pitman), Berlin: Springer-Verlag, pp. 129–188.
Raven, J. A. (1981) Nutritional strategies of submerged benthic plants: the acquisition of C, N and P by rhizophytes and haptophytes. *New Phytologist*, **88**, 1–30.
Raven, J. A. (1982) The energetics of freshwater algae: energy requirements for biosynthesis and volume regulation. *New Phytologist*, **92**, 1–20.
Raven, J. A. (1983) Cyanobacterial motility as a test of the quantitative significance of proticity transmission along membranes. *New Phytologist*, **94**, 511–519.
Raven, J. A. (1986) Physiological consequences of extremely small size for autotrophic organisms in the sea. In *Photosynthetic Picoplankton* (eds T. Platt and W. K. W. Li), *Canadian Bulletin of Fisheries and Aquatic Science* **214**, pp. 1–70.
Raven, J. A. (1988) Algae on the move. *Transactions of the Botanical Society of Edinburgh*, **45**, 167–186.
Raven, J. A. (1989) Fight or flight: the economics of repair and avoidance of photoinhibition of photosynthesis. *Functional Ecology*, **3**, 5–19.
Raven, J. A. (1991) Responses of aquatic photosynthetic organisms to increased solar UV–B. *Journal of Photochemistry and Photobiology B: Biology*, **9**, 239–244.
Raven, J. A. (1993) Energy and nutrient acquisition by autotrophic symbioses and their asymbiotic ancestors. *Symbiosis*, **14**, 33–60.
Raven, J. A. (1994a) Comparative aspects of chrysophyte nutrition with emphasis on carbon, phosphorus and nitrogen. In *Chrysophyte Algae: Ecology, Phylogeny and Development* (eds C. D. Sandgren, J. P. Smol and I. Kristiansen), Cambridge: Cambridge University Press, pp. 95–118.
Raven, J. A. (1994b) Photosynthesis in aquatic plants. In *Ecophysiology of Photosynthesis.* vol. 100 of *Studies in Ecology* (eds E. D. Schulze and M. H. Caldwell), Berlin: Springer-Verlag, pp. 299–318.
Raven, J. A. (1994c) Why are there no picoplanktonic O_2-evolvers with volumes less than 10^{-19}? *Journal of Plankton Research*, **16**, 565–580.
Raven, J. A. (1994d) The cost of photoinhibition to plant communities. In *Photoinhibition of Photosynthesis: from Molecular Mechanisms to the Field* (eds N. R. Baker and J. A. Raven), Oxford: Bios Scientific, pp. 449–465.
Raven, J. A. (1995) Costs and benefits of low osmolarity in cells of freshwater algae. *Functional Ecology*, **9**, 701–707.
Raven, J. A. (1997a) Phagotrophy in phototrophs. *Limnology and Oceanography*, **42**, 198–205.
Raven, J. A. (1997b) CO_2 concentrating mechanisms: a direct role for thylakoid lumen acidification? *Plant, Cell and Environment*, **20**, 147–154.
Raven, J. A. (1997c) The vacuole: a cost–benefit analysis. *Advances in Botanical Research*, **25**, 59–86.
Raven, J. A. (1998a) Insect and angiosperm diversity in marine environments: further comments on van der Haage. *Functional Ecology*, **12**, 977–978.
Raven, J. A. (1998b) Small is beautiful. The picophytoplankton. *Functional Ecology*, **12**, 503–513.
Raven, J. A. (1998c) Review of '*Volvox*. Molecular-Genetic Origins of Multicellularity and Cellular Differentiation'. *European Journal of Phycology*, **33**, 275–278.
Raven, J. A. (1999) Picophytoplankton. *Progress in Phycological Research*, **13**, 33–106.
Raven, J. A. and Beardall, J. (1981) Respiration and photorespiration. In *Physiological Bases*

of Phytoplankton Ecology (ed. T. Platt), *Canadian Bulletin of Fisheries and Aquatic Science*, **210**, pp. 55–82.

Raven, J. A. and Richardson, K. (1984) Dinophyte flagella: a cost–benefit analysis. *New Phytologist*, **98**, 259–276.

Rivkin, R. B., Swift, E., Biggley, W. H. and Voytek, M. A. (1984) Growth and carbon uptake by natural populations of oceanic dinoflagellates *Pyrocystis noctiluca* and *Pyrocystis fusiformis*. *Deep-Sea Research, Part A. Oceanographic Research Papers*, **31**, 353–367.

Roberts, F., Roberts, C. W., Johnson, J. J., Kyle, D. E., Krell, T., Coggins, J. R., Coombs, G. H., Milhous, W. K., Tzipori, S., Ferguson, D. J. P., Chakrabarti, D. and McLeod, R. (1998) Evidence for the shikimate pathway in apicomplexan parasites. *Nature*, **393**, 801–805.

Robinson, D. G., Hoppenrath, M., Overbeck, K., Luykx, P. and Ratajczak, R. (1998) Localization of pyrophosphatase and V-ATPase in *Chlamydomonas reinhardtii*. *Botanica Acta*, **111**, 108–122.

Round, F. E. (1981) *The Ecology of Algae*. Cambridge: Cambridge University Press.

Sand-Jensen, K., Pedérsen, O. and Geertz-Hansen, O. (1997) Regulation and role of photosynthesis in the colonial symbiotic ciliate *Ophyrydium versatile*. *Limnology and Oceanography*, **42**, 866–873.

Serrão, E. A., Pearson, G., Kautsky, L. and Brawley, S. H. (1996) Successful external fertilization in turbulent environments. *Proceedings of the National Academy of Sciences*, **93**, 5286–5290.

Sieburth, J. McN. and Johnston, P. W. (1989) Picoplankton ultrastructure: a decade of preparation for the brown tide alga, *Aureococcus anophagefferens*. In *Novel Phytoplankton Blooms Coastal and Estuarine Studies* (eds E. M. Cosper, V. M. Bricelj and E. J. Carpenter), 35, pp. 6–21.

Strom, S. C. and Morello, T. A. (1998) Comparative growth rates and yields of ciliates and heterotrophic dinoflagellates. *Journal of Plankton Research*, **20**, 571–584.

Torres de Araujo, F. F., Pires, M. A., Frankel, R. B. and Bicudo, C. E. M. (1986) Magnetite and magnetotaxis in algae. *Biophysical Journal*, **50**, 375–378.

van den Hoek, C., Mann, D. G. and Johns, H. M. (1995) *Algae – An Introduction to Phycology*. Cambridge: Cambridge University Press.

van der Haage, J. C. H. (1996) Why are there no insects and so few higher plants in the sea? New thoughts on an old problem. *Functional Ecology*, **10**, 546–547.

Vogel, S. (1981) *Life in Moving Fluids: The Physical Biology of Flow*. Boston, Mass: Willard Grant.

Walsby, A. E. (1994) Gas vesicles. *Microbiological Reviews*, **58**, 94–144.

Wilkinson, C. R. (1992) Symbiotic interactions between marine sponges and algae. In *Algae and Symbioses: Plants, Animals, Fungi, Viruses, Interactions Explored* (ed. W. Reisser), Bristol: Biopress, pp. 111–151.

Wilkinson, C. R., Cheshire, A. C., Klumpp, D. W. and McKinnon, A. D. (1988) Nutritional spectrum of animals with photosynthetic symbionts – corals and sponges. *Proceedings of the 6th International Coral Reef Symposium, Townsville, Australia, 1988*, vol. 3, pp. 27–30.

Chapter 3

Mechanisms of flagellar propulsion

Michael E. J. Holwill and Helen C. Taylor

ABSTRACT

Flagellar bending is caused by the action of molecular motors – dyneins – distributed internally along the length of the organelle. The basic internal structure of eukaryotic flagella is the axoneme, which consists of a set of nine doublet microtubules arranged cylindrically around a pair of single microtubules. A computer model of axonemal structure has been generated, based on electron micrographs of flagella prepared in different ways, and is being used to evaluate the relationships between features of the axoneme. Flagella can adopt a wide range of bend patterns, both two- and three-dimensional, with different degrees of asymmetry. Application of objective methods shows that planar bends on both smooth and hispid flagella *in vivo* consist of circular arcs separated by straight regions, suggesting that the bend shape is an intrinsic characteristic of the axoneme. The interaction of a flagellum with its liquid environment produces propulsive forces which are dominated by viscosity. Calculations of the propulsive thrust yield velocities in reasonable agreement with observation. The energy dissipated against external viscous forces can be calculated and used in assessments of motor action. During bend formation and propagation, the motor molecules, which are arranged along each doublet in two rows of composite structures known as inner and outer dynein arms, undergo cyclic activity to cause microtubule sliding. The outer dynein arms appear to control beat frequency while the inner arms influence bend symmetry. Complete arms and individual motor molecules extracted from flagella can be distributed on a glass surface and activated to transport isolated microtubules. Dynamic computer modelling is used to study the activity and mutual interaction of arms in assemblies for comparison with these *in vitro* preparations. Theoretical predictions based on separate models incorporating either random or co-ordinated arm action are in accord with experimental results, and more experimental data are needed to differentiate between them. Development of the computer models will allow the incorporation of mechanical parameters, such as the forces generated by the motors and the elastic properties of the microtubules, with the ultimate goal of constructing a functional model of the flagellar axoneme.

3.1 Introduction

The most characteristic and visible attribute of a eukaryotic flagellum is its oscillatory movement, which takes the form of a succession of bends propagated more or less regularly along the organelle. Flagellar bends may be two- or three-dimensional, with successive bends propagated with different phases and amplitudes, giving rise to movements ranging from symmetric waves to a highly-asymmetric breast-stroke action. In most flagella, bends are propagated only from the flagellar base; in some cells, however, bend propagation from the flagellar tip is normal, with reversal of the propagation direction being a common feature of the motion. Although flagellar bend patterns show considerable variety from one species to another, the basic structure of all flagella is the same, consisting of an assembly of nine microtubule doublets (bearing the motors and other significant structures) surrounding a pair of single microtubules and associated structures; this familiar 9+2 arrangement of microtubules, with a diameter of about 0.2 μm, is the flagellar axoneme. Since activity of the same arrangement of motors and other structures gives rise to a range of bend patterns, the motor mechanism itself must be controlled in different ways in different organisms.

In this chapter we will examine the propulsive forces which arise from interactions between the bending flagellum and its liquid environment and which cause either the locomotion of the flagellate or the generation of a liquid current flowing past the organism. Flagellar bending is the result of relative sliding between the microtubule doublets caused by the action of molecular motors distributed along the doublets as two rows of inner and outer arms; axonemal structures restrict the sliding in a controlled way to generate bending. We will discuss how detailed quantitative knowledge of flagellar bend patterns provides information about the *in vivo* activity of the motors and their interaction with other axonemal structures. *In vitro* studies of motor action will be described, and we will show how they may be interpreted by static and dynamic computer-modelling techniques to contribute to our understanding of motor action; these techniques allow us to predict conformational changes in an individual motor. The use of these and related techniques, suitably developed and combined with appropriate critical experimental observations, will help in the elucidation of the functions performed by the various structures within the axoneme.

3.2 Patterns of flagellar movement

Since flagellar beating results from the action of molecular motors in the axoneme, the macroscopic bend patterns provide information about the activity of the motors, their interaction with other structures to deform the axoneme and their control in bend propagation. The propulsive forces produced by flagella can be estimated by examining their hydrodynamic interaction with the fluid environment; to do this a detailed knowledge of the beat characteristics is needed. It is appropriate to summarize first the different beat patterns observed and then to consider the information that can be derived from a critical evaluation of the motion.

The range of flagellar beat patterns recorded for the flagellates is wide, with both two- and three-dimensional bend propagation being common. Beat frequencies are generally in the range 5 Hz to 100 Hz, with bend amplitudes commonly between 2 μm and 5 μm. In most species, bends are initiated at the flagellar base and propagate distally; exceptions are cells in the Trypanasomatidae, where the forward motion consists of waves propagating proximally from the flagellar tip. Under certain conditions, the Trypanasomatidae reverse their swimming direction by propagating bends from the base; occasionally bends have been observed propagating in both directions simultaneously on different regions of the same flagellum (Holwill, 1965).

While many of the Trypanasomatidae have undulating membranes which obscure the characteristics of flagellar bending, the proximally-directed bends on *Crithidia oncopelti* can be symmetric. By contrast, flagella of many cells execute asymmetric two-dimensional beat patterns; for example the biflagellate *Chlamydomonas* propagates mirror-image asymmetric bends distally along its two flagella to give the appearance of a 'breast-stroke' action (Holwill, 1966). The two flagella on some organisms, such as the dinoflagellates, have different individual beat characteristics, with, for example, one flagellum propagating planar bends and the other executing a three-dimensional movement.

It is likely that all flagellates have avoidance or tactic responses involving a change in flagellar action, demonstrating that the cells can exert considerable control over the movement of their flagella. The response takes different forms for different species. In the Trypanasomatidae the direction of wave propagation is reversed, the bends become more asymmetric and are generated at a lower frequency than during normal locomotion. In other species, the direction of bend propagation remains unchanged, but the flagella are reoriented and the bends have an altered waveform compared to that of normal motion (for example Holwill, 1994).

3.3 Bend shapes

Flagellar bends are formed and propagated by the action of the internal motor machinery. The precise shapes of bends therefore provide information about the motors and their interactions with structures in the axoneme. In early analyses (for example Machin, 1958) the waves were assumed to be sinusoidal, but Brokaw and Wright (1963) used visual curve-fitting to suggest that the waves on the longitudinal flagellum of the dinoflagellate *Ceratium* are composed of circular arcs joined by straight sections. Such a bend shape may imply that the motor mechanism responsible for bending the flagellum is capable of rapid conformational change, to give the transition between straight and curved flagellar regions, and also predicts that the relative sliding between doublets should be evenly distributed through a bend. A variation of this shape, the gradient curvature model, in which the curvature changes linearly with distance in the central region of a bend, has been considered by Eshel and Brokaw (1988). Bend initiation by a sinusoidal oscillator can produce a sine-generated curve, which has also been studied (Hiramoto and Baba, 1978; Eshel and Brokaw, 1988). The meander, a curve that minimizes elastic potential energy for bending systems under certain conditions, has been considered by Silvester and Holwill (1972). All these curves have a similar appearance (Figure 3.1), though each is distinguished easily from the cosine wave. These bend shapes do not form an exclusive set, and others, such as the catenary, are potential members. The subtle differences between the curves cannot be detected easily by eye and yet, as implied above, each has a particular significance for flagellar motor mechanisms. The bend shape also provides a critical test of the modelling procedures to be discussed in

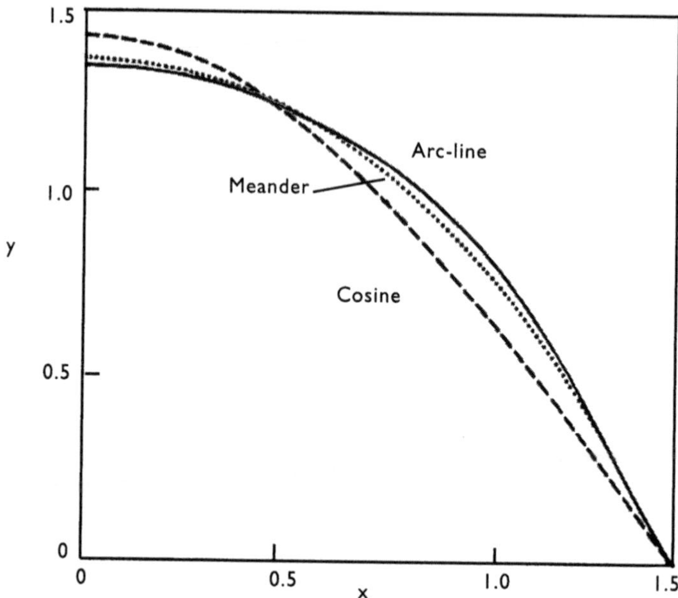

Figure 3.1 Comparison of arc-line and meander bend shapes; the cosine wave is included for comparison. The sine-generated curve is indistinguishable from the meander. The arrow on the arc-line curve indicates the transition from arc to line

section 3.5. It is therefore of value to have an objective technique for distinguishing between the wave shapes and thereby establishing the precise shape of a flagellar bend. One analytical approach is based on Fourier synthesis of the wave form and examines the relative amplitudes of the harmonics (Silvester and Holwill, 1972). For a particular bend shape, the set of coefficients is unique. Coefficients derived from flagellar images can be compared with those calculated for known analytical curves.

Using this approach the bends on the flagella of *Crithidia oncopelti* and *Ochromonas danica* have been shown to match the arc-line pattern (Johnston *et al.*, 1979; Holwill, 1996). It might be expected that the mechanical loading on a flagellum would affect its shape, but this does not appear to be the case, since the smooth flagellum of *Crithidia* bears a quite different distribution of load from the hispid flagellum of *Ochromonas* (Figure 3.2 and section 3.4); for example, the tangential stress on the flagellum near a wave crest will be high in *Ochromonas* but low in *Crithidia*. This indicates that wave shape is an intrinsic property of the axoneme itself. For a given flagellum, an increase in viscosity, and therefore a change in the viscous loading, produces a change in the wave parameters, causing the amplitude and wavelength to decrease. The ratio of amplitude to wavelength remains essentially constant, although in some spermatozoa an increase in viscosity may cause the bend angle to change (Brokaw, 1966).

These observations suggest that the axonemal structure determines a minimum radius of curvature which, in normal circumstances, is always reached, and may be due primarily to inner arm action (see section 3.5). The bending moment provided by the motor mechanism is applied rapidly, so that throughout the bend no change

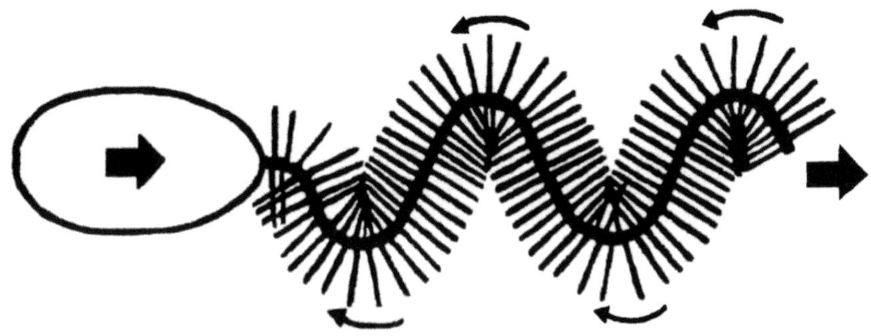

Figure 3.2 Showing the action of mastigonemes on the *Ochromonas* flagellum. Waves are propagated distally. Curved arrows indicate motion of mastigonemes; heavy arrows show direction of cell movement and of wave propagation

in curvature is observed; the regions of changing curvature occur at the leading and trailing ends of a bend.

Eshel and Brokaw (1988) used curve-fitting procedures to show that the flagella of certain sea-urchin sperm conformed to a gradient curvature model. These flagella are longer than those of *Crithidia* and *Ochromonas*, and it is possible that viscous loading has a more significant effect than for the shorter organelles; for instance, viscous bending moments would be expected to reduce curvature in the leading part of a bend and increase it in the trailing element, as is observed for the sea urchin sperm. The studies of Johnston *et al.* (1979) also revealed changes in curvature as a wave propagated, but with no clear behavioural pattern.

3.4 Interactions with the liquid environment

A moving flagellum elicits propulsive thrust from the liquid environment. From knowledge of the beat pattern and the properties of the liquid, hydrodynamic theory can be used to estimate the magnitude of the thrust and of the hydrodynamic energy dissipated. This energy, together with energy needed to overcome internal resistance provided by, for example, the elastic properties of flagellar structures, must be supplied from chemical sources within the cell.

It is well established that the forces used to propel flagellates are dominated by viscosity (for example Holwill, 1994). Forces generated by a bending flagellum have been predicted using two distinct approaches, the resistive force theory (Gray and Hancock, 1955), which is based on slender body theory (Hancock, 1953), and the boundary element method (Ramia, 1993). These techniques evaluate liquid flow and flagellar thrusts by considering mathematical singularities (for example, stokeslets) distributed along the flagellum; in resistive force theory the singularities are distributed along the flagellar centreline, while boundary element theory represents the system more appropriately by considering singularities associated with the flagellar surface. Ramia (1993) gives a detailed comparative account of the different approaches, while Lighthill (1996) has given a general review of the status of hydrodynamic theory in relation to flagellar propulsion. For most situations, the motion is not known in sufficient detail to warrant the detailed and complex calculations involved in determining the singularity distribution. For example, in microscopic observations, the cell will be swimming at variable, and unknown, distances from a surface, which will have a considerable effect on the motion.

It is usually sufficient to use the approximate approach of resistive force theory, in which force coefficients, C_N and C_L, relating to a moving cylinder are calculated; it is generally easier to perform the calculations using resistive force theory than using the singularities. The viscous force generated by a cylinder of length l moving with velocity v_L in the direction of its axis is $-C_L v_L l$, with a similar expression involving C_N for motion normal to the axis. The thrust produced by a bending flagellum is found by integrating the effects of these forces along the flagellum. Because there are no inertial effects, this thrust is balanced at every instant by the resistance to motion offered by the cell body, which is approximated by a sphere or ellipsoid, for which analytical expressions of resistance are available. Thus, if the shapes of flagellum and body are known, analytical or numerical techniques can be used to predict the hydrodynamic behaviour of the system (for example, Holwill and Miles, 1971; Holwill and Coakley, 1972; Brokaw, 1965; Chwang and Wu, 1971; Lighthill, 1996). For many organisms, the calculation of the resistance to motion is further complicated by the oscillation imposed on the cell by the flagellar undulations.

The direction of propulsion relative to that of wave propagation depends on the ratio C_N/C_L of the force coefficients. For a smooth cylinder, the limiting value of this ratio is 2; using the values proposed by Lighthill (1976) a value of approximately 1.7 is obtained. Theory predicts that if $C_N/C_L > 1$, the direction of propulsion is opposite to that of wave propagation, in agreement with observation.

Cells of *Ochromonas* and other similar flagellates move flagellum-first with waves propagating from flagellar base to tip, that is the propulsive thrust is in the direction of wave propagation. The reason for this is the presence of mastigonemes, invisible in the light microscope but revealed by the electron microscope to be cylinders a few micrometres long and 10 to 20 nm in diameter. The action of the mastigonemes can be understood by consideration of Figure 3.2. In this diagram, the mastigonemes lie in the plane of the beating flagellum, and remain normal to the flagellar surface during movement. Mastigonemes at the crest of the wave move like oars, providing thrust in a direction opposite to that of the flagellar shaft. If this opposing force is greater than that provided by the shaft, the cell will move in the direction of the wave, as observed. In terms of the resistance coefficients, theoretical considerations show that the ratio C_N/C_L must be < 1. The ratio is about 0.5 for the arrangement of mastigonemes on *Ochromonas* (Holwill and Sleigh, 1967; Holwill and Peters, 1974), and Holwill and Peters (1974) confirmed experimentally that the fluid flow pattern around the *Ochromonas* flagellum corresponds to that expected from the arrangement and motion of mastigonemes shown in Figure 3.2. If the mastigonemes lie in planes perpendicular to that of the beat, thrust reversal will not occur.

Theoretical analyses have been used to predict successfully the propulsive velocities of a wide range of flagellates. Expressions to estimate the hydrodynamic energy dissipated in maintaining a flagellar wave have also been derived. Calculations indicate that the power required to propel the cell is less than 1 per cent of that needed to maintain the oscillations (Purcell, 1997). For an individual flagellum the power requirement is of the order of 10 fW. This power is provided by dephosphorylation of ATP (see later); the efficiency with which the flagellum converts the chemical energy into mechanical energy lies in the range 40 to 90 per cent (Holwill and Satir, 1987). In reaching these values for the efficiency, the energy dissipated in overcoming elastic forces was assumed to be of the same order as that dissipated hydrodynamically.

3.5 Motor mechanisms

3.5.1 The sliding microtubule basis of bend formation

Omoto (1991) has summarized the evidence which leads to the conclusion that the axonemal microtubules slide relative to each other during flagellar bending. Since Satir (1965) provided the first evidence to support the microtubule sliding hypothesis of flagellar motion, relative sliding has been observed of doublets extruded from demembranated and reactivated axonemes of cilia (Sale and Satir, 1977) and various spermatozoa (Summers and Gibbons, 1971, 1973; Takahashi *et al.*, 1982; Woolley and Brammall, 1987). In these experiments components which maintain the integrity of the axoneme, presumably including the interdoublet links, the radial spokes and the basal body, are damaged or destroyed by the preparative procedures. Sliding between peripheral doublets in less damaged axonemes has been observed to initiate flagellar bending (Shingyoji *et al.*, 1977; Yeung and Woolley, 1984). More recently Brokaw (1989) has attached small gold beads to different microtubules in demembranated axonemes and has recorded bead movements during bending that are consistent with doublet microtubule sliding. Sale and Satir (1977) and Woolley and Brammall (1987) used optical and electron microscopy to show that sliding is unidirectional, with the dynein motors on one microtubule pushing the neighbouring microtubule tipwards. In a recent study (Yamada *et al.*, 1998), the directional characteristics of motor action have been investigated by monitoring fluorescence-labelled brain microtubules moving over rows of dynein exposed by sliding disintegration of sea urchin sperm axonemes. In the presence of ATP the motor action is polarized in the same sense as in the disintegrating axoneme, with the microtubule being driven towards the end of the doublet associated with the flagellum tip. Recent reports (Mimori and Miki-Noumura, 1994; Ishijima *et al.*, 1996) indicate that, under certain conditions, the action of dynein can be bi-directional; if this proves to be a property of *in vivo* motors, it will have important implications for the operation of the motor mechanism.

3.5.2 Action of inner and outer dynein arms to produce sliding

The hypothesis that dynein is responsible for generating the force required to slide the microtubules has been confirmed by a comprehensive series of experiments involving the gliding of microtubules over a substrate of dynein molecules. *In vitro* motility assays of purified dyneins from both rows of axonemal motors have demonstrated that microtubules can be transported not only by the intact outer dynein motors (Sale and Fox, 1988; Vale and Toyoshima, 1989; Hamasaki *et al.*, 1991, 1995) but also by many of the individual dynein isoforms from both rows of motors in *Chlamydomonas* axonemes (Sale and Fox, 1988; Moss *et al.*, 1992a, 1992b; Kagami and Kamiya, 1992; Sakakibara and Nakayama, 1998). The outer dynein motor often consists of three dynein heavy-chain molecules α, β and γ. Experiments have shown that the α subparticle alone cannot transport microtubules (Moss *et al.*, 1992a), but when it is combined with the β subparticle, translocation occurs readily (Sakakibara and Nakayama, 1998). Microtubule gliding experiments have revealed differences between the motile characteristics of the outer dynein subparticles. For instance, though microtubule gliding occurs over an activated substrate of *Chlamydomonas* β motors (Sakakibara and Nakayama, 1998), faster gliding rates are observed when the β motor from sea urchin sperm is associated with intermediate chain 1 (Sale and Fox, 1988). These faster velocities are also greater than the gliding rates produced by a substrate of the intact sea urchin outer dynein from which the fraction is derived (Sale and Fox, 1988). The γ subparticle also shows motor action (Sakakibara and Nakayama, 1998) but is absent from outer motor complexes of some species, including sea urchin sperm. At present it is not clear whether the properties of the motors depend on the species from which they are derived.

The situation is more complicated for the inner dynein motors. At least seven different isoforms of inner dynein have been identified in *Chlamydomonas* (Piperno, 1995; Gardner *et al.*, 1994) and *in vitro* motility assays of seven of these isoforms have been studied (Kagami and Kamiya, 1992). These studies revealed that all isoforms but one are capable of translocating microtubules, and of these six, five can also cause microtubule rotation. The ability of many of the inner dynein subparticles to exert torque on the microtubule may reflect a key difference in function between the two rows of dynein motors. Sliding experiments involving the disintegration of mutants of *Chlamydomonas* have demonstrated that without outer motors the sliding velocity is reduced significantly, while missing inner arm components leave the velocity unchanged (Kurimoto and Kamiya, 1991). Examination of the ability of the mutant flagella to beat and disintegrate by sliding led Kurimoto and Kamiya (1991) to suggest that though the inner motors do not control the beat frequency, they may be required to initiate bending, and hence to control the bend shape.

3.5.3 Experimental and computer studies of outer arm motor activity

In the gliding and sliding experiments discussed earlier, dynein motors were identified as the force-generating structures responsible for translocating the microtubule. These motors undergo a cycle of activity known as the mechanochemical cycle. As discussed by Omoto (1991), it has proved difficult to establish the detailed coupling between the chemical cycle, that is, the ATPase activity, and the mechanical cycle, that is, the conformational changes observed. Several research groups have obtained micrographs showing that the outer dynein motor changes configuration under different chemical conditions (Goodenough and Heuser, 1982, 1984; Avolio *et al.*, 1986; Burgess *et al.*, 1991; Burgess, 1995). Although there are no confirmed images of a motor during its force-producing phase, perhaps because this phase appears to occupy a small fraction (about 1 per cent) of the entire cycle time (but see later comments), attached arms are observed on rigor axonemes. Based on its evidence, each research group has proposed a cycle of activity. The cycles have some features in common but also show significant differences. Variations in interpretation arise in part because each of the variety of techniques used to prepare the specimens has its own individual characteristics and artifacts.

Sugrue *et al.* (1991) have used computer modelling to formulate a cycle of outer dynein activity based on micrographs from Goodenough and Heuser (1982) and Avolio *et al.* (1986). In this mechanical cycle (Figure 3.3, animated in Holwill *et al.*, 1998a), the three globular heads (α, β and γ dynein heavy-chains) observed in the micrographs are mounted on top of each other so that they bridge the interdoublet gap. The intermediate chains are folded to form a cape-like structure which links the heavy chains to the microtubule (N) on which they are mounted. During the conformational change, the heads alter their orientation such that one attaches to the neighbouring doublet (N+1). At this point (1, Figure 3.3) the arm produces a force that translocates doublet N+1 by a distance known as the step size, after which it disconnects (2, Figure 3.3) and returns (3,4,5, Figure 3.3) to a position from which the cycle can be repeated. Information relating to sea urchin sperm dynein, which lacks the γ heavy chain, indicates that, *in vitro*, the β particle makes transient contact with doublet N+1 and generates the sliding force, while the α particle is bound to microtubule N (Moss *et al.*, 1992a, 1992b).

The various stages of the mechanochemical cycle have been considered in more detail by Barkalow *et al.* (1994). The function of thin structures known as B links, which project from the dynein heads, has not yet been established, but recent evidence (Gee *et al.*, 1997) indicates that they bind to microtubules. A complete understanding of the mechanochemical cycle requires that the various stages of the enzyme reaction be correlated with the phases of the mechanical cycle; several authors (for example Brokaw and Johnson, 1989; Satir, 1989; Moss *et al.*, 1992b) have contributed towards this goal, but a complete correlation of the two cycles has yet to emerge. To understand the implications of the animated cycle shown in Holwill *et al.* (1998a) the time that the arm spends attached to the neighbouring doublet microtubule, that is, the duty phase, requires consideration.

A kinetic analysis of the results of an *in vitro* motility assay of intact outer arms translocating various lengths of microtubules at different velocities (Hamasaki *et al.*, 1995), has predicted that the duty phase could occupy just 1 per cent of the total cycle time of dynein activity. However, such a short duty phase is not supported by a recent experiment in which microtubules are transported by single arms attached to an extruded axonemal doublet (Shingyoji *et al.*, 1998). One implication of this experiment is that two heads of an individual arm interact with the translocating microtubule in such a way that continuous contact is maintained between the arm and the microtubule. The dynein motor can be considered to 'walk' along the microtubule in the manner of cytoplasmic dynein or kinesin. Further studies are needed to resolve the apparent discrepancy between the results of these experiments.

3.5.4 Co-ordination of outer arm activity

Although both inner and outer rows of dynein motors can translocate microtubules, in the axoneme each row performs a different task, with outer arms collectively controlling beat frequency and the inner row controlling the bend shape (Brokaw and

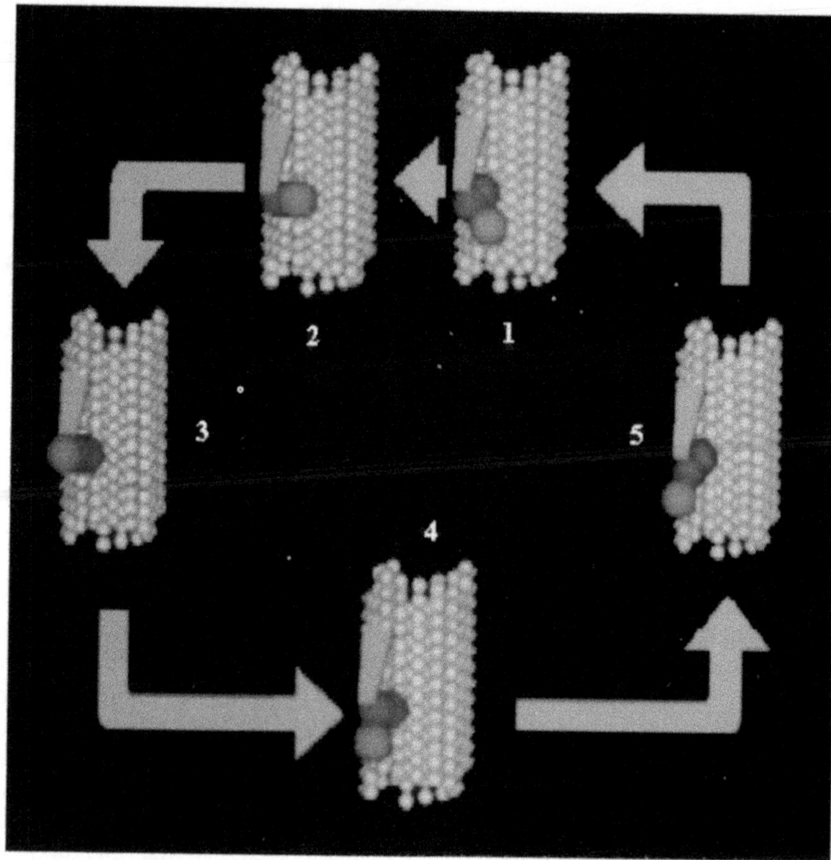

Figure 3.3 Mechanical cycle of the outer dynein arm. The arm is shown mounted on a section of microtubule which is modelled by a cylindrical arrangement of rows of small spheres representing tubulin monomers. The three dynein heavy chains are modelled by the larger spheres, with the lighter-shaded one interacting with the neighbouring microtubule, which would lie between the viewer and the arm. The cape is represented by the trapezoidal structure. The arm attaches to the neighbouring microtubule at stage 1 and produces a force which pushes the microtubule tipwards; the arm detaches at stage 2. Stages 3 to 5 prepare the motor for its next active phase

Kamiya, 1987; Kurimoto and Kamiya, 1991). The way in which these two rows of motors use the same mechanism (doublet translocation) to control different aspects of motility requires an understanding of the co-ordination of the arms in any one row. The availability of appropriate structural and experimental data has allowed us to investigate outer motor co-ordination. In a recent study (Taylor *et al.*, 1999), we have generated structures for the inner arms which reconcile previously disparate interpretations; this study, together with experimental studies of inner arm action, will allow us to investigate the co-ordination of the inner motors. In one *in vitro* motility assay of intact outer motors, dynein is laid down on a glass slide to give a random distribution of intact motors (Hamasaki *et al.*, 1995). Microtubules are dropped onto the dynein substrate and the system activated by adding ATP, whereupon the microtubules are transported in linear paths along the microtubule axis. The initial conditions suggest that the dynein motors are activated randomly, i.e. with no systematic co-ordination, to produce microtubule gliding.

In our laboratory, we (Holwill *et al.*, 1998a, 1998b; Taylor and Holwill, 1999) have developed computer simulations of the collective behaviour of outer motors in this experimental system. Using reasonable assumptions, we calculated the velocity of microtubule translocation as a function of microtubule length for a random, or stochastic, activation of the dynein arms. The simulation of stochastic activity shows the same trend as the experimental data, in that the microtubule velocity increases with microtubule length, and approaches a maximum value asymptotically. If the step size of the dynein motor is taken to be 16 nm, as suggested by Hamasaki *et al.* (1995), the magnitude of the maximum simulation velocity is considerably greater than that in the experiment. Simulations with different values for the step size and duty phase are being studied, and the indications are that the velocities predicted will be reasonably close to those observed experimentally.

The experimental data shows a high degree of scatter. By repeating the stochastic simulation several times at all microtubule lengths, the predicted velocities are also scattered, but to a significantly smaller extent than the experimental values (Holwill *et al.*, 1995). Preliminary investigations indicate that the scatter of the simulated data would be increased if the motors did not all drive the microtubule through the same distance in one cycle. In terms of motor action, this could be achieved by assuming that the angle between microtubule axis and the force generated by the dynein is variable. Given the experimental situation, this is not an unreasonable expectation. However, the linear movement of the microtubule suggests that the driven microtubule can control the dynein activity so that only the component of force along the axis is actually applied.

Feedback between the driven microtubule and the dynein motor could lead to co-ordinated, rather than random, motor activity. In one scenario, the motors are stimulated sequentially to give a wave of activity sweeping across the substrate beneath the microtubule. To simulate this situation, the system was set up to generate a constant phase difference, equal to the time occupied by the duty phase, between the cycles of neighbouring arms. With assumptions and magnitudes similar to the random case described above, the results show the same trend as that in the experiment, but as for the random simulation, the maximum velocity predicted is about an order of magnitude greater than that actually observed. By changing the step size to 8 nm, a value suggested by measurements on the other microtubule molecular motor, kinesin (Svoboda *et al.*, 1993), the maximum velocity is reduced by half. The other parameter required to predict the gliding velocities is the proportion of the cycle time occupied by the duty phase. Its value is determined from experimental data which show considerable scatter, and is therefore subject to uncertainty. If a value of 10 per cent, rather than 1 per cent, is assumed, the predicted maximum velocities for step sizes of 16 nm and 8 nm lie closer to the experimental range (Figure 3.4). The effects of varying other parameters, such as the phase difference between activity of successive arms, is currently under examination.

Within the intact axoneme it is reasonable to suggest that the activity of motor assemblies is co-ordinated systematically so that the arm action can produce a progressive bend. However, calculations show that only a small fraction of the dynein arms available is needed to achieve the microtubule sliding necessary to form a flagellar bend (for example Satir, 1998). It is therefore possible that arm activity is a random phenomenon in the axoneme, although the energy requirements also need to be considered carefully. Experiments to distinguish between these possibilities involve the disintegration of demembranated axonemes (Takahashi *et al.*, 1982) and limited data are available relating the distance of microtubule sliding to time in this situation. These data indicate that sliding occurs at a constant velocity, although one of the plots suggests that the behaviour becomes non-linear as the microtubule overlap decreases significantly. Simulations of motors activated randomly under these conditions produce an intermittent motion of the microtubule (Holwill *et al.*, 1998a, 1998b). This behaviour would not be detected with the temporal and spatial resolution of the experiments. In simulations of co-ordinated action, a linear relationship is observed between the distance moved by the microtubule and time (Holwill *et al.*, 1998a). When the overlap region was reduced to a length shorter than the activation wavelength, the microtubule movement occurred in a stepwise fashion; some experimental evidence supports this behaviour (Takahashi *et al.*, 1982), but more is required to allow an unequivocal interpretation to be made.

Studies of flagellated cells moving in media of increased viscosity indicate that the behaviour of dynein is load-dependent (Minoura and Kamiya, 1995; Brokaw, 1996). In wild-type *Chlamydomonas*, Minoura and Kamiya (1995) report that the force produced by intact flagella increases by 30-40 per cent on raising the viscosity of the

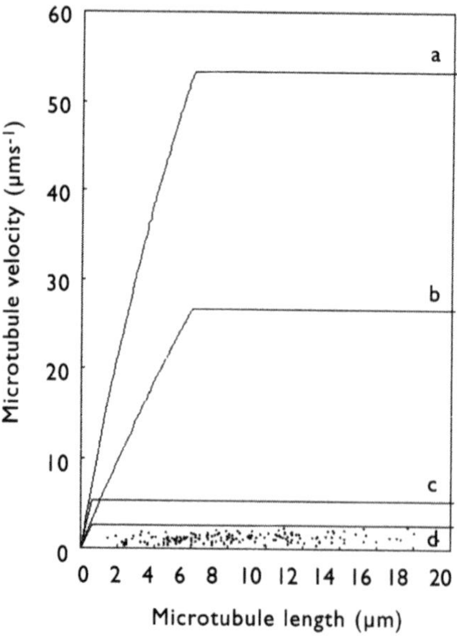

Figure 3.4 Results of simulations of gliding microtubules. The duty phase of the dynein arm is 1% of the cycle time in (a) and (b) and 10% in (c) and (d); the stepsize in (a) and (c) is 16 nm, while in (b) and (d) it is 8 nm. (. experimental data)

environment from 0.9 mPa s to 2 mPa s, but decreases with further increase in viscosity; the cells stop moving when the viscosity approaches 6 mPa s. The initial increase in force is not observed in cells which have been demembranated and reactivated with ATP, suggesting that the feedback mechanism responsible for the effect in the living flagella is removed by the preparative procedures used to demembranate the cells. Experiments with mutant cells demonstrate that the force produced by the outer dynein motors varies as the external viscous load is changed, whereas force generation by the inner dynein motors remains essentially constant under the same circumstances. These observations suggest strongly that the action of the outer motors is load dependent, whereas that of the inner motors is not. Such an interpretation fits well with the idea that the outer motors control beat frequency while the inner motors determine bend shape. In the earlier discussion of bend shape (Section 3.2), the inner motors are presumably responsible for the load-independent effects observed.

3.6 Conversion of sliding into bending

In the intact axoneme, interconnecting structures prevent free microtubule sliding so that the action of the dynein motors generates flagellar bending. At the proximal end, the doublet microtubules become triplets and form, with interlinking structures, the basal body; it is generally assumed that no microtubule sliding occurs in this region. Forces developed by the dynein motors in more distal parts of the flagellum could then cause bending, even if there were no additional links between the doublets. Observations that fragments of flagella isolated from *Crithidia oncopelti* by laser microbeam (*Goldstein et al.*, 1970) or by micromanipulation (Holwill and McGregor, 1974) are able to propagate bends indicates that the basal body is not essential for bend initiation and propagation. Fragments isolated using the laser were able to reverse the direction of propagation, as in the living cell, whereas those amputated by micromanipulation could not, and sustained waves which propagated either proximally or distally depending on the situation at the moment of amputation. The different behaviour may be associated with the different state of the exposed end produced by the two amputation procedures; the laser amputation is likely to weld the microtubules, so that no sliding can occur at the exposed end, whereas amputation by micromanipulation probably produces a cut which allows sliding to occur.

The fact that waves are formed and propagated in a coherent way along flagella indicates that sliding and the resistance to it occur in a controlled way. The observations on amputated *Crithidia* flagella, described in the preceding paragraph, demonstrate that the control mechanism resides in the flagellum itself. With two rows of motors on each doublet, and the inner and outer rows having significantly different structures and functions, the co-ordination required within each row, between the rows, and with other axonemal structures is likely to be sophisticated. The computer simulations described in section 3.5, together with integrated experiments, are directed towards understanding the nature of the co-ordination.

The absence of doublet sliding at the flagellar base leads to the phenomenon of synchronous sliding which occurs simultaneously with the asynchronous local sliding that is associated with the propagation of a bend (for example Gibbons, 1982). These two types of sliding can be understood by considering the behaviour of microtubules in a flagellum initially containing a single, plane bend (A, Figure 3.5) of angle φ_A. If the microtubules do not twist along the axoneme, the amount of relative sliding between neighbouring doublets is $d\varphi_A$, where d is the separation of the doublets parallel to the bend plane (Holwill *et al.*, 1979). As the next bend (B, Figure 3.5, angle φ_B), forms at the base with the opposite curvature to bend A, relative sliding occurs between the microtubules in the opposite direction to that which formed bend A, which is assumed to propagate without change in shape, and therefore in angle. Since the microtubules are fixed at the base, if the original bend (A) is to maintain its shape, the sliding $d\varphi_B$ required to form the new bend (B) must occur through the established bend at the same time as the local asynchronous sliding which is geometrically required for bend propagation. Since this sliding occurs along the length of the flagellum beyond the developing bend, it is known as synchronous sliding.

Bends on some sea urchin sperm flagella appear to be initiated in pairs with opposite curvature (Goldstein, 1975), so that a situation with no synchronous sliding could be envisaged; however, other cases have been analysed where synchronous sliding is clearly present (for example Brokaw, 1993, 1996). Brokaw (1993) analysed the distribution of synchronous and asynchronous sliding in sperm flagella, and concluded that bend initiation and bend propagation are independent and separable processes. The synchronous and asynchronous sliding are geometric consequences of the shape of developing and propagating bends. A mechanism whereby active dynein-microtubule interactions are responsible for bend propagation, while allowing simultaneously microtubule sliding in the direction opposite to that needed for bend propagation, has not yet been formulated, and will require an understanding of how the two sliding patterns can arise at the same time.

The observation that dynein arm action is polarized has implications for doublet sliding in the axoneme. To form a bend in one direction, the dynein arms on the doublets in only one half of the axoneme, to one side of the bend plane, are required to be active. The arms on doublets in the other half of the axoneme must be passive to allow relative basewards sliding to occur. These considerations, which are based on bend geometry, led Satir (1985) to suggest that arm activity must switch from one half of the axoneme to the other to produce coherent bend propagation. Support for such a mechanism comes from observations that *Elliptio* gill cilia can be arrested in

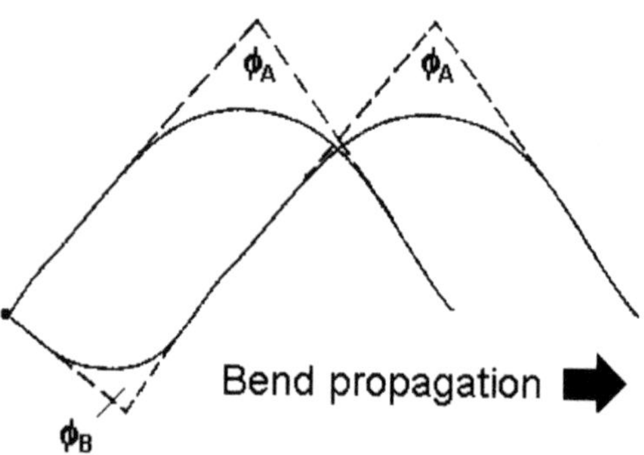

Figure 3.5 Propagation of a bend A without change in shape, and formation of bend B. For further explanation, see text

one of two configurations depending on the ionic environment (Wais-Steider and Satir, 1979; Satir, 1985) and can be forced to switch between the two by altering the ionic concentrations of calcium or vanadate. Vanadate is a potent ATP poison which has been used to inhibit flagellar motility in a range of experiments, while calcium sensitivity is a common feature of flagellar motion, with this ion being implicated in some of the avoidance responses already described (section 3.2).

The concept of a switch point is appropriate for two-dimensional flagellar bending, if the microtubules do not twist along the axoneme. As noted earlier, many flagellates propagate three-dimensional bends; such bends can be formed by dynein activity between only one pair of doublets. Bend propagation would then involve the progression of arm activity from one doublet to its neighbour continuously around the axoneme.

The smooth propagation of two- or three-dimensional bends along a flagellum requires control of the motor mechanism so that relevant torques to bend the organelle are applied with appropriate timing along its length. It is implicit that the control will include the operation of the switch, either from one half of the axoneme to the other, or from one microtubule to the next. Such control is most effectively provided through a mechanism which feeds back information about the mechanical state of the bending to the motor apparatus. One possibility is that the motors are activated by a change in curvature of the flagellum, an hypothesis proposed and developed by Brokaw in a range of papers (for example Brokaw, 1985). The principle of the mechanism is that bending in one region of a flagellum causes a change in the curvature of a neighbouring region, which is then stimulated to bend actively; this feedback is a passive process, mediated by the elasticity of the axoneme. Other control processes, which could be passive or active in nature, could rely on microtubule sliding, stimulating activity through its magnitude or velocity (for example Brokaw, 1996).

Lindemann (1994a, 1994b) has proposed a mechanism – which he refers to as the 'geometric clutch' – based on the transverse forces exerted between microtubules by the dynein arms. In a region where the dynein arms between a pair of microtubules are active, these forces will tend to draw the microtubules together, thereby favouring the attachment of further motors in the neighbouring region; the situation will progress along the pair of microtubules, resulting in propagated activity. In an extensive review, Lindemann and Kanous (1997) discuss the properties of the geometric clutch, and show that it is compatible with other suggested propagation mechanisms, as well as predicting that a flagellum will be sensitive to mechanical and other stimuli. They show how the mechanism can be used to explain both symmetric and asymmetric flagellar bending. Such an explanation is both valuable and necessary, since, as noted earlier, individual flagella are able to generate and sustain more than one type of bend pattern. The different types of movement usually occur in response to external stimuli, which may be chemical or mechanical, and it is difficult to avoid the conclusion that some form of active control system exists to change the beat form. Once the form of beat is changed, propagation could be maintained by a passive mechanism, such as the geometric clutch.

In our discussion, we have made no attempt to identify unequivocally those structures which restrict the sliding along the flagellar shaft. The interdoublet or nexin links are well placed to fulfil this role, and observations by Warner and Satir (1974) show that the radial spokes change their orientation in a way that is strongly correlated to bending. Biochemical studies linked to microtubule sliding velocity measurements have led Habermacher and Sale (1995) to conclude that inner arm dynein activity is regulated by signals transmitted by the spokes, which may therefore form an element of a mechanochemical feedback process. Another feature which has not been considered in detail is the role that might be played by the torsional strain which appears to be a property of microtubules isolated from some axonemes. In avian sperm, isolated groups of microtubules form helical ribbons (Vernon and Woolley, 1995), the pitch of which could be changed, and the change propagated on addition of ATP. The significance of this for flagellar motion is not clear, but the microtubules could be a source of elastic energy in the intact organelle.

3.7 Concluding remarks

Although we have concentrated on the behaviour of the flagellates, a considerable amount of information relating to the activity of the axoneme and its components has been derived from studies of cilia and sperm flagella. While there are some differences of detail, the axonemal structure remains remarkably constant across species and through evolution, indicating that it is well-fitted to perform its function of liquid propulsion. Just why an assembly of nine microtubules with their associated inner and outer dynein motors should be so well conserved and effective in producing a wide range of motility patterns has not yet been explained. Interdisciplinary research, some of which we have described in this paper, involving structural, biochemical, physical and computer-modelling studies, is directed towards understanding the mechanisms of motility, and may provide the answer to this question.

REFERENCES

Avolio, J., Glazzard, A. N., Holwill, M. E. J. and Satir, P. (1986) Structures attached to doublet microtubules of cilia: computer modelling of thin-section and negative-stain stereo images. *Proceedings of the National Academy of Sciences*, 83, 4804–4808.

Barkalow, K., Avolio, J., Holwill, M. E. J., Hamasaki, T. and Satir, P. (1994) Structural and geometrical constraints on the outer dynein arm *in situ*. *Cell Motility and the Cytoskeleton*, 27, 299–312.

Brokaw, C. J. (1965) Non-sinusoidal bending waves of sperm flagella. *Journal of Experimental Biology*, 43, 155–169.

Brokaw, C. J. (1966) Effects of increased viscosity on the movements of some invertebrate spermatozoa. *Journal of Experimental Biology*, **45**, 113–139.

Brokaw, C. J. (1985) Computer simulation of flagellar movement: VI Simple curvature–controlled models are incompletely specified. *Biophysical Journal*, 48, 633–642.

Brokaw, C. J. (1989) Direct measurements of sliding between outer doublet microtubules in swimming sperm flagella. *Science*, **243**, 1593–1596.

Brokaw, C. J. (1993) Microtubule sliding in reduced-amplitude bending waves of *Ciona* sperm flagella, resolution of metachronous and synchronous sliding components of stable bending waves. *Cell Motility and the Cytoskeleton*, 26, 144–162.

Brokaw, C. J. (1996) Microtubule sliding, bend initiation, and bend propagation parameters of *Ciona* sperm flagella altered by viscous load. *Cell Motility and the Cytoskeleton*, **33**, 6–21.

Brokaw, C. J. and Johnson, K. A. (1989) Dynein-induced microtubule sliding and force generation. In *Cell Movement*, vol. 1 (eds F. D. Warner, P. Satir and I. R. Gibbons), New York: Alan R. Liss, pp 191–198.

Brokaw, C. J. and Kamiya, R. (1987) Bending patterns of *Chlamydomonas* flagella: IV. Mutants with defects in inner and outer dynein arms indicate differences in dynein arm function. *Cell Motility and the Cytoskeleton,* **8**, 68–75.

Brokaw, C. J. and Wright, L. (1963) Bending waves on the posterior flagellum of *Ceratium. Science* **142**, 1169–1170.

Burgess, S. A., Dover, S. D. and Woolley, D. M. (1991) Architecture of the outer arm dynein ATPase in an avian sperm flagellum, with further evidence for the B-link. *Journal of Cell Science*, **98**, 17–26.

Burgess, S. A. (1995) Rigor and relaxed outer dynein arms in replicas of cryofixed motile flagella. *Journal of Molecular Biology,* **250**, 52–63.

Chwang, A. T. and Wu, T. Y. (1971) A note on the helical movement of micro-organisms. *Philosophical Transactions of the Royal Society of London, Series B,* **178**, 327–346.

Eshel, D. and Brokaw, C. J. (1988) Determination of the average shape of flagellar bends, a gradient curvature model. *Cell Motility and the Cytoskeleton,* **9**, 312–324.

Gee, M. A., Heuser, J. G. and Vallee, R. B. (1997) An extended microtubule binding structure within the dynein motor domain. *Nature,* **390**, 636–639.

Gardner, L. C., O'Toole, E., Perrone, C. A., Giddings, T. and Porter, M. E. (1994) Components of a 'dynein regulatory complex' are located at the junction between the radial spokes and the dynein arms in *Chlamydomonas* flagella. *Journal of Cell Biology,* **127**, 1131–1325.

Gibbons, I. R. (1982) Sliding and bending in sea urchin sperm flagella. *Symposia of the Society for Experimental Biology,* **35**, 225–287.

Goldstein, S. F. (1975) Morphology of developing bends in sperm flagella. In *Swimming and Flying in Nature* (eds T. Y. Wu, C. J. Brokaw and C. Brennan), New York: Plenum, pp. 127–132.

Goldstein S. F., Holwill, M. E. J. and Silvester, N. R. (1970) The effects of laser microbeam irradiation on the flagellum of *Crithidia (Strigomonas) oncopelti. Journal of Experimental Biology,* **53**, 401–409.

Goodenough, U. W. and Heuser, J. E. (1982) Substructure of the outer dynein arm. *Journal of Cell Biology,* **95**, 798–815.

Goodenough, U. W. and Heuser, J. E. (1984) Structural comparison of purified dynein proteins with *in situ* dynein arms. *Journal of Molecular Biology,* **180**, 1083–1118.

Gray, J. and Hancock, G. J. (1955) The propulsion of sea urchin spermatozoa. *Journal of Experimental Biology,* **32**, 802–814.

Habermacher, G. and Sale, W. S. (1995) Regulation of dynein-driven microtubule sliding by an axonemal kinase and phosphatase in *Chlamydomonas* flagella. *Cell Motility and the Cytoskeleton,* **32**, 106–109.

Hamasaki T., Barkalow K., Richmond, J. and Satir P. (1991) cAMP-stimulated phosphorylation of an axonemal polypeptide that copurifies with the 22S dynein arm regulates microtubule translocation velocity and swimming speed in *Paramecium. Proceedings of the National Academy of Sciences*, **88**, 7918–7922.

Hamasaki T., Holwill, M. E. J., Barkalow, K. and Satir, P. (1995) Mechanochemical aspects of axonemal dynein activity studied by *in vitro* microtubule translocation. *Biophysical Journal,* **69**, 2569–2579.

Hancock, G. J. (1953) Self-propulsion of microscopic organisms through liquids. *Philosophical Transactions of the Royal Society of London,* **Series A, 217**, 96–121.

Hiramoto, Y. and Baba, S. (1978) A quantitative analysis of flagella movement in echinoderm spermatozoa. *Journal of Experimental Biology,* **76**, 85–104.

Holwill, M. E. J. (1965) The motion of *Strigomonas oncopelti. Journal of Experimental Biology,* **42**, 125–137.

Holwill, M. E. J. (1966) Physical aspects of flagellar movement. *Physiological Reviews*, **46**, 696–785.

Holwill, M. E. J. (1994) Mechanical aspects of ciliary propulsion. *NATO ASI Series H*, **84**, 393–413.

Holwill, M. E. J. (1996) Bend shapes and molecular mechanisms of flagella. In *Cilia, Mucous and Mucociliary Interactions* (eds G. L. Baum, Z. Priel, Y. Roth, N. Liron and E. Ostfeld), New York: Marcel Dekker, pp, 563–568.

Holwill, M. E. J. and Miles, C. A. (1971) A hydrodynamic analysis of non-uniform flagellar undulations. *Journal of Theoretical Biology*, **31**, 25–42.

Holwill, M. E. J. and Coakley, C. J. (1972) Propulsion of micro-organisms by three-dimensional flagellar waves. *Journal of Theoretical Biology*, **35**, 525–542.

Holwill, M. E. J. and McGregor, J. L. (1974) Micromanipulation of the flagellum of *Crithidia oncopelti*. *Journal of Experimental Biology*, **60**, 437–444.

Holwill, M. E. J. and Peters, P. D. (1974) Dynamics of the hispid flagellum of *Ochromonas danica*. *Journal of Cell Biology*, **62**, 322–328.

Holwill, M. E. J. and Satir, P. (1987) Generation of propulsive forces by cilia and flagella. In *Cytomechanics* (eds J. Bereiter-Hahn and O. R. Anderson), Berlin: Springer-Verlag, pp. 120–130.

Holwill, M. E. J. and Sleigh, M. A. (1967) Propulsion by hispid flagella. *Journal of Experimental Biology*, **47**, 267–276.

Holwill, M. E. J., Cohen, H. J. and Satir, P. (1979) A sliding microtubule model incorporating axonemal twist and compatible with three-dimensional ciliary bending. *Journal of Experimental Biology*, **78**, 265–280.

Holwill, M. E. J., Foster, G., Hamasaki, T. and Satir, P. (1995) Biophysical aspects and modelling of ciliary motility. *Cell Motility and the Cytoskeleton*, **32**, 114–120.

Holwill, M., Foster, G., Guevara, E., Hamasaki, T. and Satir, P. (1998a) Computer modelling of the ciliary axoneme. *Cell Motility and the Cytoskeleton*, **39**, 337–348. (Video suppl. 5).

Holwill, M. E. J., Taylor, H. C., Guevara, E. and Satir, P. (1998b) Computer modelling: a versatile tool for the study of structure and function in cilia. *European Journal of Protistology*, **34**, 239–243.

Ishijima, S., Kubo-Irie, M., Mohri, H. and Hamaguchi, Y. (1996) Calcium-dependent bidirectional power stroke of the dynein arms in sea urchin axonemes. *Journal of Cell Science*, **109**, 2833–2842.

Johnston, D. N., Silvester, N. R. and Holwill, M. E. J. (1979) An analysis of the shape and propagation of waves on the flagellum of *Crithidia oncopelti*. *Journal of Experimental Biology*, **80**, 299–315.

Kagami, O. and Kamiya, R. (1992) Translocation and rotation of microtubules caused by multiple species of *Chlamydomonas* inner-arm dynein. *Journal of Cell Science*, **103**, 653–664.

Kurimoto, E. and Kamiya, R. (1991) Microtubule sliding in flagellar axonemes of *Chlamydomonas* mutants missing inner- or outer-arm dynein: velocity measurements on new types of mutants by an improved method. *Cell Motility and the Cytoskeleton*, **19**, 275–281.

Lighthill, J. (1976) Flagellar hydrodynamics. *SIAM Review* **18**, 161–230.

Lighthill, J. (1996) Reinterpreting the basic theorem of flagellar hydrodynamics. *Journal of Engineering and Mathematics*, **30**, 25–34.

Lindemann, C. B. (1994a) A 'Geometric Clutch' hypothesis to explain oscillations of the axoneme of cilia and flagella. *Journal of Theoretical Biology*, **168**, 175–189.

Lindemann, C. B. (1994b) A model of flagellar and ciliary functioning which uses the forces transverse to the axoneme as the regulator of dynein activation. *Cell Motility and the Cytoskeleton*, **29**, 141–154.

Lindemann, C. B. and Kanous, K. S. (1997) A model for flagellar motility. *International Review of Cytology*, **173**, 1–72.

Machin, K. E. (1958) Wave propagation along flagella. *Journal of Experimental Biology*, **35**, 796–806.

Minoura, I. And Kamiya, R. (1995) Strikingly different propulsive forces generated by different dynein-deficient mutants in viscous media. *Cell Motility and the Cytoskeleton*, **31**, 130–139.

Mimori, Y. and Miki-Noumura, T. (1994) ATP-induced sliding of microtubules on tracks of 22S dynein molecules aligned with the same polarity. *Cell Motility and the Cytoskeleton*, **27**, 180–191.

Moss, A. G., Gatti, J-L. and Witman, G. B. (1992a) The motile β/IC1 subunit of sea urchin sperm outer arm dynein does not form a rigor bond. *Journal of Cell Biology*, **118**, 1177–1189.

Moss, A. G., Sale, W. S., Fox, L. A. and Witman, G. B. (1992b) The α subunit of sea urchin sperm outer arm dynein mediates structural and rigor binding to microtubules. *Journal of Cell Biology*, **118**, 1189–1200.

Omoto, C. K. (1991) Mechanochemical coupling in cilia. *International Review of Cytology*, **131**, 255–292.

Piperno, G. (1995) Regulation of dynein activity within *Chlamydomonas* flagella. *Cell Motility and the Cytoskeleton*, **32**, 103–105.

Purcell, E. M. (1997) The efficiency of propulsion by a rotating flagellum. *Proceedings of the National Academy of Sciences*, **94**, 11307–11311.

Ramia, M. (1993) *Mathematical modelling of micro-organism locomotion*. Ph.D. thesis, University of Sydney, Australia.

Sakakibara, H. and Nakayama, H. (1998) Translocation of microtubules caused by the αβ, β and γ outer arm dynein subparticles of *Chlamydomonas*. *Journal of Cell Science*, **111**, 1155–1164.

Sale, W. S. and Fox, L. A. (1988) Isolated β-heavy chain subunit of dynein translocates microtubules *in vitro*. *Journal of Cell Biology*, **107**, 1793–1797.

Sale, W. S. and Satir, P. (1977) Direction of active sliding of microtubules in *Tetrahymena* cilia. *Proceedings of the National Academy of Sciences*, **74**, 2045–2049.

Satir, P. (1965) Studies on cilia.II. Examination of the distal region of the ciliary shaft and the role of the filaments in motility. *Journal of Cell Biology*, **26**, 805–834.

Satir, P. (1985) Switching mechanisms in the control of ciliary motility. *Modern Cell Biology*, **4**, 1–46.

Satir, P. (1989) Structural analysis of the dynein cross-bridge cycle. In *Cell Movement*, vol. 1 (eds F. D. Warner, P. Satir and I. R. Gibbons), New York: Alan R. Liss, pp. 219–234.

Satir, P. (1998) Mechanisms of ciliary motility. *European Journal of Protistology*, **34**, 267–272.

Shingyoji, C., Murakami, A. and Takahashi, K. (1977) Local reactivation of Triton-extracted flagella by iontophoretic application of ATP. *Nature*, **265**, 269–270.

Shingyoji, C., Higuchi, H., Yoshimura, M., Katayama, E. and Yanagida, T. (1998) Dynein arms are oscillating force generators. *Nature*, **393**, 711–714.

Silvester, N. R. and Holwill, M. E. J. (1972) An analysis of hypothetical flagellar waveforms. *Journal of Theoretical Biology*, **35**, 505–523.

Sugrue, P., Avolio, J., Satir, P. and Holwill, M. E. J. (1991) Computer modelling of *Tetrahymena* axonemes at macromolecular resolution. *Journal of Cell Science*, **98**, 5–16.

Summers, K. E. and Gibbons, I. R. (1971) Adenosine-triphosphate-induced sliding of tubules in tripsin-treated flagella of sea urchin sperm. *Proceedings of the National Academy of Sciences*, **68**, 3092–3096.

Summers, K. E. and Gibbons, I. R. (1973) Effects of trypsin digestion on flagellar structures and their relation to cell motility. *Journal of Cell Biology*, **58**, 618–628.

Svoboda, K., Schmidt, C. F., Schnapp, B. J. and Block, S. M. (1993) Direct observation of kinesin stepping by optical trapping interferometry. *Nature*, **365**, 721–727.

Takahashi, K., Shingyoji, C. and Kamimura, S. (1982) Microtubule sliding in reactivated flagella. *Symposia of the Society for Experimental Biology*, **35**, 159–177.

Taylor, H. C. and Holwill, M. E. J. (1999) Axonemal dynein – a natural molecular motor. *Nanotechnology*, **10**, 237–243.

Taylor, H. C., Satir, P. and Holwill, M. E. J. (1999) Assessment of inner dynein arm structure and possible function in ciliary and flagellar axonemes. *Cell Motility and the Cytoskeleton*, in press.

Vale, R. D. and Toyoshima, Y. Y. (1989) Microtubule translocation properties of intact and proteolytically digested dynein arms from *Tetrahymena* cilia. *Journal of Cell Biology*, **108**, 2327–2334.

Vernon, G. G. and Woolley, D. M. (1995) The propagation of a zone of activation along groups of flagellar doublet microtubules. *Experimental Cell Research*, **220**, 482–494.

Wais-Steider, J. and Satir, P. (1979) Effect of vanadate on gill cilia: switching mechanism in ciliary beat. *Journal of Supramolecular Structure*, **11**, 339–347.

Warner, F. D. and Satir, P. (1974) The structural basis of ciliary bend formation. Radial spoke positional changes accompanying microtubule sliding. *Journal of Cell Biology*, **63**, 35–63.

Woolley, D. M. and Brammall, A. (1987) Direction of sliding and relative sliding velocities within trypsinized sperm axonemes of *Gallus domesticus*. *Journal of Cell Science*, **88**, 361–371.

Yamada, A., Yamaga, T., Sakakibara, H. Nakayama, H. and Oiwa, K. (1998) Unidirectional movement of fluorescent microtubules on rows of dynein arms of disintegrated axonemes. *Journal of Cell Science*, **111**, 93–98.

Yeung, C-H. and Woolley, D. M. (1984) Three-dimensional bend propagation in hamster sperm models and the direction of roll in free-swimming cells. *Cell Motility and the Cytoskeleton*, **4**, 215–226.

Chapter 4

The flagellate cytoskeleton

Introduction of a general terminology for microtubular flagellar roots in protists

Øjvind Moestrup

ABSTRACT

The cytoskeleton of protists may consist of a few microtubules beneath the plasma membrane in small and apparently reduced organisms. In other species, however, it may comprise highly complex systems of microtubules and fibrous components. The 9+2 flagellum/cilium is believed to have arisen only once during the evolution of the eukaryotic cell. The thesis of the present paper is that not only the 9+2 axoneme and the 9x3 basal body but also other structures of the flagellar apparatus are homologous. A common feature of most eukaryote cells is the presence of microtubular roots, believed to play a range of functions in the cell. Roots associated with different flagella are usually different. However, the discovery of flagellar transformation, a process during which a flagellum passes through a maturation process before taking up its final position in the cell one or more generations later, has made it possible to identify the mature (final) position of each flagellum. This allows comparison of roots associated with mature flagella in different organisms. Using such information, a comparative study of the microtubular roots has been made on representatives from the major taxonomic groups of protists. It has resulted in the proposal of a common numbering (identification) system for microtubular roots, which may hopefully replace the many labelling systems presently in use in the different groups of protists.

4.1 Introduction

The cytoskeleton of eukaryotic cells comprises highly diverse systems of microtubular and fibrillar structures. Such structures occur almost universally and also in the few extant species thought to be related to the oldest eukaryotes. Cytoskeletal structures therefore probably arose early during evolution of the eukaryotic cell. The diversity of the cytoskeleton is a serious difficulty for determining which of the many structures are homologous.

Flagella and cilia of eukaryotes are based on a common 9+2 axonemal structure and the flagellar basal bodies on a common 9x3 configuration. Due to their complexity, flagella and basal bodies are considered as having evolved only once during evolution of the eukaryotic cell.

Since all cilia/flagella and basal bodies in eukaryotes appear to be homologous it is likely that this applies also to some of the structures attached to the axoneme and the basal body doublets/triplets. By identifying such homologous structures it becomes possible to determine whether organelles attached to the structures occupy homologous positions. This is a prerequisite for determining whether the structures themselves are homologous, that is, they arose only once. For example, do eyespots of dinoflagellates and green algae occupy homologous positions? Are flagellar hairs on the long flagella of cryptomonads and tripartite hairs of heterokont protists homologous?

The major taxonomic groups of eukaryotes are presently defined by a combination of biochemical, ultrastructural and sometimes genetic features. Determination of homologous structures in the different groups is difficult, however. Recently a process known as flagellar transformation has been detected which is widespread in eukaryotes and which offers a possiblity of identifying homologous flagella in different organisms (Wright *et al.*, 1980a; review in Beech *et al.*, 1991). During flagellar transformation in a biflagellate cell, one flagellum (the immature state), for example the hairy flagellum of heterokonts, changes in the next generation into another state (the mature state), in this case the smooth flagellum. The process, now discovered in many different groups of eukaryotes discussed later, enables determination of the mature state of any flagellum, provided the cell undergoes cell division (exceptions being gametes or other non-dividing cells, but see later). The situation is more complex in species with several flagella but eventually all flagella for example in a quadri/octoflagellate green alga take up the final, mature position (Moestrup and Hori, 1989) which is termed the number 1 position. The immature flagellum in a biflagellate protist occupies the number 2 position and, in cases of more flagella, these occupy the number 3, 4, etc. positions. The flagellar apparatus in some organisms consists of two or more copies of the same unit, for example in diplomonads (two apparently identical systems arranged symmetrically). This situation has been taken to its extreme in ciliates.

Organelles are not arranged randomly in the cell but take up a fixed and characteristic position in relation to the flagellar apparatus. Indeed, formation of new flagella is among the first visible processes in cell division, followed by duplication of organelles and, eventually, cytokinesis. During flagellar replication and transformation, complex processes take place, including reorganization of the flagellar roots. Roots belonging to the two flagella of a pair take up different and characteristic paths in the cell and the flagellar root systems must therefore undergo major changes when the flagellum changes from the immature to the mature state. Cytokinesis is completed when the duplicated sets of organelles rearrange relative to the replicated and transformed components of the flagellar apparatus.

In this chapter I will be discussing one of the characteristic components of the cytoskeleton, the microtubular flagellar roots. Data on flagellar root structure have accumulated since application of electron microscopy to the study of protist cells in the late 1940s, and after the finding of flagellar transformation we are now approaching a stage which allows identification of homologous cell structures in both in distantly related organisms. This is based on the assumption that the mature state of flagella in different organisms is homologous, which appears likely.

Protists in which flagellar replication has been studied include many of the main taxonomic groups and these will discussed separately.

Before proceeding further it is perhaps useful to mention that I will be using few 'formal' names of classes, divisions, and so on, this is due to the present flux in classifications caused by the ongoing combination into a single classification of groups formerly classified as plants (algae and fungi) and animals (protozoa).

4.2 Some general considerations

Numbering of individual flagella and naming of the components of the flagellar apparatus has over the years been made independently and often more or less arbitrarily in the different taxonomic groups. The aim of the present exercise has been to develop a general numbering system which may be used for most and eventually all protists. This obviously means that some of the systems presently in use will have to change.

During many years of work with the flagellar apparatus of algae I have become struck by the fact that individual flagella of many taxonomic groups possess two, and only two, microtubular roots, attached to opposite sides of the basal body. The presence of only a single root per flagellum is also relatively common, but more than two roots is exceptional. This fact may not be apparent at first sight (for example Sleigh, 1988: fig. 3) thus the flagellar apparatus of dinoflagellates was at first thought to contain only a single flagellar root, associated with the longitudinal flagellum basal body. However, in the 1980s Keith Roberts discovered that the transverse flagellum very commonly possesses two one-stranded roots, attaching to opposite sides of the basal body. Very recently Gert Hansen found a second root, also one-stranded, on the opposite side of the basal body of the longitudinal flagellum in certain thecate dinoflagellates. Two microtubular roots associated with each basal body therefore also applies to many dinoflagellates.

In cryptomonads the flagellar apparatus reconstruction offered by Roberts *et al.* (1981) and Roberts (1984) at first sight seems difficult to interpret. However, Perasso *et al.* (1992, fig. 7) showed very clearly that two microtubular roots are formed on opposite sides of the new flagellar basal bodies. They subsequently change somewhat in morphology when the immature flagellum transforms into the mature state. One root transforms only slightly, the other becomes more difficult to recognize.

While roots associated with opposite sides of each basal body are the rule in numerous organisms, euglenoids, bodonids and ciliates have two roots associated with the mature flagellum, but only a single root with the immature one. When the immature flagellum matures, two roots emerge, associated with the basal body. These findings prompt me to postulate that the bi-rooted condition arose early during eukaryote evolution and I use this as a basis for the nomenclature suggested.

The suggestions are as follows:

The mature flagellum is the number 1 flagellum (Heimann *et al.*, 1989; Moestrup and Hori, 1989).

The two roots associated with flagellum 1 are named root 1 and 2, those with flagellum 2 are named roots 3 and 4.

The roots are (usually) named in a clockwise fashion, looking down the basal body from the outside of the cell. During flagellar transformation root 3 develops into root 1 or, alternatively, root 3 is absorbed and root 1 develops in its place. Similarly root 4 develops into root 2, or root 4 is resorbed and root 2 develops in it place. In other words, root 1 and 3 are homologous, and root 4 and 2 are homologous.

The nomenclature may perhaps be extended to defining the ventral and dorsal sides of a cell, something which is sometimes done more or less arbitrarily. Thus flagellum 1 in cryptomonads is located close to what is presently called the ventral side (Perasso *et al.*, 1992) and the same applies to euglenoids. However, the opposite applies to cercomonads (discussed later). The new system may also be used to identify homologous triplets of different basal bodies. This requires identification of the triplets to which the MTOC-organizing centres responsible for root formation are attached. In the prasinophycean green alga *Pyramimonas octopus* these have been identified as triplets numbers 7–8 on one side of the basal body and triplet 3 on the other (Moestrup and Hori, 1989) (Figure 4.9). Considering that some rearrangements of the roots may take place after formation of the roots, identification may require that root ontogeny is studied in detail in each group.

This chapter is not to be considered as a complete review but as an interim report to be completed when more information becomes available.

4.3 The individual groups of protists

4.3.1 *Haptophytes*

The new numbering system agrees with that used for prymnesiophycean haptophytes (Green and Hori, 1994). Eikrem and Moestrup (1998) have recently discussed the flagellar root structure in this group and it will only be repeated briefly here.

The haptophyte protists may be grouped into two classes, the Prymnesiophyceae and the Pavlovophyceae. In the Prymnesiophyceae each flagellum typically possesses two microtubular roots (Figure 4.1).

Root 1 is often large and directed towards a chloroplast, on its way associating with a mitochondrion close to the flagellar apparatus. In coccolithophorids and a few others, R1 nucleates a 'crystal' believed to function in mitosis, perhaps as a source of tubulin for the mitotic spindle.

Root 2 is few-membered. In many species it is reduced or almost absent. In others, however, for example *Pleurochrysis* and many other coccolithophorids, it also nucleates a 'crystal'of microtubules believed to serve in mitosis. Near the origin at basal body 1, R2 often forms a trough, for example in *Pleurochrysis* (Inouye and Pienaar, 1985: fig. 18).

Root 3 is few-membered, usually four-stranded, with microtubules in two tiers.

Root 4 is 1-stranded and after leaving the flagellar apparatus joins R3. This combined root is one of the distinctive features of the Prymnesiophyceae. The two roots proceed together along the plasmalemma, reaching the second chloroplast typically present in prymnesiophyceans.

As was mentioned earlier, one of the prerequisites of the new labelling system suggested here is that R3 and R1 are homologous or occupy homologous positions and that R2 and R4 are homologous or occupy homologous positions. None of the published micrographs known to me show the early stages of root replication in haptophytes. However, the orientation of the newly replicated flagella in *Pleurochrysis carterae* (Beech *et al.*, 1988: fig. 29) strongly indicates that what I have called R3 does indeed develop at the R1 position in the transforming flagellar apparatus, and similarly with R4 and R2.

The arrangement of the roots is notably different in the Pavlovophyceae (Figure 4.2). A comparison of Green (1980: figs 30, 31: *Pavlova pinguis*) and Eikrem and Moestrup (1998, figs 19–25: *Chrysochromulina scutellum*) indicates that the posterior flagellum in the Pavlovophyceae is the equivalent of flagellum 1 in

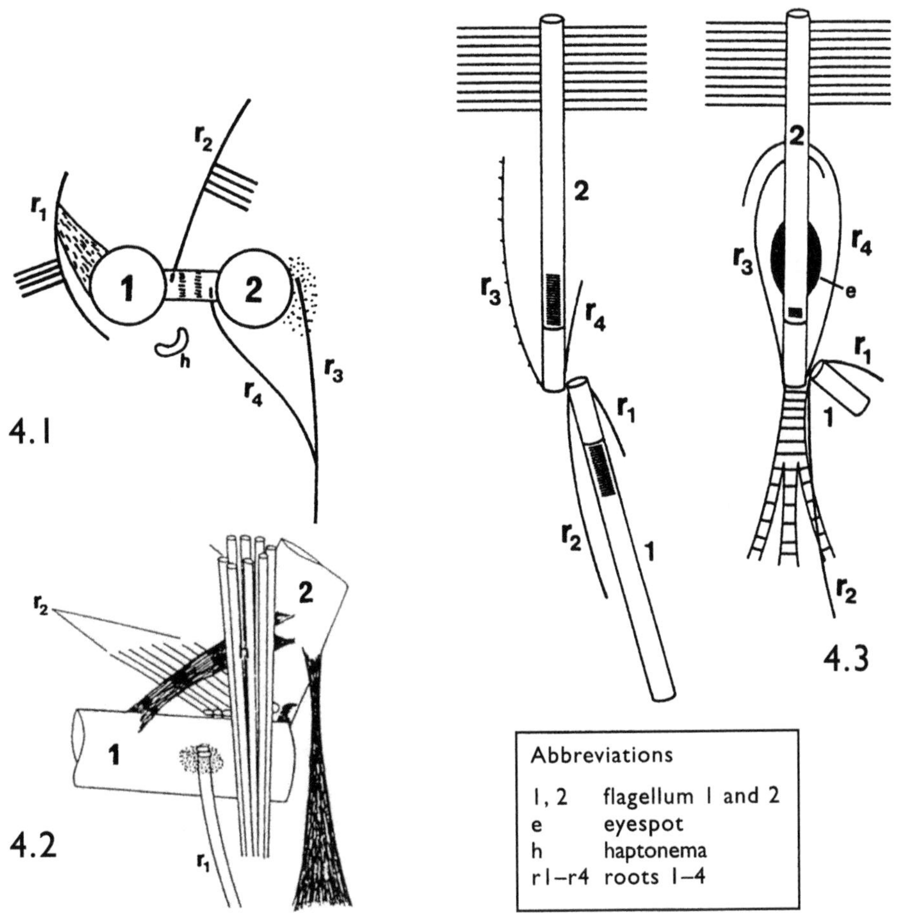

Figure 4.1 Semi-diagrammatic reconstruction of the flagellar apparatus in prymnesiophycean haptophytes

Source: Modified after Green and Hori, 1994

Figure 4.2 The flagellar apparatus in the pavlovophycean haptophytes

Source: After Green, 1980, labelling modified

Figure 4.3 The flagellar apparatus of heterokont protists, represented by the Oomycetes (left, after Andersen, 1989) and the eustigmatophyte *Vischeria stellata* (right, after Andersen, 1991 and Santos and Leedale, 1991), labelling modified

Chrysochromulina (see also Figure 4.2), and Beech *et al.* (1991) reached the same conclusion. The anterior flagellum of pavlovophyceans therefore probably changes into the posterior one at cell division. Two microtubular roots associate with the posterior flagellum 1 in *Pavlova* while anterior flagellum 2 lacks microtubular roots altogether. The orientation indicates that R1 is the narrow (two-stranded) root, and the broad root is R2. This is the opposite of the situation in the prymnesiophyceans in which R1 is the broad root. R1 of *Pavlova* passes deep into the cell in a way not seen in any root of the Prymnesiophyceae.

A fibrillar structure extends into the cell from flagellum 2, another feature by which members of the Pavlovophyceae differ from the Prymnesiophyceae. The differences in the flagellar apparatus support the inclusion of the Pavlovaales group as a separate class of protists.

4.3.2 Heterokont protists

The structure of the flagellar apparatus has been reviewed several times (for example Barr and Allan, 1985; Barr and Désaulniers, 1989; Preisig, 1989; Andersen, 1991). Flagellar transformation was studied by Wetherbee *et al.* (1988) and Beech and Wetherbee (1990). The anterior hairy flagellum, carrying two opposite rows of tripartite hairs, transforms into the smooth flagellum at cell division, that is, flagellum 1 is the smooth flagellum, the hairy flagellum is number 2.

Microtubular flagellar roots are very uniform in heterokonts, two roots attaching to each basal body (Figure 4.3).

Numbering of the root system was done by Andersen (1987), before the process of flagellar transformation was known. Andersen's system was extended to other protists by Sleigh (1995). This was done without taking flagellar homology into consideration. Andersen labelled as roots 1 and 2 the two roots associating with the anterior flagellum. Unfortunately the anterior flagellum has now been found to be the number 2 flagellum. The numbering therefore has to change as follows.

R1 is the old R4 root, a few-membered root passing under the plasmalemma, and it is sometimes absent (for example hyphochytrids: Barr and Allan, 1985; Barr and Désaulniers, 1989).

R2 is the old R3 root which in many phagotrophic heterokonts functions in feeding (Andersen, 1991; Andersen and Wetherbee, 1992). It often forms a U-shaped trough near the basal bodies, two of the microtubules subsequently extending into a loop that serves as a feeding basket. In non-phagotrophic algae it may be reduced or absent. In heterokont fungi this root is often well developed, forming a trough, for example in the thraustochytrid *Thraustochytrium aureum* (Barr and Allan, 1985: fig. 22).

In *Epipyxis pulchra* R2 was seen to serve as a microtubule-organizing centre, new microtubules extending on the surface of a nearby mitochondrion. The eyepot was located on the other side of the microtubules (Andersen and Wetherbee, 1992).

Flagellar root transformation has not been studied in heterokonts, and owing to the relative arrangement of the basal bodies to each other it is not entirely clear which of the two roots associated with basal body 2 is homologous with R1. The two basal bodies are often arranged at right or even larger angles (Barr and Allan, 1985; Andersen, 1991) which makes interpretation difficult. Species with parallel or almost parallel basal bodies have modified flagellar roots (Synurophyceae: Andersen, 1985) or they appear to lack microtubular roots (pedinellids and silicoflagellates: Zimmermann *et al.*, 1984; Moestrup and Thomsen, 1990). However, the finding of the 'x' and 'y' bars in *Mallomonas* may be useful for identification of homologous roots (Beech and Wetherbee, 1990: fig. 72). The x and y bars appear to be identical, one extending from basal body 1, the other from basal body 2. Rotation of one basal body 180 degrees brings the x bar to cover the y bar. In other words the two basal bodies are probably rotated 180 degrees with respect to each other. If this applies also to other heterokonts it identifies the R3 and R4 roots: R3 is the old R1 root, known to serve as a microtubule-organizing root in many chrysophytes, oomycetes, thraustochytrids, xanthophytes, raphidophytes and brown algae. R4 is usually few-membered. In the hyphochytrids *Rhizidiomyces* and *Hyphochytrium* the one- or two-stranded R4 nucleates microtubules (Barr and Allan, 1985; Barr and Désaulniers, 1989).

There is some variation in the structure of the flagellar apparatus in heterokonts. Thus in *Pelagomonas* (Pelagophyceae) all signs of a second flagellum are lacking (Andersen *et al.*, 1993). The only flagellum present is the hairy anterior one. This creates the unusual situation that during cell division the flagellum is not transformed. It disappears without leaving any trace while two new hairy ones (number 2 flagella) grow out (Heimann *et al.*, 1995a).

The situation is more difficult to interpret in the Synurophyceae. The three- and four-stranded fibres described by Andersen (1985) extend from the rhizoplasts, not the basal bodies. The nature of the three-stranded root in *Mallomonas splendens* was resolved by Beech and Wetherbee (1990), who demonstrated that the three-stranded roots form along a newly formed basal body and later move down to take up their position on the rhizoplast surface. Andersen (1985) interpreted the roots as R1 and R3, using the old terminology (that is R2 and R3 in the terminology introduced here). Because the new three-stranded roots form on the newly forming basal bodies, they belong to basal body 2. They nucleate numerous microtubules, making it likely that they represent R3 (new terminology). There is no information on the attachment of the other microtubular structure to the basal bodies.

4.3.3 Cryptomonads

Flagellum 1 of cryptomonads is the ventral, short flagellum, which carries a single row of short flagellar hairs. Flagellum 2 is the dorsal one, lined by two opposite rows of longer hairs (Perasso *et al.*, 1992). The root arrangement in cryptomonads is at first sight somewhat difficult to interpret, but the elegant work of Perasso *et al.* (1992), describing transformation of the flagella and the fate of the flagellar roots during transformation, has left no doubt about the identity of the three main microtubular roots (Figure 4.4). The two roots associated with flagellum 2 emerge from the transforming flagellar apparatus in a way recalling the orientation of flagella in many other eukaryotes, for example green algae and the R3/R4 roots of haptophytes. The two roots appear to be homologous with R3 and R4 of these organisms.

R3 is the broad root, comprising four microtubules in *Cryptomonas ovata* (Roberts, 1984) but twelve microtubules in *Chilomonas paramecium* (Roberts *et al.*, 1981).

R4 is the 2-stranded 'Cr' root of both *Chilomonas* and *Cryptomonas*.

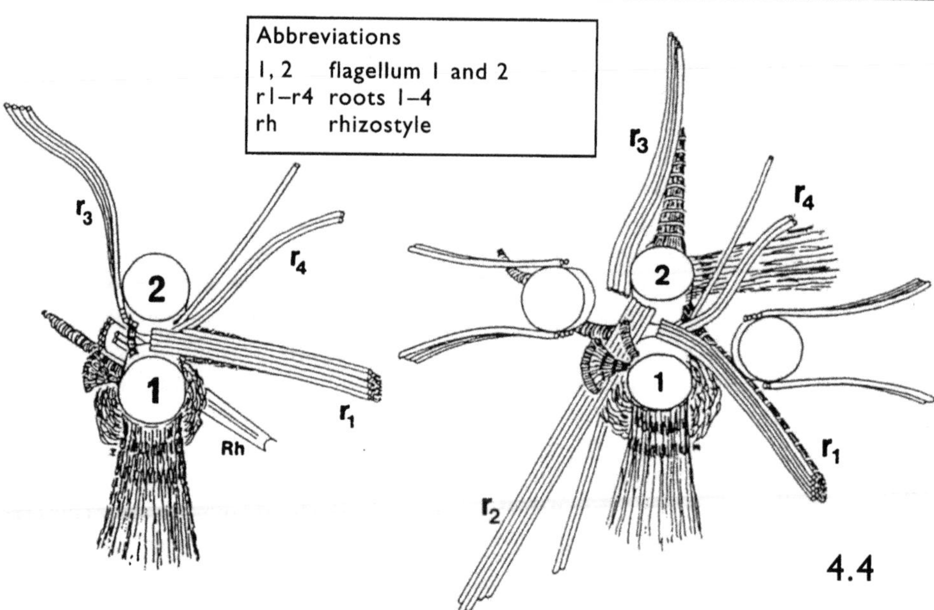

Figure 4.4 The cryptomonad *Cryptomonas ovata*, showing a diagrammatic reconstruction of the flagellar apparatus (left) and some details of flagellar replication and transformation (right)

Source: After Perasso *et al.*, 1992, labelling modified

Perasso *et al.* showed clearly that the R3 root during flagellar transformation in *Cryptomonas ovata* develops into the 'SRm', a complex comprising a five-stranded microtubular root and a transverse fibre. This is therefore the R1 root.

There is, however, some doubt about the identity of R2. The obvious candidate is the rhizostyle, a row of microtubules which in some species are associated with vanes or wings. The rhizostyle takes up slightly different paths in different species. In *Chilomonas* it passes deep into the cell, terminating near the posterior end of the cell, on the way bypassing the nucleus. In *Cryptomonas ovata* it also extends backwards but closer to the cell surface and without establishing any contact to the nucleus (Roberts *et al.*, 1981; Roberts, 1984). The two sets of figures reproduced as figures 4–6 and 7–9 by Roberts (1984) show the rhizostyle microtubules in an orientation which indicates homology to the R2 root. In *Proteomonas sulcata,* Hill and Wetherbee (1986) found differences in the flagellar apparatus of the two stages of the organism. The haplomorph was simpler than the diplomorph and there was no rhizostyle. In its place a two-stranded microtubular structure ('pr') extended from basal body 1 towards the nucleus. This structure is more like a normal flagellar root, strengthening the idea that the rhizoplast is a modified R2.

It would be interesting to examine also the ultrastructure of cryptomonads considered to be closest to the ancestral form (for example *Goniomonas*).

4.3.4 The alveolates: I Dinoflagellates

Flagellar transformation studies have shown that the transverse flagellum is abscissed or retracted during cell division and that the basal body grows out as a new longitudinal flagellum. Two new transverse flagella are produced: that is, the mature state (the number 1 flagellum) is the longitudinal flagellum, while the transverse flagellum is the number 2 flagellum (Heimann *et al.*, 1995b).

In the 1960s John Dodge and co-workers described the broad microtubular root associated with the longitudinal flagellum of dinoflagellates (generally known as the longitudinal microtubular root, LMR), and the striated fibre or root (TSR) emanating from the basal body of the transverse flagellum (literature reviewed in Moestrup, 1982; see also Figure 4.5). These structures have subsequently been found to be characteristic features of dinoflagellates, and they occur even in aberrant forms such as *Oxyrrhis marina* (Roberts, 1985). Subsequently, however, Keith Roberts discovered that the transverse flagellum carries two one-stranded roots, emanating from opposite sides of the flagellar base. One is always associated with the transverse striated fibre, while the other is microtubule-nucleating (Roberts 1991). Very recently it was discovered that in some gonyaulacoid and peridinioid dinoflagellates the longitudinal flagellum carries not only the LMR, but an additional one-stranded microtubular root is present on the opposite side of the basal body (Hansen *et al.*, 1997; Calado *et al.*, 1999). Both flagella of these dinoflagellates thus carry two microtubular roots.

Naming of the individual roots is clearest in species with almost parallel flagella, for example *Prorocentrum* (Roberts *et al.*, 1995: fig. 8). R1 is the LMR root, while the single-stranded root associated with the posterior flagellum in the thecate dinoflagellates mentioned above is R2. The microtubule-organizing root is R3, and R4 is the single-stranded root associated with the transverse fibre (TSR). During transformation of flagellum 2 into flagellum 1, R3 and R4 are resorbed. A new R1 (LMR) grows out on the side of the basal body previously taken up by the microtubule-organizing root, identifying the latter as R3 (Heimann *et al.*, 1995b: *Prorocentrum*).

Oxyrrhis marina, which in many respects deviates from typical dinoflagellates, also differs in the detailed architecture of the flagellar apparatus. R1, R3 and R4 roots are present, R1 and R3 somewhat modified, however. Additional structures such as a curved band of microtubules have no equivalent in other dinoflagellates (Roberts, 1985).

4.3.5 The alveolates: II Perkinsids

An undescribed genus of perkinsids has recently been found and the opportunity was taken to examine the ultrastructure of the flagellar apparatus of this small group of organisms (Norén *et al.*, 1999), which presently comprises one genus of free-living species, *Colponema* (Simpson and Patterson, 1996), in addition to *Perkinsus*, a parasite in oyster.

The new genus, a parasite of marine dinoflagellates, possesses two orthogonal flagella inserted laterally in grooves near the front end. The anterior flagellum carries unilateral very thin flagellar hairs in groups and the basal body is associated with two one-stranded microtubular roots that attach to opposite sides of the basal body. The roots extend along the anterior flagellar groove. The posterior flagellum associates with a single four-stranded root, extending posteriorly along a groove in which the short posterior flagellum is located. The arrangement of the flagellar roots indicates affinity to dinoflagellates, the anterior flagellum being homologous to the transverse flagellum in dinoflagellates, that is the number 2 flagellum, and the posterior flagellum being the number 1 flagellum.

The four-stranded posterior root is R1 (LMR) and the two one-stranded roots are R3 and R4, respectively. A very thin cross-banded fibre extending along R4 undoubtedly represents a homologue of the usually much more prominent transverse striated fibre of dinoflagellates (TSR). I am not aware of any studies describing in detail the flagellar root systems in apicomplexans , the other group to which perkinsids show similarity.

4.3.6 The alveolates: III Ciliates

Based on information from Pitelka (1974), Sleigh (1988) produced a drawing showing ciliary replication in *Paramecium* (Figure 4.6). In a dikinetid ciliate the basal body carrying the striated root (the posterior basal body) is the number 1 cilium (flagellum). It has two microtubular roots while the number 2 cilium has a single microtubular root. The posterior cilium is therefore the homologue of the longitudinal flagellum of dinoflagellates. Lynn (1988) illustrated a generalized 'kinetid' of a ciliate. This bears two microtubular roots associated with each basal body, as in many dinoflagellates and perkinsids.

Using the new numbering system for microtubular roots, R1 is the PT fibre ('the posterior transverse microtubular ribbon', the LMR in dinoflagellates), R2 is the PPc ('posterior postciliary microtubular ribbon'), R3 is Apc (' postciliary microtubules') and R4 is AT ('anterior transverse microtubular ribbon').

A feature of ciliates not found in other alveolates is the 'kinetodesmal fibre', a striated root extending from one of the basal body triplets. It does not associate with a microtubular root, in contrast to the transverse root of dinoflagellates and perkinsids (TSR). Cilium 2 in Lynn's drawing lacks a cross-banded root, indicating that this structure develops during ciliary transformation. In other words it is a feature of the number 1 position. In species with single rather than paired basal bodies (monokinetids) Lynn (1991: figs 3, 4) drew the basal bodies with a kinetodesmal fibre, that is they all represent basal body 1.

4.3.7 Cercomonads

Flagellar replication has not to my knowledge been studied in cercomonads. However, *Cercomonas*, *Heteromita* and a few other flagellates form a natural group together with protostelid and other myxo'mycete' protists, and flagellar replication has been studied in the myxomycete *Physarum*. The idea of a phylogenetic

(right)
Figure 4.5 Reconstruction of the flagellar apparatus in the dinoflagellate *Alexandrium catenella*
Source: After Hansen and Moestrup, 1998, labelling modified
Figure 4.6 Flagellar replication in ciliates
Source: After Sleigh, 1988, labelling modified
Figure 4.7 The flagellar apparatus in the cercomonad *Cercomonas* (Ce) and the protostelid myxomycete *Cavostelium* (Ca)
Source: After Karpov, 1997, modified

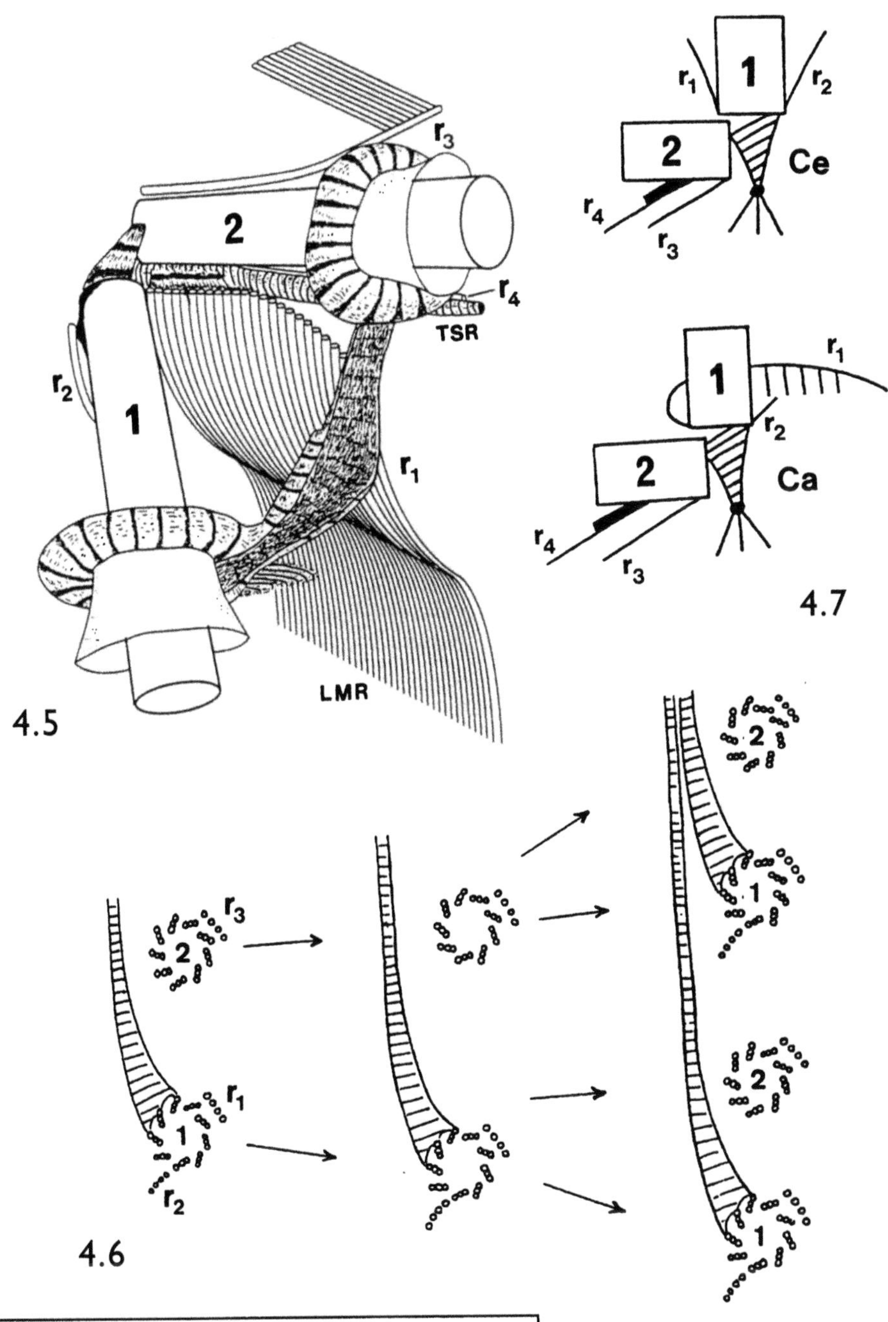

Abbreviations

1, 2	flagellum 1 and 2
LMR	longitudinal microtubular root
r1–r4	roots 1–4
TSR	transverse flagellar root

relationship between cercomonads and myxomycetes is based on ultrastructure of the flagellar apparatus but is not supported by 18S rRNA data (Cavalier-Smith, 1995). All members of these groups are biflagellate (sometimes with a barren posterior basal body). In *Physarum polycephalum* the anterior basal body remains as such in the next generation while the posterior basal body develops into an anterior one (Wright *et al.*, 1980a). Extra centrioles known as procentrioles first develop into posterior basal bodies and in the next generation into anterior ones. In other words, the anterior basal body is number 1, the posterior one number 2. This sequence of events is unusual. In all other heterotrophic groups studied so far except the choanoflagellates, the anterior flagellum develops into a posterior one at cell division.

The cercomonad/myxomycete group is characterized by the ultrastructure of its flagellar apparatus. However, I have found no micrographs illustrating flagellar root replication. As the cytoskeleton is highly complex, numbering of the roots is therefore slightly uncertain.

The anterior basal body, basal body 1, was drawn with two opposite microtubular roots in *Cercomonas* (Karpov, 1997; see Figure 4.7). Root 1 is apparently the 'dorsal root' of *Heteromita* sp. (Karpov, 1997). It possesses two microtubules in this species and extends to the anterior end of the cell, apparently nucleating a microtubular cytoskeleton. Root 2 is the 'ms' of *Heteromita* sp., a row of eight microtubules subtended by a very thin plate on the side towards the basal body (Karpov, 1997: fig. 8). This resembles the complex semicircle of microtubules illustrated in the myxomycete *Physarum polycephalum* (Wright *et al.*, 1980b: fig. 13, 'band 2'). In *Physarum* the band contains numerous microtubules, forming an outer cone around the flagellar apparatus (band 2 in Wright *et al.*, 1979 and Spiegel, 1981). In *Heteromita* root R2 extends posteriorly, running parallel to the microtubular root R3 (see below) that extends from the posterior basal body (Karpov, 1997).

Identification of the cercomonad homologue of 'band 3' in *Physarum* (Wright *et al.*, 1980b: fig. 13) is not clear.

The posterior basal body of cercomonads and myxomycetes (flagellum 2) possesses one or two microtubular roots. When two are present they associate with opposite sides of the basal body. The roots extend posteriorly in the cell, in the protostelid *Protosporangium* supporting a furrow in which the short posterior flagellum is located (Spiegel *et al.*, 1986).

Root 3 (number 5 in the old terminology) comprises a few microtubules only. In *Physarum* they were illustrated to originate as two separate groups which subsequently join into a single structure. The two-stranded part opposite R4 corresponds in its placement to a Root 3, while the identity of the additional one or two microtubules that originate on a triplet near Root 4 is uncertain.

Root 4 is a broad root associated with an unusually complex structure of parallel plates known as the ppks ('posterior para-kinetosomal structure'). It was particularly well illustrated in the protostelid *Protosporangium articulatum* (Spiegel *et al.*, 1986). Fibrillar material associated with this root in cercomonads is considered by Karpov (1997) to represent the ppks.

One of the most conspicuous components of the flagellar apparatus in the cercomonad/myxomycete cluster is the striated fibre which emanates from the base of the anterior flagellum (sometimes with a branch from the other basal body as well). It extends as a cone in the opposite direction of the anterior flagellum and terminates in a MTOC (microtubule-organizing centre). Numerous microtubules emanate from the MTOC over the nuclear surface. Together with R2 this structure indicates phylogenetic relationship between cercomonads and myxomycetes. Mir *et al.* (1984) concluded that the MTOC of *Physarum* may represent the interphase state of the mitotic centre, the microtubules (perhaps in part) contributing to the mitotic spindle.

4.3.8 Euglenoids and kinetoplastids

These groups are of particular interest, since 18S rRNA data have indicated that they originated before the major, so-called crown groups (for example, Sogin *et al.*, 1986), that is, the cytoskeleton probably originated very early. The groups will therefore be treated in more detail here. Kinetoplastids comprise the trypanosomes and the bodonids which, based on molecular evidence, appear to be sister groups (Montegut-Felkner and Triemer, 1997). Together they form a sister group to the euglenoids (Montegut-Felkner and Triemer, 1997).

Euglenoids

All euglenoid flagellates examined possess three microtubular roots (for example Moestrup, 1982; Farmer and Triemer, 1988; see also Figure 4.8). Two of these associate with the flagellum commonly known as the ventral flagellum: the short flagellum in autotrophs such as *Euglena* and *Colacium* (Surek and Melkonian, 1986; Willey and Wibel, 1987), the trailing flagellum in heterotrophs such as *Entosiphon* (Solomon *et al.*, 1987). One root associates with each side of the flagellum, and they have been termed the ventral and intermediate roots, respectively (Surek and Melkonian, 1986; Solomon *et al.*, 1987). The third, dorsal, root associates with the outside of what is known as the dorsal flagellum (for example Farmer and Triemer, 1988). This represents the long flagellum of *Euglena* and *Colacium* and the anterior flagellum of *Entosiphon* (Surek and Melkonian, 1986; Solomon *et al.*, 1987).

Based on flagellar transformation studies, Farmer and Triemer (1988) concluded that the dorsal flagellum develops into a ventral flagellum at cell division, that is, the ventral flagellum is flagellum 1. Farmer and Triemer (1988) and Brugerolle (1992) both concluded that during flagellar transformation a new root develops *de novo* on flagellum 2, opposite the dorsal root. For a short period each basal body is therefore associated with two roots. Brugerolle showed that the new root develops into a ventral root when the flagellum transforms into a flagellum 1. This identifies the dorsal root and the intermediate roots as being homologous.

Using the new terminology, the ventral root is R1, the intermediate root is R2 (as in Hilenski and Walne, 1985) and the dorsal root is R 4 (Figure 4.7). Root 3 seems to be missing in all species examined.

All three roots pass along the plasmalemma of the reservoir.

Root 1 extends to the reservoir pocket, a shallow structure opening into the reservoir near the transition between the reservoir and the canal. The reservoir pocket has been found in both autotrophic and heterotrophic species (Willey and

Wibel, 1985; Surek and Melkonian, 1986; Triemer and Farmer, 1991). In dividing cells, R1 extends into the strip separating the two new reservoirs, perhaps serving in cell division (Mignot *et al.*, 1987).

Root 2 appears to terminate on the reservoir surface and is not known to associate with any other structures (*Cryptoglena*: Owens *et al.*, 1988). However, in *Euglena* additional microtubules are added to the three-stranded root 2, and the root subsequently splits into pairs passing underneath the canal (Surek and Melkonian, 1986). R2 sometimes (always?) nucleates a set of microtubules that extend ventrally.

Root 4 terminates near the eyespot. On its way it passes very close to the main component of the reservoir/canal cytoskeleton, the so-called dorsal band (Owens *et al.*, 1988). Surek and Melkonian (1986) suggested that the dorsal band may be nucleated by R4. The two structures form a slight angle at the level where they pass each other, later becoming parallel. The dorsal band microtubules split into pairs near the transition between reservoir and canal, and the doublets subsequently extend beneath the pellicular strips (probably as microtubules 2 and 4: Mignot *et al.*, 1987). Another set of microtubules is located at right angles to the dorsal band in the canal region, encircling the canal.

Striated roots are known to extend from the flagellar apparatus to the rod organ in *Peranema trichophorum* but apparently without being associated with any of the microtubular roots (Hilenski and Walne, 1985). A striated root extending along the plasmalemma was also seen in *Eutreptiella eupharyngea* (Moestrup, 1978).

Bodonids

Members of this group are very similar to euglenoids, possessing three microtubular flagellar roots arranged as in euglenoids (Brugerolle *et al.*, 1979; Nohýnková, 1984; Surek and Melkonian, 1986). R4 nucleates a set of dorsal microtubules and, very interestingly, R2 also nucleates a band of microtubules, these microtubules extending ventrally. Root 1 passes along the cytostome, and a striated root is in contact with R1 (Nohýnková, 1984; Brugerolle, 1985; Surek and Melkonian, 1986). From the drawing by Brugerolle (1985) it appears that the anterior flagellum of the bodonid *Rhynchobodo* is flagellum 2, and the recurrent flagellum is flagellum 1 (the root on the outside of the recurrent flagellum basal body extends towards the cytostome, that is, it represents R1).

Trypanosomes

The single emergent flagellum is homologous with the anterior flagellum of bodonids (Vickerman 1990); that is, it is flagellum 2, and the barren basal body is flagellum 1. The cytoskeleton comprises a manchette of evenly spaced microtubules beneath the plasmalemma. However, fig. 11 of Vickerman (1969: *Trypanosoma congolense*) shows a four-stranded root said to originate at the basal body of the emergent flagellum. Judged from its position it could be the R4 root. Further investigations of the trypanosomatid flagellar apparatus may reveal the presence of additional roots.

4.3.9 Glaucophytes

Only *Cyanophora paradoxa* has been examined. Heimann *et al.* (1989) described the two new flagella formed as short anterior flagella, that is the long posterior flagellum is the number 1 flagellum, and the short anterior is the number 2 flagellum.

I have been unable to label the individual roots from the published electron micrographs (Mignot *et al.*, 1969).

4.3.10 Green algae

Green algae (including primitive forms belonging to the Pedinophyceae and Prasinophyceae) nearly always possess two microtubular roots associated with each basal body, a few-membered root (often comprising two microtubules) and a broad root (Figure 4.9). Green algae typically show a 180° rotational symmetry of the basal body apparatus, that is, if one basal body is rotated 180° it covers the second basal body, including its microtubular roots. In cells with more flagella, only the two central ones (numbers 1 and 2) carry microtubular roots. The 180° rotational symmetry is not exact as homologous roots of the two pair of basal bodies extend to different organelles in the cell.

Flagellar transformation was initially found in the prasinophyte *Nephroselmis.* Members of this genus are anisokont, that is, the two flagella of the biflagellate cells differ in length, and Melkonian *et al.* (1987) found that the short anterior flagellum increases in length prior to cell division, while two new short flagella appear. The long flagellum remains long. In other words, the long flagellum is number 1, the short flagellum number 2. *Nephroselmis* is unusual in having only one microtubular root associated with flagellum 2, and this root extends to the eyespot, probably determining the position of the eyespot underneath the short flagellum. Other prasinophytes lack the association between the stigma and a microtubular root, and the position of the eyespot in the cell differs, even among species belonging to the same genus. In more advanced green algae the eyespot always associates with a microtubular root, but this root extends from the opposite side of flagellum 2, compared with *Nephroselmis*; that is, it represents a different root.

Using the new terminology for microtubular roots, R1 is the (typically two-stranded) root associated with flagellum 1. In a few prasinophytes this root is multi-stranded and associates with the opening of the scale reservoir, perhaps regulating scale release. In some prasinophytes (*Cymbomonas*, *Pterosperma* and close allies), the root associates with the pouch. This structure probably served in food uptake in the heterotrophic ancestors of the green algae, and it may still do so in *Cymbomonas* (Moestrup, Inouye and Hori, unpublished observations). The root comprises an MLS in *Mesostigma*, *Cymbomonas*, *Pterosperma* and its allies, similar to the MLS of higher plants and bryophytes. The root also commonly associates with a cross-banded fibre.

R2 is a broad root extending along the plasmalemma. R3 is the two-stranded root extending from basal body 2. In *Nephroselmis* this root interconnects the flagellar apparatus and the eyespot. In another prasinophyte, *Mesostigma*, both R1 and R3 possess an MLS. R4 is the broad root from basal body 2 to the eyespot in the advanced non-charophycean green algae (Chlorophyceae, Ulvophyceae, Trebouxiophyceae). This very characteristic feature, not found in any prasinophycean

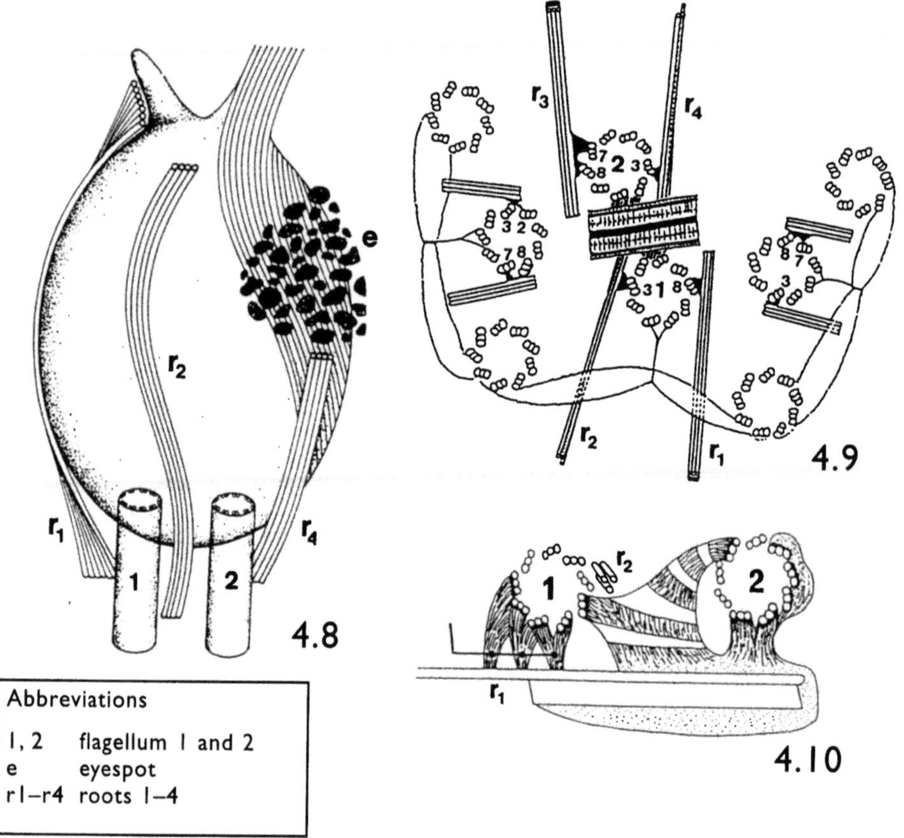

Figure 4.8 The flagellar apparatus of the euglenoid flagellate *Cryptoglena pigra*
Source: After Owens *et al.*, 1988, labelling modified
Figure 4.9 The flagellar apparatus of the prasinophycean green alga *Pyramimonas octopus*
Source: After Moestrup and Hori, 1989, labelling modified
Figure 4.10 The flagellar apparatus of the charophycean green alga *Coleochaete pulvinata*
Source: After Sluiman, 1983, labelling modified

flagellates, is likely to have arisen in the common ancestor of the advanced non-charophycean green algae.

Identification of individual flagella in charophyceans is not simple as this advanced group of green algae includes no true flagellates, only zoospores, and gametes and flagellar replication therefore does not take place. Only one of the two flagella has microtubular roots, a broad one (which may contain numerous microtubules) and, at least in some species, a few-membered one (Figure 4.10). Melkonian (1984, 1989) interpreted the root-bearing flagellum in charophyceans as the number 1 flagellum. Thus the broad, MLS-bearing root of the Charophyceae would represent R1, and the few-membered root would be R2. The basal body interpreted as basal body 2 lacks flagellar roots. There is, however, no definite proof of which flagellum in charophyceans represent flagellum 1.

The uniflagellate green algae are unusual, as flagellar transformation takes place in two different ways. *Monomastix* and *Pedinomonas* both possess a single emergent flagellum and a second, barren basal body. In *Monomastix*, a species of very uncertain taxonomic affinities, the long flagellum transforms into the barren basal body during cell division: that is, the emergent flagellum is the number 2 flagellum (Heimann *et al.*, 1989). Very surprisingly, however, in *Pedinomonas* of the class Pedinophyceae (Moestrup, 1991), the exact opposite happens. The long flagellum remains unaltered during cell division, and the barren basal body grows out into a new long flagellum (Pickett-Heaps and Ott, 1974, fig. 13; confirmed by light microscopy, Heimann *et al.*, 1989). This is an exceptional situation, and presently the only known case within the algae where a barren basal body forms a flagellum during cell division, that is, the emergent flagellum is the number 1 flagellum.

4.3.11 Choanoflagellates

Some information on flagellar replication in *Monosiga ovata* was published recently by Karpov and Leadbeater (1997). The single emergent flagellum disappears prior to cell division. Two new basal bodies appear, one next to each of the old ones, but initially of a narrower diameter. The ring of microtubules surrounding the basal body of the emergent flagellum seems to be retained at cell division. This identifies the emergent flagellum as flagellum 1. During separation of the two pairs of basal bodies prior to mitosis, a second ring develops around one of the basal bodies. Karpov and Leadbeater (1997) interpreted this basal body as the barren basal body of the interphase cell. It was therefore basal body 2 before cell division.

The microtubular system associated with the two basal bodies of choanoflagellates is unlike those of most other protists. In *Monosiga* the basal body of the emergent flagellum is surrounded by rings of opaque material, and radiating microtubules emerge from the innermost ring. They extend under the plasmalemma and apparently make contact with the fibrillar bundles that support each of the tentacles. The barren basal body apparently has no associated structures, its sole function may be to grow into an emergent flagellum in the next generation.

In *Codosiga botrytis*, examined in detail by Hibberd (1975), large numbers of microtubules radiate from the flagellar basal body, grouped in blocks. It is not possible to homologize the microtubular system of choanoflagellates and the microtubular roots of other protists. The ring of microtubules is probably a specialization of choanoflagellates, developed in connection with formation of the collar tentacles. Details of this process must await the discovery of related groups with a less derived flagellar apparatus.

4.4 Some flagellates in which flagellar replication is unknown

Numbering of flagella and associated microtubular roots is usually possible only if flagellar replication has been studied, or if the organism is closely related to a taxonomic group in which details of flagellar replication are known. From published micrographs it is, however, sometimes possible to get an idea of the identity of flagella and roots. Some examples will be given here.

Proteromonads

In Proteromonads (Brugerolle and Joyon, 1975) the presence of a helix-like structure in the flagellar transition region has been taken to indicate phylogenetic affinity to the heterokonts (Patterson, 1985, 1989). There are, however, few other similarities between the two groups, and the idea needs further support. If accepted, however, the relative orientation of the two flagella in *Proteromonas* suggests that the posterior flagellum is the homologue of the smooth flagellum in heterokonts, that is, flagellum 1. In both *Proteromonas* and *Karotomorpha* this flagellum carries two microtubular roots, shown clearly to arise on opposite sides of the basal body (Brugerolle and Joyon, 1975: fig. 32). Judging by the orientation of the basal body triplets, 'ft1' is R1 and 'ft2' is R2. They join into a 'rhizostyle' when leaving the flagellar apparatus, a structure suggested to be associated with (nucleating?) the extensive peripheral cytoskeleton in *Karotomorpha*. Flagellum 2 in *Proteromonas* bears a microtubular root that joins up with the combined R1/R2, forming a rhizostyle of three microtubular roots, a very unusual feature. The identity of the R2-associated root is not certain, indeed some of the published micrographs indicate that two microtubular roots are associated with the anterior flagellum of *Proteromonas* (ibid., 4).

Archamoebae

In a phylogenetic context Archamoebae is one of the potentially important groups. The few known members lack mitochondria, and the group may date back to before mitochondria developed endosymbiotically from bacteria. The flagellar apparatus is very unusual in several respects (Figure 4.11). There is only a single flagellum and no extra basal body (Brugerolle, 1991), an exceptional situation paralleled only in the heterokont *Pelagomonas* and in the zooflagellate *Phalansterium digitatum* (Hibberd, 1983). Perhaps flagellar transformation does not take place: the single flagellum probably disappears during cell division to be replaced with two new ones, which then distribute to each daughter cell. In *Pelagomonas* the single flagellum is the number 2 flagellum, which disappears at cell division (Heimann *et al.*, 1995a).

The single flagellum in mastigamoebids may therefore be a number 2 flagellum (cell division has not been studied). It has only a single microtubular root attached, which apparently nucleates numerous microtubules on the nuclear surface. It would be very interesting to study cell division in mastigamoebae. Mastigamoebid flagella do not beat normally, and in some genera are non-motile.

Flagellar transformation has not been studied in other mitochondria-lacking flagellates such as diplomonads, retortamonads, and parabasalians.

4.5 Some groups in which flagellar transformation does not take place

Plasmodiophoromycetes

The plasmodiophoromycete zoospore is so unusual that it deserves a comment. Barr (1981) and Barr and Allan (1982) described the root system in *Polymyxa graminis* (Figure 4.12). Each of the exceptionally long basal bodies carries two slightly different opposite roots. Most unexpectedly, however, the two pairs of roots are identical. One root consists of two microtubules that extend to and form an indentation into the plasmalemma. The other is initially one-stranded, becoming two-stranded and finally three-stranded. The microtubules extend to and subsequently run beneath the plasmalemma. This combination is unlike any other protist known to me. It somewhat resembles the arrangement of microtubular roots in green algae, but other details of the flagellar apparatus are different and it appears likely, as pointed out by Barr and Allan (1982), that the plasmodiophoromycetes have their relatives in some protozoan group.

Chytrids

These form another unusual group. Due to the large number of high quality investigations by Donald Barr and his colleagues we are better informed about the flagellar apparatus of this group than about that of most other protists. The four orders of chytrids all have uniflagellate motile cells, in which a second flagellum is represented by a very short basal body (Barr, 1981; see also Figure 4.13). In the Chytridiales a single bundle of microtubules is associated with the long flagellum basal body. Remarkable differences exist between the different orders, more so than in other classes of protists. At first sight the root system is very unusual, but the Monoblepharidales may offer a clue to the phylogeny of the chytrids. In *Monoblepharella* illustrations of the emergent flagellum basal body illustrate a series of rings around the basal body, and instead of the single bundle of microtubules, 270° of the periphery around the basal body is covered by microtubules extending from the basal body (Figure 4.13). The arrangement is, at least superficially, strongly reminiscent of the basal body and associated rings/microtubules in choanoflagellates (for example Fuller and Reichle, 1968: fig. 3). This is important since many different sequencing data have indicated a phylogenetic relationship between the 'true' fungi (at the base of which are the chytrids) and the choanoflagellates/metazoa.

Some members of the spizellomycetalean chytrids possess an MTOC located a short distance from the emergent flagellum basal body and attached to the basal body with thin fibres. Microtubules extend from the MTOC to the nuclear surface. This recalls the MTOC of the cercomonad/myxomycete line, although the details are somewhat different, and it may not reflect a phylogenetic relationship. The most aberrant flagellar apparatus is that of the Blastocladiales. Nine groups of three microtubules extend from the basal body over the nuclear surface. This type of flagellar root was derived from the spizellomycetalean type by Barr (1981). 18S rRNA data have indicated, however, that the Blastocladiellales is only distantly related to other chytrids (for example Nagahama *et al.*, 1995), and chytrids may not comprise a monophyletic group.

If the relationship between the choanoflagellates and the Monoblepharidales is true, then the emergent flagellum of chytrids is probably the number 1 flagellum and the single root is R1.

4.6 Concluding remarks

As the information given has shown, the two-rooted basal body is a very common feature in protists. In some taxonomic groups, flagellum 2 possesses only a single microtubular root and the second root develops *de novo* at cell division. When two

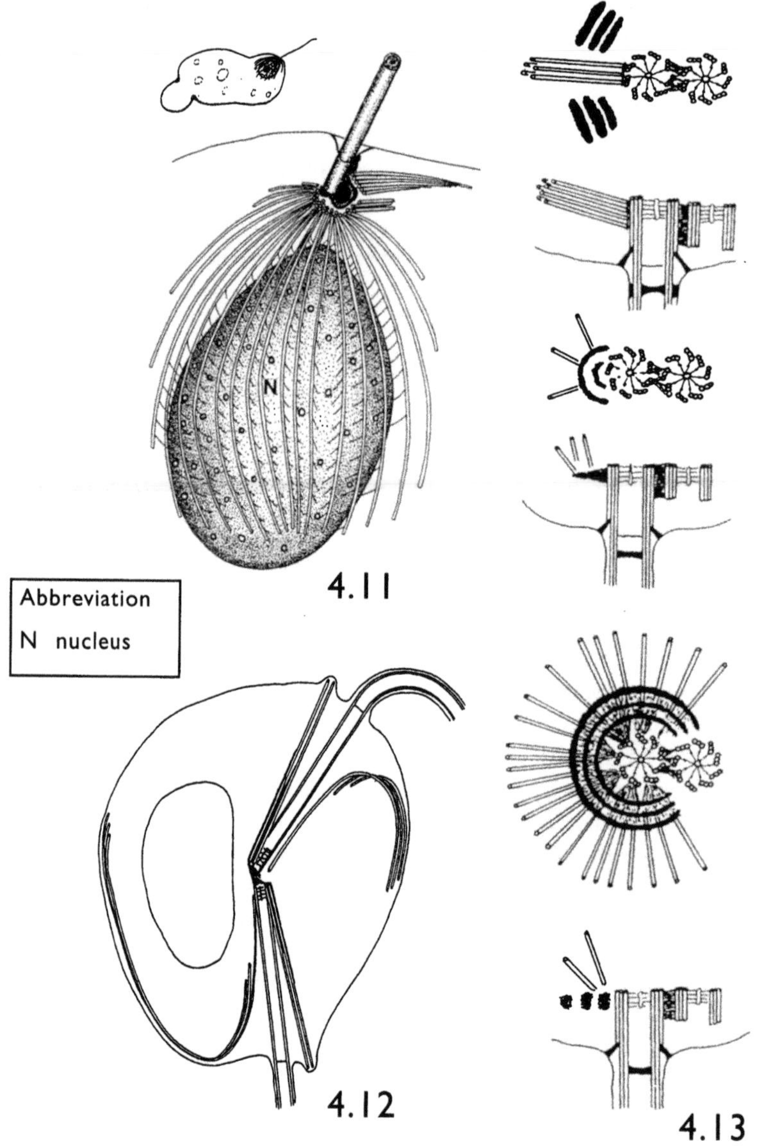

Figure 4.11 Reconstruction of the front end of the archamoeba *Pelagomonas*
Source: After Brugerolle, 1991
Figure 4.12 Diagrammatic reconstruction of the flagellar apparatus in the plasmodiophoromycete *Polymyxa*
Source: After Barr, 1981
Figure 4.13 The flagellar apparatus of selected chytrids, represented by the order Chytridiales (top pair) and the order Monoblepharidales (central pair, lower pair). The lower pair represents *Monoblepharella*
Source: After Barr, 1981

microtubular roots are present, the roots belonging to different basal bodies are usually different, and transformation of the roots therefore takes place when the roots of flagellum 2 develop into the roots of flagellum 1. In several taxonomic groups the flagellar apparatus appears modified, hampering identification of the microtubular roots. This may at first sight appear to contradict the thesis that two microtubular roots formed early during evolution of the eukaryotic cell. However, modifications of the root systems have also taken place within the major lineages of organisms, for example, the synurophyceans within the heterokonts. Is the exceptionally high diversity in the root structure of chytrids associated with the lack of flagellar transformation in this group? The functional significance of flagellar transformation is still unknown. It is notable, however, that very little variation occurs in the axoneme of protist flagellates, while sperm cells of metazoa may deviate considerably from the 9+2 structure. This is the more remarkable since the protist groups are significantly older than the metazoans. No transformation of flagella takes place in sperm and zoospores, and one may speculate that flagellar transformation causes a constraint that keeps variation of the flagellar apparatus within limits. Flagellar and root transformation are very complex processes, and faults encountered during the process may result in poorly functioning or moribund cells. We are still very poorly informed about the function of flagellar transformation today, eighteen years after its discovery.

When the data presently available are considered, the protists fall into two groups, based on the way flagellar transformation takes place. In all algal flagellates examined except one, the anterior (dorsal) flagellum transforms during cell division into a posterior (ventral) flagellum. The same applies to some protozoa, for example ciliates. However, in the cercomonad/myxomycete and the choanoflagellate groups the opposite occurs. In the former group, the posterior flagellum develops into an anterior one during flagellar transformation. In choanoflagellates, the barren basal body develops a microtubular cone during cell division, indicating that it will subsequently grow out as the emergent flagellum. Very surprisingly the same happens in a single alga, the green flagellate *Pedinomonas*. The significance of this dualism is not clear, but it supports the idea that the pedinomonad green algae are related in some unknown way to other green algae.

Future research should address flagellar transformation in additional groups of protists. Also, the MTOCs on the basal bodies from which the roots develop should be identified and examined to answer questions such as, first, what the chemical composition is of the MTOCs on opposite sides of a basal body, second, which genes are responsible for formation of the microtubular roots, and so on.

ACKNOWLEDGEMENTS

I thank Dr Niels Daugbjerg for assistance with preparation of the illustrations and Professor R. N. Pienaar for comments on the text.

REFERENCES

Andersen, R. A. (1985) The flagellar apparatus of the golden alga *Synura uvella*: four absolute orientations. *Protoplasma*, **128**, 94–106.

Andersen, R. A. (1987) Synurophyceae classis nov., a new class of algae. *American Journal of Botany*, **74**, 337–353.

Andersen, R. A. (1989) Absolute orientation of the flagellar apparatus of *Hibberdia magna* comb. nov. (Chrysophyceae). *Nordic Journal of Botany*, **8**, 653–669.

Andersen, R. A. (1991) The cytoskeleton of chromophyte algae. *Protoplasma*, **164**, 143–159.

Andersen, R. A., Saunders, G. W., Paskind, M. P. and Sexton, J. P. (1993) Ultrastructure and 18S rRNA gene sequence for *Pelagomonas calceolata* gen. et sp. nov. and the description of a new algal class, the Pelagophyceae. *Journal of Phycology*, **29**, 701–715.

Andersen, R. A. and Wetherbee, R. (1992) Microtubules of the flagellar apparatus are active during prey capture in the chrysophycean alga *Epipyxis pulchra*. *Protoplasma*, **166**, 8–20.

Barr, D. J. S. (1981) The phylogenetic and taxonomic implications of flagellar rootlet morphology among zoosporic fungi. *BioSystems*, **14**, 359–370.

Barr, D. J. S. and Allan, P. M. E. (1982). Zoospore ultrastructure of *Polymyxa graminis* (Plasmodiophoromycetes). *Canadian Journal of Botany*, **60**, 2496–2504.

Barr, D. J. S. and Allan, P. M. E. (1985) A comparison of the flagellar apparatus in *Phytophthora, Saprolegnia, Thraustochytrium*, and *Rhizidiomyces*. *Canadian Journal of Botany*, **63**, 138–154.

Barr, D. J. S. and Désaulniers, N. L. (1989) The flagellar apparatus of the Oomycetes and Hyphochytriomycetes. In *The Chromophyte Algae: Problems and Perspectives*, Systematics Association Special Volume 38 (eds J. C. Green, B. S. C. Leadbeater and W. L. Diver), Oxford: Clarendon Press, pp. 343–355.

Beech, P. L., Wetherbee, R. and Pickett-Heaps, J. D. (1988) Transformation of the flagella and associated flagellar components during cell division in the coccolithophorid *Pleurochrysis carterae*. *Protoplasma*, **145**, 37–46.

Beech, P. L. and Wetherbee, R. (1990) The flagellar apparatus of *Mallomonas splendens* (Synurophyceae) at interphase and its development during the cell cycle. *Journal of Phycology*, **26**, 95–111.

Beech, P. L., Heimann, K. and Melkonian, M. (1991) Development of the flagellar apparatus during the cell cycle in unicellular algae. *Protoplasma*, **164**, 23–37.

Brugerolle, G. (1985) Des trichocystes chez les Bodonides, un caractère phylogénétique supplémentaire entre Kinetoplastida et Euglenida. *Protistologica*, **21**, 339–348.

Brugerolle, G. (1991) Flagellar and cytoskeletal system in amitochondrial flagellates: Archamoeba, Metamonada and Parabasala. *Protoplasma*, **164**, 70–90.

Brugerolle, G. (1992) Flagellar apparatus duplication and partition, flagellar transformation during division in *Entosiphon sulcatum*. *BioSystems*, **28**, 203–209.

Brugerolle, G. and Joyon, L. (1975) Etude cytologique ultrastructurale des genres *Proteromonas* et *Karotomorpha* (Zoomastigophorea, Proteromonadida Grassé 1952). *Protistologica*, **11**, 531–546.

Brugerolle, G., Lom, J., Nohýnková, E. and Joyon, L. (1979) Comparison et évolution des structures cellulaires chez plusieurs espèces de bodonidés et cryptobiidés appartenant aux genres *Bodo*, *Cryptobia* et *Trypanoplasma* (Kinetoplastida, Mastigophora). *Protistologica*, **15**, 197–221.

Calado, A. J., Hansen, G. and Moestrup, Ø. (1999) Architecture of the flagellar apparatus and related structures in the type species of *Peridinium, P. cinctum* (Dinophyceae). *European Journal of Phycology*, **34**, 179–191.

Cavalier-Smith, T. (1995) Zooflagellate phylogeny and classification. In *The Biology of Free-*

*Living Heterotrophic Flagellates (*ed. S. A. Karpov),*Cytology*, **37**, 1010–1029.

Eikrem, W. and Moestrup, Ø. (1998) Structural analysis of the flagellar apparatus and the scaly periplast in *Chrysochromulina scutellum* sp. nov. (Prymnesiophyceae, Haptophyta) from the Skagerrak and the Baltic. *Phycologia*, **37**, 132–153.

Farmer, M. A. and Triemer, R. E. (1988) Flagellar systems in the euglenoid flagellates. *BioSystems*, **21**, 283–291.

Fuller, M. S and Reichle, R. E. (1968) The fine structure of *Monoblepharella* sp. zoospores. *Canadian Journal of Botany*, **46**, 279–283.

Green, J. C. (1980).The fine structure of *Pavlova pinguis* Green and a preliminary survey of the order Pavlovales (Prymnesiophyceae). *British Phycological Journal*, **15**, 151–191.

Green, J. C. and Hori, T. (1994) Flagella and flagellar roots. In *The Haptophyte Algae*, Systematics Association Special Volume 51 (eds J. C. Green and B. S. C. Leadbeater), Oxford: Clarendon Press, pp. 47–71.

Hansen, G., Moestrup, Ø. and Roberts, K. R. (1997) Light and electron microscopical observations on *Protoceratium reticulatum* (Dinophyceae). *Archiv für Protistenkunde*, **147**, 381–391.

Hansen, G. and Moestrup, Ø. (1998) Fine structural characterization of *Alexandrium catenella* (Dinophyceae), with special emphasis on the flagellar apparatus. *European Journal of Phycology*, **33**, 281–291.

Heimann, K., Reize, I. B. and Melkonian, M. (1989) The flagellar developmental cycle in algae: flagellar transformation in *Cyanophora paradoxa* (Glaucocystophyceae). *Protoplasma*, **148**, 106–110.

Heimann, K., Andersen, R. A. and Wetherbee, R. (1995a) The flagellar development cycle of the uniflagellate *Pelagomonas calceolata* (Pelagophyceae). *Journal of Phycology*, **31**, 577–583.

Heimann, K., Roberts, K. R. and Wetherbee, R. (1995b) Flagellar apparatus transformation and development in *Prorocentrum micans* and *P. minimum* (Dinophyceae). *Phycologia*, **34**, 323–335.

Heimann, K., Benting, J., Timmermann, S. and Melkonian, M. (1989) The flagellar development cycle in algae. Two types of flagellar development in uniflagellated algae. *Protoplasma*, **153**, 14–23.

Hibberd, D. J. (1975) Observations on the ultrastructure of the choanoflagellate *Codosiga botrytis* (Ehr.) Saville-Kent with special reference to the flagellar apparatus. *Journal of Cell Science*, **17**, 191–219.

Hibberd, D. J. (1983) Ultrastructure of the colonial colourless zooflagellates *Phalansterium digitatum* Stein (Phalansteriida ord. nov.) and *Spongomonas uvella* Stein (Spongomonadida ord. nov.). *Protistologica*, **19**, 523–535.

Hilenski, L. L. and Walne, P. L. (1985) Ultrastructure of the flagella of the colorless phagotroph *Peranema trichophorum* (Euglenophyceae). II. Flagellar roots. *Journal of Phycology*, **21**, 125–134.

Hill, D. R. A. and Wetherbee, R. (1986) *Proteomonas sulcata* gen. et sp. nov. (Cryptophyceae), a cryptomonad with two morphologically distinct and alternating forms. *Phycologia*, **25**, 521–543.

Inouye, I. and Pienaar, R. N. (1985) Ultrastructure of the flagellar apparatus in *Pleurochrysis* (class Prymnesiophyceae). *Protoplasma*, **125**, 24–35.

Karpov, S. A. (1997) Cercomonads and their relationship to the myxomycetes. *Archiv für Protistenkunde*, **158**, 297–307.

Karpov, S. A. and Leadbeater, B. S. C. (1997) Cell and nuclear division in a freshwater choanoflagellate, *Monosiga ovata* Kent. *European Journal of Protistology*, **33**, 323–334.

Lynn, D. H. (1988) Cytoterminology of cortical components of ciliates: somatic and oral kinetids. *BioSystems*, **21**, 299–307.

Lynn, D. H. (1991) Implications of recent descriptions of kinetid structure to the systematics of the ciliated protists. *Protoplasma*, **164**, 123–142.

Melkonian, M. (1984) Flagellar apparatus ultrastructure in relation to green algal classification. In *Systematics of the Green Algae,* Systematics Association Special Volume 27 (eds D. E. G. Irvine and D. M. John), London and Orlando: Academic Press, pp. 73–120.

Melkonian, M. (1989) Flagellar apparatus ultrastructure in *Mesostigma viride* (Prasinophyceae). *Plant Systematics and Evolution*, **164**, 93–122.

Melkonian, M., Reize, I. B. and Preisig, H. R. (1987) Maturation of a flagellum/basal body requires more than one cell cycle in algal flagellates: studies on *Nephroselmis olivacea* (Prasinophyceae). In *Algal Development, Molecular and Cellular Aspects* (eds W. Wiessner, D. G. Robinson and R. C. Starr), Berlin, Heidelberg, New York, Tokyo: Springer, pp. 102–113.

Mignot, J. P., Brugerolle, G. and Bricheux, G. (1987) Intercalary strip development and dividing cell morphogenesis in the euglenid *Cyclidiopsis acus. Protoplasma*, **139**, 51–65.

Mignot, J. P., Joyon, L. and Pringsheim, E. G. (1969) Quelques particularités structurales de *Cyanophora paradoxa* Korsch., protozoaire flagellé. *Journal of Protozoology*, **16**, 138–145.

Mir, L., Wright, M. and Moisand, A. (1984) Variations in the number of centrioles, the number of microtubule organizing centers and the percentage of mitotic abnormalities in *Physarum polyxcephalum* amoebae. *Protoplasma*, **120**, 20–35.

Moestrup, Ø. (1978) On the phylogenetic validity of the flagellar apparatus in the green algae and other chlorophyll a and b containing plants. *BioSystems*, **10**, 117–144.

Moestrup, Ø. (1982) Flagellar structure on algae: a review with new observations particularly on the Chrysophyceae, Phaeophyceae (Fucophyceae), Euglenophyceae and *Reckertia. Phycologia*, **21**, 425–528.

Moestrup, Ø. (1991) Further studies of presumedly primitive green algae, including the description of Pedinophyceae class. nov. and *Resultor* gen. nov. *Journal of Phycology*, **27**, 119–133.

Moestrup, Ø. and Hori, T. (1989) Ultrastructure of the flagellar apparatus in *Pyraminonas octopus* (Prasinophyceae). II. Flagellar roots, connecting fibres, and numbering of individual flagella in green algae. *Protoplasma*, **148**, 41–56.

Moestrup, Ø. and Thomsen, H. A. (1990) *Dictyocha speculum* (Silicoflagellata, Dictyochophyceae), studies on armoured and unarmoured stages. *Biologiske Skrifter*, **37**, 1–57.

Montegut-Felkner, A. E. and Triemer, R. E. (1997) Phylogenetic relationships of selected euglenoid genera based on morphological and molecular data. *Journal of Phycology*, **33**, 512–519.

Nagahama, T., Sato, H., Shimazu, M. and Sugiyama, J. (1995) Phylogenetic divergence of the entomophthoralean fungi: evidence from nuclear 18S ribosomal RNA gene sequences. *Mycologia*, **87**, 203–209.

Nohýnková, E. (1984) A new pathogenic *Cryptobia* from freshwater fishes: a light and electron microscope study. *Protistologica*, **20**, 181–195.

Norén, F., Moestrup, Ø. and Rehnstam-Holm, A-S. (1999) *Parvilucifera infectans* Norén et Moestrup gen. et sp. nov. (Perkinsozoa phylum nov.): a parasitic flagellate capable of killing toxic microalgae. *European Journal of Protistology*, **35**, 233–254.

Owens, K. J., Farmer, M. A. and Triemer, R. E. (1988) The flagellar apparatus and reservoir/canal cytoskeleton of *Cryptoglena pigra* (Euglenophyceae). *Journal of Phycology*, **24**, 520–528.

Patterson, D. J. (1985) The fine structure of *Opalina ranarum* (family Opalinidae): opalinid phylogeny and classification. *Protistologica*, **21**, 413–428.

Patterson, D. J. (1989) Stramenopiles: chromophytes from a protistan perspective. In *The Chromophyte Algae: Problems and Perspectives*, The Systematics Association Special Volume 38 (eds J. C. Green, B. S. C. Leadbeater and W. L. Diver), Oxford: Clarendon Press, pp. 357–379.

Perasso, L., Hill, D. R. A. and Wetherbee, R. (1992) Transformation and development of the flagellar apparatus of *Cryptomonas ovata* (Cryptophyceae) during cell division. *Protoplasma*, **170**, 53–67.

Pickett-Heaps, J. D. and Ott, D. W. (1974) Ultrastructural morphology and cell division in *Pedinomonas. Cytobios*, **11**, 41–58.

Pitelka, D. R. (1974) Basal bodies and root structures. In *Cilia and Flagella*, (ed. M. A. Sleigh), London and New York: Academic Press, pp. 437–469.

Preisig, H. (1989) The flagellar base ultrastructure and phylogeny of chromophytes. In *The Chromophyte Algae: Problems and Perspectives*, The Systematics Association Special Volume 38 (eds J. C. Green, B. S. C. Leadbeater and W. L. Diver), Oxford: Clarendon Press, pp. 167–187.

Roberts, K. R. (1984) Structure and significance of the cryptomonad flagellar apparatus. I. *Cryptomonas ovata* (Cryptophyta). *Journal of Phycology*, **20**, 590–599.

Roberts, K. R. (1985) The flagellar apparatus of *Oxyrrhis marina* (Pyrrhophyta). *Journal of Phycology*, **21**, 641–655.

Roberts, K. R. (1991) The flagellar apparatus and cytoskeleton of dinoflagellates: organization and use in systematics. In *The Biology of Free-living Heterotrophic Flagellates*, The Systematics Association, Special Volume 45 (eds D. J. Patterson and J. Larsen), Oxford: Clarendon Press, pp. 285–302.

Roberts, K. R., Heimann, K. and Wetherbee, R. (1995) The flagellar apparatus and canal structure in *Prorocentrum micans* (Dinophyceae). *Phycologia*, **34**, 313–322.

Roberts, K. R., Stewart, K. D. and Mattox, K. R. (1981) The flagellar apparatus of *Chilomonas paramecium* (Cryptophyceae) and its comparison with certain zooflagellates. *Journal of Phycology*, **17**, 159–167.

Santos, L. M. A. and Leedale, G. F. (1991) *Vischeria stellata* (Eustigmatophyceae): ultrastructure of the zoospores, with special reference to the flagellar apparatus. *Protoplasma*, **164**, 160–167.

Simpson, A. G. B. and Patterson, D. J. (1996) Ultrastructure and identification of the predatory flagellate *Colpodella pugnax* Cienkowski (Apicomplexa) with a description of *Colpodella turpis* n. sp. and a review of the genus. *Systematic Parasitology*, **33**, 187–198.

Sleigh, M. A. (1988) Flagellar root maps allow speculative comparisons of root patterns and of their ontogeny. *BioSystems*, **21**, 277–282.

Sleigh, M. A. (1995) Progress in understanding the phylogeny of flagellates. *Cytology*, **37**, 985–1009.

Sluiman, H. J. (1983) The flagellar apparatus of the zoospore of the filamentous green alga *Coloechaete pulvinata*: absolute configuration and phylogenetic significance. *Protoplasma*, **115**, 160–175.

Sogin, M. L., Elwood, H. J. and Gunderson, J. H. (1986) Evolutionary diversity of eukaryotic small-subunit rRNA genes. *Proceedings of the National Academy of Sciences USA*, **83**, 1383–1387.

Solomon, J. A., Walne, P. L. and Kivic, P. A. (1987) *Entosiphon sulcatum* (Euglenophyceae): flagellar roots of the basal body complex and reservoir region. *Journal of Phycology*, **23**, 85–98.

Spiegel, F. W. (1981) Phylogenetic significance of the flagellar apparatus in protostelids (Eumycetozoa). *BioSystems*, **14**, 491–499.

Spiegel, F. W., Feldman, J. and Bennett, W. E. (1986) Ultrastructure and development of the amoebo-flagellate cells of the protostelid *Protosporangium articulatum*. *Protoplasma*, **132**, 115–128.

Surek, B. and Melkonian, M. (1986) A cryptic cytostome is present in *Euglena*. *Protoplasma*, **133**, 39–49.

Triemer, R. E. and Farmer, M. A. (1991) An ultrastructural comparison of the mitotic apparatus, feeding apparatus, flagellar apparatus and cytoskeleton in euglenoids and kinetoplastids. *Protoplasma*, **164**, 91–104.

Vickerman, K. (1969) The fine structure of *Trypanosoma congolense* in its bloodstream phase. *Journal of Protozoology,* **16**, 54–69.

Vickerman, K. (1990) Phylum Zoomastigina Class Kinetoplastida. In *Handbook of Protoctista* (eds L. Margulis, J. O. Corliss, M. Melkonian and D. J. Chapman), Boston: Jones and Bartlett, pp. 215–238.

Wetherbee, R., Platt, S. J., Beech, P. L. and Pickett-Heaps, J. D (1988) Flagellar transformation in the heterokont *Epipyxis pulchra* (Chrysophyceae): direct observations using image enhanced light microscopy. *Protoplasma*, **145**, 47–54.

Willey, R. L. and Wibel, R. G. (1985) The reservoir cytoskeleton and possible cytostomal homologue in *Colacium* (Euglenophyceae). *Journal of Phycology*, **21**, 570–577.

Willey, R. L. and Wibel, R. G. (1987) Flagellar roots and the reservoir cytoskeleton of *Colacium libellae* (Euglenophyceae). *Journal of Phycology*, **23**, 283–288.

Wright, M., Moisand, A. and Mir, L. (1979) The structure of the flagellar apparatus of the swarm cells of *Physarum polycephalum. Protoplasma*, **100**, 231–250.

Wright, M., Moisand, A. and Mir, L. (1980a) Centriole maturation in the amoebae of *Physarum polycephalum. Protoplasma*, **105**, 159–160.

Wright, M., Mir, L. and Moisand, A. (1980b) The structure of the pro–flagellar apparatus of the amoebae of *Physarum polycephalum*: relationship to the flagellar apparatus. *Protoplasma*, **103**, 69–81.

Zimmermann, B., Moestrup, Ø. and Hällfors, G. (1984) Chrysophyte or heliozoon: ultrastructural studies on a cultured species of *Pseudopedinalla* (Pedinellales ord. nov.), with comments on species taxonomy. *Protistologica*, **20**, 591–612.

Chapter 5

Molecular aspects of the centrin-based cytoskeleton in flagellates

Peter E. Hart and Jeffrey L. Salisbury

ABSTRACT

The small calcium-binding protein, centrin, is ubiquitious among flagellates and ciliates, and appears to play significant roles in flagellar behaviour, feeding and cortical patterning. In this chapter, we present a current review of the molecular aspects of centrin. Numerous flagellate and ciliate centrin cDNAs have been identified that encode homologous proteins of ~20 kDa. All of these centrin proteins consist of four calcium-binding EF-hand domains, a highly variable amino terminal domain that is rich in hydrophobic and positively charged residues, and a short carboxy terminus. Only a few centrin genes have been characterized in lower eukaryotes. In general, centrin genes are similar to other genes in a given species with regard to codon bias, the presence of introns, the length of the untranslated regions, and the presence of internal eliminated sequences. However, the number of centrin-encoding genes varies greatly between species, raising the intriguing question as to the role of centrin microdiversity in some flagellates and ciliates. Also, the position of introns is highly variable in centrin genes, and does not conform to the boundaries of the functional EF-hand domains; this suggests a complex evolutionary history for centrin. Phylogenetic analysis demonstrates the existence of three distinct centrin clades: an algal clade, a ciliate clade and a distantly related clade consisting of yeast and one of the *Giardia* sequences. The phylogenetic data, taken together with the structural data and genomic analysis, indicate a high level of centrin diversity. Understanding this centrin diversity will be critically important to elucidating the mechanisms of flagellar behaviour, feeding, and cortrical patterning in flagellates and ciliates. Furthermore, understanding these processes in lower eukaryotes should contribute significantly to an understanding of centrosome function in higher eukaryotes.

5.1 Introduction

Complex fibre systems organize and anchor the flagellar basal apparatus in unicellular flagellates (Pitelka, 1969; Brown *et al.*, 1976; Melkonian, 1980; Goodenough, 1989). Certain of these fibres show contractile properties, and in many cases contraction is induced by elevated cytoplasmic levels of free Ca^{2+} (Melkonian, 1989). Contractile fibres associated with the flagellar apparatus function to orient basal bodies, position the flagellar basal apparatus relative to the nucleus or cortical cell structures, and in certain cases to sever or break microtubules (McFadden *et al.*,

1987; Salisbury *et al.*, 1987; Sanders and Salisbury, 1989, 1994). In some ciliates and flagellates, such as *Paramecium* and *Holomastigotoides*, cortical fibre systems function in contraction of the cell body (Klotz *et al.*, 1997; Lingle and Salisbury, 1997). There is a considerable variety in architecture and organization of these contractile fibre systems (Figure 5.1). Nonetheless, centrin, a small calcium-binding protein has emerged as a common feature of many contractile fibres associated with flagellar basal bodies and the cell cortex in flagellates and ciliates. Centrin was first identified as a component of flagellar roots in the marine green alga *Tetraselmis* (Salisbury and Floyd, 1978; Salisbury *et al.*, 1984). Subsequently, centrin has been identified in unicells of diverse phylogenies, as well as in pericentriolar material of centrosomes and mitotic spindle poles of animal cells (Salisbury *et al.*, 1986; Salisbury, 1995, 1998).

5.2 Centrin: protein structure comparisons

Centrin is a small (~20 kDa) calcium-binding protein (Salisbury *et al.*, 1984) consisting of four EF-hand domains (Kretsinger, 1975), a highly variable amino terminal region, and a short carboxy terminus. For unicellular flagellates and ciliates, centrins range in size from 161 to 181 amino acids, with most of the differences attributed to a variable amino-terminal domain (Table 5.1, Figure 5.2). All centrins show an overall acidic isoelectric point which is due to the preponderance of negatively charged residues in the EF-hands. Common features unique to the centrin subfamily include the pattern of hydrophobic and charged residues in the amino-terminal region, and a potential phosphorylation site and penultimate F or Y residue at the carboxy terminus. Centrins also characteristically (yeast and *Giardia* centrins are exceptions) contain no histidine, cysteine or tryptophan residues.

5.2.1 The calcium-binding EF-hand domains

The most notable feature of the centrin molecule is the calcium-binding EF-hand domain that is present in four copies. EF-hands are helix-loop-helix structures (the helices are denoted E and F) first identified in parvalbumin by Kretsinger (Kretsinger, 1975) as the canonical Ca^{2+}-binding domain of cytoplasmic calcium-binding proteins. Each EF-hand consists of thirty amino acids with the first nine residues comprising the e-helix, the last nine residues comprising the f-helix and the intervening twelve residues making up the loop (Moncrief *et al.*, 1990). Calcium ion binding is thought to be coordinated within this loop, directly or indirectly via water molecules, by oxygen atoms located on the side chains of residues, or in some cases the carbonyl oxygen (Moncrief *et al.*, 1990). $^{45}Ca^{2+}$ overlay and gel-shift assays directly demonstrated that centrin is capable of calcium-binding (Salisbury *et al.*, 1984; Baron *et al.*, 1992).

Furthermore, biochemical studies of recombinant algal centrin suggest that all four of the EF-hands bind calcium, albeit with different affinities for the ion (Weber *et al.*, 1994). Analysis of the crystal structure of EF-hand calcium-binding proteins and molecular modeling of centrin (Figure 5.3) suggests a dumbbell-shaped molecular configuration. This configuration consists of two globular domains, each with two EF-hands, separated by a flexible central helix region composed of the e-helix of the second EF-hand and the f-helix of the third EF-hand (Kretsinger *et al.*,

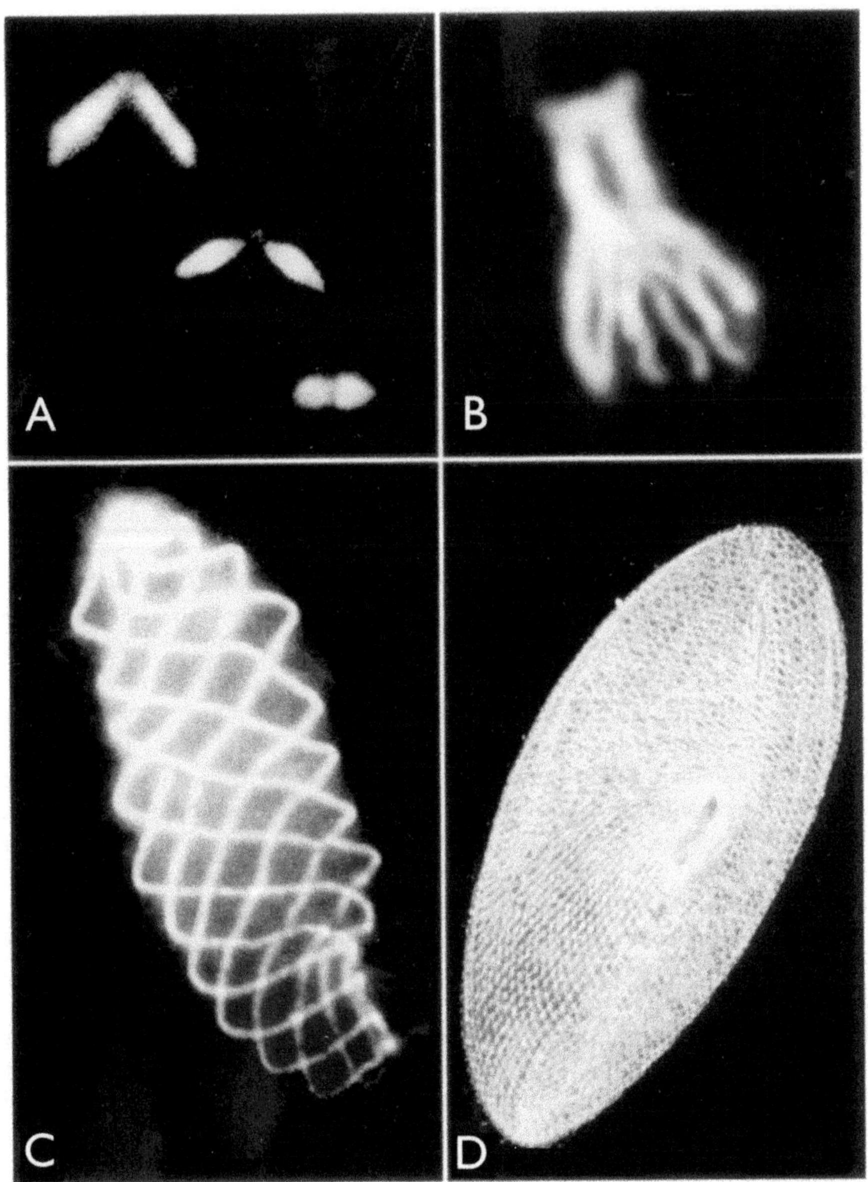

Figure 5.1 Centrin-containing contractile fibres show a diverse range of architecture and structure in flagellates and ciliates. Several examples of indirect immunofluorescence of centrin localization are illustrated here. A) A collage of three distinct stages of Ca^{2+}-induced flagellar root contraction in *Tetraselmis striata* is shown in this image. B) Basal body, descending fibres, and fimbria of the centrin-based fibre system of *Chlamydomonas* is shown here. C) Five flagellar bands of the termite gut endosymbiont *Holomastigotoides sp.*spiral just beneath the plasma membrane. One of the bands has released from the cell apex and is being drawn toward the distal half of the cell. D) *Paramecium* centrin is localized in a complex infraciliary lattice, as well as in basal bodies and in the cord, a filamentous structure associated with the oral apparatus

Source: Figure 5.1D courtesy of Professor Janine Beisson

Table 5.1 Summary of the properities of predicted centrin proteins. All sequences are full-length with the exception of the *Paramecium* ICL1b and ICL1c centrins

Organism	Full-length protein			Amino terminal region		
	Amino acids	MW (Da)	pKa	Amino acids	hydrophobic residues	(+)charged residues
Chlamydomonas	169	19,459	4.52	28	7	7
Dunaliella	169	19,397	4.64	28	6	8
Scherffelia	168	19,334	4.57	27	7	7
Naegleria	172	19,608	4.99	32	6	6
Giardia lamblia	176	20,236	4.53	34	12	8
Giardia intestinalis	161	18,685	4.49	20	6	5
Paramecium ICL1a	181	20,298	4.28	40	18	5
Paramecium ICL1b	>181	>20,336	~4.28	>40	≥17	≥5
Paramecium ICL1c	>182	>20,437	~4.28	>41	≥17	≥5
Paramecium ICL1d	181	20,312	4.28	40	18	5
Yeast	161	18,748	4.22	23	6	3

1986; Babu *et al.*, 1988; Heidorn and Trewhella, 1988). Analysis of centrin mutations in *Chlamydomonas* demonstrates that the flexible central helix region is critically important for centrin function (Taillon *et al.*, 1992). A G-to-A point mutation converting residue 101 from glutamic acid to lysine imposes a positive charge at this site, and results in a failure of the assembly of centrin-based fibres, and in an inappropriate basal body segregation phenotype (Taillon *et al.*, 1992).

A model explaining the calcium-sensitive behaviour of centrin-containing fibres can be proposed, based on these studies and others on centrin, calmodulin and troponin C (Heidorn and Trewhella, 1988; Persechini and Kretsinger, 1988; Putkey *et al.*, 1988; Seeholzer and Wand, 1989; VanBerkum *et al.*, 1990; Mehler *et al.*, 1991). In this model, calcium binding within the loop of the EF-hands results in the presentation of hydrophobic pockets at the ends of the central linking helix. The normally dynamic thermal bending of the central helix region becomes constrained when centrin–centrin or centrin–target protein interactions become stabilized by intermolecular interactions involving these hydrophobic pockets. Compaction of the overall molecular structure follows (Wiech *et al.*, 1996). Because each of the molecular components of the contractile fibres has yet to be identified, the precise mechanism of contraction remains unclear. However, ultrastructural investigations suggest that twisting and 'supercoiling' of centrin-containing fibres themselves operate in the fibre shortening process (Salisbury and Floyd, 1978; Salisbury, 1998). Centrin–target protein interactions have been identified in yeast where a calcium-dependent interaction between CDC31p and KAR1p has been shown to be required for spindle pole body duplication (Biggins and Rose, 1994; Vallen *et al*, 1994; Geier *et al.*, 1996).

5.2.2 The amino-terminal region

The amino-terminal regions of centrins from different species share little sequence homology, and vary in length from twenty to greater than forty amino acids (Table 5.1). None the less, the amino terminal region does share a conserved pattern of positively charged and hydrophobic residues. Although no role has been attributed to

```
Consensus              ...................................1.-E...EI.EAF.LFD..2.2.1D.+ELK
S.cerevisiae                              MSKNRSSLQSGPLNSELLEEQKQEIYEAFSLFDMNNDGFLDYHELK
G.intestinalis                               MARKPNARSTEPDLTEEQKHEIREAFDLFDSDGSGRIDVKELK
G.lamblia                       MNRAAIAAGKPSGSISTGKPRRKTRAEVSEEMKHEIREAFDLFDADRSGRIDFHELK
C.reinhardtii                          MSYKAKTVVSARRDQKKGRVGLTEEQKQEIREAFDLFDTDGSGTIDAKELK
D.salina                               MSYRKTVVSARRDQKKGRVGGLTEEQKQEIREAFDLFDTDGSGTIDAKELK
N.gruberi                          MQKYGSKKIGATSATSSNKQKVQIELTDEQRQEIKEAFDLFDMDGSGKIDAKELK
S.dubia                                 MSYRKAASARRDKAKTRSAGLTEEQKQEIREAFDLFDTDGSGTIDAKELK
P.tetraurelia (ICL1a)  MARRGQQPPP--QQAPPAQKNQTGKFNPAEFVKPGLTEEEVLEIKEAFDLFDTDGTQSIDPKELK
P.tetraurelia (ICL1b)  <ARRGQQPPPQQQQAPP-QKNQAGKFNPAEFVKPGLTEEEVLEIKEAFDLFDTDGTQSIDPKELK
P.tetraurelia (ICL1c)  <ARRGQQPPPQQQQAPPTQKNQAGKFNPAEFVKPGLTEEEVLEIKEAFDLFDTDGTQSIDPKELK
P.tetraurelia (ICL1d)  MARRGQQPPP--QQAPPAQKNQTGKFNPAEFVKPGLTEEEVLEIKEAFDLFDTDGTQSIDPKELK

Consensus              1AM..LGF-1....1..11...D..2...1...-F..1M..+1..+D...-1.+1F.LFD.-..G
S.cerevisiae           VAMKALGFELPKREILDLIDEYDSEGRHLMLYDDFYIVMGEKILKRDPLDEIKRAFQLFDDDHIG
G.intestinalis         VAMRALGFEPKREELKRMIAEVDTSGSGMIDLNDFFRIMTAKMAERDSREEILKAFRLFDEDDTG
G.lamblia              VAMRALGFDVKKEEIQRIMNEYDRDQLGEITFQDFEEVMIEKISNRDPTEEILKAFRLFDDDATG
C.reinhardtii          VAMRALGFEPKKEEIKKMISEIDKDGSGTIDFEEFLTMMTAKMGERDSREEILKAFRLFDDDNSG
D.salina               VAMRALGFEPKKEEIKKMIADIDKAGSGTIDFEEFLQMMTSKMGERDSREEIIKAFKLFDDDNTG
N.gruberi              VAMRALGFEPKKEEIKKMISGID-NGSGKIDFNDFLQLMTAKMSEKDSHAEIMKAFRLFDEDDSG
S.dubia                VAMRALGFEPKKEEIKKMIADIDKDGSGTIDFEEFLQMMTAKMGERDSREEIMKAFRLFDDDETG
P.tetraurelia (ICL1a)  AAMTSLGFEAKNQTIYQMISDLDTDGSGQIDFAEFLKLMTARISERDSKADIQKVFNLFDSERAG
P.tetraurelia (ICL1b)  AAMTSLGFEAKNQTIYQMISDLDTDGSGQIDFAEFLKLMTARISERDSKADIQKVFNLFDSERAG
P.tetraurelia (ICL1c)  AAMTSLGFEAKNQTIYQMISDLDTDGSGQIDFAEFLKLMTARISERDSKADIQKVFNLFDSERAG
P.tetraurelia (ICL1d)  AAMTSLGFEAKNQTIYQMISDLDTDGSGQIDFAEFLKLMTARISERDSKADIQKVFNLFDSERAG

Consensus              .131+.L++VAKELGE21.-.EL..MI..1D......1...+F..I.........
S.cerevisiae           KISIKNLRRVAKELGETLTDEELRAMIEEFDLDGDGEINENEFIAICTDS
G.intestinalis         KISFKNLKKVAKELGENLTDEEIQEMIDEADRDGDGEINEEEFLRIMRRTSLY
G.lamblia              RISLKNLRRVAKELSENISDEELLAMIQEFDRDGDGEIDEEDFIAILRSTSAFS
C.reinhardtii          TITIKDLRRVAKELGENLTEEELQEMIAEADRNDDNEIDEDEFIRIMKKTSLF
D.salina               FITLKNLKRVAKELGENLTDEELQEMTDEADRNGDGQIDEDEFYRIMKKTSLF
N.gruberi              FITFANLKRVAKDLGENMTDEELREMIEEADRSNQGQISKEDFLRIMKKTNLF
S.dubia                KISFKNLKRVAKELGENMTDEELQEMIDEADRDGDGEVNEEEFFRIMKKTSLF
P.tetraurelia (ICL1a)  VVTLKDLRKVAKELGETMDDSELQEMIDRADSDGDAQVTFEDFYNIMTKKTFA
P.tetraurelia (ICL1b)  VITLKDLRKVAKELGETMDDSELQEMIDRADSDGDAQVTFEDFYNIMTKKTFA
P.tetraurelia (ICL1c)  VITLKDLRKVAKELGETMDDSELQEMIDRADSDGDAQVTFEDFYNIMTKKTFA
P.tetraurelia (ICL1d)  VITLKDLRKVAKELGETMDDSELQEMIDRADSDGDAQVTFEDFYNIMTKKTFA
```

Figure 5.2 Protein sequence alignment of full-length flagellate centrins. The following centrin sequences are considered: *Saccharomyces cerevisiae* CDC31 (M14078), *Giardia intestinalis* (U59300), *Giardia lamblia* (U42428), *Chlamydomonas reinhardtii* (X12364), *Dunaliellla salina* (U53812), *Scherffelia dubia* (X69220), *Naegleria gruberi* (U21725), and *Paramecium tetraurelia* ICL1b (U35397) and ICL1c (U35396). The nearly full-length *Paramecium* ICL1a (U35344) and ICL1d (U76540) centrins are also included. Conservative substitutions are represented as follows: (-) = negatively charged residues (D and E), (+) = positively charged residues (K, R, and H), (1) = hydrophobic residues (I, L, A, V, P, M, F, and W), (2) = uncharged/polar residues (S, T, Y, C, G, Q, R, and N), and (3) = hydroxyl-containing residues (S, T, and Y). EF-hands are indicated (helices are underlined and loops are double underlined)

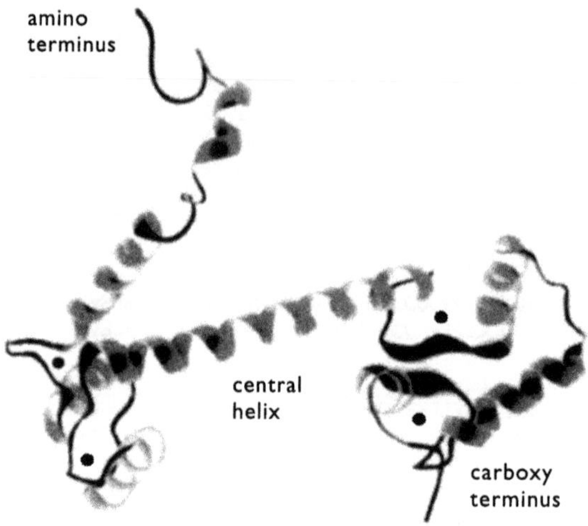

Figure 5.3 Molecular model of centrin protein structure based on a comparison with other calcium-binding proteins and computer modelling. The model illustrates a protein consisting of two calcium-binding globular domains and a central flexible connecting helix. The sites of four bound Ca^{2+} ions are represented by black dots

the amino-terminal region, these conserved features suggest preservation throughout evolution of common functional properties for this part of the protein. The four centrins of *Paramecium* possess the most divergent amino-terminal regions. The amino-terminal regions of *Paramecium* centrins are particularly long, and are unique with regard to the presence of numerous proline and glutamine residues. In *Paramecium*, multiple centrins showing microheterogeneity may be required to generate the complex geometry of the infraciliary lattice (Madeddu *et al.*, 1996; Klotz *et al.*, 1997; Ruiz *et al.*, 1998).

5.2.3 The carboxy terminal region

Centrin was first identified as a phosphoprotein, and phosphorylation was correlated with centrin-fibre extension (Salisbury *et al.*, 1984; Martindale and Salisbury, 1990). Centrins display sequence conservation in their extreme carboxy terminus, which often contains the consensus [K/R] [K/R] TSL [F/Y]. This sequence conforms to a consensus for protein kinase A (PKA) phosphorylation with the serine serving as the phosphoacceptor. We have previously identified this serine as a *bona fide* phosphoacceptor for human centrin (Lingle *et al.*, 1998). Variations of the carboxy-terminal sequence, that resemble the consensus but lack the serine residue, are found in *Naegleria* (KKTNLF), and *Paramecium* (KKTFA) centrins. It is possible that the threonine residue may serve as a phosphoacceptor in these proteins. Interestingly, one of the *Giardia* sp. centrins contains the consensus (RRTSLY) while the other *Giardia* sp. centrin contains a relatively divergent carboxy terminus (STSAFS). None the less, the putative phosphoacceptor serine is retained in both *Giardia* sp. centrins. We have preliminary immunological evidence that *Giardia* centrin is phosphorylated at the

RRTSLY carboxy-terminus site during mitosis and excystment (unpublished observations Lingle, Gillan, and Salisbury). The yeast CDC31p shows the most divergent carboxy-terminal sequence, lacking this phosphorylation site.

5.3 Centrin: gene and cDNA sequence comparisons

Centrin cDNAs have been sequenced for yeast (Baum *et al.*, 1986, 1988), mammals (Lee and Huang, 1993; Ogawa and Shimizu, 1993; Errabolu *et al.*, 1994; Middendorp *et al.*, 1997), plants (Zhu *et al.*, 1992), and the following ciliates and flagellates: *Chlamydomonas reinhardtii* (Huang *et al.*, 1988), *Dunaliella salina* (Ko and Lee, 1996), *Micromonas pusilla* (Steinkoetter and Melkonian, 1996), *Tetraselmis striata* (Bhattacharya *et al.*, 1993), *Spermatozopsis similis* (Bhattacharya *et al.*, 1993), *Scherffelia dubia* (Bhattacharya *et al.*, 1993), *Pterosperma cristatum* (Steinkoetter and Melkonian, 1996), *Naegleria gruberi* (Levy *et al.*, 1996), *Paramecium tetraurelia* (Madeddu *et al.*, 1996; Vayssie *et al.*, 1997), *Giardia* sp. (Meng *et al.*, 1996), and *Entodinium caudatum* (Eschenlauer *et al.*, 1998). Several of these sequences are partial, but all of the cDNA sequences available minimally encompass the four EF-hands.

All of the full-length flagellate sequences (*Naegleria, Dunaliella, Chlamydomonas, Scherffelia, Paramecium* and *Giardia*) begin with ATG methionine codons with *Naegleria, Dunaliella* and *Chlamydomonas* having the consensus translation initiation sequence (that is, an A at −3) (Kozak, 1986). The stop codons vary: *Naegleria, Dunaliella* and *Chlamydomonas* have TAA stop codons while *Paramecium, Giardia* and *Scherffelia* have TGA stop codons. Among the ciliates and flagellates, multiple centrin cDNAs have only been identified for *Paramecium, Entodinium* and *Giardia* sp., while among mammalian species multiple centrins appears to be the norm (mice and humans each contain at least three centrin genes) (Lee and Huang, 1993; Errabolu *et al.*, 1994; Middendorp *et al.*, 1997). Therefore, individual ciliate and flagellate species may, first, only posses a single centrin (as does yeast); second, show restricted expression of centrin(s) during specific developmental stages (as is the case for *Naegleria*); or third, have multiple centrin genes (as seen in *Paramecium*). It is unknown why some species contain multiple centrins. Do different centrins have distinct functions, or are centrins functionally redundant? The yeast centrin, along with mouse and human centrin 3 (Middendorp *et al.*, 1997), appear to be quite divergent from the centrins of most flagellate species. It may be that the unique structural properties of basal bodies / centrioles and contractile fibres requires distinct centrins.

The function of centrin, and its analogues spasmin and calmyonemin, in ciliate and flagellate species remains unclear, but several systems may provide some clues as to the diversity of centrin function. Calcium-modulated centrins fibres appear to have been adapted for use in a variety of structures requiring contractile behaviour, including flagellar excision, basal body positioning, and feeding. In *Chlamydomonas*, centrin-based fibres play a role in the excision of flagella in response to physiological stress (Sanders and Salisbury, 1994), and in the positioning of basal bodies within the cell (Wright *et al.*, 1989). In *Paramecium*, centrin is an essential component of the infraciliary lattice, a structure involved in the control of basal body distribution in the epiplasm (Klotz *et al.*, 1997). Recently, Vigues and co-workers (1999) have identified centrin-like filaments closely associated with the microtubule-based cytopharyngeal apparatus (feeding basket) of the ciliates *Nassula* and *Furgasonia*. It has been suggested that centrin filaments in *Nassula* and *Furgasonia*, and possibly *Entodinium*, are involved in the contraction of the cytopharyngeal apparatus that is required for the ingestion of food (Vigues *et al.*, 1999; Vigues and David, 1994). It is also possible that distinct centrins possess unique functions during discrete developmental stages. For example, the centrin analog, spasmin, has been identified as a major protein of the contractile spasmoneme in the trophont stage of some peritrich ciliates such as *Vorticella* and *Carchesium* (Ochiai *et al.*, 1988; Buhse, 1998). Additionally, expression of centrin is restricted to the differentiating flagellate stage of the life history of *Naegleria* (Levy *et al.*, 1998). It should also be pointed out that particular centrin species may show cellular functions distinct from those associated contractile fibre systems (Paoletti *et al.*, 1996).

Paramecium may represent a particularly useful system to investigate centrin function. Four centrin genes have been identified in *Paramecium*, but as many as twenty genes may exist (Madeddu *et al.*, 1996). *Paramecium* may require multiple centrins of subtly different structure in order appropriately to assemble the complex architecture of the infraciliary lattice. The four known centrin genes are co-expressed in *Paramecium*. Microinjection of high copy number of plasmids containing only the coding region of one centrin gene into the *Paramecium* somatic macronucleus leads to a marked reduction in the expression of the corresponding endogenous gene(s) (Ruiz *et al.*, 1998). Clonal descendents of microinjected cells show complete or partial disruption of infraciliary lattice structure. These experiments provide the first evidence that the members of the centrin gene family in *Paramecium* are required for assembly of the infraciliary lattice and that they are not functionally redundant (Ruiz *et al.*, 1998).

5.3.1 Structural features of centrin genes

Among the ciliates and flagellates, centrin genomic sequence information has only been reported for *Chlamydomonas*, *Paramecium* and *Giardia*. While they contain many features common to eukaryotic genes in general, the centrin genes of the different species share few specific structural features. Lee and co-workers (Lee *et al.*, 1991) were the first to identify and characterize the genomic sequence for *Chlamydomonas* centrin. This gene contains seven exons (106, 36, 50, 167, 42, 75, and 508 bp in length), separated by six introns that, as is generally the case for calcium-binding protein genes, do not occur at the boundaries of the EF-hand encoding sequence (Table 5.2). The intron/exon boundaries are consistent with the eukaryotic consensus, and conform to the GT/AG rule (Mount, 1982) (Table 5.2). The 5'-untranslated region (UTR) is 59 bp in length while the 3'-UTR is 184 bp in length and contains a TGTAA polyadenylation signal. The sequence immediately upstream of the translation start site contains an element that shares structural features with the canonical Hogness–Goldberg TATA box, and could serve as a RNA polymerase II recognition site. A CCAAT box typically located upstream of the TATA element is not present in this gene.

Table 5.2 Summary of centrin gene intron properities

Organism	*Intron*	*Class*	*Size (bp)*	*Position*	*Splice site sequence*
Chlamydomonas	1	II	117	E-helix of EF-hand 1	CCT/GTAGGT...ATGCAG/C
	2	0	231	loop of EF-hand 1	AAG/GTACGC...TAACAG/G
	3	0	503	E-helix of EF-hand 2	GAG/GTGAGA...CCTCAG/A
	4	0	111	loop of EF-hand 3	AAG/GTGAGC...TCACAG/G
	5	I	144	linker of EF-hand 3 and 4	CTG/GTGAGT...ACGCAG/A
	6	II	155	F-helix of EF-hand 4	CCG/GTGAGT...CAACAG/G
Paramecium ICL1a	1	II	23	loop of EF-hand 1	CGG/GTATTT...ATCCAGG/C
	2	0	26	E-helix of EF-hand 3	TAA/GTATAT...TAATAG/A
Paramecium ICL1b	1	II	25	loop of EF-hand 1	TGG/GTATAA...TATTAG/T
	2	0	28	E-helix of EF-hand 3	TAA/GTACAT...TTCTAG/A
Paramecium ICL1c	1	II	25	loop of EF-hand 1	TGG/GTATGT...TCTTAG/T
	2	0	28	E-helix of EF-hand 3	CAA/GTATCT...CATTAG/A
Paramecium ICL1d	1	II	24	loop of EF-hand 1	CGG/GTATTA...ATTAAG/C
	2	0	26	E-helix of EF-hand 3	TAA/GTATAA...TAATAG/A
Giardia intestinalis	Intronless				
Giardia lamblia	Intronless				

Note: All flagellate centrins for which genomic sequence data are available are considered. Class refers to the position of the introns as defined by Sharp (1981). Class 0 introns lie between codons, class I introns lie between the first and second base of a codon, and class II introns lie between the second and third base of a codon. The eukaryotic consensus for intron splice site sequence is [A/C]AG/GT[A/G]AGT...[T/C][T/C]N[T/C]AG/G

The genome of *Paramecium* contains a relatively large centrin gene family; it is estimated that up to twenty centrin genes may exist in this organism (Madeddu *et al.*, 1996; Vayssie *et al.*, 1997). Thus far, four centrin genes (ICL1a, b, c, and d) have been identified in the transcriptionally active macronucleus of *Paramecium tetraurelia* (Madeddu *et al.*, 1996; Vayssie *et al.*, 1997). The macronuclear genes, ICL1a, b, and c, and the micronuclear gene, ICL1d, each contain two introns that are conserved between the four genes with respect to their location, but vary slightly in size (Table 5.2) (Madeddu *et al.*, 1996). The sequence of the intron/exon boundaries of the first and second introns contain a high degree of conservation between genes, and show partial conservation with the general eukaryotic consensus (Table 5.2).

All four *Paramecium* centrin genes, have been identified in the macronucleus as well as the germline micronucleus suggesting that the diversity of centrin genes is derived from gene duplication events, and not from the gene rearrangements that occur during macronuclear differentiation (Vayssie *et al.*, 1997). The micronuclear ICL1b gene contains a 75 bp germline-specific internal eliminated sequence (IES) inserted in the second intron (Vayssie *et al.*, 1997). Furthermore, the IES in ICL1b is flanked by TA dinucleotides, and contains conserved inverted terminal repeats (5'- TAYAGYNR- 3') as defined by Klobutcher and Herrick (Klobutcher and Herrick, 1995).

In contrast to ICL1b, the micronuclear ICL1d does not contain an IES within the coding sequence. However, a 28 bp IES is located 1,057 bp upstream of the coding sequence (Vayssie *et al.*, 1997). The codon usage observed for ciliates is also conserved in the macronuclear centrin genes (for example, TAA and TAG are stop codons in mammalian system, but encode glutamine in *Paramecium*). The *Paramecium* centrin genes also conform to the trend in ciliate genomes of being A+T rich with coding sequences having a relatively lower A+T content (for example, the coding sequence of ICL1d is 65 per cent A+T and the surrounding genomic sequence is 78.5 per cent A+T) (Prescott, 1994).

Two *Giardia* centrin genes (reported as *G. intestinalis* and *G. lamella*) have been identified (Meng *et al.*, 1996; Sanchez and Muller, 1996, Accession U59300), and both genes are intronless. The *G. lamblia* centrin appears to be a single copy gene, and contains a 486 bp open reading frame, a 1 or 2 bp 5'-UTR, and a 27 bp 3'-UTR (Meng *et al.*, 1996). The predicted size of the transcriptional unit is consistent with northern blots demonstrating a transcript of ~0.6 kb (Meng *et al.*, 1996). The 3'-UTR contains a polyadenylation signal (TGTAAA). The 5' promoter region of the gene contains a palendromic octet (TAAATTTA) at –2 to –9 and an A+T rich region at –20 to –42. These features have been previously described for *Giardia* genes and may serve regulatory roles (Holberton and Marshall, 1995). The intronless nature of the gene, the short UTRs, and potential regulatory motifs in the sequence upstream of the initiation codon are all features that appear to be characteristic of *Giardia* genes, and these features are not unique to *Giardia* centrin genes.

5.3.2 Phylogenetic analysis

We have used parsimony analysis (PAUP* 4.0d64 by David Swofford) to determine the phylogenetic relationships of the ciliate and flagellate centrins. Figure 5.4 is a cladogram based on the protein sequence of the flagellates for which full-length sequences are available; the analysis also includes two *Paramecium* sequences that are nearly full-length. We have also analyzed protein sequences based on the amino acids spanning the four EF-hands, and found essentially the same features seen in Figure 5.4. Several phylogenetic relationships are evident. First, the algal centrins form a distinct clade. Second, the *Paramecium* centrins form a separate clade with the *Naegleria* and *G. intestinalis* centrins being closely related. This indicates that centrins of ciliate species are distinct from those of flagellate species. Within the *Paramecium* centrins, ICL1a and ICL1d appear to be most related, and ICL1b appears to be most divergent. Third, the G. *lamblia* centrin and CDC31p form the most divergent clade, which also includes mouse and human centrin 3 (Middendorp *et al.*, 1997). Given that CDC31p is the only centrin in the genome of *S. cerevisiae*, this divergent group of centrins may delineate the modern representatives of an ancestral centrin. Finally, the *Giardia* centrins appear to be relatively divergent from each other; this divergence may indicate a functional difference between centrins in *Giardia*.

5.4 Conclusion

Centrin is a component of contractile-fibre systems of diverse flagellate species. The identification of centrin encoding cDNAs from numerous flagellates has allowed the description of some common physical properties as well as several conserved sequence features of the molecule. Characterization of centrin encoding genes has only appeared for a few flagellate species.

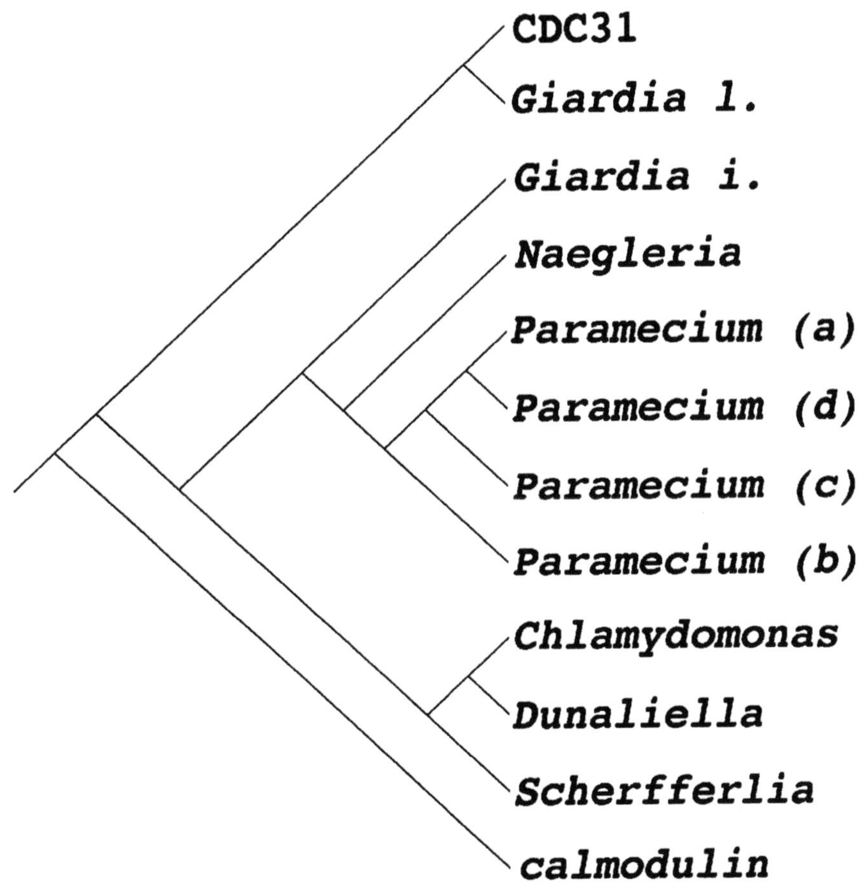

Figure 5.4 Cladogram illustrating parsimony analysis of relationships between flagellate centrins based on amino acid sequences

Outstanding questions that a number of laboratories are currently investigating include the following. What precisely is the molecular organization of centrin-containing contractile fibres? What is the nature of the contractile mechanism? What is the role of phosphorylation in regulating centrin-fibre contraction and extension? What are the distinguishing features of centrin associated with basal bodies and centrioles versus contractile fibre and the pericentriolar lattice? Does centrin have functions that are not tightly linked to centrosome and flagellar apparatus behaviour?

REFERENCES

Babu, Y. S., Bugg, C. E. and Cook, W. J. (1988) Structure of calmodulin refined at 2. 2 A resolution. *Journal of Molecular Biology*, **204**, 191–204.

Baron, A. T., Greenwood, T. M., Bazinet, C. W. and Salisbury, J. L. (1992) Centrin is a component of the pericentriolar lattice. *Biology of the Cell*, **76**, 383–388.

Baum, P., Furlong, C. and Byers, B. (1986) Yeast gene required for spindle pole body duplication, homology of its product with Ca^{2+}-binding proteins. *Proceedings of the National Academy of Sciences USA*, **83**, 5512–5516.

Baum, P., Yip, C., Goetsch, L. and Byers, B. (1988) A yeast gene essential for regulation of spindle pole duplication. *Molecular and Cellular Biology*, **8**, 5386–5397.

Bhattacharya, D., Steinkotter, J. and Melkonian, M. (1993) Molecular cloning and evolutionary analysis of the calcium-modulated contractile protein, centrin, in green algae and land plants. *Plant Molecular Biology*, **23**, 1243–1254.

Biggins, S. and Rose, M. D. (1994) Direct interaction between yeast spindle pole body components, Kar1p is required for Cdc31p localization to the spindle pole body. *Journal of Cell Biology*, **125**, 843–852.

Brown, D. L., Massalski, A. and Patenaude, R. (1976) Organization of the flagellar apparatus and associate cytoplasmic microtubules in the quadriflagellate alga *Polytomella agilis*. *Journal of Cell Biology*, **69**, 106–125.

Buhse, H. E. (1998) *Vorticella*, a cell for all seasons. *Journal of Eukaryotic Microbiology*, **45**, 469–474.

Errabolu, R., Sanders, M. A. and Salisbury, J. L. (1994) Cloning of a cDNA encoding human centrin, an EF-hand protein of centrosomes and mitotic spindle poles. *Journal of Cell Science*, **107**, 9–16.

Eschenlauer, S. C., McEwan, N. R., Calza, R. E., Wallace, R. J., Onodera, R. and Newbold, C. J. (1998) Phylogenetic position and codon usage of two centrin genes from the rumen ciliate protozoan, *Entodinium caudatum*. *FEMS Microbiology Letters*, **166**, 147–154.

Geier, B. M., Wiech, H. and Schiebel, E. (1996) Binding of centrins and yeast calmodulin to synthetic peptides corresponding to binding sites in the spindle pole body components Kar1p and Spc110p. *Journal of Biological Chemistry*, **271**, 28366–28374.

Goodenough, U. W. (1989) Cilia, flagella and the basal apparatus. *Current Opinion in Cell Biology*, **1**, 58–62.

Heidorn, D. B. and Trewhella, J. (1988) Comparison of the crystal and solution structures of calmodulin and troponin C. *Biochemistry*, **27**, 909–915.

Holberton, D. V. and Marshall, J. (1995) Analysis of consensus sequence patterns in *Giardia* cytoskeleton gene promoters. *Nucleic Acids Research*, **23**, 2945–2953.

Huang, B., Mengersen, A. and Lee, V. (1988) Molecular cloning of cDNA for caltractin, a basal body-associated Ca^{2+}-binding protein, Homology in its protein sequence with calmodulin and the yeast CDC31 gene product. *Journal of Cell Biology*, **107**, 133–140.

Klobutcher, L. A. and Herrick, G. (1995) Consensus inverted terminal repeat sequence of Paramecium IESs, resemblance to termini of Tc1-related and *Euplotes* Tec transposons. *Nucleic Acids Research*, **23**, 2006–2013.

Klotz, C., Garreau de Loubresse, N., Ruiz, F. and Beisson, J. (1997) Genetic evidence for a role of centrin-associated proteins in the organization and dynamics of the infraciliary lattice in *Paramecium*. *Cell Motility and the Cytoskeleton*, **38**, 172–186.

Ko, J. H. and Lee, S. H. (1996) Nucleotide sequence of a cDNA (Accession no. U53812) encoding a caltractin–like protein from *Dunaliella salina*. *Plant Physiology*, **112**, 445.

Kozak, M. (1986) Point mutations define a sequence flanking the AUG initiator codon that modulates translation by eukaryotic ribosomes. *Cell*, **44**, 283–292.

Kretsinger, R. (1975) Hypothesis, calcium-modulated proteins contain EF-hands. In *Calcium Transport in Contraction and Secretion* (ed. E. Carafoli), Amsterdam: North Holland, pp. 469–478.

Kretsinger, R. H., Rudnick, S. E. and Weissman, L. J. (1986) Crystal structure of calmodulin. *Journal of Inorganic Biochemistry*, **28**, 289–302.

Lee, V. D. and Huang, B. (1993) Molecular cloning and centrosomal localization of human caltractin. *Proceedings of the National Academy of Sciences USA*, **90**, 11039–11043.

Lee, V. D., Stapleton, M. and Huang, B. (1991) Genomic structure of *Chlamydomonas* caltractin. Evidence for intron insertion suggests a probable genealogy for the EF-hand superfamily of proteins. *Journal of Molecular Biology*, **221**, 175–191.

Levy, Y., Lai, E., Remillard, S. and Fulton, C. (1998) Centrin is synthesized and assembled into basal bodies during *Naegleria* differentiation. *Cell Motility and the Cytoskeleton*, **40**, 249–260.

Levy, Y., Lai, E., Remillard, S., Heintzelman, C. and Fulton, C. (1996) Centrin is a conserved protein that forms diverse associations with centrioles and MTOCs in *Naegleria* and other organisms. *Cell Motility and the Cytoskeleton*, **33**, 298–323.

Lingle, W. L., Lutz, W. H., Ingle, J. N., Maihle, N. J. and Salisbury, J. L. (1998) Centrosome hypertrophy in human breast tumors, implications for genomic stability and cell polarity. *Proceedings of the National Academy of Sciences USA*, **95**, 2950–2955.

Lingle, W. L. and Salisbury, J. L. (1997) Centrin and the cytoskeleton of the protist *Holomastigotoides. Cell Motility and the Cytoskeleton*, **36**, 377–390.

Madeddu, L., Klotz, C., Le Caer, J. P. and Beisson, J. (1996) Characterization of centrin genes in *Paramecium. European Journal of Biochemistry*, **238**, 121–128.

Martindale, V. E. and Salisbury, J. L. (1990) Phosphorylation of algal centrin is rapidly responsive to changes in the external milieu. *Journal of Cell Science*, **96**, 395–402.

McFadden, G. I., Schulze, D., Surek, B., Salisbury, J. L. and Melkonian, M. (1987) Basal body reorientation mediated by a Ca^{2+}-modulated contractile protein. *Journal of Cell Biology*, **105**, 903–912.

Mehler, E. L., Pascual-Ahuir, J. L. and Weinstein, H. (1991) Structural dynamics of calmodulin and troponin C. *Protein Engineering*, **4**, 625–637.

Melkonian, M. (1980) Ultrastructural aspects of basal body associated fibrous structures in green algae, a critical review. *BioSystems*, **12**, 85–104.

Melkonian, M. (1989) Centrin-mediated motility, a novel cell motility mechanism in eukaryotic cells. *Botanica Acta*, **102**, 3–4.

Meng, T. C., Aley, S. B., Svard, S. G., Smith, M. W., Huang, B., Kim, J. and Gillin, F. D. (1996) Immunolocalization and sequence of caltractin/centrin from the early branching eukaryote *Giardia lamblia. Molecular and Biochemical Parasitology*, **79**, 103–108.

Middendorp, S., Paoletti, A., Schiebel, E. and Bornens, M. (1997) Identification of a new mammalian centrin gene, more closely related to *Saccharomyces cerevisiae* CDC31 gene. *Proceedings of the National Academy of Sciences USA*, **94**, 9141–9146.

Moncrief, N., Kretsinger, R. and Goodman, M. (1990) Evolution of EF-hand calcium-modulated proteins. I. Relationships based on amino acid sequences. *Journal of Molecular Evolution*, **30**, 522–562.

Mount, S. (1982) A catalogue of splice junction sequences. *Nucleic Acids Research*, **10**, 459–472.

Ochiai, T., Kato, M., Ogawa, T. and Asai, H. (1988) Spasmin-like proteins in various ciliates revealed by antibody to purified spasmins of *Carchesium polypinum. Experientia*, **44**, 768–771.

Ogawa, K. and Shimizu, T. (1993) cDNA sequence for mouse caltractin. *Biochimica et Biophysica Acta*, **1216**, 126–128.

Paoletti, A., Moudjou, M., Paintrand, M., Salisbury, J. L. and Bornens, M. (1996) Most of centrin in animal cells is not centrosome-associated and centrosomal centrin is confined to the distal lumen of centrioles. *Journal of Cell Science*, **109**, 3089–3102.

Persechini, A. and Kretsinger, R. H. (1988) The central helix of calmodulin functions as a flexible tether. *Journal of Biological Chemistry*, **263**, 12175–12178.

Pitelka, D. A. (1969). Fibrillar systems in protozoa. *Research in Protozoology*, **3**, 279–388.

Prescott, D. M. (1994) The DNA of ciliated protozoa. *Microbiological Reviews*, **58**, 233–267.

Putkey, J. A., Ono, T., VanBerkum, M. F. and Means, A. R. (1988) Functional significance of the central helix in calmodulin. *Journal of Biological Chemistry*, **263**, 11242–11249.

Ruiz, F., Vayssie, L., Klotz, C., Sperling, L. and Madeddu, L. (1998) Homology-dependent gene silencing in *Paramecium. Molecular Biology of the Cell*, **9**, 931–943.

Salisbury, J. L. (1995) Centrin, centrosomes, and mitotic spindle poles. *Current Opinion in Cell Biology* **7**, 39–45.

Salisbury, J. L. (1998) Roots. *Journal of Eukaryotic Microbiology*, **45**, 28–32.

Salisbury, J. L., Aebig, K. W. and Coling, D. E. (1986) Isolation of the calcium-modulated contractile protein of striated flagellar roots. *Methods in Enzymology*, **134**, 408–414.

Salisbury, J. L., Baron, A., Surek, B. and Melkonian, M. (1984) Striated flagellar roots, isolation and partial characterization of a calcium-modulated contractile organelle. *Journal of Cell Biology*, **99**, 962–970.

Salisbury, J. and Floyd, G. (1978) Calcium-induced contraction of the rhizoplast of a quadriflagellate green alga. *Science*, **202**, 975–977.

Salisbury, J. L., Sanders, M. A. and Harpst, L. (1987) Flagellar root contraction and nuclear movement during flagellar regeneration in *Chlamydomonas reinhardtii. Journal of Cell Biology*, **105**, 1799–1805.

Sanders, M. A. and Salisbury, J. L. (1989) Centrin-mediated microtubule severing during flagellar excision in *Chlamydomonas reinhardtii. Journal of Cell Biology*, **108**, 1751–1760.

Sanders, M. A. and Salisbury, J. L. (1994) Centrin plays an essential role in microtubule severing during flagellar excision in *Chlamydomonas reinhardtii. Journal of Cell Biology*, **124**, 795–805.

Seeholzer, S. H. and Wand, A. J. (1989) Structural characterization of the interactions between calmodulin and skeletal muscle myosin light chain kinase, effect of peptide (576–594)G binding on the Ca^{2+}-binding domains [published erratum appears in *Biochemistry*, 28,7974]. *Biochemistry*, **28**, 4011–4020.

Steinkoetter, J. and Melkonian, M. (1996). Centrin. In *Guidebook to Calcium-Binding Proteins* (ed. M. R. Celio), Oxford: Oxford University Press.

Taillon, B. E., Adler, S. A., Suhan, J. P. and Jarvik, J. W. (1992) Mutational analysis of centrin, an EF-hand protein associated with three distinct contractile fibres in the basal body apparatus of *Chlamydomonas. Journal of Cell Biology*, **119**, 1613–1624.

Vallen, E. A., Ho, W., Winey, M. and Rose, M. D. (1994) Genetic interactions between CDC31 and KAR1, two genes required for duplication of the microtubule organizing center in *Saccharomyces cerevisiae. Genetics*, **137**, 407–422.

VanBerkum, M. F., George, S. E. and Means, A. R. (1990) Calmodulin activation of target enzymes. Consequences of deletions in the central helix. *Journal of Biological Chemistry*, **265**, 3750–3756.

Vayssie, L., Sperling, L. and Madeddu, L. (1997) Characterization of multigene families in the micronuclear genome of *Paramecium tetraurelia* reveals a germline specific sequence in an intron of a centrin gene. *Nucleic Acids Research*, **25**, 1036–1041.

Vigues, B. and David, C. (1994) Calmyonemin, identification and distribution throughout the cell cycle in *Entodinium bursa* (ciliate). *Biology of the Cell*, **82**, 121–127.

Vigues, B., Blanchard, M. and Bouchard, P. (1999) Centrin-like filaments in the cytopharyngeal apparatus of the ciliates *Nassula* and *Furgasonia*, evidence for a relationship with microtubular structures. *Cell Motillity and the Cytoskeleton*, **43**, 72–81.

Weber, C., Lee, V., Chazin, W. and Huang, B. (1994) High level expression in *Escherichia coli* and characterization of the EF-hand calcium-binding protein caltractin. *Journal of Biological Chemistry*, **269**, 15795–15802.

Wiech, H., Geier, B. M., Paschke, T., Spang, A., Grein, K., Steinkotter, J., Melkonian, M. and Schiebel, E. (1996) Characterization of green alga, yeast, and human centrins. Specific subdomain features determine functional diversity. *Journal of Biological Chemistry*, **271**, 22453–22461.

Wright, R. L., Adler, S. A., Spanier, J. G. and Jarvik, J. W. (1989) Nucleus-basal body connector in *Chlamydomonas,* evidence for a role in basal body segregation and against essential roles in mitosis or in determining cell polarity. *Cell Motillity and the Cytoskeleton*, **14**, 516–526.

Zhu, J. K., Bressan, R. and Hasegawa, P. (1992) An *Atriplex nummularia* cDNA with sequence relatedness to the algal caltractin gene. *Plant Physiology*, **99**, 1734–1735.

Chapter 6

The cell surface of flagellates

Burkhard Becker

ABSTRACT

An overview on the structure and function of flagellate cell surfaces is presented. Cell surfaces can be simple, that is, consisting only of the plasma membrane. However, in most cases cell surfaces are complex and consist of a plasma membrane and associated extracellular and/or intracellular material. A new classification scheme for cell surfaces of flagellates is presented and illustrated with various examples, and the evolution of the various cell surface types is discussed.

6.1 Introduction

The cell surface forms the border between the living and the non-living. The simplest cell surface consists of a single biomembrane, the plasma membrane (PM) which is involved in many basic functions of a cell. For example, the PM represents a permeability barrier for many solutes and it is therefore the site of uptake and excretion/secretion of various compounds. Other functions often associated with the plasma membrane are cell signalling and cell–cell interaction.

Many organisms have developed complex cell surfaces: in other words, additional intra- and/or extracellular material is associated with the plasma membrane. For these complex systems, additional functions can be attributed to the cell surface. It is often, together with the cytoskeleton, responsible for the shape of the cell, and cell walls allow cells to survive in hypertonic media without special water secretion systems. In addition, cell surfaces protect cells against pathogens, mediate cell adhesion, and can be a storage place for various compounds.

All functional aspects mentioned so far have been studied in various groups of flagellate organisms. The following is an incomplete list of examples:

1 Endocytosis, secretion and plasma membrane transporters have been studied in various groups of flagellates (for example, endocytosis in kinetoplastids: Duszenko and Seyfang, 1993; glucose transporter in the PM of kinetoplastids, Tetaut *et al.*, 1997; secretion in diplomonads, kinetoplastids, ciliates and chlorophytes was reviewed by Becker and Melkonian, 1996, exocytosis in ciliates, Glas-Albrecht *et al.*, 1991).
2 Cell signalling has been studied in many groups (see Kawai and Kreimer, Chapter 7 in this volume, for light perception as an example); osmosensing has

been studied in *Dunaliella* (for example Zelasny *et al.*, 1995). Cell–cell interactions have been studied in kinetoplastids (for example, host–cell interactions, see Chapter 10 by Vickerman, this volume) and *Chlamydomonas* (mating, for example, Wilson *et al.*, 1997).

3 The role of extracellular material as a storage compartment has been addressed in *Emiliania*, where coccoliths can function as a source of dissolved inorganic carbon for the cells (Sekino *et al.*, 1996).

A special problem of flagellate cells is that every type of cell covering should not affect cell motility. A complete rigid covering of the entire cell including the flagella would obviously interfere with flagella-mediated cell motility. The development of cell walls, forming a rigid, homogenous but porous extracellular matrix, therefore requires holes to allow the flagella to protrude into the medium. In addition, these holes have to be exactly positioned so that cell motility is effective. This is obviously difficult to achieve, since a cell wall like structure is only found twice among flagellates.[1]

In many organisms the flagellate cell is not the major life form, but is restricted to certain stages of a life-cycle, often involved in sexual (gametes) or asexual (zoospores) reproduction. In this review I will restrict myself to the cell surfaces of organisms where the flagellate form is the major life form.

It is difficult to define which parts of a cell constitute the cell surface. In its literal meaning, the term cell surface describes just the border area between a cell and the environment. However, even many intracellular structures have been considered historically to be part of the cell surface. In the context of this review, I define the cell surface as the border between the living and the non-living (permeability barrier), which consists of the plasma membrane and associated intra- and/or extracellular material. Not discussed in this review in detail are some intracellular structures with a fixed spatial relationship to the cell surface, such as the cytoskeleton, peripheral ER and extrusive organelles. Specialized cell surfaces of flagellates like the flagellar or the haptonematal surfaces are also not dealt with.

6.2 Nomenclature

The terminology and nomenclature used to describe cell surfaces is confusing. Homologous structures have been given different names by protozoologists, phycologists and mycologists (for example amphiesma, alveolus and inner membrane system have been used to describe flattened vesicles underlying the plasma membrane in ciliates, dinoflagellates and apicomplexans respectively), whereas the same name (such as theca) has been applied to unrelated structures (theca of, for example, dinoflagellates, diatoms and some prasinophytes).

A collection of terms used to describe the cell surface of protists can be found in the excellent review by Preisig *et al.* (1994), who collected 100 terms including thirty-four synonyms. These authors also made recommendations to avoid these problems. However, their definitions of glycocalix and extracellular matrix are not in agreement with the ones used in the biomedical literature (for example Cooper, 1997; Lodish *et al.*, 1995). In the following, I will use both terms as in the biomedical literature:

glycocalix is a carbohydrate coat covering the cell surface, formed by the carbohydrate side chains of the plasma membrane glycoproteins and glycolipids; and extracellular matrix is the material surrounding the cell, which can consist of polysaccharides, glycoproteins, proteoglycans, and also inorganic components. 'Extracellular matrix' defined in this way includes every extracellular covering of a cell.

6.3 Basic types of cell surfaces

Cell surfaces can be grouped into a few different basic types. In the following I introduce a new classification scheme (Figure 6.1) and present some examples for each cell surface type. Terms defined by Preisig *et al.* (1994) are given in italics. The natural occurrence of the various types is also given. In some groups, more than one type of cell surface is present. For instance although most green algae possess an extracellular matrix, the cell surface of a few green flagellates consists only of a plasma membrane, for example *Spermatozopsis similis* (Preisig and Melkonian, 1984). Such organisms I consider as exceptions of the general surface type accounted in a given group of organisms.[2] Only the major cell surface types of any particular group are presented. In addition, many species secrete large amounts of mucilage which can form a kind of extracellular matrix (for example the euglenophyte *Trachelomonas* is enclosed by a mineralized lorica made of mucistrands impregnated with manganese (West and Walne, 1980)), or stalks, which the cells use to adhere themselves to a substratum.

6.3.1 Type I: simple plasma membrane (Figure 6.1)

The type I cell surface consists of a simple or modified *plasma membrane*, which can be covered with a thick *glycocalix* forming a *fuzzy coat* visible after staining in the transmission electron microscope. Type 1 cell surfaces can be found in various groups of organisms. As an example, Figure 6.2A shows an electron micrograph of a *Trichomonas* cell. The plasma membrane is clearly visible, and no type of additional intracellular or extracellular material can be observed. In some protists like *Trypanosoma*, however, the PM is covered by a thick (15nm) glycocalix. In *Trypanosoma* the glycocalix is made of glycosyl-phosphatidyl-inositol anchored glycoproteins (see Vickerman, Chapter 10 this volume).

This type of cell surface occurs in many flagellates, which are most often placed in today's phylogenetic trees at the root of the eukaryotic cell evolution.[3] Included are pelobionts, retortamonads, diplomonads, oxymonads, parabasalians, kinetoplastids, pseudociliates, opalinids and raphidophytes.

6.3.2 Type II: cell surfaces with additional intracellular material (Figure 6.1)

Only one eukaryotic group possesses a type II cell surface. In the euglenoids the PM is underlayed by internal protein complexes ordered in tile-like overlapping stripes (epiplasm, membrane skeleton: see Figure 6.2B). The membrane skeleton is associated with cisternae of ER and microtubules which form together the cell cortex or membrane skeleton complex. This type of cell surface is called *pellicle*. The pellicle

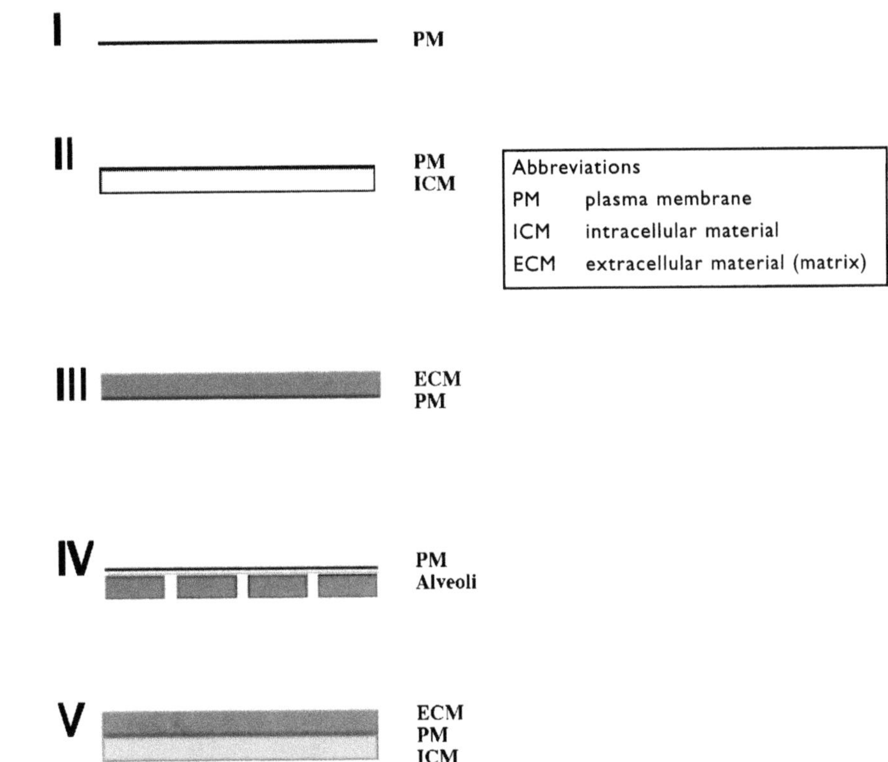

Figure 6.1 Schematic drawing of basic types of cell surfaces among flagellates

of *Euglena* has been characterized in some detail (Bouck and Ngô, 1996). The major plasma membrane protein of *Euglena* (IP39) functions as the membrane anchor for the two major epiplasmic proteins (articulin 80 and 86) which are probably embedded in an epiplasmic matrix, consisting of proteins crosslinked by disulphide bridges (Bouck and Ngô, 1996).

6.3.3 Type III: cell surfaces with additional extracellular material (extracellular matrix) (Figure 6.1)

In many groups of organisms (part of the so-called crown group of eukaryotic cell evolution) an extracellular matrix is present.[4] These organisms belong to the heterokont group (chrysophytes, dictyochophytes, bicosoecids, proteomonads), and also include the prymnesiophytes, pseudodendromonads, thaumatomonads, chlorophytes and choanoflagellates. Extracellular matrices occur in various forms and include *cases, cell walls, coccoliths, costae, loricae, mucilage, scales, sheaths* and *thecae* of prasinophytes. Figure 6.3 shows some examples for type III surfaces.

Usually the extracellar matrix consists of carbohydrates, for example prasinophyte scales (Becker *et al.*, 1994) or prymnesiophyte cellulose-containing scales (Brown and Romanovicz, 1976), but inorganic material (calcium carbonate or silica) and glycoproteins are also present. In some cases, inorganic material or glycoproteins can

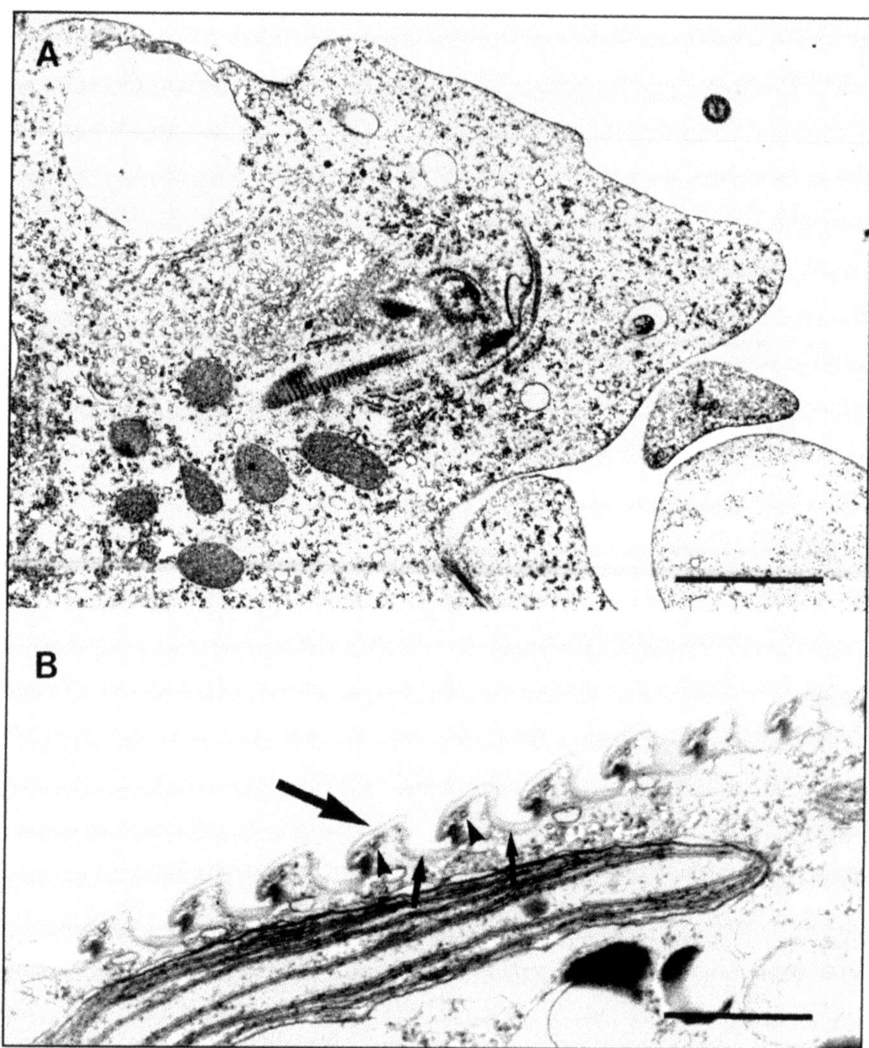

Figure 6.2

A The cell surface of *Trichomonas vaginalis* (scale bar = 1 μm)

Source: Courtesy of G. Brugerolle

B The pellicle of *Euglena mutabilis*, large arrow plasma membrane, small arrow epiplasm, arrow head microtubule (scale bar = 0.5 μm)

Source: Courtesy of M. Melkonian

(right)

Figure 6.3 Type III cell surface

A and B *Synura petersenii*

A Thin section showing the cell case and several intracellular scales residing in vesicles can be seen (scale bar = 1 μm)

B Whole mount of isolated scales (scale bar = 1 μm)

C Thin section showing the cell surface of *Nephroselmis olivacea* (scale bar = 0.5 μm). The different scale types on the flagellar and body surface can be seen

D The case (arrow) of the choanoflagellate *Monosiga* (scale bar = 1 μm)

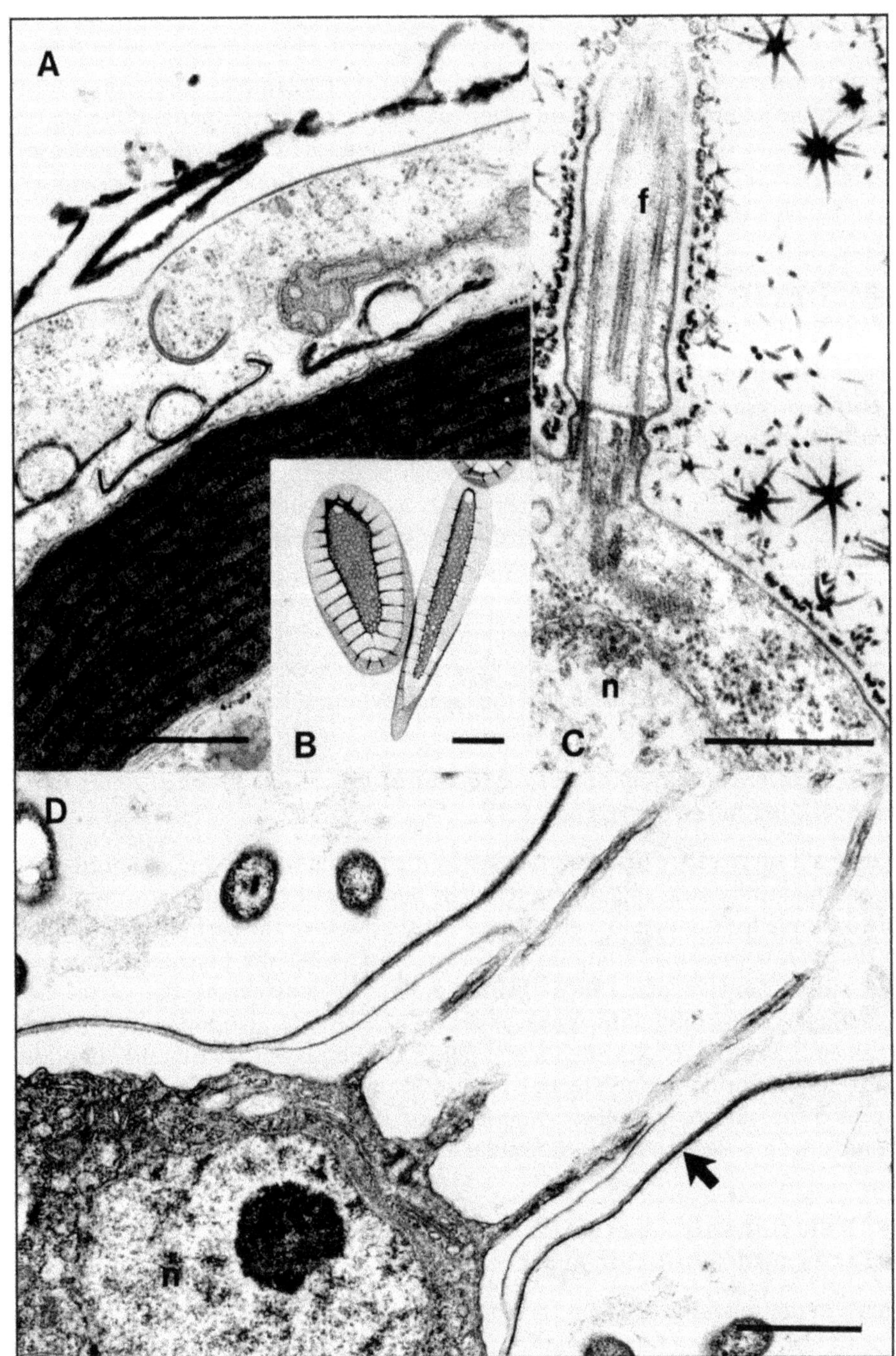

General abbreviations:

n nucleus

f flagellum

Sources: A, B and D courtesy of G. Brugerolle

C courtesy of M. Melkonian

account for most of the extracellular material. For example, coccoliths consists of calcium carbonate crystallized on a polysaccharide matrix (Faber and Preisig, 1994), the scales of *Mallomonas* consists of silica deposited on a glycoprotein matrix (Ludwig *et al.*, 1996), and the cell wall of *Chlamydomonas* is made entirely of glycoproteins (Adair and Snell, 1990).

6.3.4 Type IV: cell surfaces with additional intracellular material in vesicles (Figure 6.1)

In alveolates and glaucocystophytes, just internal to the plasma membrane is a system of flattened vesicles, which is sometimes underlain by epiplasm. The cisternae are called *alveoli* in ciliates, *amphiesma* vesicles in dinoflagellates, and *inner membrane complex* in apicomplexans. Plasma membrane, alveoli and epiplasm are called together *cell cortex* in ciliates (Figure 6.4). Whether the structurally similar *lacuna* system in glaucocystophytes is a homologous or analogous structure cannot be decided at present. Thecate dinoflagellates contain one or several cellulose plates in the amphiesmal vesicles (Morrill and Loeblich, 1983). Similarly, some ciliates have plates in the *alveoli*, which can consist of glycoproteins, polysaccharides or calcified organic material (for example Kloetzel, 1991; Williams *et al.*, 1989), and in *Cyanophora paradoxa* (glaucocystophytes), scales (plates) are present in the lacuna system (Heimann *et al.*, 1997)

6.3.5 Type V: cell surfaces with additional intracellular and extracellular material (Figure 6.1)

Type V cell surfaces are made of the plasma membrane with additional proteinaceous intracellular and extracellular material. The *periplast* of cryptomonads represents the only example for this type of cell surface. The stiffness of the periplast is derived from nearly gapless protein-containing plates accompanying the plasma membranes on both sides (Brett *et al.*, 1994) (Figure 6.5A). In most cases the plates have a hexagonal or rectangular appearance (Figure 6.5B). Roundish or polygonal intramembranous particles anchor the plates to the plasma membrane. In addition, extracellular fibrillar material and rosette scales (Santore, 1983) can be present. Recently, the surface periplast component of *Komma caudata* was shown to self-assemble from a secreted high-molecular mass protein (Perasso *et al.*, 1997).

6.4 Relationships between the various types of cell surfaces and evolutionary trends

What are the relations between the various types of cell surfaces? It is very difficult to draw some definite conclusions. However, every non-cytoplasmic structure is a product of the secretory pathway, and the intracellular membrane system of the alveolates and the glaucocystophytes can be considered as non-discharged secretory vesicles. It is therefore not surprising that these 'vesicles' occasionally contain typical secretory products like glycoproteins and polysaccharides (for example cellulose). It is much more difficult to determine whether the intracellular materials used to strengthen the cell surface (epiplasmins) by various flagellates are related. However,

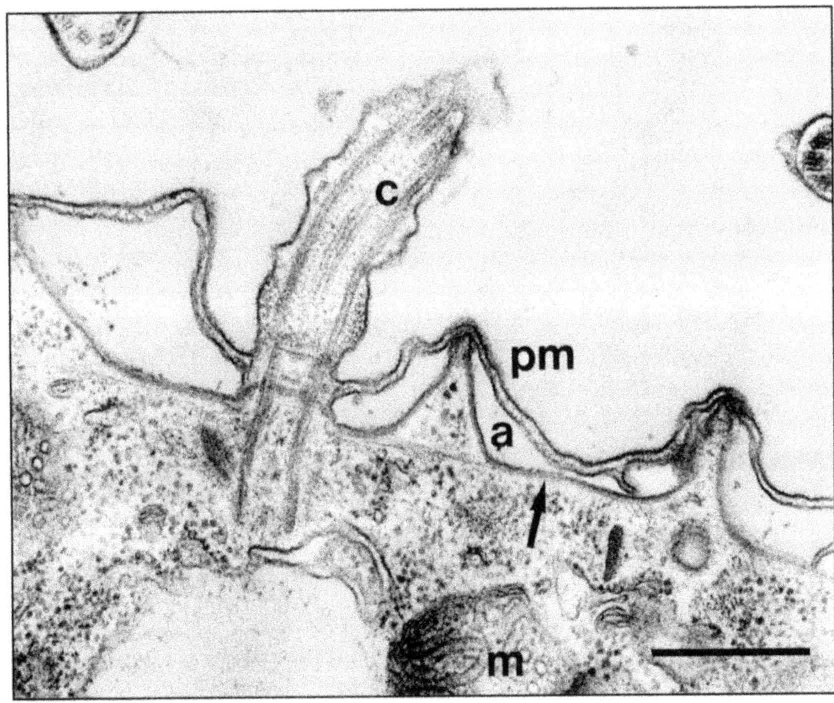

Figure 6.4 The cell cortex of the ciliate *Paramecium tetraurelia*. pm plasma membrane, a alveolus, arrow epiplasm, c cilium, m mitochondrion. (scale bar = 0.5 μm)

Source: Courtesy of G. Brugerolle

recent molecular data on some articulins (epiplasmins) of euglenoids and ciliates might be a first indication that this is the case (see later).

Cell surfaces show a great variety among the flagellates, and it is difficult to determine trends in the evolution of cell surfaces within the eukaryotes. However, when the data is put into a phylogenetic context a non-random distribution is observed. Figure 6.6 shows the distribution of the types of cell surfaces among flagellates. The upper part of the diagram shows the topology of the phylogenetic tree obtained by 18S RNA sequence comparisons (Sogin, 1997). For simplicity some minor groups of organisms have been omitted. With the exception of the euglenoids which possess a type II surface, simple surfaces (type I) are present in the early lines of eukaryotic cell evolution. This could reflect the original eukaryotic condition. In almost every scenario of the evolution of eukaryotes, the 'ureukaryote' had a simple surface. This is considered to be necessary for the phagotrophy required for the uptake of bacteria, which later became mitochondria and chloroplasts (Cavalier-Smith, 1991, 1993; Margulis, 1996). It is considered either that this 'ureukaryote' is a flagellate, or that flagella evolved very soon (Cavalier-Smith, 1991, 1993; Knoll, 1992; Margulis, 1996; Vellai *et al.*, 1998).

The other types of cell surfaces (types III, IV and V) are found only among crown group organisms (Figure 6.6). As has been indicated, all alveolates are characterized by a type IV cell surface, which suggests that this type of surface was already developed in the last common ancestors of the various alveolates. In contrast, within the heterokonts,

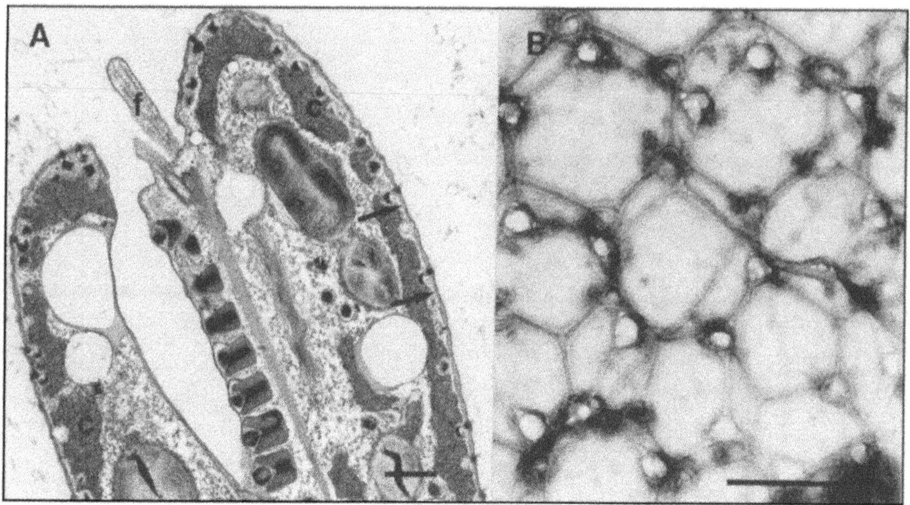

Figure 6.5 The cell surface of cryptomonads
A *Chilomonas paramaecium* thin section (scale bar = 1 µm; arrows indicate the periplast plates)
Source: Courtesy of G. Brugerolle
B *Cryptomonas* sp., whole-mount stain of detergent extracted cells showing the hexagonal pattern of the periplast (scale bar = 1 µm)
Source: Courtesy of K. Hoef-Embden

green algae and choanoflagellates, type I and type III cell surfaces are found. This might indicate that type III surfaces evolved several times independently in these groups. Heterokonts and green algae acquired chloroplasts during their evolution, which changed the host cell metabolism considerably. Carbohydrates became an abundant organic material, which might have allowed the development of type III surfaces as protection against pathogens. Remarkably, only few algal viruses seem able to penetrate a cell wall (Lee *et al.*, 1998) and infect naked spores or gametes.[5] In addition, many polysaccharides derived from algal extracellular matrices were shown to possess antiviral activity (Witvrouw and de Clerq, 1997). Thus an extracellular matrix might have been developed to protect cells against viruses or other pathogens.

Based on the current phylogenetic trees, green algae, glaucophytes and cryptophytes are sister groups, but all three evolutionary lines possess different types of cell surfaces. This could indicate that the three types of cell surfaces evolved after the last common ancestor of these organisms.

(right)
Figure 6.6 Evolutionary trends of types of cell surfaces among flagellates
Above: The topology of an 18S RNA phylogenetic tree
Source: Modified from Sogin, 1997
Below: The major type of cell surface present in various flagellate groups is indicated with an + or the name used for the structure. The grey shade indicates crown groups and the white windows heterokont protists and alveolates, respectively.
In some flagellate groups additional types of cell surfaces are found (e.g. a lorica is present in the euglenophyte *Trachelomonas* (West and Walne, 1980)) which are for simplicity not included in the table

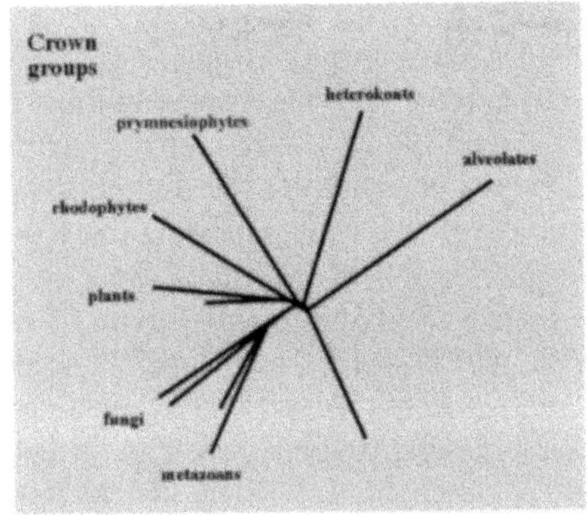

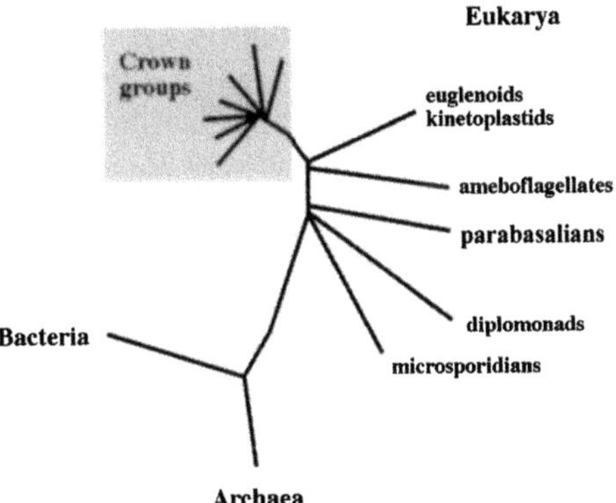

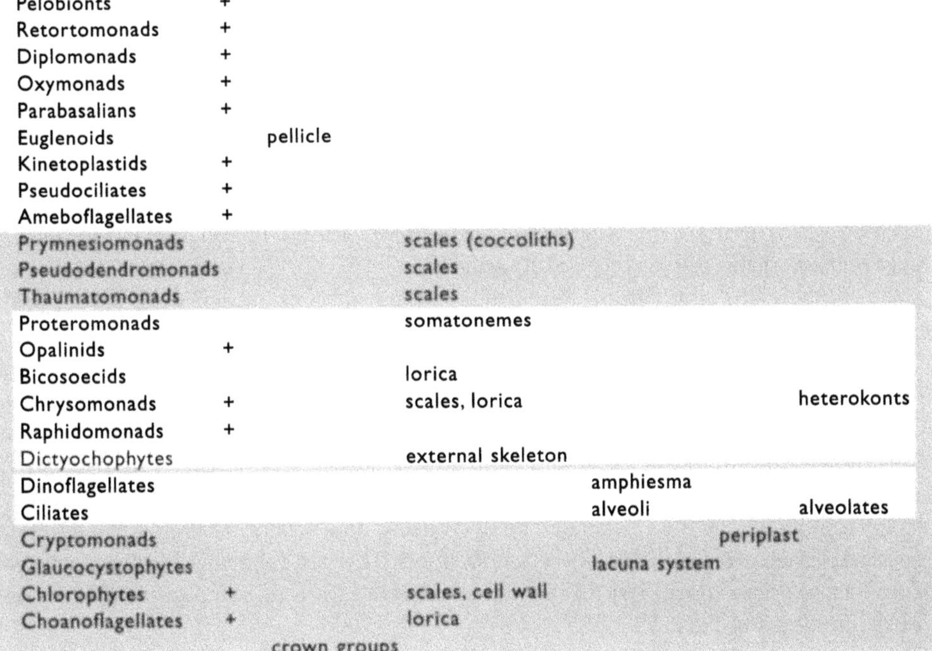

Type of cell surface

	I	II	III	IV	V	
Pelobionts	+					
Retortomonads	+					
Diplomonads	+					
Oxymonads	+					
Parabasalians	+					
Euglenoids		pellicle				
Kinetoplastids	+					
Pseudociliates	+					
Ameboflagellates	+					
Prymnesiomonads			scales (coccoliths)			
Pseudodendromonads			scales			
Thaumatomonads			scales			
Proteromonads			somatonemes			
Opalinids	+					
Bicosoecids			lorica			
Chrysomonads	+		scales, lorica			heterokonts
Raphidomonads	+					
Dictyochophytes			external skeleton			
Dinoflagellates				amphiesma		
Ciliates				alveoli		alveolates
Cryptomonads					periplast	
Glaucocystophytes				lacuna system		
Chlorophytes	+		scales, cell wall			
Choanoflagellates	+		lorica			
		crown groups				

However, are these considerations correct? Most bacteria and Archaea possess a plasma membrane covered by extracellular material, and most flagellates are capable of forming cysts in which the protoplast is covered by an extracellular matrix, thus forming a complex cell surface. For example, the diplomonad *Giardia* forms cysts where the protoplast is covered by a cell wall consisting of carbohydrates and protein components (Luján *et al.*, 1997). Other examples are chrysophytes which form endogenous cysts, covered by a wall predominantly made of silica (Hibberd, 1977), and the cysts of cryptophytes, which possess a cell wall most likely made of polysaccharides (Lichtlé, 1979). The hypnospores and zygotes of dinoflagellates develop by shedding large parts of the amphiesma, yielding a structure called pellicle (Morrill and Loeblich, 1983; Höhfeld and Melkonian, 1992). The structure and the development of the pellicle of the dinoflagellates is still controversial (Morrill and Loeblich, 1983; Höhfeld and Melkonian, 1992), but in both models, extracellular material covers the plasma membrane. According to Höhfeld and Melkonian (1992), the plasma membrane is reinforced by an epiplasmic layer, which is missing in the model of Morrill and Loeblich (1983). In *Euglena* the cells do not form real cysts. However, resting non-motile cells can secrete large amounts of mucilage, which covers the total surface (Triemer, 1980). Thus it seems possible that organisms reflecting the various types of cell surfaces have the intrinsic capability to produce an extracellular matrix.

Furthermore, recent results indicate that the cell surfaces in euglenoids, trypanosomes, ciliates and dinoflagellates might be related in both structural and molecular terms (Bouck and Ngô, 1996). For example, epiplasmic components, which are used to strengthen the cell surface in euglenoids, ciliates and dinoflagellates, are immunologically related. Sequence information available on articulins from *Euglena* (Marrs and Bouck, 1992) and the ciliate *Pseudomicrothorax* (Huttenlauch *et al.*, 1995) indicates that both might form a new protein family. In a recent study, Huttenlauch *et al.* (1998) showed that epiplasmins and articulins are two different protein families, which can both be found in the epiplasm of *Euglena gracilis*, the dinoflagellate *Amphidinium carterae*, and various ciliates. Glycosyl-phosphatidyl-inositol-linked variable surface antigens are present in ciliates and trypanosomes (Bouck and Ngô, 1996) and a surface associated membrane system (alveoli in the case of alveolates, cortical ER in euglenoids and trypanosomes) functions in the storage of calcium (Bouck and Ngô, 1996). If these cell surfaces are really homologous structures, this might have great implications for our understanding of the evolution of the cell surfaces of flagellates.

Interestingly, the assumption that the 'ureukaryote' had no cell wall has recently been challenged by Vellai *et al.* (1998). These authors argue that the energy metabolism of free living bacteria and archaebacteria requires a cell wall. Therefore, complete loss of a cell wall could not occur before the endosymbiosis between an α-purple bacterium, which became later the mitochondrion, and the 'ureukaryote' was established (Vellai *et al.*, 1998). Furthermore, whether the amitochondriate eukaryotes like microsporidia, *Giardia* and *Trichomonas* are early branching eukaryotes has been questioned (Embley and Hirt, 1998). Recent phylogenetic analysis using data sets other than small-subunit ribosomal RNA indicated that these amitochondriates are related to different crown groups, rather than being deep branching as depicted in trees based on small-subunit RNAs (Embley and Hirt, 1998).

Currently, we have not enough information to develop a conclusive model for the evolution of eukaryotic cell surfaces. However, progress in our understanding of the molecular components of the cell surface will help to reveal this topic. Important questions to be addressed are, for example, do articulins from ciliates and *Euglena* really form a protein family, and how widely are articulins distributed within flagellates and other eukaryotes? Cellulose is used by various eukaryotic groups and bacteria (Delmer and Amor, 1995), but did the capacity to synthesize cellulose evolve independently in the various groups, or was it derived from a common ancestor? Only work on such questions might finally lead to a conclusive model of the evolution of eukaryotic cell surfaces.

ACKNOWLEDGMENTS

The author would like to thank G. Brugerolle and K. Hoef-Emden for supplying micrographs, and M. Melkonian for supplying micrographs and critical reading of the manuscript.

NOTES

1 Whether the theca of prasinophytes and the chlamys of chlamydomonadalean taxa are real cell walls can be questioned, since not the whole cell is surrounded by the cell wall and they do not confer osmotic stability to the cell.
2 Based on phylogenetic analysis *S. similis* lost its extracellular matrix during evolution, thus the simple surface type is a secondary development (M. Melkonian, personal communication).
3 The classification and nomenclature of eukaryotes in this review is based on the *Handbook of Protoctista* (eds. Margulis *et al.*, 1990). Phylogenetic trees are based on ssu ribsomal RNA sequences.
4 The crown group is an unresolved assemblage of mitochondriate eukaryotes in phylogenetic trees (based on 18S rRNA), including animals, fungi, green plants, red algae, alveolates and heterokonts. The term crown group was introduced by Knoll (1992).
5 At least in brown algae, viruses do not penetrate a cell wall, and the penetration of the host wall by the *Chorella* virus has been discusses as an exception (Lee *et al.*, 1998). Viruses have been frequently observed in other algal groups too; however, the mode of infection was in most cases not established.

REFERENCES

Adair, W. S. and Snell, W. J. (1990) *The Chlamydomonas reinhardtii* cell wall: structure, biochemistry and molecular biology. In *Organization and Assembly of Plant and Animal Extracellular Matrix* (eds W. S. Adair and M. Mecham), San Diego: Academic Press, pp. 15–84.

Becker, B. and Melkonian, M. (1996) The secretory pathway of protists: spatial and functional organization and evolution. *Microbiological Reviews,* **60**, 697–721.

Becker, B., Marin, B. and Melkonian, M. (1994) Structure, composition and biogenesis of prasinophyte cell coverings. *Protoplasma,* **181**, 233–244.

Bouck, G. B. and Ngô, H. (1996) Cortical structure and function in euglenoids with reference to trypanosomes, ciliates and dinoflagellates. *International Review of Cytology,* **169**, 267–318.

Brett, S. J., Perasso, L. and Wetherbee, R. (1994) Structure and development of the cryptomonad periplast: a review. *Protoplasma,* **181**, 106–122.

Brown, R. M. Jr and Romanovicz, D. K. (1976) Biogenesis and structure of the Golgi-derived cellulosic scales in *Pleurochrysis*. I. Role of the endomembrane system in scale assembly and exocytosis. *Applied Polymer Symposium*, **28**, 537–585.

Cavalier-Smith, T. (1991) The evolution of cells. In *Evolution of Life* (eds S. Osawa and T. Honjo), Tokyo: Springer-Verlag, pp. 271–304.

Cavalier–Smith, T. (1993) Kingdom Protozoa and its 18 phyla. *Microbiological Reviews*, **57**, 953–994.

Cooper, J. M. (1997) *The Cell: A Molecular Approach*. Washington, D.C.: ASM Press.

Delmer, D. P. and Amor, Y. (1995) Cellulose biosynthesis. *Plant Cell*, **7**, 987–1000.

Duszenko, M. and Seyfang, A. (1993) Endocytosis and intracellular transport of variant surface glycoproteins in trypanosomes. *Advances in the Cell Biology of Membranes*, **2**, 227–258.

Embley, T. M. and Hirt, R. P. (1998) Early branching eukaryotes? *Current Opinion in Genetics and Development*, **8**, 624–629.

Faber, W. W. Jr and Preisig. H.–R. (1994) Calcified structures and calcification in protists. *Protoplasma*, **181**, 78–105.

Glas-Albrecht, R., Kaesberg, R., Knoll, G., Allmann, K., Pape, R. and Plattner, H. (1991) Stimulus-secretion coupling in *Paramecium* cells. *European Journal of Cell Biology*, **55**, 3–16.

Heimann, K., Becker, B., Harnisch, H., Mukherjee, K. D. and Melkonian, M. (1997) Biochemical characterization of plasma membrane vesicles of *Cyanophora paradoxa*. *Botanica Acta*, **110**, 401–410

Hibberd, D. J. (1977) Ultrastructure of cyst formation in *Ochromonas tuberculata* (Chrysophyceae). *Journal of Phycology*, **13**, 309–320.

Höhfeld, I. and Melkonian, M. (1992) Amphiesmal ultrastructure of dinoflagellates: a re-evaluation of pellicle formation. *Journal of Phycology*, **28**, 82–89.

Huttenlauch, I., Geisler, I., Plessmann, U., Peck, R. K., Weberm, W. and Stick, R. (1995). Major epiplasmatic proteins of ciliates are articulins: cloning, recombinant expression, and structural characterization. *Journal of Cell Biology*, **130**, 1401–1412.

Huttenlauch, I., Peck, R. K. and Stick, R. (1998) Articulins and epiplasmins: two distinct classes of cytoskeletal proteins of the membrane skeleton in protists. *Journal of Cell Science*, **111**, 3367–3378.

Kloetzel, J. A. (1991) Identification and properties of plateins, major proteins in the cortical alveolar plates of *Euplotes*. *Journal of Protozoology*, **38**, 392–401.

Knoll, A. H. (1992) The early evolution of eukaryotes: a geological perspective. *Science*, **256**, 622–627.

Lee, A. M., Ivy, R. G. and Meints, R. H. (1998). The DNA polymerase gene of a brown algal virus: structure and phylogeny. *Journal of Phycology*, **34**, 608–615.

Lichtlé, C. (1979). Effects of nitrogen deficiency and light of high intensity on *Cryptomonas rufescens* (Cryptophyceae). I. Cell and photosynthetic apparatus transformations and encystment. *Protoplasma*, **101**, 283–299.

Lodish, H., Baltimore, D., Berk, A., Zipursky, S. L., Matsudaira, P. and Darnell, J. (1995) *Molecular Cell Biology*. New York: W. H. Freeman.

Ludwig, M., Lind, J. L., Miller, E. A. and Wetherbee, R. (1996) High molecular mass glycoproteins associated with the siliceous scales and bristles of *Mallomonas splendens* (Synurophyceae) may be involved in the cell surface development and maintenance. *Planta*, **199**, 219–228.

Luján, H. D., Mowatt, M. R. and Nash, T. E. (1997) Mechanisms of *Giardia lamblia* differentiation into cysts. *Microbiology and Molecular Biology Reviews*, **61**, 294–304.

Margulis, L. (1996). Archaeal-eubacterial mergers in the origin of eukarya: phylogenetic classification of life. *Proceedings fo the National Academy of Science USA*, **93**, 1071–1076.

Margulis, L., Corliss, J. O., Melkonian, M. and Chapman, D. J. (1990) *Handbook of Protoctista*. Boston: Jones and Bartlett.

Marrs, J. A. and Bouck. G. B. (1992) The two major membrane skeletal proteins articulins of *Euglena gracilis* define a novel class of cytoskeletal proteins. *Journal of Cell Biology*, **118**, 1465–1475.

Morrill, L. C. and Loeblich III, A. R. (1983) Ultrastructure of the dinophyte amphiesma. *International Review of Cytology*, **82**, 151–180.

Perasso, L., Ludwig, M. and Wetherbee, R. (1997) The surface periplast component of the protist *Komma caudata* self-assembles from a secreted high-molecular mass polypeptide. *Protoplasma,* **200**, 186–197.

Preisig, H. R.and Melkonian, M. (1984) A light and electron microscopical study of the green flagellate *Spermatozopsis similis* spec. nova. *Plant Systematics and Evolution*, **146**, 57–74.

Preisig. H. R., Anderson, O. R., Corliss, J. O., Moestrup, Ø., Powell, M. J., Roberson, R. W. and Wetherbee, R. (1994) Terminology and nomenclature of protist cell surfaces structures. *Protoplasma,* **181**, 1–28.

Santore, U. J. (1983) Flagellar and body scales in the Cryptophyceae. *British Phycological Journal,* **18**, 239–248.

Sekino, K., Kabayashi, H. and Shiraiwa, Y. (1996) Role of coccoliths in the utilization of inorganic carbon by a marine unicellular coccolithophorid *Emiliana huxleyi*: a survey using intact cells and protoplasts. *Plant Cell Physiology*, **37**, 123–127.

Sogin, M. L. (1997) History assignment: when was the mitochondrion founded? *Current Opinion in Genetics and Development*, **7**, 792–799.

Tetaut, E., Barrett, M. P., Bringaud, F. and Baltz, T. (1997) Kinetoplastid glucose transporters. *Biochemical Journal,* **325**, 569–580.

Triemer, R. E. (1980). Role of golgi apparatus in mucilage production and cyst formation in *Euglena gracilis* (Euglenophyceae). *Journal of Phycology*, **16**, 46–52.

Vellai, T., Takacs, K. and Vida, G. (1998). A new aspect to the origin and evolution of eukaryotes. *Journal of Molecular Evolution*, **46**, 499–507.

West, L. K. and Walne, P. L. (1980). *Trachelomonas hispida* var. *coronata* (Euglenophyceae): III. Envelope elemental composition and mineralization. *Journal of Phycology*, **16**, 582–591.

Williams, N. E., Honts, J. E., Lu, Q., Olson L. C. and Moore, K. C. (1989). Identification and localization of major cortical proteins in the ciliated protozoan, *Euplotes eurystomus*. *Journal of Cell Science*, **92**, 433–439.

Wilson, N. F., Foglesong, M. F. and Snell, W. J. (1997) The *Chlamydomonas* mating type plus fertilization tubule, a prototypic cell fusion organelle: isolation, characterization and *in vitro* adhesion to mating type minus gametes. *Journal of Cell Biology,* **137**, 1537–1553.

Witvrouw M. and De Clercq, E. (1997). Sulfated polysaccharides extracted from sea algae as potential antiviral drugs. *General Pharmacology*, **29**, 497–511.

Zelasny, A. M., Shaish, A. and Pick, U. (1995). Plasma membrane sterols are essential for sensing osmotic changes in the halotolerant alga *Dunaliella. Plant Physiology*, **109**, 1395–1403.

Chapter 7

Sensory mechanisms

Phototaxes and light perception in algae

Hiroshi Kawai and Georg Kreimer

ABSTRACT

Most photosynthetic flagellates have evolved various photo-orientation mechanisms after the acquisition of chloroplasts. For phototactic responses, eyespot apparatuses (EAs) are usually involved in light perception. As the endosymbiosis of chloroplasts occurred multiple times, the EAs are polyphyletic. We classified known EAs of algae into the following types considering the phylogenetic origins and the fine structural features: first, Chlorophyte type; second, Cryptophyte type; third, Euglenophyte type; fourth, Phaeophyte type; fifth, Eustigmatophyte type; sixth, Dinoflagellate type (five subtypes). In addition we discuss and summarize aspects of signalling elements and flagellar responses involved in photoresponses in the green algae.

7.1 Introduction

A large proportion of flagellates, including some non-photosynthetic flagellates, zooids of some fungi, and other protists, show various kinds of photo-orientation responses, movements induced by light stimulation (Doughty, 1991; Kuhlmann, 1998 for reviews). For these organisms, it is essential to detect strong UV irradiance and escape from it to avoid lethal damage. Moreover, for photosynthetic flagellates and motile reproductive cells of benthic algae, it is also essential to detect the direction and intensity of the irradiation in order to move to places where favourable light conditions for photosynthesis are available.

Light-induced responses are generally classified into the following three types (Diehn *et al.*, 1977): first, photokinesis; second, photophobic or photoshock response; third, phototactic response (true phototaxis).

Photokinesis is a response in which the rate of movement or the frequency of directional changes is affected by the absolute magnitude of the stimulating light. Photokinesis is known in a number of photosynthetic flagellates; however, detailed investigations are mainly restricted to gliding movements of non-flagellate cells (Häder and Hoiczyk, 1992). In most cases, action spectra of photokinetic responses imply that photosynthetic pigments (for example chlorophyll *a*, β-carotene and phycobilins) are involved in photoreception. Therefore, the photokinetic responses are generally considered to have evolved along with the differentiation of photosynthetic pigments.

Photophobic responses are caused by a rapid change in light intensity, irrespective of the direction of light. Typical responses consist of a stop response followed by a change of the direction of movement. The photoreceptive pigments of photophobic responses are generally not photosynthetic pigments, but carotenoids and flavins common to the phototactic responses (discussed later).

In phototaxis, in contrast to photokinesis and photophobic response, the orientation of the movement is affected by the direction and intensity of the stimulating light. An orientation of the cell towards the light is classified as positive, and away from it as negative, phototaxis. As the origin of photosynthetic protists is polyphyletic (Bhattacharya *et al.*, 1992; Melkonan, 1996), the origin of their photoreceptive structures and photoreceptors also reflects numerous parallel evolutionary processes. Whereas in some groups phototactic and photophobic responses are most likely initiated by excitation of a single photoreceptor (for example green algae; Kröger and Hegemann, 1994), others use presumably several different photoreceptors for these responses. This review will be focused on the diversity and mechanisms of the phototactic responses in flagellate algae.

7.2 Diversity of action spectra and photoreceptive pigments

The first element involved in photo-orientation responses is a receptor with its chromophoric group, which absorbs the light and initiates signal transduction events leading finally to the movement response. In elucidating the receptor mechanisms and identifying the photoreceptive pigments, action spectroscopy is a primary technique (Lipson, 1995). The action spectrum (relative effectiveness of each stimulating light wavelength for the response rate) theoretically agrees with the absorption spectrum of the photoreceptive pigment. Therefore, comparison of an action spectrum with the absorption spectrum of known pigments can reveal candidates for photoreceptive pigments of a particular photomovement. However, distortion of the spectra by screening pigments accumulated in the photoreceptive structures may occur (Foster and Smyth, 1980). Thus in some cases, threshold action spectra are necessary to identify the photoreceptive pigments.

Watanabe (1995) classified the reported action spectra of photomovements and photomorphogenetic responses into the following five types (Figure 7.1): first, 'UV-B C type', for which the pigment(s)/receptor(s) are not yet identified; second, '(UV-B C), UV-A, Blue type', for which flavins and pterins are discussed as plausible chromophores of the putative receptor(s); third, 'Green Yellow type', for which rhodopsins are established as the photoreceptors; fourth, 'Red/Far-Red type', for which phytochrome is the photoreceptor; and fifth, high irradiance responses ('HIR type'), where the involvement of phytochromes is also suggested.

Although there remain some major algal groups from which no action spectrum data are available (for example Haptophyta, Chlorarachniophyta), the published data suggest the following characteristics. First, in the cyanobacteria most of the classified types are observed. Second, most algal groups exhibit '(UV-B C), UV-A, Blue type' responses. Third, 'UV-B C type' responses are currently only reported for Prasinophyceae, a class of the Chlorophyta. Fourth, 'Green Yellow type' responses occur in Chlorophyta and Cryptophyta, although the latter have peaks at longer

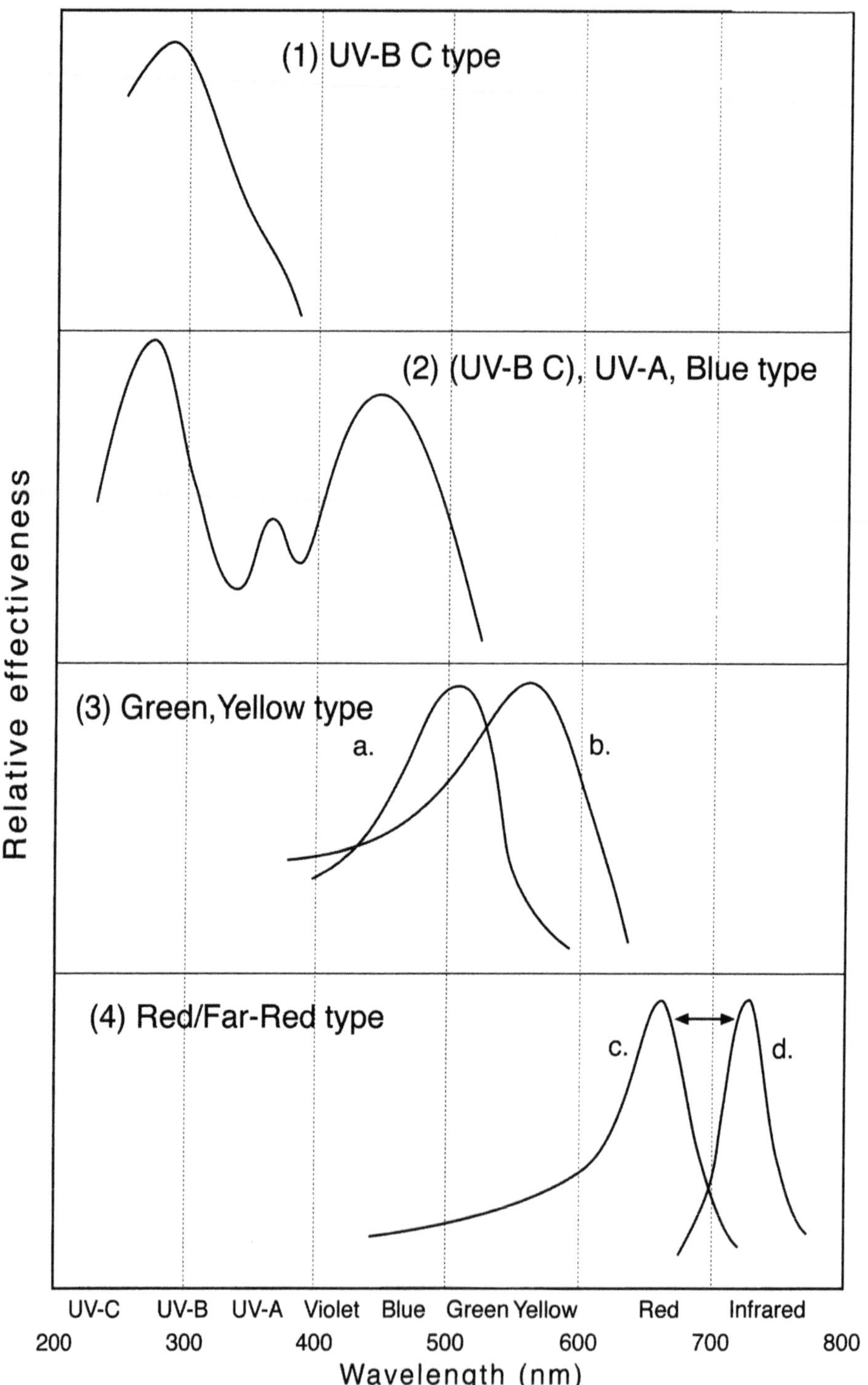
(1) UV-B C type
(2) (UV-B C), UV-A, Blue type
(3) Green, Yellow type
a.
b.
(4) Red/Far-Red type
c.
d.
Relative effectiveness
UV-C UV-B UV-A Violet Blue Green Yellow Red Infrared
200 300 400 500 600 700 800
Wavelength (nm)

wavelengths (yellow) than the former (blue to green). Fifth, 'Red/Far-Red type' responses are known in some Chlorophyta and Dinophyta. However, Euglenophyta also show sensitivity in this wavelength range in their photomorphogenesis responses (Watanabe, 1995). As the '(UV-B C), UV-A, Blue type' responses (often also described as blue-light responses) are seen in most groups, they are considered as a universal photoreceptive system that evolved in a very early stage of evolution. However, parallel evolutionary processes cannot be excluded until all receptors of this type are identified on a molecular basis. In contrast, 'Green Yellow type' responses and 'Red/Far-Red type' responses are probably due to more recently evolved receptor systems, possibly as a consequence of the chloroplast-acquisition by endosymbiotic events in some lineages. Since red/infrared light hardly penetrates in underwater habitats beyond a few metres depth, this type should have evolved in the algal groups that entered shallower water areas.

7.3 Structural diversity, classification and function of eyespot apparatuses

The structures used for photoreception in phototaxis include at least one photoreceptor in a well-defined location, often in conjunction with light-modifying structures. When combined, these are termed eyespot apparatuses (EAs). EAs are diverse in both structure and function, reflecting the independent evolution of a variety of advanced photoreception systems.

7.3.1 Basic functional demands

In general, EAs must operate over a wide range of light intensities and incidences, and they must discriminate sufficiently between the wavelengths used by the photoreceptor(s) and the fluctuating, diffuse background illumination. EAs can be regarded as optimized directional antennas, which take advantage of basic physical principles such as absorption, interference reflection, and polarization, to enable the cells to determine the ambient light conditions. The light-modifying structures of the EA (the eyespot *sensu stricto*, discussed later) are mainly used for contrast enhancement at the location of the photoreceptor. Structures with high refractive indices and good light-absorbing properties such as pigmented lipid globules are apparently widely utilized for such functions. A close match between the spectral properties of the eyespot and those of the photoreceptor(s) is thus a prerequisite for the performance of these optical devices (Foster and Smyth, 1980; Kreimer, 1994; Hegemann and Harz, 1998).

(left)
Figure 7.1 Major types of photomovement action spectra
(1) UV-B C type
(2) (UV-B C), UV-A, Blue type
(3) Green, Yellow type: a, with major peak in green region; b, with major peak in yellow region
(4) Red/Far-Red type: c and d indicate reversible major peaks
Source: Simplified after Watanabe, 1995. The HIR type in Watanabe's classification has been omitted

As linear progression in most flagellate algae is combined with simultaneous rotation, the mechanisms used must also be compatible with the actual displayed motility pattern; that is, signal generation, processing, response, and return to a new excitable state must occur within the time window determined by the rotational speed of the cell during locomotion. Most flagellate algae appear to take advantage of cell rotation to generate a periodically modulated light signal, which is ultimately used to obtain directional information from the regular scans of the environment. Both the shape of the modulated signal and the duration of light/dark periods depend on several parameters (Hegemann and Harz, 1998). This concept of periodic shading/illumination of the photoreceptor in conjunction with helical movement was proposed early by several authors (Foster and Smyth, 1980; Nultsch and Häder, 1988).

What are the characteristics common to most algal EAs, which therefore appear to be essential for the outlined minimal functional demands? The following attributes are observed in the majority of the EAs.

First, carotenoid-rich lipid globules form the most conspicuous part of the EA, and are commonly termed the stigma or eyespot. Their arrangement varies from irregularly packed to highly ordered hexagonal domains forming the eyespot *sensu stricto*.

Second, usually they are single structures in peripheral positions, most often oriented roughly perpendicular to the axis of the swimming path. When more than one EA is permanently present, an array with a common directionality is formed.

Third, their location is exactly defined with respect to either the flagellar apparatus or the plane of the flagellar beat. Pronounced associations between the EA and microtubules or microtubular roots/bands during interphase are well documented, and apparently ensure consistent EA localization.

7.3.2 Classification, ultrastructure and function in signal modulation

The grouping and classification of EAs has been conducted on the basis of various characteristics: for example, first, the nature of photoreceptive pigments and/or types of action spectra; second, the functional properties of the eyespot *sensu stricto*; and third, ultrastructural features. As the range of photobiological chromophores in nature is small and conservative (Lipson and Horwitz, 1991), a grouping based on the putative chromophores results in only three major lineages: those which use either photosynthetic accessory pigments, retinal pigments, or cryptochromes (Foster and Smyth, 1980).

In addition, three major lineages have been recognized on the basis of eyespot function in light signal modulation (Kreimer, 1994). This grouping refers to the absorbing and reflective properties of the eyespot *sensu stricto*. Dodge proposed two ultrastructural classification systems, one applied for all algal groups (Dodge, 1973), and the other only for dinoflagellates (Dodge, 1984).

The evolution of different EAs probably has occurred after the acquisition of chloroplasts by endosymbiotic events (primary and secondary) in each algal phylogenetic lineage. Based on current knowledge, such endosymbioses took place several times, and hence EAs are apparently polyphyletic. Probably they also reflect different intracellular adaptive requirements after acquisition of photosynthesis. Thus, in principle, any grouping of EAs based on superficial similarities cannot reflect phylogenetic relationships. Here we modify Dodge's classification of EA types considering the features mentioned and the evolutionary backgrounds. The basic types are: first, Chlorophyte type; second, Cryptophyte type; third, Euglenophyte type; fourth, Phaeophyte ype; fifth, Eustigmatophyte type; and sixth, Dinoflagellate type including five subtypes.

Chlorophyte type (Figure 7.2A)

This type corresponds to the 'Type A' of Dodge (1973), but excludes the cryptophycean type. These EAs are not closely associated with a flagellum. The functional EA consists of one to several, usually highly ordered, layers of carotenoid-rich lipid globules within the chloroplast, and specialized, intramembrane particle-rich areas of the chloroplast envelope and the plasma membrane overlying the globules (the eyespot membranes (Melkonian and Robenek, 1984)). The retinal-based photoreceptor is most likely located in the plasma membrane patch of the EA (Deininger *et al.*, 1995). The eyespot membranes are often attached to each other by a fuzzy fibrillar material. The globules are laterally tightly associated and are hexagonally close-packed. This association is even maintained to a high degree in isolated EAs (Kreimer *et al.*, 1991a). The average globule diameter varies between 80 and 130 nm. They are enriched in specific carotenoids (Grung *et al.*, 1994). Depending on the number of globule layers, it is possible to differentiate single, double and multilayered eyespots. The spacing of the layers is extremely constant. As will be outlined shortly, the fixed position of the photoreceptor relative to the eyespot, the absorption properties and the spacing of the eyespot layers are important for the contrast-enhancing mechanism. Each globule layer is usually subtended by a thylakoid. Several subtypes have been introduced based on globule-layer arrangement and association with thylakoids (Melkonian and Robenek, 1984).

The shape of the chlorophyte EA can range from ovoid (most common) to comma-shaped, and its area ranges from about 0.3 μm^2 to 10 μm^2 in different species. EAs exhibit different surface geometry and can cause a slight protrusion of the cell surface. They are located at the cell surface in the anterior, median (most often) or posterior half of the cell. Although the EA can be positioned in the plane of the flagellar beat, a clockwise displacement by 20–45° from this plane is more common. Prominent associations with the flagellar root system are well known (Melkonian and Robenek, 1984).

Cell rotation in conjunction with light absorption by cellular constituents leads, depending on the deviation from the light source, to a periodically modulated photoreceptor excitation. The modulation pattern is strongly affected by different parameters such as EA position and the scan and tracking angles (Foster and Smyth, 1980; Hegemann and Harz, 1998). The eyespot greatly enhances the precision of the phototactic orientation by increasing the front-to-back contrast (Morel-Laurens and Feinleib, 1983; Kreimer *et al.*, 1992). Determination of the threshold for rhodopsin-triggered photocurrents established total contrast values of up to eight (Harz *et al.*, 1992). The pigment compositions of isolated EAs and highly purified globules underline these shielding properties (Grung *et al.*, 1994; Backendorf and Kreimer,

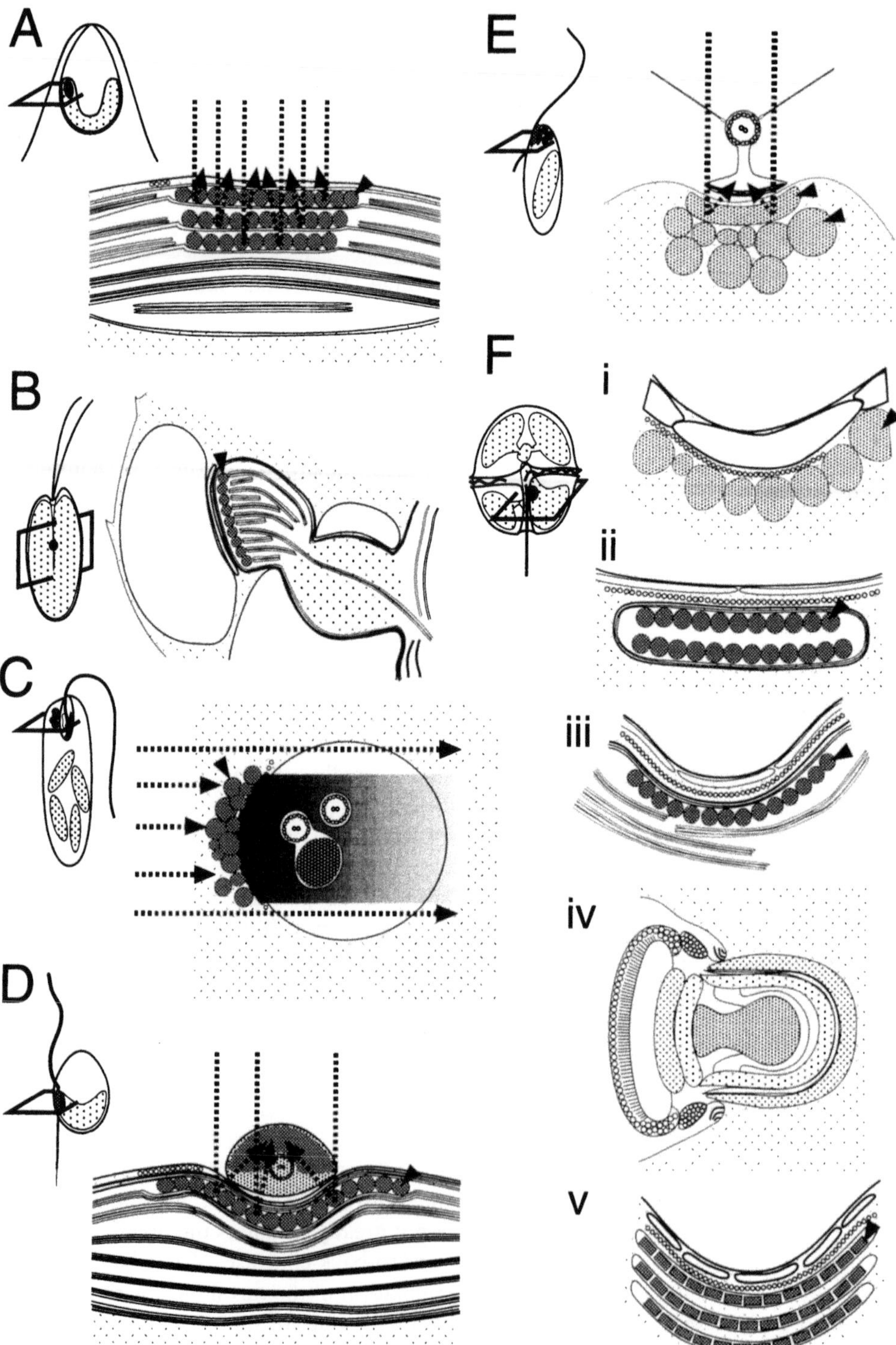
A
B
C
D
E
F
i
ii
iii
iv
v

unpublished results). High contrast is achieved by the eyespot through its combined function as absorbance screen and interference reflector. The exactly-spaced alternating layers of high and low refractive indices are the basis for its function as a quarter-wave interference reflector (Foster and Smyth, 1980). Light intensity and illumination time at the location of the photoreceptor are increased by positive interference reflection for phototactically active light striking the EA from outside. Reflection thus extends the sensitivity to lower fluence rates. Maximal reflection for green light is observed when the EA surface is roughly perpendicular to the light source, and reflection decreases with changing angles of light incidence. For light falling through the cell on the EA, absorption by cell constituents and the eyespot, as well as back reflection, will largely suppress excitation of the photoreceptor.

These properties have been analyzed by confocal reflection microscopy in different green algae, and the results confirmed the theory of Foster and Smyth (Kreimer and Melkonian, 1990; Kreimer *et al.*, 1992; Kreimer, 1994). Eyespot reflection peaks at 540 to 550 nm (Schaller and Uhl, 1997). This coincides well with the EA absorption properties (discussed earlier).

An intact eyespot is important for the overall spectral characteristics and reflectance of the directional antenna (Kreimer *et al.*, 1992). General reflectance of the EA is also affected in a complex manner by a series of additional parameters, for example the overall geometry of the EA, the angle of light incidence, and the pitch angle of the helical swimming path. The latter itself is affected in some species by the light intensity. The pattern of reflectance changes has recently been calculated for some of these parameters, and points to interesting properties of the eyespot under these conditions (Hegemann and Harz, 1998). Of special importance are the calculated effects on colour modulation at low angles of light incidence. Briefly, reflectance with a good colour match to rhodopsin was calculated for perpendicular light incidence. However, at low angles the eyespot reflectance is predicted to exhibit a blue shift, whereas the reflectance of light of increasing wavelengths should be reduced. As was suggested by these authors, the results may well explain both the prominent blue-side peak as well as the sharp red drop in many action spectra of *C. reinhardtii*. Colour modulation is thus of importance for action spectroscopy. It may, however, also be important for a photocycle of the receptor (Kreimer, 1994). If a

(left)
Figure 7.2 Various types of eyespot apparatuses (EAs)
A Chlorophyte type
B Cryptophyte type
C Euglenophyte type
D Phaeophyte type
E Eustigmatophyte type
F Dinoflagellate types:
i *Woloszynskia coronata* type
ii Chromophyte-symbiont type
iii Peridinin-containing-group type
iv *Warnowia* type
v *Gymnodinium natalense* type
Dotted arrows indicate stimulation lights (reflecting in A, D and E, and shading in C)
Arrowheads indicate eyespot globules

cycle is indeed present, it must apparently be coupled both to the rotational speed of the cells and to the light-modifying properties of the eyespot.

Cryptophyte type (Figure 7.2B)

This type of EA is found only in some members of the Cryptophyta, although members without EAs are also phototactic (Erata *et al.*, 1995). The EA of the Cryptophyceae is located in a conical lobe of the chloroplast (Dodge, 1973). The thylakoids in the eyespot are oriented perpendicular to the globule layer. It has been suggested that the photoreceptor in Cryptophyceae is a phycoerythrin and/or a phycocyanin (Watanabe and Furuya, 1974; Foster and Smyth, 1980), but additional pigment(s) appear to be involved in their photoresponses (Erata *et al.*, 1995). Experimental data about the exact mechanisms by which these EAs function are lacking. However, rotational motion of the cells is again essential for the reaction pattern (Watanabe and Furuya, 1982).

Euglenophyte type (Figure 7.2C)

The EAs of Euglenophyta are composed of a paraxonemal body (PAB) and carotenoid-rich eyespot globules located in the cytoplasm. The eyespot globules are membrane-bound. No hexagonal packing of the globules is observed. Their diameters range between 240 nm and 1200 nm (Rosati *et al.*, 1991). The exact position of the eyespot globules is somewhat variable among species, but is invariably roughly opposite the PAB. The PAB is usually located on the emergent locomotive flagellum, 2–3 µm above the flagellar base, and appears to be fixed by filaments of the paraxonemal rod (Rosati *et al.*, 1991). It has an ovoid shape and forms hook- or nose-like projections, with a unique orthogonal paracrystalline organization composed of monoclinic or slightly distorted hexagonal unit cells. The PAB shows dominant fluorescence compared to the flagellum (Benedetti and Checucci, 1975; Sineschekov *et al.*, 1994).

Indications for a photocycle of a pigment in the PAB have recently been obtained (Barsanti *et al.*, 1997). The photoreceptor(s) under discussion are flavins/pterins and a rhodopsin (Kreimer, 1994). The eyespot globules of all euglenoids analyzed so far do not significantly reflect blue-green light, whereas the PAB weakly reflects light (Kreimer, 1994). The receptor for phototactic orientation is most likely located in the PAB. However, different locations for additional photoreceptors must also be considered, because colourless euglenoids missing a PAB also exhibit phobic responses. Also in euglenophytes, complex interactions between the EA and the flagellar/basal body apparatus and associated microtubular roots are thought to be involved in the organization and positioning of the cytoplasmic globules (Dodge, 1973; Rosati *et al.*, 1991).

Phaeophyte type (Figure 7.2D)

This type corresponds the type B EAs of Dodge (1973) found in Chrysophyceae, Xanthophyceae and the motile reproductive cells of many Phaeophyceae. Similar structures are also found in some Haptophyta (Prymnesiophyta), and they have been classified in the same group by previous authors (Dodge, 1973; Kreimer, 1994). The presence of flagellar fluorescent substances in one of the isokont flagella in Haptophyta might support their close relationship (Kawai, 1988, 1992). However, considering the distant phylogenetic relationships revealed by recent molecular data (Bhattacharya *et al.*, 1992), the EAs of Haptophyta (Prymnesiophyta) might have a different phylogenetic origin.

The eyespot is found within a chloroplast in close association with a flagellum (Dodge, 1973; Kawai, 1992). It is usually the shorter posterior flagellum that bears a PAB. The PAB is either positioned in a depression in the eyespot or located together with the eyespot in a groove in the cell surface. Its dimensions and that of the eyespot depression point to a relatively tight fit of both structures. This spatial relationship lends support to the idea that both structures form the functional EA.

The phototactic zoospores and gametes of Phaeophyceae exhibit a green flagellar fluorescence in the posterior flagella throughout its length (Müller *et al.*, 1987; Kawai, 1988, 1992). Similar fluorescence occurs also in the corresponding flagella of many Chrysophyceae, Xanthophyceae (exceptionally in both anterior and posterior flagella in *Botrydiopsis intercedens*), Synurophyceae and Haptophyta (Kawai and Inouye, 1989).

Microspectrofluorometry of detached flagella of phaeophycean zooids showed that both flavin and pterin-like substances are localized in the PAB (Kawai *et al.*, 1996). Analysis of isolated flavins from *Scytosiphon lomentaria* (Phaeophyceae) revealed the presence of 4′,5′-Cyclic FMN (Riboflavin-4′,5′-cyclic phosphate) and a still unidentified photochemically unstable flavin (Yamano *et al.*, 1996). In *Ochromonas danica* (Chrysophyceae), Walne *et al.* (1995) reported the presence of retinal material in isolated PAB preparations. Although the photoreceptor has not yet been conclusively localized in these algae, the autofluorescence data strongly suggest the PAB as its most probable location.

Typically the eyespot is composed of a single layer of carotenoid-rich globules situated directly underneath the chloroplast envelope of a normal or reduced chloroplast. Association with microtubules/microtubular bands and roots occurs, and it has been suggested that this is involved in correct eyespot positioning (Maier, 1997a). No pronounced hexagonal packing is observed, and the globule diameter varies (100 to 500 nm). As in the chlorophyte type, the chloroplast envelope and plasma membrane are closely attached in this area. The PAB contains granular material. This material is neither highly ordered nor paracrystalline. In some Chrysophyceae it is, however, clearly layered. In male gametes of *Ectocarpus siliculosus* (Phaeophyceae) the PAB is asymmetrically structured (Maier, 1997a). Some members (Synurophyceae) possess two PABs and lack an eyespot. Here, however, only the PAB of the non-mastigoneme-bearing flagellum exhibits autofluorescence and contains electron-dense material (Andersen, 1987; Kawai and Inouye, 1989)

Whereas the shape of the PAB is relatively uniform in the Xanthophyceae and Phaeophyceae and the thickest part faces the eyespot, round to wedge/anchor-shaped PABs are found in the Chrysophyceae. Again, peculiar reflective properties have been reported for some of these EAs (Kawai *et al.*, 1990; Kreimer *et al.*, 1991b; Kreimer, 1994). However, due to the presumably dispersed location of the photo-receptor in the PAB of a freely movable flagellum, eyespot function can not relate to quarter-wave interference reflection. Here a strong focusing of the reflected light on the PAB by the concave eyespot surfaces has been observed. Signal strength and shape are again strongly affected by the angle of light incidence. Partial engulfment of the PAB in the depression of the eyespot surface probably results in an additional increase of EA directivity. Phaeophycean zooids require cell rotation for normal phototactic responses (Kawai *et al.*, 1990).

Eustigmatophyte type (Figure 7.2E)

Eustigmatophyceae are members of the Chromophyta *sensu stricto* based on molecular phylogenetic data (Andersen *et al.*, 1998). However, the flagella and EAs of their zoospores have unique features (Hibberd, 1990). Although the EA has morphological similarities with the euglenophyte type, clear differences in the structure of the eyespot and the PAB exist between the two types. In the eustigmatophycean zoospores the eyespot is prominent and occupies almost the whole anterior part of the cell. The PAB is always located at the very base of the long flagellum, and has a T-shaped appearance in transverse sections. Numerous eyespot globules of variable diameters (not membrane-bound) are located in the cytoplasm. There is a large D-shaped globule in the middle of the globules, and its crescent surface faces toward the large surface of the T-shaped PAB. The bulk of the eyespot globules do not reflect blue/green light. Only signals presumably originating from the D-globule were observed. Confocal microscopic analysis supports the view that the concave surface of this globule focuses the light on the PAB. Reflection here also is thus presumably used for enhancement of photoreceptor illumination (Santos *et al.*, 1996). The photoreceptor is not yet known, but its location is very likely the PAB. The PAB contains layered electron-dense material with a paracrystalline appearance at the large surface. This material is completely shielded by the D-globule and the other globules. In some species, this region of the PAB/eyespot shows a weak green autofluorescence (Hibberd, 1990; Santos *et al.*, 1996). No further functional analyses are currently available on this type.

Dinoflagellate types (Figure 7.2F)

EAs are not common, but are found in certain species of Dinophyta. In contrast to those of the algal groups already discussed, the EAs of the Dinophyta show considerable variation. The presence of multiple EA types is most likely due to the fact that the dinoflagellate cells inherited the eyespots from various photosynthetic symbionts having different types of eyespots (Dodge, 1984). However, despite considerable morphological variations, their EAs also show similarities. They are usually located on the ventral side adjacent to the sulcus, or the junction of it with the girdle, just beneath the base of the longitudinal flagellum. The position within the sulcus is somewhat variable. A PAB is missing and the location of the photoreceptor is not yet known. However, the longitudinal flagellum and the eyespot are always in close proximity, suggesting that either the flagellar base and the eyespot, or the overlying plasma membrane and the eyespot, form the functional unit. The EAs reveal structural similarities to the eyespot types already described, and can be subdivided into five types.

WOLOSZYNSKIA CORONATA TYPE (FIGURE 7.2F I)

This type of EA is found in *Woloszynskia coronata* (Crawford and Dodge, 1974) and a species of *Glenodinium* (Kreimer, 1994). Here densely packed carotenoid-rich globules are located in the cytosol in close association with large microtubular strands. The globules are not membrane-bound. Their function has not been studied.

CHROMOPHYTE-SYMBIONT TYPE (FIGURE 7.2F II)

This type of EA is so far found only in the dinoflagellates with a chromophyte (perhaps diatom) endosymbiont (Dodge and Crawford, 1969; Tomas and Cox, 1973; Dodge, 1984; Horiguchi and Pienaar, 1991, 1994a). They are independent structures outside chloroplasts, and are surrounded by a triple-membrane envelope. The globules show a somewhat variable hexagonal packing density with an average diameter of 70 to 100 nm. The eyespot of *Peridinium foliaceum* is double-layered. However, multiple layers are often formed due to the peculiar eyespot structure. In *P. balticum*, multiple layers are generally observed. The eyespot surface is concave towards the flagellum. Consequently a strong focusing on the base of the longitudinal flagellum is observed (Kreimer, 1999). Regarding the origin of this type of EA, it is suggested that the *Peridinium* cells originally had a peridinin-containing-group type of EA (discussed later) within a normal chloroplast. After acquisition of a new endosymbiont of chromophyte origin, the original chloroplasts became reduced, and only the eyespot-containing part was retained and further reduced (Dodge, 1984; Horiguchi *et al.*, 1999). In favour of this hypothesis is the observation that this type of eyespot is surrounded by three layers of membranes, like the chloroplasts possessing the peridinin-containing type eyespot.

PERIDININ-CONTAINING-GROUP TYPE (FIGURE 7.2F III)

This type of EA is more or less common in dinoflagellates, and has morphological similarities with the chlorophyte type. It is found in some species of *Gymnodinium* (Schnepf and Deichgräber, 1972) and *Woloszynskia* (Dodge, 1984; Kreimer, 1994). Here a single layer of somewhat smaller lipid globules (30–110 nm), subtended by a thylakoid, is situated directly underneath the chloroplast envelope. Also here a close hexagonal packing is observed. These eyespots reflect blue/green light and probably use interference reflection in conjunction with absorption for signal modulation (Kreimer, 1994).

WARNOWIA TYPE (FIGURE 7.2F IV)

EAs of this type are complex organelles often also called ocelli, and are found in members of the family Warnowiaceae, oceanic heterotrophic dinoflagellates. As a unique exception among the algal EAs, they possess a simple refractive lens system (hyalosome), which presumably acts as a focusing system for light collection on the melanosome. A detailed structural description is given by Greuet (1987).

GYMNODINIUM NATALENSE TYPE (FIGURE 7.2F V)

This type of eyespot was recorded relatively recently in a tide pool dinoflagellate, *Gymnodinium natalense* (Horiguchi and Pienaar, 1994b). It consists of several layers of hemi-cylindrical sheets including regularly arranged crystalline bricks. Each sheet is membrane-bound. Although no functional aspects are known, the ultrastructure of this eyespot is extremely regular, and suggests a function as reflector. It may focus the light on the flagellum. Analysis of the eyespot ontogeny has revealed that the major components are synthesized in the chloroplast and are transported towards their final position in the sulcal region. Also here a broad microtubular band is observed in front of the eyespot. A similar eyespot was recently recorded in *Amphidinium lacustre*; however, in the latter the chloroplast is not involved in the formation of the bricks constituting the eyespot (Calado *et al.*, 1998). Their phylogenetic relationship and the function are still unclear.

7.4 Signal transduction elements and flagellar responses

Signalling elements and flagellar responses are currently best studied in the green algae. We therefore here restrict our review to this group. In contrast to the flagellar signalling machinery (Witman, 1993), the molecular characterization of elements expected to facilitate and regulate such a finely tuned and highly adaptable response as phototaxis is in its infancy. However, various recent approaches allow a preliminary assessment of this cascade(s).

7.4.1 Flagellar responses

Green algae respond to various external stimuli with well defined changes in the beat pattern of their flagella. Differences in the responses, however, exist between unicellular and colonial forms. Here only unicellular green algae are considered. For recent review of the multicellular Volvocales see Hoops (1997). Light signals are processed rapidly (< 140 ms) and, depending on the intensity, lead to two basic differential responses (Witman, 1993). Taxis is based on subtle differences in the normal asymmetric, breaststroke-type beat pattern of the flagella. In contrast, the stop or shock response is characterized by a transient switch to a symmetrical, undulating waveform that causes the cell to move backward. Both beat changes and general motility strongly depend on extracellular Ca^{2+} and thus finally on transmembrane Ca^{2+} fluxes.

The presence of voltage-gated Ca^{2+} channels in the flagellar membrane of unicellular green algae has been shown by the suction-pipette technique on whole cells. Upon photoshock at least two inward-directed flagellar currents, carried mainly by Ca^{2+}, are triggered in an all-or-none manner by the preceding currents induced by activation of the rhodopsin. These channels appear to be evenly distributed along the flagella. As can be deduced from effects of Ba^{2+}, at least one of these channels appears to be regulated down by increased flagellar matrix Ca^{2+} (see for example, Litvin *et al.*, 1978; Beck and Uhl, 1994; Yoshimura, 1996; Hegemann, 1997). The link between these currents and the photoshock response was established by simultaneous current and optical analysis on single cells (Holland *et al.*, 1997). Effective triggering of flagellar beat reversal by electrical stimulation indicates that the channels involved in this response are voltage-gated (Yoshimura *et al.*, 1997). Whereas the shock response lasts for about 500 to 1000 ms, dim flashes trigger only changes in flagellar beat amplitude and frequency for 20 to 80 ms (Holland *et al.*, 1997). Recording of flagellar currents under these conditions is not yet possible. Apparently only very few flagellar channels are activated under low light conditions.

Only recently the first green algal flagellar channels have been characterized at the single channel level (Hill *et al.*, 2000). A dominating low conductance, Ca^{2+}-permeable channel is probably involved in the peculiar mechanoshock response previously characterized for *Spermatozopsis similis* (Kreimer and Witman, 1994) and not in the photoresponses. In addition to voltage, its open probability is controlled by Ca^{2+}. Recently, four *Chlamydomonas* mutants (*ppr1* to *4*), which appear to be defective only in channel(s) involved in generation of the flagellar currents observed during photoshock, have been isolated. These mutants are not affected in respect to currents originating from the eyespot region or in the phototactic response (Matsuda *et al.*, 1998). This observation further supports the view that different flagellar channels are activated during tactic and shock responses. Hill *et al.* (2000) also briefly reported reconstitution of other flagellar ion channels. Future analyses of these channels in conjunction with the analysis of phototaxis mutants like *ptx 2* and *ptx 8*, which are probably affected in the activation of voltage-gated flagellar Ca^{2+} channels (Pazour *et al.*, 1995) and the *ppr*-mutants, will help to further unravel ion channel function within the flagellar regulatory network.

During undisturbed forward swimming of most green algae, the flagella move in an asymmetric, breaststroke-type beat pattern consisting of an effective and recovery stroke. In free-swimming cells of *C. reinhardtii,* in the absence of a stimulus, the beat frequencies of both flagella are similar. Cells captured with micropipettes, however, exhibit slightly slower (25–40 per cent) frequencies of the *cis* flagellum, that is, the flagellum close to the EA. Analysis of permeabilized cells and isolated axonemes have shown that a differential Ca^{2+}-sensitivity of the axonemes is the basis for this behaviour. At elevated free Ca^{2+} concentrations (10^{-7} to 10^{-6} M) the frequency of the *cis* axoneme is slower, whereas at resting Ca^{2+} levels (10^{-8} to 10^{-9} M) the *cis* axoneme dominates (Kamiya and Witman, 1984; Rüffer and Nultsch, 1987; Holland *et al.*, 1997). A part of the frequency differences is caused by components of the outer dynein arm (Takada and Kamiya, 1997). Upon illumination, *C. reinhardtii* responds with waveform and frequency changes in opposite ways for the *cis* and *trans* flagella. Periodic illumination of the EA during cell rotation thus produces a signal, presumably Ca^{2+}-based (see later), that leads transiently to a more powerful stroke of one flagellum. The flagellar responses to step-up or step-down stimuli are complementary in such a way that they allow the cells a smooth phototactic steering towards or away from the light source (Rüffer and Nultsch, 1991). Thus the developmental difference between the flagella – the *cis* flagellum is usually the immature one – is reflected in a differential sensitivity towards one of the most central signalling molecules in cells.

Often cells exhibit both positive and negative phototactic responses; that is, they must switch the response of the flagella depending on the stimulus intensity. Moreover, some strains of *C. reinhardtii* show, at a constant light intensity, positive taxis for the first few seconds and then switch to negative phototaxis (for example Zacks *et al.*, 1993). It is not yet understood how *cis/trans* flagellar responses are reversed. Adaptational phenomena might be involved in the reversal. It has been shown that the sign of the phototaxis can be reversed from positive to negative by red background illumination, probably by affecting the membrane excitability (Takahashi and Watanabe, 1993). In addition, Rüffer and Nultsch (1991) demonstrated that step-up/step-down responses of the flagella, which are widely accepted as forming the basis of phototactic steering, could be reversed by pre-illumination treatments.

The main Ca^{2+}-sensitive elements responsible for the change in flagellar bending and beat frequency differences are components of the axoneme (Kamiya and Witman, 1984; McFadden *et al.*, 1987). Some elements involved in regulation of the differential axonemal activity have been identified. In the *Chlamydomonas* mutant *ptx1, cis/trans* axonemal differentiation is missing. The *ptx1* mutant is defective in phototaxis but not photoshock, and lacks two 75 kDa axonemal proteins (Horst and Witman, 1993; Rüffer and Nultsch, 1997). Asymmetrical phosphorylation of a 138 kDa component of the *f* dynein complex may represent an additional Ca^{2+}-sensitive element involved in phototactic turning (King and Dutcher, 1997). In the model suggested by these authors, Ca^{2+}-regulated phosphorylation/dephosphorylation of *f* dynein would only occur in the *trans* flagellum. Different Ca^{2+}-modulated type-1 and calcineurin-like phosphatases necessary for this model are present in axonemal preparations (Tash, 1989; Habermacher and Sale, 1996). Other Ca^{2+}-binding proteins such as calmodulin and centrin are also present in flagella (Witman, 1993). Their functions within the light-induced flagellar responses are, however, still not elucidated. Because Ca^{2+} is also intricately involved in regulating the adaptation and sign of phototactic behaviour (Morel-Laurens, 1987), highly linked signalling cascades can be expected. Several phototaxis mutants that are probably defective in the flagella Ca^{2+} responses (Pazour *et al.*, 1995) will help to further elucidate these signalling pathways.

7.4.2 Signal transduction elements in the EA region

Following rhodopsin excitation, opening of ion channels in the EA region is triggered, which in turn initiates signal spread towards the flagella and the different flagellar responses. In most studies, high-intensity flash excitation has been used to analyze the electrical processes downstream of the photoreceptor. It is well established that the electrical events differ with light intensity. The first detectable inward current originates from the EA region and is called the photoreceptor current (P current). This current is graded with light intensity and is mainly carried by Ca^{2+}. Other ions, however, also contribute (Litvin *et al.*, 1978; Holland *et al.*, 1996). High light intensities trigger the P current extremely rapidly. This has led to the suggestion that rhodopsin and the ion channel form a complex (Sineshchekov *et al.*, 1990) or even that the rhodopsin itself forms the ion channel (Deininger *et al.*, 1995).

However, the situation appears to be different in response to low light stimuli. Here existence of a presumably biochemically-amplified current component saturating at low intensities has been deduced from a biphasic dependence of the current amplitudes (Sineschekov and Govorunova, 1999). Recent kinetic analysis of photocurrents in *C. reinhardtii* under low light conditions supports this view. These data point to the possibility that bleaching of a single rhodopsin may activate more than one Ca^{2+}channel in the eyespot region (P. Hegemann and R. Uhl, personal communication). However, it is not yet possible to differentiate between direct rhodopsin-channel interaction and the involvement of additional signalling elements.

The presence of heterotrimeric G-proteins in EA preparations of both *C. reinhardtii* and *S. similis* has been reported. In *S. similis*, different approaches point to the coupling of green light and Ca^{2+}-modulated heterotrimeric G-proteins to the rhodopsin (Hegemann and Harz, 1993; Calenberg *et al.*, 1998). In analogy to other visual systems, G-proteins might thus represent good candidates for such a coupling. Alternatively, they also may be involved in light adaptation by direct control of open probabilities of ion channels, a mechanism known in many systems. For example, some types of voltage-gated Ca^{2+}-channels are inhibited by G-proteins. However, they may also be involved in other rhodopsin-controlled processes. Thus the postulated biochemical amplification cascade under steady-state low light conditions (Kreimer, 1994; Sineshchekov and Govorunova, 1999), although gathering increasing experimental support, still awaits further conclusive experimental evidence.

Signalling processes initiated by flash excitation of the rhodopsin now turn out to be more complex than was initially suggested (Harz *et al.*, 1992). Evidently in the region of the EA, rhodopsin activation triggers additional current components, as can be seen in some published current traces. In addition, Braun and Hegemann (personal communication) have recently identified a new long-lasting current component. This current, following the fast inactivating P current, also originates in the eyespot region. It is not caused by a decrease in the intracellular pH, although it was only resolved after acidification of the bath solution. Its function is not yet clear.

Present current analyses are mainly limited to high-intensity flash excitation, which triggers the shock response. Thus no conclusions about signalling cascades under low light conditions can be drawn from these studies. Although flash excitation represents a rather artificial system, periodic stimulation of the photoreceptor at low intensities and frequencies of about 1–2 Hz could mimic the light signal received by the photoreceptor in a free-swimming cell. Only a few measurements have yet been conducted under the more complex situation of continuous illumination (Sineshchekov and Govorunova, 1999).

Future analysis of the signalling cascade of phototaxis should therefore include not only P current analysis under low intensities, but also current analysis under continuous illumination. However, as green algal photoresponses are affected in a complex manner not only by many extracellular factors, but also by intracellular processes such as photosynthesis and the stage of the cell cycle (Nultsch, 1979; Takahashi and Watanabe, 1993), the analysis will be extremely complex.

Assignment of signalling elements to the cascade initiated by excitation of the rhodopsin is further complicated by complex adaptational processes present in the green algae. Rapid adaptation occurs to bright flashes as well as to longer stimuli (Sineshchekov *et al.*, 1990; Zacks *et al.*, 1993; Zacks and Spudich, 1994; Govorunova *et al.*, 1997) and is also known for flagellar responses induced by mechanical stimulation (Kreimer and Witman, 1994; Yoshimura, 1996). Again the mechanisms involved are not yet known. Govorunova *et al.* (1997) showed that light-induced desensitization is a consequence of reduced membrane excitability rather than of photoreceptor bleaching. These authors point to the importance of depolarization-activated K^+ efflux as an effective way for controlling the cell's photosensitivity. Ca^{2+} appears to be involved in a complex manner in the overall control of the cells photosensitivity.

The kinetics of desensitization by preillumination point to two competitive processes: a light-induced decrease in sensitivity, and a mechanism by which sensitivity is maintained (Zacks *et al.*, 1993). Recovery of full photophobic sensitivity from a completely adapted status is slow and has a $t0.5$ of about one minute (Zacks *et al.*, 1993). In contrast, adaptational mechanisms to changing light conditions that do not trigger a photophobic response are rapid (Zacks and Spudich, 1994). Mechanisms involved in desensitization to phobic stimuli under continuous illumination probably involve modifications of either the receptor and/or ion channels involved in signal transmission. In vertebrate vision, Ca^{2+} plays a central role in adaptational processes; for example, Ca^{2+}-modulated phosphorylation/dephosphorylation of rhodopsin is one of the key elements in this process. It is not yet known whether there are similar mechanisms involved in green algal light adaptation. Rapid and strong changes in protein phosphorylation occurs upon changes in the concentration of free Ca^{2+} between 10^{-8} and 10^{-7} M in isolated EAs (Linden and Kreimer, 1995). Functional analyses were, however, not conducted in this study.

What signalling elements from the EA region are known at the molecular level? Currently, the gene of the photoreceptor is the only characterized element. A series of analyses from different laboratories, initiated by the work of Foster *et al.* (1984), has finally lead to the identification of a rhodopsin as the photoreceptor in green algae. The rhodopsin genes from *Chlamydomonas reinhardtii* and *Vischeria carteri* are very similar in structure (Deininger *et al.*, 1995; Hegemann, 1997). Biochemical and molecular biological analyses also point to the presence of a similar protein in *Spermatozopsis similis* (Calenberg *et al.*, 1998; Rupprich Jakubzik and Kreimer, unpublished results). No homology is observed with the halobacterial opsins. Only the chromophore used (all-*trans*, 6-S-*trans* retinal) and the light-induced isomerization (13-*trans* to *cis*) resemble the archaebacterial system. In some regions of the sequence, homologies to motifs of invertebrate opsins are evident. However, the overall homology is very low, and the two known green algal opsins are characterized by principal differences from all other known opsins. They are highly charged, truncated at both termini, and only four transmembrane domains can be clearly identified (Deininger *et al.*, 1995; Hegemann, 1997).

Although, as was discussed earlier, strong biochemical evidence points to coupling of a G-protein to the rhodopsin, the algal opsin sequences do not allow any clear conclusion with respect to putative reaction partners. All eukaryotic opsins so far known belong to the group of G protein-coupled receptors (GPCR). They possess a DRY/ERY consensus motif involved in G-protein binding. This motif is missing in green algae. However, in other subfamilies of GPCRs this motif is also absent (Birnbaumer and Birnbaumer, 1995). Thus lack of this motif *per se* does not exclude interactions with G-proteins. Also, in the sequence of the first identified putative plant GPCR, only the arginine residue is conserved of this motif (Plakidou-Dymock *et al.*, 1998). Clearly more information is needed about the different components of the signalling cascade initiated upon rhodopsin activation, to provide further insights into this early-developed rhodopsin-based visual system.

ACKNOWLEDGEMENTS

Thanks are given for continuing support by the Deutsche Forschungsgemeinschaft (G.K.). Special thanks are given to Dr E. Henry for improving the English of this paper, Drs P. Hegemann and R. Uhl for sharing unpublished work and sending preprints of their work, Drs T. Horiguchi, B. Mann and M. Melkonian for helpful advice.

REFERENCES

Andersen, R. A. (1987) Synurophyceae classis nov., a new class of algae. *American Journal of Botany*, **74**, 337–53.

Andersen, R. A., Robyn, W. B., Potter, D. and Sexton, J. P. (1998) Phylogeny of the Eustigmatophyceae based upon 18S rDNA, with emphasis on *Nannochloropsis*. *Protist*, **149**, 61–74.

Barsanti, L., Passarelli, V., Walne, P. L. and Gualtieri, P. (1997) *In vivo* photocycle of the *Euglena gracilis* photoreceptor. *Biophysical Journal*, **72**, 545–553.

Beck, C. and Uhl, R. (1994) On the localization of voltage-sensitive calcium channels in the flagella of *Chlamydomonas reinhardtii*. *Journal of Cell Biology*, **125**, 1119–1125.

Benedetti, P. A. and Checucci, A. (1975) Paraflagellar body (PFB) pigments studied by fluorescence microscopy in *Euglena gracilis*. *Plant Science Letter*, **4**, 47–51.

Bhattacharya, D., Medlin, L., Wainwright, P. O., Ariztia, E. V., Biseau, C., Stickel, S. K. and Sogin, M. L. (1992) Algae containing chlorophylls *a* + *c* are paraphyletic: molecular evolutionary analysis of the Chromophyta. *Evolution*, **46**, 1801–1817.

Birnbaumer, L. and Birnbaumer, M. (1995) G proteins in signal transduction. In *Biomembranes: Signal Transduction across Membranes*, vol. 3, (ed. M. Shinitzky), Weinheim: VHC-Verlagsgesellschaft, pp. 153–252.

Calado, A. J., Craveiro, S. C. and Moestrup, Ø. (1998) Taxonomy and ultrastructure of a freshwater, heterotrophic *Amphidinium* (Dinophyceae) that feeds on unicellular protists. *Journal of Phycology*, **34**, 536–554.

Calenberg, M., Brohsonn, U., Zedlacher, M. and Kreimer, G. (1998) Light- and Ca^{2+}-modulated heterotrimeric GTPases in the eyespot apparatus of a flagellate green alga. *The Plant Cell*, **10**, 91–103.

Crawford, R. M. and Dodge, J. D. (1974) The dinoflagellate genus *Woloszynskia*. II. The fine structure of *W. coronata*. *Nova Hedwigia*, **22**, 699–719.

Deininger, W., Kröger, P., Hegemann, U., Lottspeich, F. and Hegemann, P. (1995) Chlamyrhodopsin represents a new type of sensory photoreceptor. *EMBO Journal*, **14**, 5849–5858.

Diehn, B., Feinleib, M., Haupt, W., Hildebrand, E., Lenci, F. and Nultsch, W. (1977) Terminology of behavioral responses of motile microorganisms. *Photochemistry and Photobiology*, **26**, 559–560.

Dodge, J. D. (1973) *The Fine Structure of Algal Cells*. London: Academic Press.

Dodge, J. D. (1984) The functional and phylogenetic significance of dinoflagellate eyespots. *BioSystems*, **16**, 259–267.

Dodge, J. D. and Crawford, R. M. (1969) Observations on the fine structure of the eyespot and associated organelles in the dinoflagellate *Glenodinium foliaceum*. *Journal of Cell Science*, **5**, 479–493.

Doughty, M. J. (1991) Mechanism and strategies of photomovement in protozoa. In *Biophysics of Photoreceptors and Photomovements in Microorganisms*, NATO ASI Series, Ser. A: Life Sciences (eds F. Lenci, F. Ghetti, G. Colombetti, D-P. Häder and P-S. Song), New York/London: Plenum Press, pp. 73–101.

Erata, M., Kubota, M., Takahashi, T., Inouye, I. and Watanabe, M. (1995) Ultrastructure and phototactic action spectra of two genera of cryptophyte flagellate algae, *Cryptomonas* and *Chroomonas*. *Protoplasma*, **188**, 258–266.

Foster, K. W. and Smyth, R. D. (1980) Light antennas in phototactic alga. *Microbiological Reviews*, **44**, 572–630.

Foster, K. W., Saranak, J., Patel, N., Zarilli, G., Okabe, M., Kline, T. and Nakanishi, K. (1984) A rhodopsin is the functional photoreceptor for phototaxis in the unicellular eukaryote *Chlamydomonas*. *Nature*, **311**, 756–759.

Govorunova, E. G., Sineshchekov, O. A. and Hegemann, P. (1997) Desensitization and dark recovery of the photoreceptor current in *Chlamydomonas reinhardtii*. *Plant Physiology*, **115**, 633–642.

Greuet, C. 1987. Complex organelles. In *The Biology of Dinoflagellates* (ed. F. J. R. Taylor), Oxford: Blackwell, pp. 119–142.

Grung, M., Kreimer, G., Calenberg, M., Melkonian, M. and Liaaen-Jensen, S. (1994) Carotenoids in the eyespot apparatus of the flagellate green alga *Spermatozopsis similis*: adaptation to the retinal-based photoreceptor. *Planta*, **193**, 38–43.

Häder, D-P. and Hoiczyk, E. (1992) Gliding motility. In *Algal Cell Motility* (ed. M. Melkonian), New York/London: Chapman and Hall, pp. 1–38.

Habermacher, G. and Sale, W. S. (1996) Regulation of flagellar dynein by an axonemal type-I phosphatase. *Chlamydomonas*. *Journal of Cell Science*, **109**, 1899–1907.

Harz, H., Nonnengässer, C. and Hegemann, P. (1992) The photoreceptor current of the green alga *Chlamydomonas*. *Philosophical Transactions of the Royal Society London Series B*, **338**, 39–52.

Hegemann, P. (1997) Vision in microalgae. *Planta*, **203**, 265–274.

Hegemann, P. and Harz, H. (1993) Photoreception in *Chlamydomonas*. In *Signal Transduction: Prokaryotic and Simple Eukaryotic Systems* (eds J. Kurjan and B. L. Taylor), San Diego: Academic Press, pp. 279–307.

Hegemann, P. and Harz, H. (1998) How microalgae see the light. In *Microbial Responses to Light and Time* (eds M. X. Caddick, S. Baumberg, A. Hodgson and M. K. Phillip-Jones), Society for General Microbiology Symposium 56, Cambridge: Cambridge University Press, pp. 95–105.

Hibberd, D. J. (1990) Phylum Eustigmatophyta. In *Handbook of Protoctista* (eds L. Margulis, J. O. Corliss, M. Melkonian and D. J. Chapman), Boston: Jones and Bartlett, pp. 326–333.

Hill, K., Hemmler, R., Calenberg, M., Kreimer, G. and Wagner, R. (2000) A Ca^{2+}- and voltage-modulated flagellar ion channel is a component of the mechanoshock response in the unicellular green alga *Spermatozopsis*, in press.

Holland, E-M., Braun, F-J., Nonnengässer, C., Harz, H. and Hegemann, P. (1996) The nature of rhodopsin-triggered photocurrents in *Chlamydomonas*. I. Kinetics and influence of divalent cations. *Biophysical Journal*, 70, 924–931.

Holland, E. M., Harz, H., Uhl, R. and Hegemann, P. (1997) Control of phobic behavioral responses by rhodopsin-induced photocurrents in *Chlamydomonas*. *Biophysical Journal*, **73**, 1395–1401.

Hoops, H. J. (1997) Motility in the colonial and multicellular Volvocales: structure, function, and evolution. *Protoplasma*, **199**, 99–112.

Horiguchi, T., Kawai, H., Kubota, M., Takahashi, T. and Watanabe, M. (1999) Phototactic responses of four marine dinoflagellates with different types of eyespot and chloroplast. *Phycological Research*, **47**, 101–107.

Horiguchi, T. and Pienaar, R. N. (1991) Ultrastructure of a marine dinoflagellate, *Peridinium quinquecorne* Abe (Peridiniales) from South Africa with particular reference to its chrysophyte endosymbiont. *Botanica Marina*, **34**, 123–131.

Horiguchi, T. and Pienaar, R. N. (1994a) Ultrastructure of a new marine sand-dwelling dinoflagellate, *Gymnodinium quadrilobatum* sp. nov. (Dinophyceae) with special reference to its endosymbiotic alga. *European Journal of Phycology*, **29**, 237–245.

Horiguchi, T. and Pienaar, R. N. (1994b) Ultrastructure and ontogeny of a new type of eyespot in dinoflagellates. *Protoplasma*, **179**, 142–150.

Horst, C. J. and Witman, G. B. (1993) *ptx1*, a nonphototactic mutant of *Chlamydomonas*, lacks control of flagellar dominance. *Journal of Cell Biology*, **120**, 733–741

Kamiya, R. and Witman, G. B. (1984) Submicromolar levels of calcium control the balance of beating between the two flagella in demembranated models of *Chlamydomonas*. *Journal of Cell Biology*, **98**, 97–107.

Kawai, H. (1988) A flavin-like autofluorescent substance in the posterior flagellum of golden and brown algae. *Journal of Phycology*, **24**, 114–117.

Kawai, H. (1992) Green flagellar autofluorescence in brown algal swarmers and their phototactic responses. *Botanical Magazine, Tokyo*, **105**, 171–184.

Kawai, H. and Inouye, I. (1989) Flagellar autofluorescence in forty-four chlorophyll c-containing algae. *Phycologia*, **28**: 222–227.

Kawai, H., Müller, D. G., Fölster, E. and Häder, D-P. (1990) Phototactic responses in the gametes of a brown alga, *Ectocarpus siliculosus*. *Planta*, **182**, 292–297.

Kawai, H., Nakamura, S., Mimuro, M., Furuya, M. and Watanabe, M. (1996) Microspectrofluorometry of the autofluorescent flagellum in phototactic brown algal zoids. *Protoplasma*, **191**, 172–177.

King, S. J. and Dutcher, S. K. (1997) Phosphoregulation of an inner dynein arm complex in *Chlamydomonas reinhardtii* is altered in phototactic mutant strains. *Journal of Cell Biology*, **136**, 177–191.

Kreimer, G. (1994) Cell biology of phototaxis in flagellate algae. *International Review in Cytology*, **148**, 229–310.

Kreimer, G. (1999) Reflective properties of different eyespot types in dinoflagellates. *Protist*, **150**, 311–323.

Kreimer, G. and Melkonian, M. (1990) Reflection confocal laser scanning microscopy of eyespots in flagellated green algae. *European Journal of Cell Biology*, **53**, 101–111.

Kreimer, G. and Witman, G. B. (1994) Novel touch-induced, Ca^{2+}-dependent phobic response in a flagellate green alga. *Cell Motility and the Cytoskeleton*, **29**, 97–109.

Kreimer, G., Brohsonn, U. and Melkonian, M. (1991a) Isolation and partial characterization of the photoreceptive organelle for phototaxis of a flagellate green alga. *European Journal of Cell Biology*, **55**, 318–327

Kreimer, G., Kawai, H., Müller, D. G. and Melkonian, M. (1991b) Reflective properties of the stigma in male gametes of *Ectocarpus siliculosus* (Phaeophyceae) studied by confocal laser scanning microscopy. *Journal of Phycology*, **27**, 268–276.

Kreimer, G., Overländer, C., Sineshchekov, O. A., Stolzis, H., Nultsch, W. and Melkonian, M. (1992) Functional analysis of the eyespot in *Chlamydomonas reinhardtii* mutant *ey627, mt–*. *Planta*, **188**, 513–521.

Kröger, P. and Hegemann, P. (1994) Photophobia and phototaxis in *Chlamydomonas* are triggered by a single rhodopsin photoreceptor. *FEBS-Letters*, **341**, 5–9.

Kuhlmann, H-W. (1998) Photomovements in ciliated protozoa. *Naturwissenschaften*, **85**, 143–154.

Linden, L. and Kreimer, G. (1995) Calcium modulates rapid protein phosphorylation/dephosphorylation in isolated eyespot apparatuses of the green alga *Spermatozopsis similis*. *Planta*, **197**, 343–351

Lipson, E. D. (1995) Action spectroscopy: methodology. In *CRC Handbook of Organic Photochemistry and Photobiology* (eds W. M. Horspool and P-S. Song), CRC Press, pp. 1257–1266.

Lipson, E. D. and Horwitz, B. A. (1991) Photosensory reception and transduction. In *Sensory Receptors and Signal Transduction, Modern Cell Biology, vol. 10* (eds J. L. Spudich and B. H. Satir), New York: Wiley-Liss, pp. 1–64.

Litvin, F. F., Sineschekov, O. A. and Sineschekov, V. A. (1978). Photoreceptor electric potential in the phototaxis of the alga *Haematococcus pluvialis*. *Nature*, **271**, 476–478.

Maier, I. (1997a) The fine structure of the male gametes of *Ectocarpus siliculosus* (Ectocarpales, Phaeophyceae). I. General structure of the cell. *European Journal of Phycology*, **32**, 241–253.

Maier, I. (1997b) The fine structure of the male gametes of *Ectocarpus siliculosus* (Ectocarpales, Phaeophyceae). II. The flagellar apparatus. *European Journal of Phycology*, **32**, 255–266.

Matsuda, A., Yoshimura, K., Sineshchekov, O. A., Hirono, M., and Kamiya, R. (1998). Isolation and characterization of novel *Chlamydomonas* mutants that display phototaxis but not photophobic response. *Cell Motility and the Cytoskeleton*, **41**, 353–362.

McFadden, G. I., Schulze, D., Surek, B., Salisbury, J. L. and Melkonian, M. (1987). Basal body reorientation mediated by a Ca^{2+}-modulated contractile protein. *Journal of Cell Biology*, **105**, 903–912.

Melkonian, M. (1996) Phylogeny of photosynthetic protists and their plastids. *Verhandlungen der Deutschen Zoologischen Gesellschaft*, **89**, 71–96.

Melkonian, M. and Robenek, H. (1984) The eyespot apparatus of flagellated green algae: a critical review. In *Progress in Phycological Research, vol. 3* (eds F. E. Round and D. J. Chapman), Bristol: Biopress, pp. 193–268.

Morel-Laurens, N. (1987) Calcium control of phototactic orientation in *Chlamydomonas reinhardtii*: Sign and strength of response. *Photochemistry and Photobiology*, **45**, 119–128.

Morel-Laurens, N. and Feinleib, M. E. (1983) Photomovement in an eyeless mutant of *Chlamydomonas*. *Photochemistry and Photobiology*, **37**, 189–194.

Müller, D. G., Maier, I. and Müller, H. (1987) Flagellum autofluorescence and photoaccumulation in heterokont algae. *Photochemistry and Photobiology*, **46**, 1003–1008.

Nultsch, W. (1979) Effect of external factors on phototaxis of *Chlamydomonas reinhardtii*. III. Cations. *Archives of Microbiology*, **123**, 93–99.

Nultsch, W. and Häder, D-P. (1988) Photomovement in motile microorganisms. II. *Photochemistry and Photobiology*, **47**, 837–869.

Pazour, G. J., Sineshchekov, O. A. and Witman, G. B. (1995) Mutational analysis of the phototransduction pathway of *Chlamydomonas reinhardtii*. *Journal of Cell Biology*, **131**, 427–440.

Plakidou-Dymock, S., Dymock, D. and Hooley, R. (1998) A higher plant seven-transmembrane receptor that influences sensitivity to cytokinins. *Current Biology*, **8**, 315–24.

Rosati, G., Verni, F., Barsanti, L., Passarelli, V. and Gualtieri, P. (1991) Ultrastructure of the apical zone of *Euglena gracilis*: photoreceptors and motor apparatus. *Electron Microscopical Reviews*, 4, 319–342.

Rüffer, U. and Nultsch, W. (1987) Comparison of the beating of *cis-* and *trans-*flagella of *Chlamydomonas* cells held on micropipettes. *Cell Motility and the Cytoskeleton*, **7**, 87–93.

Rüffer, U. and Nultsch, W. (1991) Flagellar photoresponses of *Chlamydomonas* cells held on micropipettes: II Change in flagellar beat pattern. *Cell Motility and the Cytoskeleton*, **18**, 269–278.

Rüffer, U. and Nultsch, W. (1997) Flagellar photoresponses of *ptx1*, a nonphototactic mutant of *Chlamydomonas. Cell Motility and the Cytoskeleton*, **37**, 111–119.

Santos, L. M. A., Melkonian, M. and Kreimer, G. (1996) A combined reflection confocal laser scanning, electron and fluorescence microscopy analysis of the eyespot in zoospores of *Vischeria* spp. (Eustigmatales, Eustigmatophyceae). *Phycologia*, **35**, 299–307.

Schaller, K. and Uhl, R. (1997) A microspectrophotometric study of the shielding properties of eyespot and cell body in *Chlamydomonas. Biophysical Journal*, **73**, 1573–1578.

Schnepf, E. and Deichgräber, G. (1972) Über den Feinbau von Theka, Pusule und Golgi-Apparat bei dem Dinoflagellaten *Gymnodinium* spec. *Protoplasma*, **74**, 411–425.

Sineshchekov, O. A. and Govorunova, E. G. (1999) Rhodopsin-mediated photosensing in green flagellated algae. *Trends in Plant Sciences*, **4**, 58–63.

Sineshchekov, O. A., Litvin, F. F. and Keszthelyi, L. (1990) Two components of photoreceptor potential in phototaxis of the flagellated green alga *Haematococcus pluvialis. Biophysical Journal*, **57**, 33–39.

Sineshchekov, V. A., Geiss, D., Sineshchekov, O. A., Galland, P. and Senger, H. (1994) Fluorometric characterization of pigments associated with isolated flagella of *Euglena gracilis*: evidence for energy migration. *Journal of Photochemistry and Photobiology*, **23**, 225–237.

Takada, S. and Kamiya, R. (1997) Beat frequency difference between the two flagella of *Chlamydomonas* depends on the attachment site of outer dynein arms on the outer-doublet microtubules. *Cell Motility and the Cytoskeleton*, **36**, 68–75.

Takahashi, T. and Watanabe, M. (1993) Photosynthesis modulates the sign of phototaxis of wild-type *Chlamydomonas reinhardtii. FEBS-Letters*, **336**, 516–520.

Tash, J. S. (1989) Protein phosphorylation: the second messenger signal transducer of flagellar motility. *Cell Motility and the Cytoskeleton*, **14**, 332–339.

Tomas, R. N. and Cox, E. R. (1973) Observations on the symbiosis of *Peridinium balticum* and its intracellular alga. I. Ultrastructure. *Journal of Phycology*, **9**, 304–323.

Walne, P. L., Passarelli, V., Lenzi, P., Barsanti, L. and Gualtieri, P. (1995) Isolation of the flagellar swelling and identification of retinal in the phototactic flagellate, *Ochromonas danica* (Chrysophyceae). *Journal of Eukaryotic Microbiology*, **42**, 7–11.

Watanabe, M. (1995) Action spectroscopy: photomovement and photomorphogenesis spectra. In *CRC Handbook of Organic Photochemistry and Photobiology* (eds W. M. Horspool and P-S. Song), CRC Press, pp. 1276–1288.

Watanabe, M. and Furuya, M. (1974) Action spectrum of phototaxis in a cryptomonad alga, *Cryptomonas* sp. *Plant Cell Physiology*, **15**, 413–420.

Watanabe, M. and Furuya, M. (1982) Phototactic behaviour of individual cells of *Cryptomonas* sp. in response to continuous and intermittent light stimuli. *Photochemistry and Photobiology*, **35**, 559–563.

Witman, G. B. (1993) *Chlamydomonas* phototaxis. *Trends in Cell Biology*, **3**, 403–408.

Yamano, K., Saito, H., Ogasawara, Y., Fujii, S., Yamada, H., Shirahama, H. and Kawai, H. (1996) The autofluorescent substance in the posterior flagellum of swarmers of the brown alga *Scytosiphon lomentaria. Zeitschrift für Naturforschung (c)*, **51**, 155–159.

Yoshimura, K. (1996) A novel type of mechanoreception by the flagella of *Chlamydomonas. Journal of Experimental Biology*, **199**, 295–302.

Yoshimura, K., Shingyoji, C. and Takahashi, K. (1997). Conversion of beating mode in

Chlamydomonas flagella induced by electric stimulation. *Cell Motility and the Cytoskeleton*, **36**, 236–245.

Zacks, D. N., Derguini, F., Nakanishi, K. and Spudich, J. L. (1993) Comparative study of phototactic and photophobic receptor chromophore properties in *Chlamydomonas reinhardtii*. *Biophysical Journal*, **65**, 508–518.

Zacks, D. N. and Spudich, J. L. (1994) Gain setting in *Chlamydomonas reinhardtii*: mechanism of phototaxis and the role of the photophobic response. *Cell Motility and the Cytoskeleton*, **29**, 225–230.

Chapter 8

Trophic strategies

Michael A. Sleigh

ABSTRACT

Flagellates display all of the basic trophic strategies seen among both autotrophic and heterotrophic eukaryotes. Photoautotrophic flagellates employ a diversity of different pigment combinations in energy capture for photosynthesis. Their wide size range enables them to exploit habitats that are rich or poor in mineral nutrients. Autotrophic members of several groups have established successful and important symbiotic relationships with other protists and higher animals. Many of the studied examples of photosynthetic flagellates are auxotrophs, depending on an external supply of one or more vitamins.

Heterotrophic flagellates are mostly phagotrophs, but osmotrophs are found in several groups, especially among parasites and other symbionts. The organic requirements of osmotrophs vary widely from only a simple energy source to a long list of molecules representing a wide range of the main groups of compounds used in living cells. An osmotrophic flagellate may use active processes to transport food molecules across its surface membrane, either over the general body surface or at a restricted area; uptake of larger soluble molecules by pinocytosis may likewise occur over the whole body surface or only at a specialized site. Phagotrophs may catch and ingest living prey, or feed on debris or fragments of other living organisms.

The main classes of living prey are bacteria, protozoa and algae, including diatoms and motile flagellates. The different methods used to capture these prey include passive diffusion feeding, which depends on movement of the prey, and active interception processes, in which either the predator moves around seeking individual prey items, or the predator collects suspended particles from a water current which it has created. Prey capture often involves the use of extrusive organelles.

Many flagellates are mixotrophs combining photoautotrophy with some form of heterotrophy. The availability of mineral nutrients determines the success of autotrophs, while the availability of suitable prey or dissolved organic molecules determines the distribution and success of heterotrophs. The response of flagellates to fluctuating nutrients varies; both autotrophs and heterotrophs may produce cysts when nutrients are scarce, but other forms may sharply reduce metabolism and produce small cells to survive periods of shortage.

8.1 Introduction, terminology and general principles of flagellate nutrition

Living organisms require energy to drive the chemical reactions necessary for maintenance, growth and other activities characteristic of life. This energy is obtained by two routes, autotrophy and heterotrophy. Light energy is absorbed by photo-autotrophic organisms, and is used to drive reactions in which inorganic molecules are combined to make energy-rich organic molecules in the process of photosynthesis. These molecules may later be broken down to provide the energy needed to power the cellular processes of these organisms. Certain bacteria, known as chemo-autotrophs, gain energy for synthesis of organic molecules from the promotion of inorganic chemical reactions. Heterotrophic organisms take in organic molecules that have been synthesized by autotrophs and break them down to release the energy they need for living. Flagellates may employ either photo-autotrophy or heterotrophy or both of these methods of obtaining energy, and are second only to bacteria among all groups of organisms in the diversity of variant ways used to obtain energy.

Flagellates that are pure autotrophs require no organic molecules from their environment because they are able to synthesize all of the classes of organic molecule needed for their metabolism and growth from simple inorganic molecules and mineral ions, with the aid of energy from photosynthesis. In many cases, however, photo-autotrophic flagellates have found it possible to gain from their environment one or several members of certain classes of essential organic molecules such as vitamins, which are only required in small quantities, and have given up synthesis of these molecules. Autotrophs with such needs for trace quantities of specific organic molecules are referred to as auxotrophs.

Some heterotrophs are able to synthesize most of the molecules they need, but the majority of heterotrophs can synthesize few if any of the smaller organic molecules which form the building blocks of larger molecules involved in their metabolism and in the formation of cell structures. These are therefore dependent on the uptake from their environment of many classes of organic molecule, for use in synthesis of their body structures and enzymes, as well as for the energy needed both for this synthesis and for other essential processes of life. Many flagellates ingest particles of food, including other whole organisms, into internal vacuoles, within which the particles are digested enzymically to release small organic molecules that can be absorbed into the cell cytoplasm. This process of ingestion and subsequent digestion is known as phagocytosis, and organisms which perform it are phagotrophs. Other heterotrophic flagellates live in environments which contain soluble small organic molecules that can be taken in through the cell membrane without further digestion. Organisms that absorb such small molecules from their external environment practice osmotrophic (or saprophytic/saprozoic) nutrition, and organisms using this method of food uptake are osmotrophs (or saprotrophs).

It it often difficult to determine whether an organism is a phagotroph or an osmotroph, for example if it feeds on larger soluble organic molecules that are digested in small (pinocytotic) vacuoles that are not easily seen. It is likely that almost any flagellate may be an opportunistic osmotroph, if suitable molecules are present in its environment, but some are capable of very active uptake by use of specialized transport systems in their membranes. Many flagellates practice both autotrophy and heterotrophy (either phagotrophy or osmotrophy), simultaneously or at different times, and are referred to as mixotrophs.

The cells of all living organisms must contain proteins constructed from about twenty amino acids, carbohydrates based on various hexose and pentose sugars, lipids in which a range of fatty acids attach to specific linking molecules and nucleic acids in which purine and pyrimidine bases are combined with phosphate and ribose sugars, as well as smaller amounts of compounds required as co-factors for enzymes or other molecules. A phagotrophic heterotroph which ingests other whole organisms should obtain all of the classes of small molecules in this list from the food it eats; an ability to synthesize any of them will therefore not be required by such an organism. It is true that the small molecules in the food may not be in the proportions it requires, so that some ability to interconvert molecules within a class can be important. The range of dissolved organic molecules available to an osmotroph may be less balanced, or even quite restricted, and such heterotrophs may show a capacity to synthesize a greater range of small organic molecules, even approaching that shown by autotrophs.

It is the purpose of this chapter to survey the ways in which flagellates exploit these various methods of satisfying their needs for energy and nutrients: their trophic strategies, both physiological variants associated with nutrition, and ecological variants associated with maximizing the use of resources in particular environments. The main modes of nutrition will be considered first, and then some consideration will be given to ecological features that relate to these. Certain aspects of the trophic strategies of flagellates are discussed in other chapters in this book, notably changes adopted by flagellates which have become parasites (Vickerman, Chapter 10), grazing and mixotrophy practiced by flagellates involved in the microbial loop (Laybourn-Parry and Parry, Chapter 11) and links between feeding activities and exploitation of ecological niches by flagellates (Arndt, Chapter 12).

8.2 Autotrophic nutrition among flagellates

The driving force for the process of photosynthesis is the absorption of light energy by a selection of chlorophyll and other pigments, molecules of which are embedded in, or associated with, the thylakoid membranes of photoautotrophs. These pigment molecules are grouped into light-harvesting (or antennal) complexes arranged around reaction centres. The energy possessed by photons of light absorbed by these pigment molecules is passed to a molecule of chlorophyll at the reaction centre, and is there used to energize electrons at one of two stages of photosynthesis. In photosystem II reaction centres, electrons derived from water are given sufficient energy to reduce intermediate electron acceptors. Some of the energy of these electrons is used to drive processes which phosphorylate ADP to ATP as the electrons are passed through a chain of cytochrome molecules, before their energy is boosted at a reaction centre of photosystem I to a level sufficient to reduce NADP molecules. The reduced NADP and ATP are used in the 'light-independent' stages of photosynthesis to fix carbon dioxide in the synthesis of carbohydrate molecules.

Electromagnetic radiations fall into a wide range of wavelengths. A narrow band of wavelengths between about 400 nm (violet) and 700 nm (red) constitutes the spectrum of light visible to the human eye. Photons with shorter wavelengths have higher energy; infra-red radiation, at say 1000 nm or longer, has too little energy to drive biological reactions, while ultra-violet radiation, at say 300 nm or shorter, has so much energy that it can be intensely damaging, but solar radiation in this range is fortunately mostly absorbed by molecules in the earth's atmosphere. Different photosynthetic pigments absorb light photons most strongly at different wavelengths, and the range of pigments present determines the active spectrum of the process in a particular organism. A single pigment may have only one or two narrow peaks of absorption (Table 8.1), but by combining several pigments together in each light harvesting complex, it is possible to make use of energy available over a wider range of the spectrum.

All flagellates contain chlorophyll *a*, but different groups of flagellates vary in their associated pigments (Table 8.2); the green chlorophytes and euglenids contain chlorophyll *b*, while the brown chrysophytes and dinoflagellates have chlorophyll *c* instead. Light harvesting complexes of photosystem I also have carotene molecules, while those of photosystem II have xanthophylls (oxygen-containing carotenoids); again, different flagellate groups have different members of these pigment classes.

The flattened thylakoid sacs, within whose membranes chlorophyll and most related pigments are incorporated, occur in the cytoplasmic matrix of prokaryote cells. In flagellates, as in all other eukaryotes, these thylakoids are enclosed within chloroplast membanes, reflecting the generally accepted ancestry of chloroplasts from symbiotic prokaryotes (see Farmer and Darley, Chapter 15, this volume). A few flagellates are photo-autotrophs because they live in association with cyanobacteria such as Glaucophytes (see Kies, 1980). Other flagellates may establish rather temporary symbiotic associations with ingested photosynthetic eukaryotes, or their chloroplasts (see Gaines and Elbrächter, 1987; Laybourn-Parry and Eccleston-Parry, Chapter 11 this volume). Most photo-autotrophic flagellates belong to groups which have long-established chloroplasts, derived either from prokaryote symbionts (where

Table 8.1 Wavelengths of characteristic absorption peaks of some photosynthetic pigments found in flagellates

Type of pigment	*Solvent*	*Wavelength(s) of peak(s) (nm)*	
Chlorophyll *a*	diethyl ether	420	662
Chlorophyll *b*	diethyl ether	455	644
Chlorophyll *c*	diethyl ether	444	626
α-carotene	hexane	420 440 470	
β-carotene	hexane	425 450 480	
lutein	ethanol	425 445 470	
fucoxanthin	hexane	425 450 475	
phycoerythrin (cryp)	water	274 310	556
phycocyanin (cryp)	water	270 350	583 625 643

Notes:
Peaks may shift somewhat on extraction in organic solvents
cryp denotes cryptophyte
Source: Data mainly selected from Govindjee and Braun, 1974

there are two enclosing chloroplast membranes) or from eukaryote symbionts (where there are four, or sometimes three, membranes) (see Farmer and Darley, Chapter 15 this volume).

Chloroplasts differ not only in the types of pigments they contain and the number of surrounding membranes, but also in the arrangement of the thylakoids within the chloroplast. Phycobilic proteins (phycoerythrin and phycocyanin), characteristic of cyanobacteria, but also present in red algae and cryptophyte flagellates, are water-soluble pigments, and are confined within small vesicles called phycobilisomes associated with the thylakoids. In prokaryotes and red algae the phycobilisomes are found on the surfaces of the single, separate thylakoid sacs, but in cryptophytes the thylakoids occur in pairs with phycobilisomes in the cavity within the thylakoids. In other flagellates all of the pigments are lipophilic and occur within membranes, the thylakoids being grouped in threes in most cases, but in chlorophytes they are found in larger stacks resembling the grana within chloroplasts of land plants.

Light energy absorbed by pigments and used as described above to synthesize ATP and reduced NADP in the chloroplast cytoplasm around the thylakoids is subsequently stored as chemical bond energy in larger molecules. Although it is usual to consider that the primary result of the fixation of carbon dioxide in the cytoplasmic stroma of the chloroplast by the Calvin cycle is the formation of hexose sugars, various intermediates in this cycle can be siphoned off and used in the synthesis of amino acids, fatty acids and other components needed to fulfil the needs of the autotrophic cell for structural, storage, genetic and enzymic or other molecules involved in metabolism.

It is likely that many autotrophic flagellates that synthesize all of the required major classes of organic molecules are in fact auxotrophs which make use of external sources of certain minor organic molecules (vitamins), usually one or more of vitamin B_{12}, thiamine and biotin (Droop, 1957; Provasoli and Carlucci, 1974). A pure autotroph can be grown in a simple inorganic medium axenically, in the complete absence of any other organism, including any parasitic or symbiotic prokaryote. The nature of the organic molecule(s) required by an auxotroph which grows axenically in pure culture has to be determined by trial and error. Thus, the chlorophyte *Chlamydomonas moewusii* and the dinoflagellate *Peridinium inconspicuum* are pure autotrophs with no requirement for any organic compound, but *Euglena gracilis* and *Peridinium foliaceum* are auxotrophs whose only organic requirement is vitamin B_{12}, *Peridinium*

Table 8.2 Principal photosynthetic pigments typical of some flagellate groups

Group	*Chlorophyll*	*Carotene*	*Xanthophylls etc.*
Euglenophyta	*a,b*	*b*	Diadinoxanthin (Neoxanthin)
Cryptophyta	*a,c*	*a*,(*b*)	Alloxanthin (Diatoxanthin) Phycocyanin, Phycoerythrin
Dinophyta	*a,c*	*b*	Peridinin (Diadinoxanthin, Dinoxanthin)
Chrysophyta	*a,c*	*b*	Fucoxanthin, Diatoxanthin
Xanthophyta	*a,c*	*b*	Diadinoxanthin, Diatoxanthin, Heteroxanthin
Chlorophyta	*a,b*	*b*	Lutein, Violaxanthin, Neoxanthin
Prymnesiophyta	*a,c*	*b*	Fucoxanthin, Diatoxanthin

limbatum needs only thiamine, and *Amphidinium carterae* needs vitamin B_{12}, biotin and thiamine (Provasoli, 1958; Holt and Pfiester, 1981).

In general, members of the Chrysophyta, Cryptophyta, Dinophyta and Euglenophyta are more commonly auxotrophs, and members of the Chlorophyta and Xanthophyta are more commonly pure autotrophs (Cox, 1980). The vitamin most commonly required by auxotrophic flagellates is B_{12}.

While the fixation of carbon dioxide can lead to the formation of carbohydrates and fatty acids without additional nutrient input, nitrogen compounds, usually nitrate or ammonium ions, are required by phototrophs to synthesize amino acids, and both nitrogen and phosphorus compounds to synthesize nucleotides. Iron limitation has been shown to control rates of phytoplankton production and biomass in the Pacific Ocean (Martin *et al.*, 1994), and it has been found that grazing protozoa may play a role in making refractory forms of iron available to phytoplankton (Barbeau *et al.*, 1996). Other elements like sulphur, magnesium and zinc are required in small amounts for metabolic purposes, and some flagellates need silica for skeletons or loricas.

The need for mineral nutrients, particularly N and P, has a strong influence on the distribution and abundance of different flagellates. In eutrophic waters, such as those enriched with nutrients by upwelling or pollution, it is possible for the flagellates, even the large species, to multiply and form blooms. For this to happen, there must be a seeding of the bloom-producing organism into the nutrient-rich water from cysts or through transport by currents, and the autotrophs must multiply more quickly than any grazing microzooplankton that prey upon them. In general, the predators of small flagellates can respond more quickly than predators of large flagellates to an increase in numbers of their prey, so the blooms of larger forms may be long-lasting.

However, the production of toxic products by some bloom-forming species may also give them an advantage in competition with other species. The 'red-tides' of various dinoflagellates are probably the most spectacular and best known of blooms caused by flagellates (Taylor and Seliger, 1979; Steidinger and Baden, 1984; Taylor and Pollingher, 1987). Extensive blooms of prymnesiophytes such as toxic species of *Chrysochromulina* or the glutinous colonial *Phaeocystis* also develop in coastal waters, and massive, long-lived blooms of the coccolithophorids *Emiliania huxleyi* or *Coccolithus pelagicus* can occur at rather lower levels of nutrients (see for example Tarran *et al.*, 1999).

At the other end of the nutrient scale, in the oligotrophic tropical oceans, the minute cyanobacterium *Prochlorococcus* dominates the primary producers in surface waters, but the larger cyanobacterium *Synechococcus* and a number of species of picoeukaryotes, mostly flagellate chrysophytes (such as Pelagophyceae, only 2–3 µm in size: Andersen *et al.*, 1993), chlorophytes, prasinophytes and prymnesiophytes, tend to occur in deeper water where they contribute to the 'deep chlorophyll maximum' and can take advantage of nutrients moving upward from the ocean depths, in spite of the very low light levels. These small cells have a large surface-to-volume ratio, and so compete effectively for scarce nutrients with any larger cells. The smallest known eukaryote cell is a prasinophyte (*Ostreococcus taurii*), found in nutrient-deficient Mediterranean tide pools (Courties *et al.*, 1994), which measured only 0.97 x 0.70 µm by electron microscopy, and is of a size likely to be found in extremely oligotrophic waters. The oligotrophic oceans also contain phototrophic flagellates which gain mineral nutrients by virtue of another special adaptation: those (principally dinoflagellates) which live in symbiosis with heterotrophs such as radiolarians, contributing energy-rich carbon compounds to the host, but gaining recycled N and P from it.

Autotrophic flagellates from several groups are found within the cells of a wide diversity of animal and protist groups (see review in Smith and Douglas, 1987). The most widespread of these symbionts are dinoflagellates of the marine genus *Symbiodinium,* usually referred to as *S. microadriaticum* (see Trench, 1987). These occur in many radiolarians, foraminiferans, ciliates, hydroids, sea anemones, corals and clams, although the dinoflagellates in some radiolarians, cnidarians and a flatworm have been found to be species of *Amphidinium* (Taylor, 1974) or *Zooxanthella.* These dinoflagellates usually live within endocytic vacuoles within host cells, where they are somehow recognized and protected from attack by host cell enzymes, although in clams the symbionts are found in haemal spaces. Photosynthetic activity of the dinoflagellate produces an excess of photosynthate which is released as organic compounds like glucose, alanine or glycerol that are taken up by the host; in return the flagellate gains N and P, principally as ammonium and phosphate ions, from the host (Trench, 1987). The symbionts living in cells of corals and foraminiferans are believed to enhance the deposition of calcium carbonate in the coral skeleton/shell of the host when they remove carbon dioxide for photosynthesis (Goreau, 1961).

Symbiotic flagellates of other groups have received less attention, but include prasinophytes which have been found in the marine flatworm *Convoluta roscoffensis* (Holligan and Gooday, 1975) and in some radiolaria, and some acantharians have been reported to contain prymnesiophyte symbionts. The planktonic ciliate *Mesodinium* (*Myrionecta*) *rubrum* contains a well-integrated cryptophyte symbiont, whose photosynthetic activity is very high, with the result that ciliate 'red tides' are frequent in coastal locations in summer (Lindholm, 1985; Stoecker *et al.,* 1991).

8.3 Osmotrophic flagellates

Some flagellates grow in media containing only the very simplest organic molecules (Droop, 1974), and can certainly take these in through their surface membrane without the need for vacuole formation. Thus the volvocid flagellate *Polytoma uvella* is like a colourless chlamydomonad; its energy needs are satisfied by organic molecules as simple as acetate, and it does not use organic sources of nitrogen or vitamin-type molecules, so that it retains an almost complete synthetic machinery, except for photosynthesis. The colourless cryptophyte flagellate *Chilomonas paramecium* can be cultured in a medium whose only organic components are acetate and thiamine, although it can use a range of organic acids, including fatty acids and ethanol, as energy sources, and amino acids as a nitrogen source, if available. While these, like the euglenid acetate flagellate *Astasia,* are obligate osmotrophs, facultative osmotrophy is probably much more widespread, being found both among forms which are primarily phagotrophic and among mixotrophs that use osmotrophy as a supplement or alternative to photosynthesis. Thus, various strains of *Euglena gracilis*, which can be fully functional photo-autotrophs, can alternatively survive on ethanol, acetate or even some carbohydrates as an organic energy source, and can use but do not require amino acids; they have a requirement for vitamin B_{12} in either nutritional mode.

It is interesting that some phototrophic flagellates require acetate as a carbon source because they are unable to assimilate carbon dioxide (Droop, 1974). Chrysophyte flagellates like *Poteriochromonas malhamensis* grow osmotrophically on sugars rather than on acetate (Moestrup and Andersen, 1991). Among dinoflagellates, *Crypthecodinium cohnii* has been cultured axenically on histidine, acetate, biotin and thiamine (Provasoli and Gold, 1962) and *Gyrodinium lebourae* has been grown in seawater containing glucose, galactose and acetate (Lee, 1977).

Obligate osmotrophic flagellates may be assumed to require fairly high concentrations of the organic molecules which provide their energy source. To capture such molecules from their environment they are likely to be in strong competition with bacteria, whose much higher surface-area-to-volume ratio, diverse exo-enzymes and specialized permease uptake systems give them a substantial advantage. The decomposition of organic matter promoted by bacterial exoenzymes, from whose products many bacteria may have selected the larger and more attractive organic molecules, will produce large amounts of compounds like acetate and ethanol. Flagellates like *Polytoma, Chilomonas* and *Euglena gracilis* inhabit, and may bloom in, water containing decaying plant or animal remains, but such places usually provide only transient habitats requiring dispersal to find new organically rich sites.

Parasitic flagellates, even if derived from phagotrophic ancestors, tend to become osmotrophs if their parasitic habitats provide suitable organic molecules in plenty. Thus the parasitic kinetoplastid flagellate *Crithidia fasciculata*, which lives in body fluids of insects, derives energy from carbohydrate, and has a requirement for ten amino acids, a purine and eight vitamin-type substances in order to synthesize its other organic needs, but can absorb all of these through its surface membrane from the gut fluids of the insect. It seems likely that other flagellate parasites have needs at least as complex as these, whether they inhabit the gut, the circulatory system, tissue fluids or are intracellular in vacuolar cavities or the cytoplasm of host tissues. The uptake of organic molecules through the surface membranes of such flagellates involves mechanisms similar to those used by the host cells themselves, including transporters for sugars and amino acids (Zilberstein, 1993) and receptor-mediated endocytosis, such as that localised within the flagellar pocket of trypanosomatids (De Souza, 1989; Russell, 1994). Some parasitic flagellates also synthesize enzymes (such as the protease gp63 (Russell, 1994)), which may be either bound to the surface membrane of the parasite or secreted to break down host molecules extracellularly before absorption. More details of mechanisms involved in osmotrophic nutrition of flagellate parasites are given by Vickerman (Chapter 10, this volume).

The diversity of parasitic flagellates is probably most extreme among the dinoflagellates, yet these are poorly known (Cachon and Cachon, 1987). Those which attach to the outside of other plant or animal cells by some sort of stalk and quickly suck out the contents are probably better regarded as phagotrophs (see Section 8.4); these forms commonly retain chloroplasts and are presumably mixotrophic. Others establish root-like contacts with host cells or tissues, and remain there, feeding through their stalk-like connection until they reproduce. Generally the original contact is made by a tentacle or pseudopodium, emerging from the parasite's sulcus region, which penetrates the host surface (cuticle, egg-case or other protection), ramifies among the host cells, and often goes on to penetrate the host cell membrane to have direct contact with host cell cytoplasm. This parasite root system presumably extracts nutrients from the host cells osmotrophically, although phagotrophic endocytosis of host cell contents is not excluded. These dinoflagellates parasitize other protists, including other dinoflagellates, cnidarians, annelids, crustaceans, appendicularians and even fish, both eggs and epidermis, and can be economically important.

8.4 Phagotrophic flagellates

The ingestion of particulate food into food vacuoles and its subsequent digestion by lysosomal enzymes before the soluble organic molecules released can be absorbed into the cytoplasm is probably a primitive feature of eukaryote cells. Examples of such phagotrophic nutrition remain in most flagellate groups, although some of those whose ancestors acquired photo-autotrophic symbionts and became pure autotrophs have developed cell walls which prevent phagocytosis, for example chlorophytes.

Pinocytosis, involving fairly large molecules capable of passing through cell walls but not through cell membranes, can presumably occur within cell walls of chlorophytes, as in fungi. Many, if not most, flagellates without cell walls probably also use pinocytosis to take up large soluble molecules, either over large areas of naked surface or through specialized areas where the surface is covered by a pellicle. Where pinocytotic uptake of larger soluble molecules is followed by digestion by lysosomal enzymes, the process is not in principle distinguishable from phagocytosis, although the food is not visible, and its intake may involve receptor mediated selection. Pinocytotic nutrition is not easy to study; it has not received the attention from researchers that it deserves, and much remains to be discovered about its extent and importance, although it is clear that many parasites, particularly, depend on it (see Vickerman, Chapter 10 this volume).

The types of food particle ingested by phagotrophic flagellates range from large insoluble molecules and cell fragments, through bacterial and protistan cells of all types, to fragments of animal and plant tissues. Protists, and in particular flagellates, play major roles in both herbivory and bacterivory in pelagic food webs (Sherr and Sherr, 1994). The smaller free-living forms are generally bacterivores, although many of them will also take small algae or other protists. The larger forms mostly feed upon protists, and are more usually algivorous herbivores than predators on protozoa. The nature of the food within this wide range of diets determines which feeding strategy and which of the different mechanisms can be used to capture and manipulate the food. These strategies are probably best considered by describing the main classes of feeding mechanisms (Laybourn-Parry, 1984; Fenchel, 1987; Sanders, 1991).

The particles which form the food of phagotrophic flagellates may be either suspended in water or associated with surfaces, and particles in either category may be alive and actively moving or inactive, though not necessarily dead. Actively moving prey organisms, usually swimming ones, may be captured on adhesive surfaces of stationary flagellate predators, by a mechanism described as 'diffusion feeding' (Fenchel, 1984, 1986). More commonly, activity of the flagellate either creates 'feeding currents' of the water in which prey are suspended, so as to bring food particles into contact with the body ('suspension feeders'), or moves itself around to make contact with food particles ('hunting feeders' or raptors). These two forms of 'interception feeding' (Fenchel, 1984, 1986) are here regarded as the principal categories of flagellate feeding.

Among larger organisms it is often practicable to subdivide interception feeding in a slightly different way into two subclasses: filter feeding by mechanically straining suspended particles which are typically considerably smaller than the predator, and (raptorial) direct interception by capturing relatively larger particles, often after active pursuit. Such subdivision is less easy in flagellates. Interception of suspended food particles from a water current flowing around a flagellate cell may take place at a restricted site or over much of the cell body, which may be extended into projecting tentacles to increase the surface for interception. Whether there is ever well-defined filtration, of the type found in such invertebrates as bivalve molluscs and in ciliate protozoa, remains to be proven among the flagellates. The diversity of feeding mechanisms will be illustrated with a selection of examples. The capture of prey by interception or diffusion feeders, but especially in raptors, is often aided by the release of the contents of extrusomes, varying from simple glutinous adhesive to complex anchoring devices (Hausmann, 1978).

Where prey numbers are high, and flagellate predators are able to reproduce rapidly, it is possible for the populations of phagotrophic flagellates to achieve concentrations comparable with those of the blooms of autotrophic forms. These 'blooms' of phagotrophs tend to be short lived in most habitats, because various competing predatory species (as well as ciliates and some metazoans) usually replace one another in succession. The feeding/growth strategies of the flagellate predators themselves are also interesting. On encountering a rich food supply, many flagellates will feed quickly and grow to many times their usual biomass before dividing (for example *Cafeteria*: Zubkov and Sleigh, 2000). If enough food is still available they will continue to feed and grow actively, keeping the body size above that of the original flagellate. Once food concentrations fall below a critical level, some flagellate species will multiply several times to produce numerous small flagellates which can disperse in search of more food, or may encyst; *Paraphysomonas* is an example of this type (Zubkov and Sleigh, 1995). Starved *Ochromonas* have been shown to produce small cells whose respiration is only 2–5 per cent of that of growing cells (Fenchel, 1982b). Small cells of many dinoflagellates produce gametes, and the zygote resulting from fertilization often forms a resting cyst (Pfiester and Anderson, 1987). These species are particularly well-adapted to the feast-and-famine existence experienced by microbes (Koch, 1971; Fenchel, 1987). Other flagellates, like *Pteridomonas* or *Bodo*, are better equipped to exploit the steady background levels of prey normally present in the environment, since they are able to continue feeding at lower prey concentrations (Zubkov and Sleigh, 1995, 2000); they can still grow and multiply in times of plenty, but only outcompete those forms capable of very rapid multiplication when food is less plentiful.

8.4.1 Motile raptors and other bulk feeders

Feeding in which the predator hunts for its prey by moving around occurs in both the pelagic realm and on surfaces. Well-known examples of pelagic predators are found among dinoflagellates. *Oxyrrhis* is an unusual dinoflagellate in various aspects of cell structure, but moves its two flagella in a manner that is fairly typical of the group, with one longitudinal flagellum directed posteriorly and a transverse one which winds around the body, though not in a prominent groove. The transverse flagellum sweeps through a large volume of water as it undulates, and in doing so it makes contact with prey organisms which adhere to the flagellum. These prey may be bacteria or algae. If algal cells like *Isochysis* or *Dunaliella* are caught, the *Oxyrrhis* rotates actively to draw the flagellum close to the body surface, and the prey cell is engulfed into a food vacuole in the posterior half of the body (Tarran, 1991). The ingestion of bacteria probably only occurs when many prey are adhering to the flagellum.

Larger prey such as diatoms or ciliates may be ingested through the sulcus region of more conventional dinoflagellates like *Gymnodinium* or *Gyrodinium*, often with the aid of a lobed pseudopodium (Biecheler, 1952; Popovsky, 1982). Even some thecate dinoflagellates are capable of engulfing whole cells, for example *Peridinium gargantua* ingests other thecate dinoflagellates (Biecheler, 1952). The large naked dinoflagellate *Noctiluca* enjoys a more varied diet, since it captures bacteria, algae, protozoa and even small metazoa using a prehensile tentacle, and engulfs its prey into the vacuolated cytoplasm through a well-defined cytostome (Uhlig, 1972, cited by Elbrächter, 1991).

Planktonic algae, generally diatoms, also form the food of the thecate dinoflagellate *Protoperidinium* (Gaines and Taylor, 1984; Jacobson and Anderson, 1986). In this case the prey may be larger than the predator. A suitable diatom, or even a diatom filament, encountered by the protruding 'peduncle' of the dinoflagellate is held by the peduncle, which expands to form a fine cytoplasmic veil, or pallium, which is extended around the prey. This forms a food vacuole within which the cytoplasm of the diatom is digested. Species in a number of other genera feed in a similar manner (Gaines and Elbrächter, 1987)

A comparable peduncle is used by *Katodinium fungiforme* to predate upon algae larger than itself using a different mechanism called myzocytosis; in this case the predator sucks fluids from the prey through the tubular, microtubule-lined peduncle into food vacuoles within the cell body of the dinoflagellate for digestion (Spero, 1982). *Dinophysis* has been found to feed on tintinnid ciliates in the same way (Elbrächter, 1991), and *Protoodinium* sucks cytoplasmic materials from the cells of cnidarians through a 'hollow tube' supported by sheets of microtubules (Cachon and Cachon, 1987). These and other variants of phagotrophy in dinoflagellates feeding on motile and filamentous protists, as well as upon metazoa, have been reviewed by Gaines and Elbrächter (1987) and Elbrächter (1991).

A feeding structure with superficial similarity to the peduncle of some dinoflagellates, and which digests material outside the cell body of the predator, is found in several species of *Pirsonia* which penetrate a diatom cell with a pseudopodium to digest host cell cytoplasm *in situ* (Schnepf and Schweikert, 1997). *Pirsonia* was once regarded as a dinoflagellate genus (for example by Elbrächter, 1991), but ultrastructural features suggest that they are heterokont (stramenopile) flagellates (Schweikert and Schnepf, 1997).

On a rather different scale, there are pelagic predators among the prymnesiophytes (haptophytes), perhaps better called 'fishing feeders' than hunting feeders! Here the cells swim using two anterior homodynamic flagella, from between which extends another filiform appendage called the haptonema. This is supported by a bundle of microtubules having a different arrangement to those of the flagellar axoneme. The haptonema is often quite long, and its name reflects its adhesive properties. Food particles have been observed to adhere to the mobile haptonema of *Chrysochromulina hirta* and to be moved along the surface of this organelle to accumulate with other particles before being passed to the posterior pole of the cell, where they are engulfed into food vacuoles (Kawachi *et al.*, 1991). The types of food particle reported to be ingested by haptophytes include bacteria, small chrysophytes, diatoms, green algae and a red alga (Green, 1991).

There are various frequent pelagic flagellates which are believed to catch their food while swimming in the open water, but which have seldom been studied alive, so that their feeding mechanism is uncertain. The biflagellate *Leucocryptos* looks superficially like a cryptophyte, but has been observed with cyanobacteria (*Synechococcus*) lodged in its presumed cytostome and ingested within food vacuoles (Zubkov, personal communication). Another pelagic biflagellate of uncertain affinities, *Telonema*, which swims by being pushed along by its two flagella, is a phagotroph which feeds on bacteria, algal cells and other flagellates (Tong, 1997), apparently ingested using anterior pseudopodia (Patterson and Zolffel, 1991). *Cryptaulax marina*, whose structure suggests that it is probably a euglenozoan, is an active predator on nanophytoplankton in Southampton Water, where it has been found to ingest centric diatoms, chrysophytes, cryptophytes, dinoflagellates, the euglenid *Eutreptiella marina* and haptophytes, as well as other *Cryptaulax* cells (Tong, 1997).

The capture of food particles from the fluid contents of the gut by flagellates endobiotic in insects is probably more like a pelagic situation than a surface-creeping one. The hind-gut of termites and wood-roaches provides a habitat for numerous parabasalian flagellates, both trichomonads and hypermastigids, as well as oxymonads. The anaerobic requirements of these flagellates probably explains our limited physiological knowledge about them; it seems likely that they are mostly osmotrophs, but some large hypermastigids like *Trichonympha* and *Joenia* certainly engulf wood fragments into food vacuoles at the naked posterior end of the body. The free-living trichomonad *Pseudotrichomonas keilini* ingests bacteria into food vacuoles at the general body surface, and pinocytosis also occurs here (Brugerolle, 1991). Parasitic trichomonads such as *Trichomonas vaginalis* also contain food vacuoles enclosing bacteria and host cells, which appear to be formed at almost any part of the cell surface (Rendón-Maldonado *et al.*, 1998). Parasitic kinetoplastids spend part of their life in the gut of insects, as well as within circulating fluids and often within cells of their hosts; pinocytosis appears to take place in the flagellar pocket region, and if phagocytosis ever occurs in these forms it presumably also takes place in this region (Vickerman, Chapter 10 this volume).

Surface-inhabiting predatory flagellates which actively hunt for their prey during gliding migrations are found among the Euglenozoa, both euglenids and bodonid kinetoplastids (Triemer and Farmer, 1991a). Structural features associated with phagotrophy in euglenids were described by Triemer and Farmer (1991b). Food particles are taken into the cell at an anterior or ventral cytostome. The mouths of bacterivorous forms range from a simple pocket reinforced with microtubules in *Petalomonas*, to a complex 'siphon' supported by three large bundles of microtubules in *Entosiphon*. *Peranema* is a large euglenoid predator, familiar for its gliding motion and for the large rod-like ingestive organelles that enable it to eat other protists like *Euglena*, which may be only a little smaller than itself, as well as smaller eukaryotes and bacteria. Several species of *Bodo*, which are also familiar for their characteristic form of gliding, ingest single bacteria through a cytostome which is similar to that of *Petalomonas* (Triemer and Farmer, 1991a; Vickerman, 1991). The lips of the cytostome are often extended onto a rostrum or proboscis (Burzell, 1973), which enables such bodonids as *Rhynchomonas* to pick single bacteria from surfaces. *Caecitellus* (Patterson *et al.*, 1993) is a flagellate of uncertain affinities which inhabits sediments and particle surfaces; this small cell ingests bacteria ventrally at a lateral mouth enclosed in a microtubular band, which appears to be lowered over individual bacterial prey on surfaces (Tong, 1997).

Various mastigamoebids move over surfaces and through sediments, some of them more by amoeboid than flagellar locomotion, and ingest bacteria, algae and detritus from surfaces by the formation of food vacuoles at almost any part of the cell surface - apparently in the uroid region in *Pelomyxa palustris* (Mylnikov, 1991). An amoeboid form of ingestion, at a naked ventral region which is capable of pseudopod formation, is apparently also used by such surface-dwelling flagellates as *Cercomonas* (Patterson and Zölffel, 1991), and is probably usual in thaumatomastigids (Tong, 1997).

8.4.2 Suspension-feeding flagellates

Flagellates which intercept particles from water currents which they create by the beating of their flagella are also found in both pelagic and benthic realms. The efficiency of these feeding currents is enhanced if the cell is anchored to some larger structure (Lighthill, 1976; Higdon, 1979), so most of the flagellates concerned attach themselves by a secreted stalk to a particle of suspended or sedimented debris, or to another organism, or secrete a substantial lorica which may serve the same function. The hydrodynamics of direct-interception feeding have been considered recently by Monger and Landry (1990).

The loricate acanthoecid choanoflagellates are important members of the true marine plankton; members of the aloricate choanoflagellate families Codosigidae and Salpingoecidae are also found in the plankton, and there are members of all three families which live attached to substrata. All appear to be bacterivores. The flagellum of these collar flagellates arises at the centre of the collar, and drives water away from the cell as it beats. Many authors (such as Lapage, 1925; Sleigh, 1964; Fenchel, 1982a) have assumed that at least part of the water propelled by the flagellum is drawn through the collar, which was believed to act as a filter to extract bacteria from the water. However, the spaces between adjacent filaments of the collar are so narrow that it seems doubtful whether the flagellar current is strong enough to overcome the viscous resistance that water would encounter in passing through the collar (Leadbeater and Blake, personal communication). Food particles intercepted by the cell are seen adhering on the outside of the collar (as correctly illustrated in the colourful frontispiece of Saville Kent's (1880) monograph), and are carried down the collar filaments to the cell body, where they are ingested near the base of the filaments.

The presence of mastigonemes on the (usually) longer flagellum of heterokont flagellates reverses the flow of water around the beating flagellum and enables these cells to draw a current of water down the flagellum towards its base (Holwill and Sleigh, 1967). Different beat patterns are associated with somewhat different modes of particle capture (Sleigh, 1964, 1991). Large amplitude waves are used to create a broad current flow around the cells of the pedinellids *Pteridomonas* and *Actinomonas,* and particles in the flow which make contact with one of the arms may adhere to it, and be carried by a membrane-transport mechanism to the cell body where they are ingested. Many particles that are intercepted by the arms, especially any larger and more vigorous swimming organisms, fail to be retained; the mesh of the 'filter' formed by the arms diverges to be much greater than the typical size of the particles ingested.

Two types of flagellar beat pattern can direct particles in the water current to the body surface near the base of flagella that bear mastigonemes. The flagellum of *Bicosoeca* (*Poteriodendron*) *petiolata* carries many waves of small amplitude (Sleigh, 1964), and funnels a narrow stream of water towards a cytostome region of the cell surface enclosed by a loop of microtubules derived from a flagellar root (Moestrup and Thomsen, 1976). The small number of waves on the longer flagellum of *Ochromonas* or *Spumella* increases in amplitude towards the tip and funnels a converging stream of water towards the flagellar base (Sleigh, 1964, as *Monas*). When a potential food particle makes contact with the flagella of such chrysophytes, the longer flagellum generally stops beating briefly and the shorter flagellum participates in trapping the food particle from the water current. Wetherbee and Andersen (1992) described the way the two flagella of *Epipyxis pulchra* then hold the particle for a short time, following which it is either rejected or held near the cell surface for ingestion. In *E. pulchra*, trapped food particles are engulfed into food vacuoles by means of a feeding basket formed within a loop of microtubules of the type referred to above, which expands and contracts as the microtubules of the flagellar roots slide along one another in response to particle capture (Andersen and Wetherbee, 1992). In most of these forms, bacteria probably predominate among the ingested food particles, but small flagellates and diatoms will also be taken by many forms, as in the common genus *Paraphysomonas* (Caron *et al.*, 1986); some forms are cannibalistic (Fenchel, 1987).

While the methods of flagellar feeding just described are best suited to capturing suspended particles from open water, similar methods of particle collection are seen in a diversity of largely bacterivorous benthic flagellates which collect particles from the water around and between sediment particles. *Cafeteria* is a small aloricate bicosoecid, found on suspended particles as well as benthic surfaces, which attaches itself to a substratum with one short flagellum, while the other short flagellum draws particles to a cytostome in a ventral depression (Fenchel and Patterson, 1988). Flagellates belonging to various groups have two flagella (for example jakobids: O'Kelly, 1993; the phagotrophic cryptomonad *Goniomonas*: McFadden *et al.*, 1994) or four (such as *Percolomonas*: Fenchel and Patterson, 1986), one or more of which is closely associated with a ventral cytostomial groove into which particles are swept by flagellar beating; the parasitic retortamonads are very similar (Brugerolle, 1991). In *Apusomonas* the cytostome is at the rear of a ventral groove in the proboscis, but is again associated with a flagellum which lies in this groove (Karpov and Zhukov, 1986).

8.4.3 Diffusion feeding

Diffusion feeding, in which passive organisms capture motile prey, seems to be rare in flagellates. The flagellate example usually quoted is the pedinellid *Ciliophrys* (Sanders, 1991). This organism has granular arms, like those of a centrohelid heliozoan, but more numerous, and a single flagellum, but the flagellum is almost always inactive or only very slowly moving. Bacteria and small protists which swim into the array of arms adhere to them, and are carried to the cell body for ingestion into food vacuoles. The dinoflagellate *Noctiluca* also has a technique for diffusion feeding, which involves the secretion of strands of mucoid slime in which prey become entangled (Paranjape and Gold, 1982). Sometimes swarms of the flagellates produce a cloud of sticky slime which sinks slowly through the water entangling prey as it sinks (Uhlig, 1983); the *Noctiluca* later ingest slime and captured prey together into food vacuoles.

8.5 Conclusions

Flagellates exhibit the full range of trophic strategies seen among eukaryotes, albeit at a microscopic scale. These strategies help these versatile protists to exploit all of the main moist habitats on earth available to either free-living or parasitic organisms. In doing so they contribute substantially to primary production of the planet, both as free-living and endosymbiotic photoautotrophs. They play a major role in carbon flux through the first stages of aquatic food webs, and through their feeding and metabolism recycle significant amounts of mineral nutrients in soils, freshwater and the seas. They live as ectoparasites and endoparasites in plants and animals, and in the latter they may occupy the gut, tissue spaces, cell vacuoles and even the cytoplasm itself, causing important diseases of man and his domestic animals.

REFERENCES

Andersen, R. A., Saunders, G. W., Paskind, M. P. and Sexton, J. P. (1993) Ultrastructure and 18S rRNA gene sequence for *Pelagomonas calceolata* gen. et sp. nov. and the description of a new algal class, the Pelagophyceae classis nov. *Journal of Phycology*, **29**, 701–715.

Andersen, R. A. and Wetherbee, R. (1992) Microtubules of the flagellar apparatus are active during prey capture of the chrysophycean alga *Epipyxis pulchra*. *Protoplasma*, **166**, 8–20.

Barbeau, K., Moffett, J. W., Caron, D. A., Croot, P. L. and Erdner, D. L. (1996) Role of protozoan grazing in relieving iron limitation of phytoplankton. *Nature*, **380**, 61–64.

Biecheler, B. (1952) Recherches sur les péridiniens. *Bulletin Biologique de la France et de la Belgique*, Suppl., **36**, 1–149.

Brugerolle, G. (1991) Cell organization in free-living amitochondriate heterotrophic flagellates. In *The Biology of Free-Living Heterotrophic Flagellates* (eds D. J. Patterson and J. Larsen), Oxford: Clarendon Press, pp. 133–148.

Burzell, L. A. (1973) Observations on the proboscis-cytopharynx complex and flagella of *Rhynchomonas metabolica* Pshenin 1964 (Zoomastigophora, Bodonidae). *Journal of Protozoology*, **20**, 385–393.

Cachon, J. and Cachon, M. (1987) Parasitic dinoflagellates. In *The Biology of Dinoflagellates* (ed. F. J. R. Taylor), Oxford: Blackwell, pp. 571–610.

Caron, D. A., Goldman, J. C. and Dennett, M. R. (1986) Effect of temperature on growth, respiration, and nutrient regeneration by an omnivorous microflagellate. *Applied and Environmental Microbiology*, **52**, 1340–1347.

Courties, C., Vaquer, A., Troussellier, M., Lauter, J., Chétiennot-Dinet, M. J., Neveux, J., Machado, C. and Claustre, H. (1994) Smallest eukaryote organism. *Nature*, **370**, 255.

Cox, E. R. (ed.) (1980) *Phytoflagellates*. New York: Elsevier/North Holland.

De Souza, W. (1989) Components of the cell surface of trypanosomatids. *Progress in Protistology*, **3**, 87–184.

Droop, M. R. (1957) Auxotrophy and organic compounds in the nutrition of marine phytoplankton. *Journal of General Microbiology*, **16**, 286–293.

Droop, M. R. (1974) Heterotrophy of carbon. In *Algal Physiology and Biochemistry* (ed. W. D. P. Stewart), Oxford: Blackwell, pp. 530–559.

Elbrächter, M. (1991) Food uptake mechanisms in phagotrophic dinoflagellates and classification. In *The Biology of Free-living Heterotrophic Flagellates* (eds D. J. Patterson and J. Larsen), Oxford: Clarendon Press, pp. 303–312.

Fenchel, T. (1982a) Ecology of heterotrophic microflagellates. I. Some important forms and their functional morphology. *Marine Ecology Progress Series*, **8**, 211–223.

Fenchel, T. (1982b) Ecology of heterotrophic microflagellates. III. Adaptations to heterogeneous environments. *Marine Ecology Progress Series*, **9**, 25–33.

Fenchel, T. (1984) Suspended marine bacteria as a food source. In *Flows of Energy and Materials in Marine Ecosystems* (ed. M. J. R. Fasham), New York: Plenum Press, pp. 301–315.

Fenchel, T. (1986) Protozoan filter feeding. *Progress in Protistology*, **1**, 65–113.

Fenchel, T. (1987) *The Ecology of Protozoa: The Biology of Free-living Phagotrophic Protists*. Madison, Wisconsin: Science Technical.

Fenchel, T. and Patterson, D. J. (1986) *Percolomonas cosmopolitanus* (Ruinen) n. gen., a new type of filter feeding flagellate from marine plankton. *Journal of the Marine Biological Association of the United Kingdom*, **66**, 465–482.

Fenchel, T. and Patterson D. J. (1988) *Cafeteria roenbergensis* nov. gen., nov. sp., a heterotrophic microflagellate from marine plankton. *Marine Microbial Food Webs*, **3**, 9–19.

Gaines, G. and Elbrächter, M. (1987) Heterotrophic nutrition. In *The Biology of Dinoflagellates* (ed. F. J. R. Taylor), Oxford: Blackwell, pp. 224–268.

Gaines, G. and Taylor, F. J. R. (1984) Extracellular digestion in marine dinoflagellates. *Journal of Plankton Research*, **6**, 1057–1061.

Goreau, T. F. (1961) Problems of growth and calcium deposition in reef corals. *Endeavour*, **20**, 32–39.

Govindjee and Braun, B. Z. (1974) Light absorption, emission and photosynthesis. In *Algal Physiology and Biochemistry* (ed. W. D. P. Stewart), Oxford: Blackwell, pp. 346–390.

Green, J. C. (1991) Phagotrophy in prymnesiophyte flagellates. In *The Biology of Free-living Heterotrophic Flagellates* (eds D. J. Patterson and J. Larsen), Oxford: Clarendon Press, pp. 401–414.

Hausmann, K. (1978) Extrusive organelles in protists. *International Review of Cytology*, **52**, 197–276.

Higdon, J. J. L. (1979) The generation of feeding currents by flagellar motions. *Journal of Fluid Mechanics*, **94**, 305–330.

Holligan, P. M. and Gooday, G. W. (1975) Symbiosis in *Convoluta roscoffensis*. *Symposia of the Society for Experimental Biology*, **29**, 205–227.

Holt, J. R. and Pfiester, L. A. (1981) A survey of auxotrophy in five freshwater dinoflagellates (Pyrrhophyta). *Journal of Phycology*, **17**, 415–416.

Holwill, M. E. J. and Sleigh, M. A. (1967) Propulsion by hispid flagella. *Journal of Experimental Biology*, **47**, 267–276.

Jacobson, D. M. and Anderson D. M. (1986) Thecate heterotrophic dinoflagellates: feeding behaviour and mechanisms. *Journal of Phycology*, **22**, 249–258.

Karpov, S. A. and Zhukov, B. F. (1986) Ultrastructure and taxonomic position of *Apusomonas proboscidea* Alexeieff. *Archiv für Protistenkunde*, **131**, 13–26.

Kawachi, M., Inouye, I., Maeda, O. and Chihara, M. (1991) The haptonema as a food-capturing device: observations on *Chrysochromulina hirta* (Prymnesiophyceae). *Phycologia*, **30**, 563–573.

Kent, W. S. (1880–1) *A Manual of the Infusoria*. London: Bogue.

Kies, L. (1980) Morphology and systematic position of some endocyanomes. In *Endocytobiology, Endosymbiosis and Cell Biology vol. 1* (eds W. Schwemmler and H. E. A. Schenk), Berlin: W. de Gruyter, pp. 7–19.

Koch, A. L. (1971) The adaptive responses of *Escherichia coli* to a feast and famine existence. *Advances in Microbial Physiology*, **6**, 147–217.

Lapage, G. (1925) Notes on the choanoflagellate *Codosiga botrytis* Ehrbg. *Quarterly Journal of Microscopical Science*, **69**, 471–508.

Laybourn-Parry, J. (1984) *A Functional Biology of Free-living Protozoa*. London: Croom Helm.

Lee, R. E. (1977) Saprophytic and phagocytic isolates of the colourless heterotrophic dinoflagellate *Gyrodinium lebouriae* Herdman. *Journal of the Marine Biological Association of the United Kingdom*, **57**, 303–315.

Lighthill, J. (1976) Flagellar hydrodynamics. *SIAM Review*, **18**, 161–230.

Lindholm, T. (1985) *Mesodinium rubrum* – a unique photosynthetic ciliate. *Advances in Aquatic Microbiology*, **3**, 1–48.

Martin, J. H. and 43 others (1994) Testing the iron hypothesis in ecosystems of the equatorial Pacific Ocean. *Nature*, **371**, 123–129.

McFadden, G. I., Gilson, P. R. and Hill, D. R. A. (1994) *Goniomonas*: rRNA sequences indicate that this phagotrophic flagellate is a close relative of the host component of cryptomonads. *European Journal of Protistology*, **29**, 29–32.

Moestrup, Ø. and Andersen, R. A. (1991) Organization of heterotrophic heterokonts. In *The Biology of Free-Living Heterotrophic Flagellates* (eds D. J. Patterson and J. Larsen), Oxford: Clarendon Press, pp. 333–360.

Moestrup, Ø. and Thomsen, H. A. (1976) Fine structural studies on the flagellate genus *Bicoeca*. *Protistologica*, **12**, 101–120.

Monger, B. C. and Landry, M. R. (1990) Direct-interception feeding by marine zooflagellates: the importance of surface and hydrodynamic forces. *Marine Ecology Progress Series*, **65**, 123–140.

Mylnikov, A. P. (1991) Diversity of flagellates without mitochondria. In *The Biology of Free-living Heterotrophic Flagellates* (eds D. J. Patterson and J. Larsen), Oxford: Clarendon Press, pp. 149–158.

O'Kelly, C. J. (1993) The jakobid flagellates: structural features of *Jakoba, Reclinomonas* and *Histiona* and implications for the early diversification of eukaryotes. *Journal of Eukaryote Microbiology*, 40, 627–636.

Paranjape, M. A. and Gold, K. (1982) Cultivation of marine pelagic protozoa. *Annales de l'Institut Océanographique, Paris,* 58(S), 143–150.

Patterson, D. J., Nygaard, K., Steinberg, G. and Turley, C. M. (1993) Heterotrophic flagellates and other protists associated with oceanic detritus throughout the water column in the mid North Atlantic. *Journal of the Marine Biological Association of the United Kingdom*, **73**, 67–95.

Patterson, D. J. and Zölffel, M. (1991) Heterotrophic flagellates of uncertain taxonomic position. In *The Biology of Free-living Heterotrophic Flagellates* (eds D. J. Patterson and J. Larsen), Oxford: Clarendon Press, pp. 427–475.

Pfiester, L. A. and Anderson, D. M. (1987) Dinoflagellate reproduction. In *The Biology of Dinoflagellates* (ed. F. J. R. Taylor), Oxford: Blackwell, pp. 611–648.

Popovsky, J. (1982) Another case of phagotrophy by *Gymnodinium helveticum* f. *achroum* Skuja. *Archiv für Protistenkunde*, **125**, 73–78.

Provasoli, L. (1958) Nutrition and ecology of protozoa and algae. *Annual Review of Microbiology*, **12**, 279–308.

Provasoli, L. and Carlucci, A. F. (1974) Vitamins and growth regulators. In *Algal Physiology and Biochemistry* (ed. W. D. P. Stewart), Oxford: Blackwell, pp. 741–787.

Provasoli, L. and Gold, K. (1962) Nurtition of the American strain of *Gyrodinium cohnii. Archives of Microbiology*, **42**, 196–203.

Rendón-Maldonado, J. G., Espinosa-Cantellano, M., Gonzáles-Robles, A. and Martínez-Palomo, A. (1998) *Trichomonas vaginalis*: *in vitro* phagocytosis of lactobacilli, vaginal epithelial cells, leukocytes and erythrocytes. *Experimental Parasitology*, **89**, 241–250.

Russell, D. G. (1994) Biology of the *Leishmania* surface: with particular reference to the surface proteinase gp63. *Protoplasma*, **181**, 191–201.

Sanders, R. W. (1991) Trophic strategies among heterotrophic flagellates. In *The Biology of Free-living Heterotrophic Flagellates* (eds D. J. Patterson and J. Larsen), Oxford: Clarendon Press, pp. 21–38.

Schnepf, E. and Schweikert, M. (1997) *Pirsonia*, phagotrophic nanoflagellates incertae sedis, feeding on marine diatoms: host recognition, attachment and fine structure. *Archiv für Protistenkunde*, **147**, 361–371.

Schweikert, M. and Schnepf, E. (1997) Light and electron microscopical observations on *Pirsonia punctigerae* spec. nov., a nanoflagellate feeding on the marine centric diatom *Thalassiosira punctigera. European Journal of Protistology*, **33**, 168–177.

Sherr, E. B. and Sherr, F. B. (1994) Bacterivory and herbivory: key roles of phagotrophic protists in pelagic food webs. *Microbial Ecology*, **28**, 223–235.

Sleigh, M. A. (1964) Flagellar movement of the sessile flagellates *Actinomonas, Codonosiga, Monas* and *Poteriodendron. Quarterly Journal of Microscopical Science*, **105**, 405–414

Sleigh, M. A. (1991) Mechanisms of flagellar propulsion. A biologist's view of the relation between structure, motion, and fluid mechanics. *Protoplasma*, **164**, 45–53.

Smith, D. C. and Douglas, A. E. (1987) *The Biology of Symbiosis*. London: Edward Arnold.

Spero, H. J. (1982) Phagotrophy in *Gymnodinium fungiforme* (Pyrrophyta): the peduncle as an organelle of ingestion. *Journal of Phycology*, **18**, 356–360.

Steidinger, K. A. and Baden, D. G. (1984) Toxic marine dinoflagellates. In *Dinoflagellates* (ed. D. L. Spector), Orlando: Academic Press, pp. 201–261.

Stoecker, D. K., Putt, M., Davis, L. H. and Michaels, A. E. (1991) Photosynthesis in *Mesodinium rubrum*: specific measurements and comparison to community rates. *Marine Ecology Progress Series*, **73**, 245–252.

Tarran, G. A. (1991) *Aspects of the Grazing Behaviour of the Marine Dinoflagellate* Oxyrrhis marina, *Dujardin*. Unpublished Ph.D. thesis, University of Southampton.

Tarran, G. A., Zubkov. M. V., Sleigh, M. A., Burkill, P. H. and Yallup, M. (2000) Microbial community structure and standing stocks in the NE Atlantic in June and July of 1996. *Deep-Sea Research II,* (in press).

Taylor, D. L. (1974) Symbiotic marine algae: taxonomy and biological fitness. In *Symbiosis in the Sea* (ed. W. B. Vernberg) , Columbia: University of South Carolina Press, pp. 245–262.

Taylor, D. L. and Seliger, H. H. (eds) (1979) *Toxic Dinoflagellate Blooms.* Amsterdam: Elsevier/North Holland.

Taylor, F. J. R. and Pollingher, U. (1987) Ecology of dinoflagellates. In *The Biology of Dinoflagellates* (ed. F. J. R. Taylor), Oxford: Blackwell, pp. 398–529.

Tong, S. M. (1997) Heterotrophic flagellates and other protists from Southampton Water, UK. *Ophelia*, **47**, 71–131.

Trench, R. K. (1987) Dinoflagellates in non-parasitic symbioses. In *The Biology of Dinoflagellates* (ed. F. J. R. Taylor), Oxford: Blackwell, pp. 530–570.

Triemer, R. E. and Farmer, M. A. (1991a) An ultrastructural comparison of the mitotic apparatus, feeding apparatus, flagellar apparatus and cytoskeleton in euglenoids and kinetoplastids. *Protoplasma*, **164**, 91–104.

Triemer, R. E. and Farmer, M. A. (1991b) The ultrastructural organization of the heterotrophic euglenids and its evolutionary implications. In *The Biology of Free-living Heterotrophic Flagellates* (eds D. J. Patterson and J. Larsen), Oxford: Clarendon Press, pp. 185–204.

Uhlig, G. (1982) Entwicklung von *Noctiluca miliaris. Institut für wissenschaftlichen Film Göttingen*, **C 879**, 1–15.

Uhlig, G. (1983) Untersuchungen 'zum red-tide' Phänomen von *Noctiluca miliaris. Jahresbericht der Biologischen Anstalt Helgoland*, 44–48.

Vickerman, K. (1991) Organization of the bodonid flagellates. In *The Biology of Free-living Heterotrophic Flagellates* (eds D. J. Patterson and J. Larsen), Oxford: Clarendon Press, pp. 159–176.

Wetherbee, R. and Andersen, R. A. (1992) Flagella of the chrysophycean alga play an active role in prey capture and selection. Direct observations on *Epipyxis pulchra* using image enhanced video microscopy. *Protoplasma,* **166**, 1–7.

Zilberstein, D. (1993) Transport of nutrients and ions across membranes of trypanosomatid parasites. *Advances in Parasitology*, **32**, 261–291.

Zubkov, M. V. and Sleigh, M. A. (1995) Bacterivory by starved marine heterotrophic nanoflagellates of two species that feed differently, estimated by uptake of dual radioactive-labelled bacteria. *FEMS Microbiology Ecology*, **17**, 57–66.

Zubkov, M. V. and Sleigh, M. A. (2000) Growth of protozoa on bacteria deposited on surfaces, (in press).

Chapter 9

Amitochondriate flagellates

Guy Brugerolle and Miklōs Müller

ABSTRACT

Amitochondriate organisms are present in several eukaryotic lineages and usually inhabit anoxic and hypoxic environments. They are free-living, symbiotic or parasitic. Several lineages of flagellates comprise exclusively amitochondriate species: metamonads (retortamonads, two genera; diplomonads, ten genera; and oxymonads, eleven genera), pelobionts (five genera), and parabasalids (trichomonads, forty-six genera; and hypermastigids, thirty-four genera). Some other groups contain the occasional amitochondriate species. The ultrastructure of many species has been well explored. Such studies have defined the characteristic features of each group, but many genera remain to be studied. No organelles of energy metabolism have been recognized in metamonads and in pelobionts, while parabasalids have hydrogenosomes. Metabolic biochemistry has been studied only in a few, primarily parasitic, species of amitochondriate flagellates. Amitochondriate energy metabolism is either not compartmentalized, as in diplomonads (type I), or shows a hydrogenosome/cytosol compartmentation, as in trichomonads (type II). It remains to be established whether other amitochondriate flagellate groups also belong to these types. Molecular studies indicate that the absence of mitochondria in diplomonads and trichomonads is secondary. Phylogenetic reconstructions based on ribosomal RNA (rRNA) and certain protein sequences are in agreement with a monophyly of diplomonads and parabasalids. The status of the other amitochondriate flagellate groups has not been resolved yet. The evolutionary emergence of amitochondriate groups in eukaryotic phylogeny remains an open question.

9.1 Introduction

Although mitochondria are considered as universal attributes of eukaryotic cells, protists of several lineages lack morphologicaly recognizable mitochondria and can be regarded as amitochondriate. Among flagellates some lineages are entirely amitochondriate with or without hydrogenosomes. Other lineages contain occasional species with hydrogenosomes or modified mitochondria.

Individual lineages of amitochondriate flagellates are characterized by diverse sets of cytological features. Particularly important is the flagellar apparatus consisting of the flagella, their basal bodies and associated roots, which together form the flagellar/basal body cytoskeleton. Some of these structures consist of proteins also found in other eukaryotes while other structures and their proteins are restricted to the groups considered here.

Most amitochondriate protists are sensitive to oxygen and are found predominantly in anaerobic or hypoxic ecosystems, either free-living in the bottom layers of various water-bodies or as parasites in anaerobic sites in their hosts, for example certain parts of their digestive system.

The absence of mitochondria raises the question of energy production in these organisms and of the existence of metabolic features necessary for adaptation to their anoxic habitats and for survival at low pO_2. Our knowledge of the hydrogenosome and its relationships to mitochondria has considerably progressed in the last years.

In studies on the molecular phylogeny of protists, rRNA and proteins provided significant information complementary to morphological data helping to establish the evolutionary relationships of individual groups within the major lineages. Molecular studies also raised the hope that they will decypher the order of emergence of the different amitochondriate flagellate groups in the eukaryotic tree and resolve the interesting question whether these flagellates are primitively amitochondriate and descended directly from the first eukaryotes before the a-proteobacterial endosymbiosis gave rise to mitochondria.

9.2 Groups, features and intragroup relationships

9.2.1 Groups

Amitochondriate flagellates include genera of the metamonad grouping, (retortamonads (Order Retortamonadida), diplomonads (Order Diplomonadida) and oxymonads (Order Oxymonadida)), of the parabasalid lineage (trichomonads (Order Trichomonadida) and hypermastigids (Order Hypermastigida)) and of the pelobionts (Class Pelobiontidea) as well as some genera which belong to known mitochondriate flagellate groups (Figure 9.1; Table 9.1, page 181) (Brugerolle, 1991a, b; Cavalier-Smith, 1993, 1999).

9.2.2 Features

Metamonads

This taxonomic grouping is based on few shared morphological characters such as an arrangement of basal bodies/flagella in two pairs with a recurrent flagellum. Mitochondria and developed Golgi stacks are consistently absent (Brugerolle, 1991a). They are probably not monophyletic, but molecular phylogeny should help in resolving this question.

The retortamonads comprise two genera (Figure 9.1), *Chilomastix*, which has four basal bodies with flagella, and *Retortamonas*, which has four basal bodies of which only two develop a flagellum. Both parasitic and free-living, they capture their prey by a cytostome/cytopharynx containing a modified flagellum (Figures 9.2 a, c) (Brugerolle, 1991a). Major features of the group established from EM studies of several parasitic species are essentially confirmed by a recent EM study of the free-

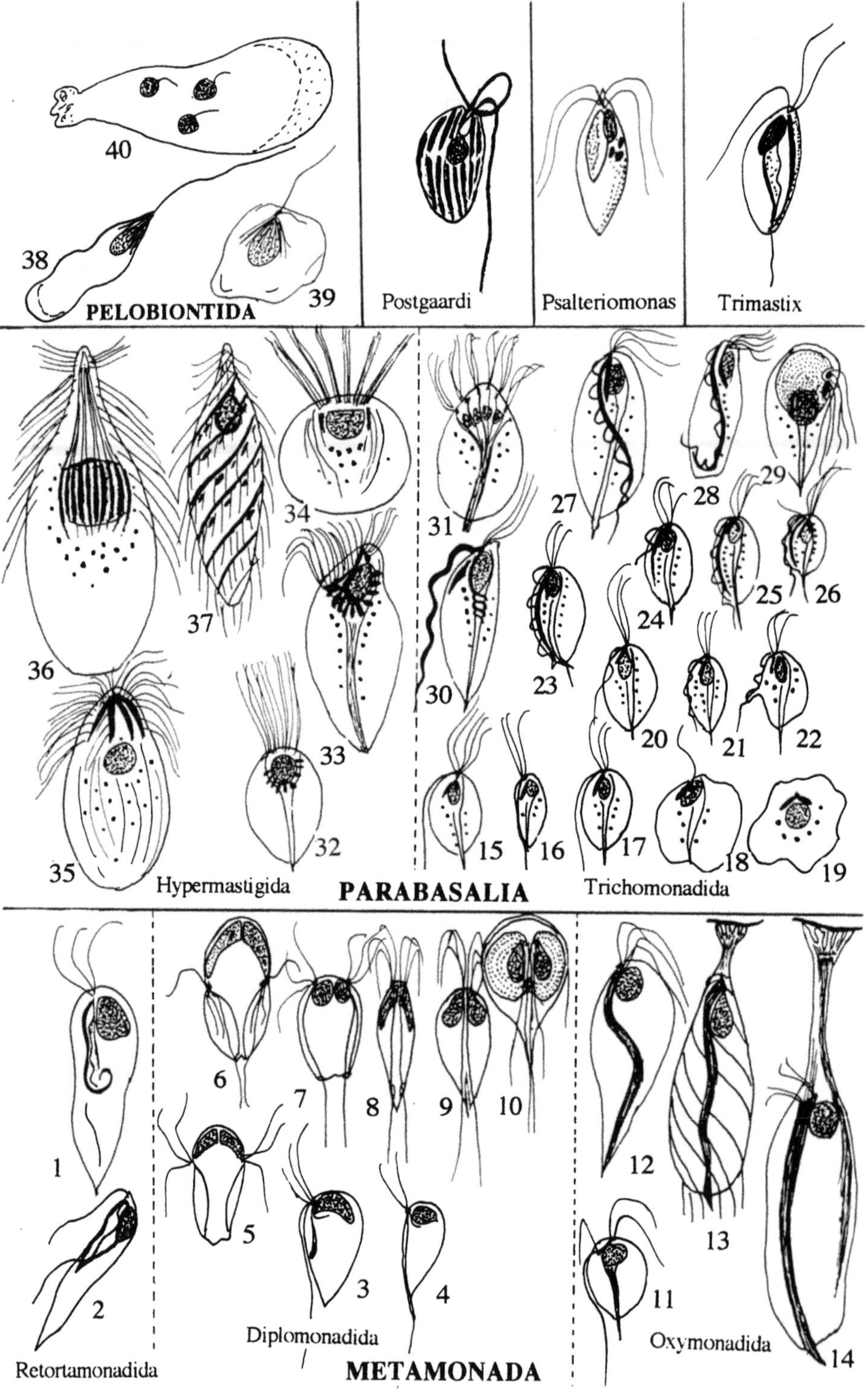
40
38
39
PELOBIONTIDA
Postgaardi
Psalteriomonas
Trimastix
36
37
34
33
35
32
31
30
27
28
29
23
24
25
26
20
21
22
15
16
17
18
19
Hypermastigida
PARABASALIA
Trichomonadida
1
2
6
7
8
9
10
5
3
4
12
13
14
11
Retortamonadida
Diplomonadida
METAMONADA
Oxymonadida

living *Chilomastix cuspidata* (Bernard *et al.*, 1997). A corset of interlinked microtubules underlies the cell membrane. The four basal bodies are grouped in two pairs: the posterior one corresponds to the basal body of the modified recurrent flagellum located in the cytostomal groove (Figures 9.2 a, b). Two microtubular roots originate from basal bodies and reinforce the cytostomal lips; the right one is associated with a striated or paracrystalline structure.

The diplomonads (about ten genera, Figure 9.1) have either a single flagellar apparatus composed of four flagella/basal bodies associated with the nucleus (Enteromonadina: *Trimitus, Enteromonas*) or have two nuclei each of which is associated with a flagellar apparatus (Diplomonadina: *Trepomonas, Hexamita, Spironucleus, Octomitus, Giardia*). They take up food through a cytostome/cytopharynx (Hexamitinae: *Trepomonas, Hexamita, Spironucleus*) or by pinocytosis (Giardiinae: *Octomitus, Giardia*) (Figure 9.3a). Most species of the genera *Trepomonas, Hexamita* are free-living but some *Trepomonas* and *Hexamita* are parasitic. All species of *Giardia, Octomitus* and *Spironucleus* are parasitic and some of them are pathogenic. Major features of the group were revealed by EM studies (Brugerolle, 1991a and Figure 9.3).

In each set of flagella, basal bodies are arranged in two pairs and one gives rise to the recurrent flagellum. Three microtubular roots are attached to basal bodies: a supranuclear, an infranuclear and a cytostomal fibre; the latter together with a striated lamina forms the armature of the cytostome/cytopharynx when present, or accompanies axonemes of the recurrent flagella (Figures 9.3 b, c, d, e). Recent ultrastructural studies on *Spironucleus* of fishes, *S. vortens, S. barkhanus* and *S. torosus,* have confirmed previous studies (Sterud, 1998). Many free-living species of *Trepomonas, Hexamita* and parasitic *Spironucleus* remain to be studied by EM. Progress in cultivation of these genera such as *Hexamita* will make biochemical

(left)
Figure 9.1 Genera of amitochondriate flagellate groups:
Retortamonadida (1 *Chilomastix*, 2 *Retortamonas*)
Diplomonadida (3 *Enteromonas*, 4 *Trimitus*, 5 *Trigonomonas*, 6 *Trepomonas*, 7 *Hexamita*, 8 *Spironucleus*, 9 *Octomitus*, 10 *Giardia*)
Oxymonadida (11 *Monocercomonoides*, 12 *Saccinobaculus*, 13 *Pyrsonympha*, 14 *Oxymonas*)
Parabasalia Trichomonadida:
Monocercomonadidae (15 *Hexamastix*, 16 *Tricercomitus*, 17 *Monocercomonas*, 18 *Histomonas*, 19 *Dientamoeba*, 20 *Hypotrichomonas*, 21 *Pseudotrichomonas*, 22 *Ditrichomonas*)
Trichomonadidae (23 *Tritrichomonas*, 24 *Trichomonas*, 25 *Tetratrichomonas*, 26 *Pentatrichomonas*, 27 *Trichomitopsis*, 28 *Pentatrichomonoides*)
Cochlosomatidae (29 *Cochlosoma*)
Devescovinidae (30 *Devescovina*)
Calonymphidae (31 *Calonympha*)
Parabasalia Hypermastigida (32 *Lophomonas*, 33 *Joenia*, 34 *Kofoidia*, 35 *Staurojoenina*, 36 *Trichonympha*, 37 *Spirotrichonympha*)
Pelobiontida (38 *Mastigamoeba*, 39 *Mastigina*, 40 *Pelomyxa*)
the euglenoid *Postgaardi mariagerensis*
the percolozoan *Psalteriomonas vulgaris*
Trimastix convexa (of unknown affinity)
Sources: Various (G. Brugerolle)

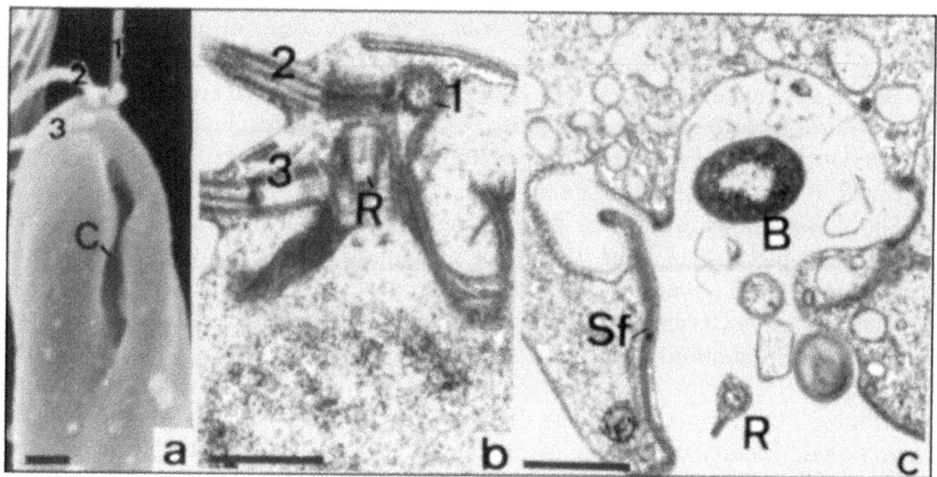

Figure 9.2 Retortamonads

a Scanning EM view of *Chilomastix*, showing subapical insertion of the three flagella (1, 2, 3) and cytostomal opening (C).

b Basal bodies arranged in two pairs (1, 2) and (3, R); recurrent flagellum (R).

c Transverse section showing the striated fibres (Sf) bordering the cytostome, the modified flagellum (R) and bacteria (B) being phagocytosed. Bar: 1 μm

Source: G. Brugerolle

studies possible. *Giardia* displays an unusual ventral disc whose cytoskeleton is composed of microtubules and associated ribbons made of a, b, g isoforms of giardins (Figure 9.3 f) (Nohria *et al.*, 1992). Centrin is present in *Giardia* (Meng *et al.*, 1996). During cell division the parental ventral disc of *Giardia* disintegrates before two new ones assemble and separate, joining the two flagellar apparatuses in the sister cells (Kulda and Nohýnková, 1995).

The oxymonads (about eleven genera, Figure 9.1) have four flagella separated in two distant pairs and a crystalline axostyle, which is contractile in some species. All are parasitic or symbiotic in the intestine of various animals. Some are present in diverse hosts such as *Monocercomonoides*, others are restricted to lower termites and wood-roaches. They take up their food by pinocytosis or phagocytosis. Features of the group have been determined by EM studies of the less complex *Monocercomonoides*, of several termite symbionts: *Pyrsonympha, Streblomastix* (Brugerolle, 1991a) and more recently of *Oxymonas* (Brugerolle and König, 1997) (Figure 9.4). The two pairs of basal bodies lie separated and are linked to a preaxostylar structure (Figures 9.4 a, b, e). They possess a crystalline axostyle composed of several staggered rows of microtubules linked by two types of cross-bridges

(right)
Figure 9.3 Diplomonads

a Organization and suggested phylogeny of the main genera based on proteins and rRNA sequences (?= no sequence)

b Basal body and fibre arrangement close to the nucleus (N): basal body pairs (1–2) and (3–R), microtubular roots: supra-nuclear (Snf), infranuclear (Inf, Inf'), cytostomal fibre (Cf) and striated lamina (Sl)

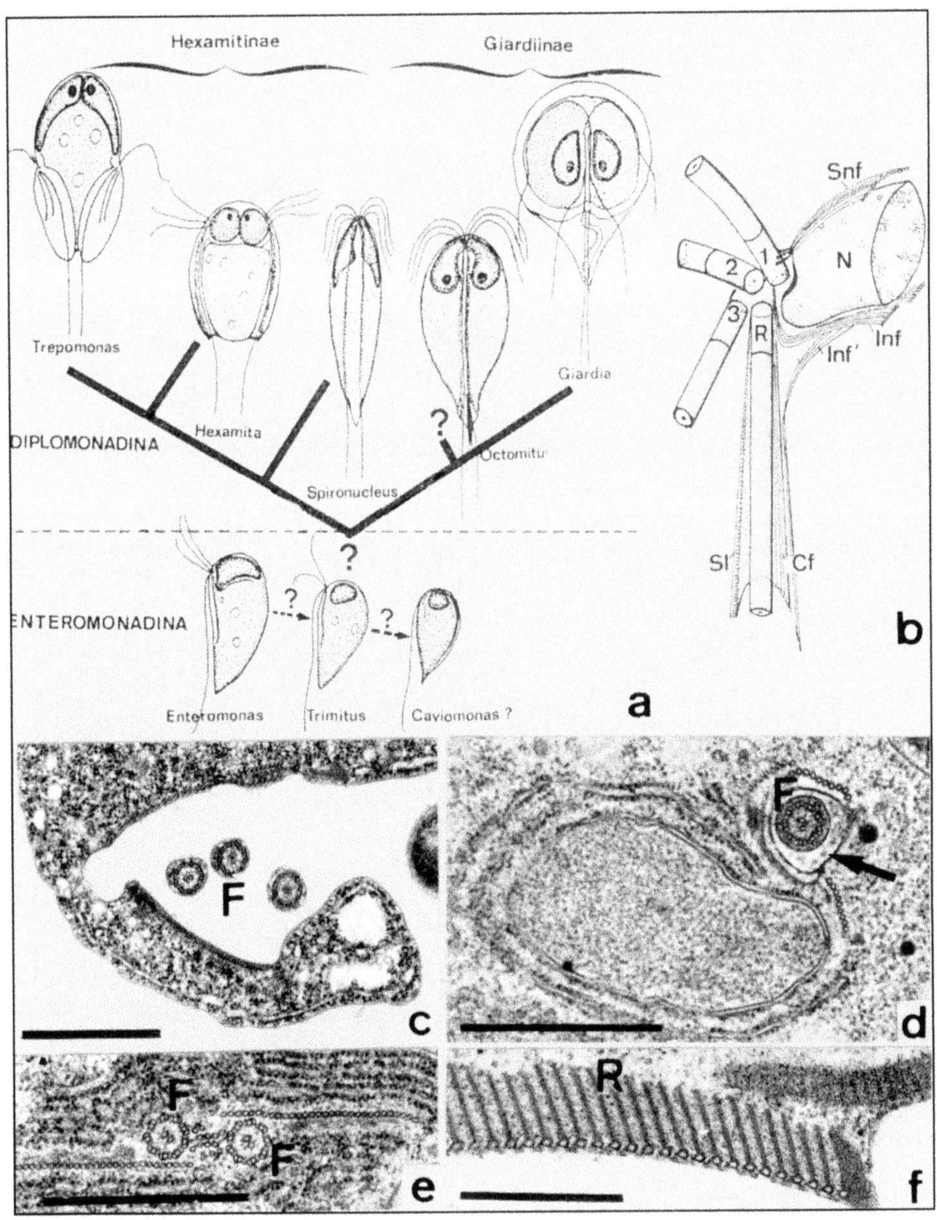

c Cytostome of *Trepomonas* containing 3 flagella (F)
d Cytostomal channel of *Hexamita* (arrow) containing one flagellum (F)
e Recurrent axonemes (F) inside the cytoplasm in *Octomitus*
f Transverse section of *Giardia* ventral disc showing giardin ribbons (R)
Bar: 1 μm

Source: G. Brugerolle

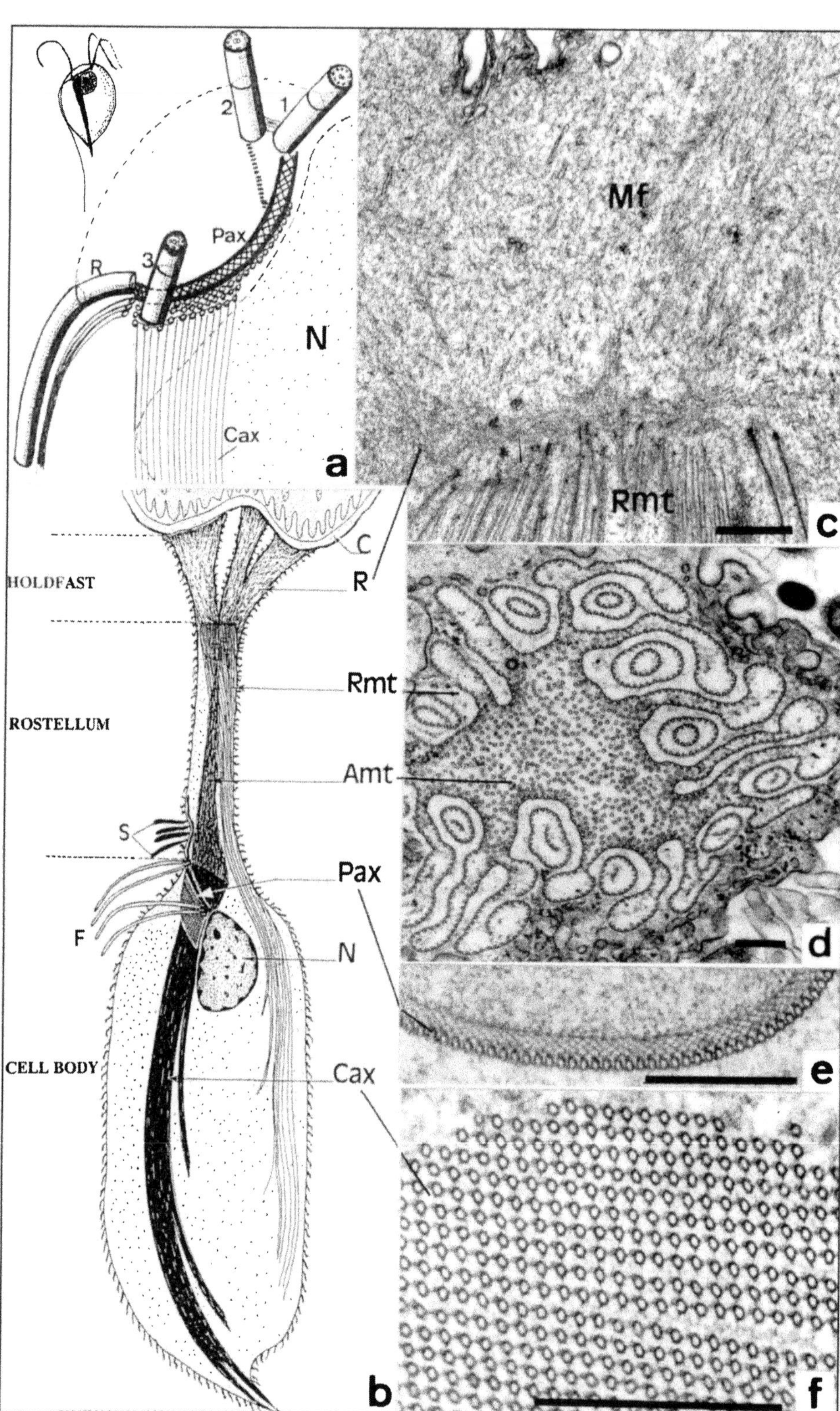

2
1
Pax
R
3
N
Cax
a
Mf
Rmt
c
C
HOLDFAST
R
Rmt
ROSTELLUM
Amt
S
Pax
F
N
d
CELL BODY
Cax
e
b
f

(Figure 9.4 f). One of these is dynein, which serves as the motor for these contractile axostyles. Most of the termites species such as *Pyrsonympha* and *Oxymonas* possess an anterior holdfast apparatus with which the parasite attaches to the lining of the host intestine. This structure consists of rhizoids filled with microfilaments (Figure 9.4 c). The *Oxymonas* cell extends into a long anterior rostellum of varying size containing many microtubule ribbons and terminates by the holdfast (Figures 9.4 b, d).

Parabasalids

Parabasalids (about eighty genera, Figure 9.1) have hydrogenosomes and no conventional mitochondria (Figure 9.6). This phylum comprises the trichomonads with less than six flagella and the hypermastigids with numerous flagella (up to 1,000). They share a set of characteristics: a typical arrangement of basal bodies, tubular axostyle, Golgi stack supported by a fibre forming the parabasal apparatus, and division by pleuromitosis, which is characterized by an external mitotic spindle (Figure 9.5). They feed by pinocytosis and phagocytosis. Only few genera are free-living; most are intestinal parasites of various animals or symbionts of insects, or live in the urogenital tract of humans and animals.

The Order *Trichomonadida* (about forty-six genera, Figure 9.1), comprising five families, is defined on the basis of morphological characters: Monocercomonadidae without a costa and an undulating membrane (UM), Trichomonadidae with a costa and a UM, Devescovinidae with a cresta and a thick recurrent flagellum. Calonymphidae correspond to a polymonad state of devescovinids. The trichomonad affinities of the last family, the Cochlosomatidae, has been confirmed by ultrastructural data (Pecka *et al.,* 1996). While molecular phylogeny strongly supported the monophyly of the trichomonads, it also has shown that some groups are artificial, such as Monocercomonadidae which turned out to be polyphyletic.

The Order Hypermastigida (about thirty-four genera, Figure 9.1) comprises flagellates parasitic or symbiotic in termites and wood-roaches. They have retained the basic characters of trichomonads and they have multiplied the number of flagella, axostyles, parabasal apparatuses and developed other structures which are as yet incompletely described at the ultrastructure level. Some families, the Joeniidae, Lophomonadidae and Rhizomastigidae show close relationships with

(left)
Figure 9.4 Oxymonads
- a Basal body and fibre arrangement in *Monocercomonoides*: the two pairs of basal bodies (1,2) and (3,R) are separated by the preaxostylar lamina (Pax) from which arises the crystalline axostyle (Cax)
- b *Oxymonas* organization: holdfast rhizoids (R) attached to the gut cuticle (C), rostellum, basal bodies and flagella (F), preaxostyle (Pax) and crystalline axostyle (Cax), spirochaetes (S)
- c Microfibril rich holdfast rhizoids
- d Rostellum with ribbons of microtubules (Rmt) and axial free microtubules (Amt)
- e Preaxostyle (Pax)
- f Crystalline axostyle (Cax)

Bar: 1 µm

Source: G. Brugerolle

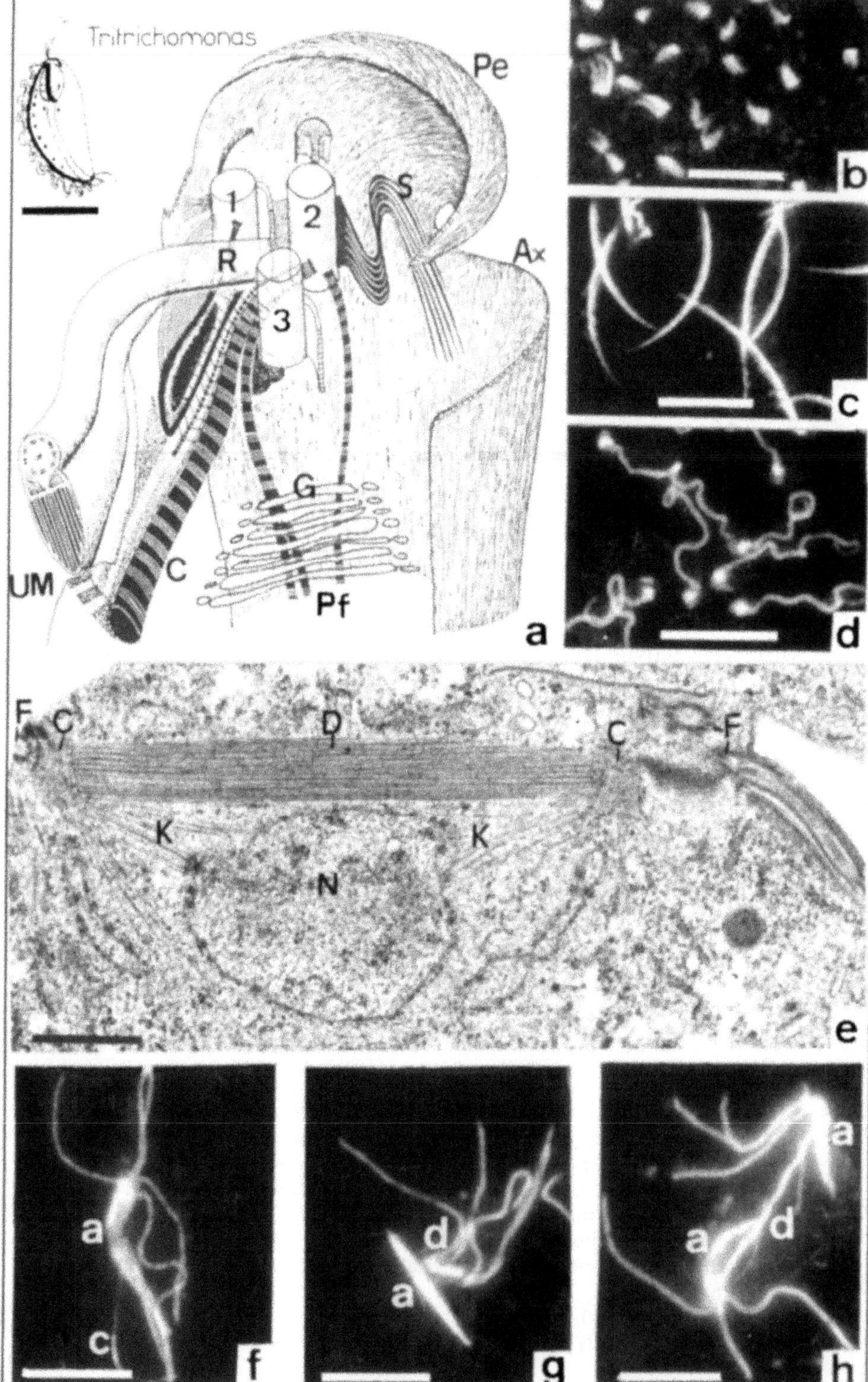
Tritrichomonas
Pe
S
1
2
R
3
Ax
G
C
UM
Pf
a
b
c
d
F
C
D
C
F
K
K
N
e
a
c
f
d
a
g
a
a
d
h

Devescovinidae, their putative trichomonad ancestors. Families of the two suborders Trichonymphina and Spirotrichonymphina, however, are greatly modified as compared to trichomonads (Hollande and Carruette-Valentin, 1971). Both ultrastructural and molecular phylogenetic studies are handicapped by difficulties in cultivating these organisms or in separating the species which occur as mixed populations of often more than five genera in the same host.

Features of trichomonads are well known (Brugerolle, 1991a) (Figure 9.5 a). Basal bodies 1, 2, 3 of the anterior flagella are typically arranged around the basal body R of the recurrent flagellum. These four privileged basal bodies bear appendages such as the sigmoid fibres (S) connecting basal body 2 to the pelta-axostyle non-contractile microtubular root. Two striated parabasal fibres support Golgi stacks forming the parabasal apparatus and the striated costa underlies the UM. Recently some genera revealed unexpected structures. In *Pentatrichomonoides* the pelta-axostyle forms a peripheral corset that is only open in the direction of the UM (Brugerolle *et al.*, 1994). EM study of *Cochlosoma* has revealed characteristic features of trichomonads and a peculiar adhesive disc reminiscent of the ventral disc of *Giardia*, which is composed of pelta microtubules associated with ribbons of tubular fibrils of 8 nm in diameter (Pecka *et al.*, 1996). In *Pseudotrypanosoma* the microtubules of the pelta are also reinforced by similar ribbons but do not form a disc (Brugerolle, unpublished).

The protein composition of the striated parabasal and costa fibres of trichomonads has been studied by immuno-biochemical techniques (Viscogliosi and Brugerolle, 1993a, 1994a; Brugerolle and Viscogliosi, 1994). Their main constituents are a novel group of striated fibre-forming proteins of 100 to 135 kDa showing fibre- and genus-dependent differences (Figure 9.5 c). Gene sequences of these proteins indicate some homology with myosin II (Viscogliosi, unpublished). This brings to mind the contractile costa of *Pseudotrypanosoma* and *Trichomitopsis* (Amos *et al.*, 1979). The sigmoid fibres consist of proteins of 45 and 55 kDa in *Tritrichomonas* species (Figure 9.5 b) (Viscogliosi and Brugerolle, 1993b). The lamellar UM of *Trichomitus* has been isolated and an antibody recognizes a protein of 25 kDa within

(left)
Figure 9.5 Trichomonads
- a Basal body and fibre organization: arrangement of basal bodies (1,2,3) around the basal body of the recurrent flagellum (R); sigmoid fibres (S), microtubules of the axostyle-pelta (Ax, Pe), striated parabasal fibre (Pf) supporting the Golgi (G), costa fibre (C) associated with the undulating membrane (UM) in *Tritrichomonas*. Bar: 10µm
- b Antibody decoration of sigmoid fibres
- c Antibody decoration of costa
- d Antibody decoration of UM
- e Pleuromitosis in *Trichomonas*: centrosomal structure (C) linked to the flagella (F), paradesmosis (D), kinetochore microtubules (K), nucleus (N). Bar: 1µm
- f Antibody decoration of flagella, axostyle (a) and costa (c) in interphase cell
- g Beginning of mitosis: parental axostyle breakage (a), short paradesmose (d)
- h Elongation of paradesmosis (d) new axostyles (a,a') and completion of flagellar apparatuses

Figure a inset Bar: 10 µm; Figures b, c, d, f, g, h Bar: 5µm

Sources: Figures a and e from G. Brugerolle; Figures b, c, d, f, g, h from E. Viscogliosi and G. Brugerolle

it (Germot *et al.*, 1996) while several proteins have been revealed in the rail-like UM of *Tritrichomonas* species (Figure 9.5d) (Viscogliosi and Brugerolle, 1993b). Striated parabasal fibres of trichomonads have protein homologues in the parabasalids *Pseudotrichonympha* and *Holomastigotoides* (Brugerolle and Viscogliosi, 1994). Anti-centrin antibody labels microfibrillar links between basal bodies in *Holomastigotoides* (Lingle and Salisbury, 1997). Actin has been identified as a component of microfilaments present in pseudopods, phagocytic cups and adhesive ectoplasm in *Trichomonas vaginalis* (Brugerolle *et al.*, 1996). This species possess about ten divergent actin genes encoding proteins similar in size and pI to muscle actin (Bricheux and Brugerolle, 1997).

Division in parabasalids is pleuromitotic and characterized by a pole-to-pole bundle of microtubules or paradesmosis and external kinetochore microtubules (Figure 9.5 e) (Brugerolle, 1991a). Morphogenesis of trichomonads monitored with antibodies has shown that the parental axostyle disassembles from its anterior end and two new ones arise around the two sets of partitioned basal bodies (Figures 9.5 f, g, h). The parental costa in contrast seems permanent; a new one is formed in the sister flagellar apparatus. Subsequent to the elongation of the paradesmosis, the parental flagella separate in two sets while addition of new flagella completes each apparatus in the sister cells (Viscogliosi and Brugerolle, 1994b).

A conspicuous double-membrane-bounded organelle present in all parabasalids is the hydrogenosome of about 200–400 nm in diameter containing moderately electron dense matrix (Müller, 1993; Benchimol *et al.*, 1996) (Figure 9.6). Often a dense spherical core is present in the matrix and between the membranes a flat, membrane-bounded vesicle can be seen (Figure 9.6 b). No DNA has been detected in these organelles. They multiply by division as suggested by frequently seen dumb-bell shaped organelles (Figure 9.6 a), but recent observations indicate also a multiplication by formation of intra-organellar septa. In certain species, endosymbiotic prokaryotes (probably methanogenic archebacteria), can be seen in close proximity of the hydrogenosomes indicating a syntrophic association (Figure 9.6 c), and are frequently surrounded by glycogen particles (Figure 9.6 d) (see Fenchel and Finlay, 1995; Martin and Müller, 1998). Hydrogenosomes play a significant role in energy metabolism (Müller, 1993). The biogenesis of this organelle is very similar to that of mitochondria (Bradley *et al.*, 1997; Müller, 1997a). Extensive molecular data are compatible with the notion that parabasalid hydrogenosomes and mitochondria derive from a common a-proteobacterial ancestor that arose from the original endosymbiotic event (see Biagini *et al.*, 1997a; Müller, 1997a; Martin and Müller, 1998), although the origin of significant metabolic and structural differences between the two organelles needs to be elucidated.

Pelobionts

Pelobionts (about four genera, Figure 9.1) are also named Archamoebae (Cavalier-Smith, 1993). Species of these amoeboid protists bear one basal body/flagellum associated with the nucleus by a cone of microtubules; they do not have mitochondria, hydrogenosomes, or a developed Golgi apparatus (Figures 9.7 a, b) (Brugerolle, 1991a, 1991b). The recent data on the ultrastructure of *Mastigamoeba*

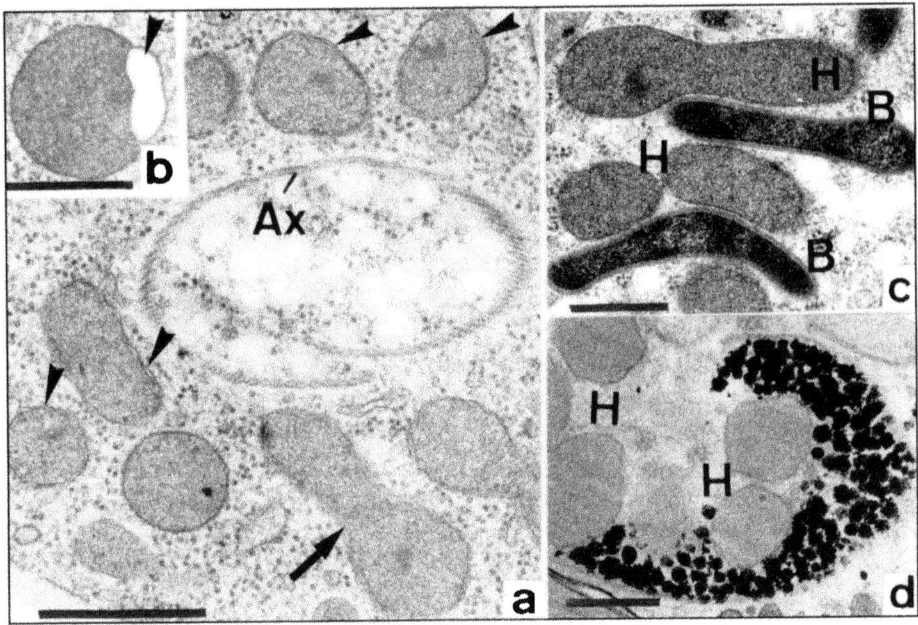

Figure 9.6 Hydrogenosomes of trichomonads

a Sections of hydrogenosomes around the axostyle (Ax) showing the two closely apposed membranes (arrowheads)

b Peripheral lens-like bleb (arrowhead)

c Bacteria (B) intercalated between hydrogenosomes (H) in *Monocercomonas lacertae*

d Concentration of glycogen particles around hydrogenosomes in *Tritrichomonas foetus*

Bar: 1 μm Figures a, c, d and 0.5 μm Figure b

Source: G. Brugerolle

schizophrenia (Simpson *et al.*, 1997) are in conformity with this organization. The group comprises genera bearing one flagellum such as *Mastigamoeba, Mastigina* and *Mastigella* and a genus which has multiple nuclei and flagella, *Pelomyxa*. They live in anoxic or microaerobic water or are endocommensal such as *Mastigina hylae*. EM studies indicate that they form a monophyletic lineage (Brugerolle, 1991a, 1991b).

Other amitochondriate flagellates

Some genera or species which belong to mitochondriate groups/lineages of flagellates but have transformed mitochondria and live in microaerobic habitats need to be mentioned here: the heterolobosean or percolozoan amoeboflagellate *Psalteriomonas* (Broers *et al.*, 1993), the euglenoid *Postgaardi mariagerensis* (Simpson *et al.*, 1996), and *Trimastix convexa*, a flagellate unrelated to any known group (Brugerolle and Patterson, 1997) (Figure 9.1).

Ribosomal RNA sequence supports the heterolobosean or percolozoan affinities of *Psalteriomonas* (Weekers *et al.*, 1997), which suggests that mitochondria can lose cristae and transform in hydrogenosome-like organelles.

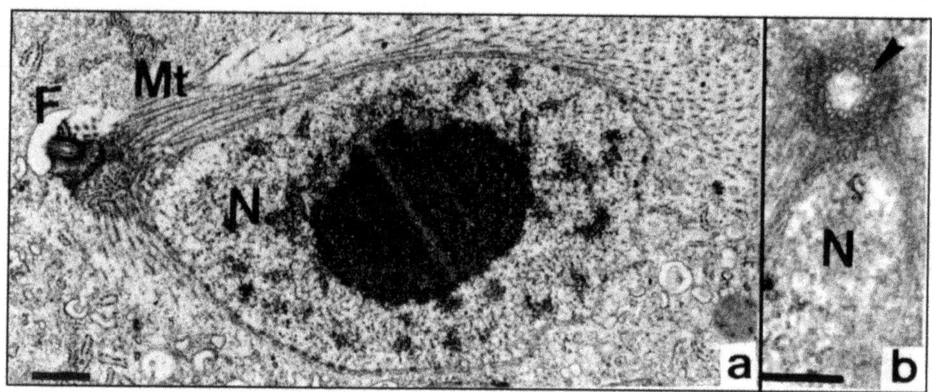

Figure 9.7 Pelobionts
a Microtubule cone (Mt) arising at the base of the unique basal body/flagellum (F) covering the nucleus (N) in *Mastigina hylae*
b Basal body triplets (arrowhead) and microtubule cone covering the nucleus (N) in *Mastigamoeba. sp.*
Bar: 1 µm and 0.2 µm

Source: G. Brugerolle

Cytological features did not reveal clear relationships for *Trimastix convexa* (Brugerolle and Patterson, 1997), while *Postgaardi mariagerensis* (Simpson *et al.*, 1996) is a *bona fide* euglenid.

9.2.3 Intragroup relationships

Morphological, primarily electron microscopic (EM), data gave a seemingly solid basis for elucidation of the relationships of genera within the major groups and led to well-supplemented proposals of intragroup evolutionary relationships. The advent of molecular phylogeny, primarily analysis of small subunit rRNA sequences, provided an independent approach to this problem. While much of the information obtained confirmed earlier notions, some aspects were put in a new light.

Among metamonads, there are no published sequence data for retortamonads, but rRNA of *Chilomastix*, recently established in culture, is being sequenced. In diplomonads, ribosomal RNA sequences are known from *Giardia, Hexamita, Spironucleus* and *Trepomonas*, but not yet from *Octomitus*. Sequences of rRNA (Leipe *et al.*, 1993; Branke *et al.*, 1996; Cavalier-Smith and Chao, 1996), glyceraldehyde-3-phosphate dehydrogenase (GAPDH) (Rozario *et al.*, 1996), but not of the elongation factor EF-1a (Hashimoto *et al.*, 1997), reveal a deep dichotomy with *Giardia* species on one side and the series *Spironucleus, Hexamita* and *Trepomonas* on the other side (Figure 9.3 a). This indicates that the lineage leading to *Giardia*, which lacks a cytostome and has acquired a ventral disc structure adapted to the parasitic lifestyle, has diverged very early. The three other genera seem to have conserved and progressively developed a cytostome in relation to the prey capture and free-living lifestyle. This scenario proposed on the basis of molecular data could represent true relationships. The long branch leading to *Giardia*, however, might have placed this genus at the root artificially. Further data and analysis might resolve this contradiction. These first molecular results show that phylogenetic relationships in this group are more complex than suggested by a previous hypothesis of an evolution from free-living genera to parasitic ones, with a progressive reduction of the cytostomal apparatus (Brugerolle, 1991a; Siddall *et al.*, 1992). The direction of the passage between the monomonad state, represented by *Enteromonas* and *Trimitus*, and the diplomonad state, is another intriguing but unresolved question.

Oxymonads are difficult to cultivate and there are few sequences available. The sequence of the elongation factor EF-1a of *Dinenympha* (*Pyrsonympha*) has been published (Moriya *et al.*, 1998) but symbiotic species of termites can be screened and rRNA sequences are currently obtained.

There are few sequences currently available to decide whether retortamonads, diplomonads and oxymonads are monophyletic or closely related to each other and whether the metamonad grouping must be conserved in systematics.

In parabasalids, new insights on the evolutionary relationships within this lineage were given by sequences of rRNA (Viscogliosi *et al.*, 1993; Berchtold and König, 1995; Gunderson *et al.*, 1995; Silberman *et al.*, 1996; Dacks and Redfield, 1998; Ohkuma *et al.*, 1998; Keeling *et al.*,1998), GAPDH (Viscogliosi and Müller, 1998) and superoxide dismutase (Viscogliosi *et al.*, 1996). However these results must be compared with morphological studies to suggest a possible scenario of the cytoskeleton and protein evolution in this group. The latest phylogenetic tree (Keeling *et al.*, 1998), analyzing thirty-six rRNA sequences of parabasalids comprising several sequences of hypermastigids, is the most complete (Figure 9.8). We must notice that many sequences are not assigned to known species and that thirty-four genera of hypermastigids are only represented by *Trichonympha* and probably by *Spirotrichonympha* and *Joenina*. This analysis reveals a large cluster named Trichomonadidae comprising all trichomonads which have a lamellar UM and a costa with a mesh-like (B type) structure, such as *Trichomonas* and close relatives *Tetratrichomonas, Pentatrichomonas, Pseudotrypanosoma* and *Pentatrichomonoides,* and trichomonads which have a lamellar UM but no costa such as *Pseudotrichomonas* and *Ditrichomonas*. Possibly *Trichomitus* and *Hypotrichomonas* will belong to this group too, as well as *Cochlosoma* which has a lamellar UM and a similar costa. Remarkably *Tritrichomonas* and *Monocercomonas* are not part of this cluster.

A dichotomy into Trichomonadidae, separating *Tritrichomonas* from other members of the trichomonadidae family, has been noted by both morphological and molecular phylogeny (Viscogliosi *et al.*, 1993). Another cluster named Devescovinidae/Calonymphidae is evident, and confirms morphological observations that calonymphids are devescovinid polymastigotes. This cluster also contains a symbiont of *Porotermes* which is probably a joeniid, *Joenina pulchella*.

EM studies have shown the existence of a 'missing link' genus *Projoenia* between the Devescovinidae and Joeniidae (Lavette, 1970). This is one of the evolutionary links between trichomonads and a first group of hypermastigids such as Joeniidae, Lophomonadidae, Rhizomastigidae and genera such as *Deltotrichonympha* and *Koruga* which have a morphology and a morphogenesis close to that of trichomonads (Hollande and Carruette-Valentin, 1971; Brugerolle, 1999). The Trichonymphidae cluster comprises *Trichonympha* and their close relatives, spirotrichonymphids. This

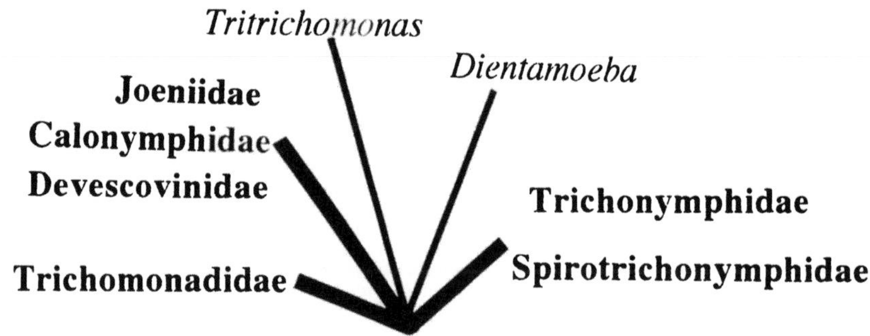

Figure 9.8 Broad phylogenetic relationships of parabasalids based on rRNA sequences analysis by Keeling *et al.*, 1998 (modified by G. Brugerolle). Three main clusters were seen: Trichomonadidae, Devescovinidae/Calonymphidae to which a first group of hypermastigids comprising Joeniidae could be associated, and Trichonymphidae/Spirotrichonymphidae the second group of hypermastigids. Several genera such as *Tritrichomonas* and *Dientamoeba* do not cluster with these major groups

second group of hypermastigids that divides symmetrically, conserving their basal bodies/flagella and attached fibres, is the most distant from the trichomonads. This cluster could accomodate most of the trichonymphids and spirotrichonymphids (Hollande and Carruette-Valentin, 1971).

The major difference from the earlier evolutionary hypothesis based on morphology is the splitting of the Monocercomonadidae to which trichomonads with a simple flagellar apparatus were assigned. Many of these genera, such *Dientamoeba*, have lost their flagellar apparatus or fibres such as costa and UM, for example *Monocercomonas* (Viscogliosi *et al.*, 1993). Obviously the family Monocercomonadidae is not monophyletic, and a revision of the taxonomy is necessary, but this will only be possible after information has been obtained from most of the genera representing the different families. Surprisingly, hypermastigids such as Trichonymphidae, considered as the highest level of the evolution in this parabasalid lineage, emerge at the base of the tree. The phylogeny of this lineage of about eighty genera, if ever completed, could present more surprises.

Ribosomal RNA sequences of several pelobiont genera have been obtained, but phylogenetic relationships within this group are not resolved yet.

9.3 Habitats/ecology

Free-living and parasitic amitochondrial flagellates live in anaerobic or microaerobic habitats. Most species perish in fully oxygenated environments but some tolerate media containing dissolved oxygen or can even multiply in them (Table 9.1).

9.3.1 Free-living forms (Table 9.2)

They are found in niches with little or no oxygen and/or in sites rich in organic matter, such as sediments of stagnant reservoirs, water treatment plants, soils

Table 9.1 Comparison of oxygen affinities and O_2 inhibition thresholds in amitochondriate flagellates, 'anaerobic' ciliates and amoebae

Organism	*Habitat*	*In situ O_2 (μM)*	*Apparent K_m O_2 (μM)*	*O_2 inhibition threshold (μM)*
Diplomonads				
Hexamita sp.	Free-living limnic	0–30	12.97	100
Giardia lamblia	Human small intestine (lumen)	0–60	6.4	80
Giardia muris	Mouse small intestine	?	2.0	15
Parabasalids				
Trichomonas vaginalis	Human vagina	13–56	3.2	19
Tritrichomonas foetus	Bovine uro-genital tract	?	1.6	?
Entamoebids				
Entamoeba histolytica	Human small intestine (lumen)	0–60	5.5	16
Ciliates				
Eudiplodinium maggii	Rumen	<0.25–3.2	5.2	5
Plagiopila frontata	Marine sediment	<2.6	2.6	<10

Source: simplified from Biagini *et al.*, 1997b

fertilized with manure, anoxic ponds and fjords. Most of the free-living genera have been described before (see section 9.1) and their anaerobic ecosystems/habitats have been extensively studied (Fenchel and Finlay, 1995). A detailed study (Fenchel *et al.*, 1995) of an anoxic fjord showed the presence of amitochondrial flagellates such as the retortamonad *Chilomastix cuspidata*, the diplomonads *Trepomonas agilis* and *Hexamita inflata*, the pelobiont *Mastigamoeba* sp. and the euglenoid *Postgaardi mariagerensis*. They were found in the anaerobic zone of the water column between 15 to 26 m in depth, where CH_4, S^{2-} and NH_4^+ were present. The living conditions of the free-living diplomonad *Hexamita* sp. have been studied in some detail (Biagini *et al.*, 1997b). This organism has been isolated from a lake, was cultivated in the presence of bacteria, and finally established in axenic culture in a medium used for parasitic trichomonads. *Hexamita* consumes O_2 and tolerates O_2 under a threshold of 100 μM. A comparison of various 'anaerobic' protists, including *Giardia*, several trichomonads, rumen ciliates and 'anaerobic' free-living ones as well as *Entamoeba histolytica*, shows *Hexamita* sp. to be the most aerotolerant (Table 9.1).

9.3.2 Parasitic groups (Table 9.2)

They comprise more genera than the free-living ones and usually occupy anaerobic sites in their hosts. Most species are endocommensal, some can be qualified as symbiotic (termite flagellates), and some are pathogenic parasites (some trichomonads and diplomonads) (Grassé, 1952; Kulda and Nohynková, 1978). Many live in the hindgut of animals that have an extended bowel, where bacteria, protozoa and fungi together establish thriving fermentative ecosystems (McBee, 1977) which make significant contributions to the nutrition of the host. The hindgut of fishes, amphibia, reptiles and birds accommodates endocommensal

Table 9.2 Distribution of amitochondriate flagellates between free-living and parasitic life styles

	Retortamonads (2 genera)	*Diplomonads (10 genera)*	*Oxymonads (11 genera)*	*Parabasalids (80 genera)*	*Pelobionts (4 genera)*
Free-living	Chilomastix	Trepomonas Hexamita Trigonomonas Gyromonas?		Pseudotricho-monas Ditricho-monas	Mastigamoeba Mastigina Mastigella Pelomyxa
Parasitic	Chilomastix Retortamonas	Giardia Spironucleus Octomitus Trepomonas Hexamita ? Enteromonas Trimitus Caviomonas?	all genera e.g. Polymastix Monocer-comonoides Saccinobaculus Pyrsonympha Streblomastix Oxymonas	78 genera e.g. Trichomonas Tritrichomonas Monocer-comonas Histomonas Dientamoeba Joenia Trichonympha	Mastigina Mastigamoeba

retortamonads, diplomonads, oxymonads and trichomonads. The caecum of Equidae and rodents, for example the guinea-pig, hamster, marmot, ground squirrel and many others, also contains numerous amitochondriate flagellates. Many insects harbour amitochondriate flagellates in their hindgut, for example the mole-cricket accomodates retortamonads, many plant-eating melolonthoid larvae (*Cetonia, Oryctes*) contain oxymonads (*Monocercomonoides, Polymastix*), and the tipulid and some trichopterid larvae harbour retortamonads and trichomonads.

The most successful ecosystem that contains amitochondriate flagellates is found in lower termites (Grassé, 1952; Honigberg, 1970) and wood-roaches (*Cryptocercus*) (Cleveland *et al.*, 1934). All termite families, except Termitidae, accommodate bacteria and symbiotic flagellates, mainly oxymonads and parabasalids, in a paunch of the hindgut and rely on them for cellulose fermentation. Flagellates can represent up to 16 per cent of the total body weight. The hindgut content is anaerobic and the flagellates are highly sensitive to aerobiosis. Termites can be defaunated by twenty-four-hour exposure to a 100 per cent oxygen atmosphere. Termite flagellates have no cytostome and feed by phagocytosis and pinocytosis (McBee, 1977). The cytoplasm of most of them is filled with vacuoles containing pieces of wood in various stages of digestion. Digestion of cellulose has been demonstrated in cultures of certain flagellate species with the formation of acetate, carbon dioxide and H_2 as fermentative endproducts. Acetate in turn serves as a nutrient for the termite (Honigberg, 1970; Odelson and Breznak, 1985).

The rumen is another well-studied anoxic ecosystem composed of prokaryotes (both archebacteria and bacteria), protists and fungi (Bauchop, 1977). Among protozoa, ciliates are dominant but amitochondriate flagellates, such as retortamonads (*Chilomastix*), diplomonads (*Octomitus*), oxymonads (*Monocercomonoides*), and trichomonads (*Monocercomonas*) are also present. The redox potential of the rumen fluid is between -250 to -400mV, reflecting anaerobiosis, and its pH is between 6 to 7. Dissolved gases represent major end-products of the metabolism: 60–70 per cent CO_2 and 30-40 per cent CH_4. The rumen serves as a fermenter, in which cellulose, hemicelluloses and pectins are degraded to glucose, further metabolized via glycolysis to pyruvate and volatile fatty acids; this process is accompanied by ATP formation (Thivend *et al.*, 1985). Flagellates represent only a small portion of the biomass compared with prokaryotes, ciliates and fungi (Bauchop, 1977), thus their contribution to this ecosystem, so far not studied, might be small.

Some amitochondriate flagellates are much-studied pathogenic parasites. In view of their excellent coverage in the literature, we mention only the most important species. These include the diplomonads, *Giardia* spp. in humans, wild and domestic animals (dogs), *Spironucleus muris* in mice (Kulda and Nohýnková, 1978, 1995), and *Hexamita/Spironucleus* of salmonids (Sterud, 1997). Among the intestinal trichomonads only the bird parasites *Trichomonas gallinae* and *Histomonas meleagridis* are pathogenic (Honigberg, 1978). *Trichomonas vaginalis* and *Tritrichomonas foetus* are pathogens of the human and bovine genito-urinary tract respectively (Honigberg, 1978). Most of these flagellates form resistant cysts, which serve as transmission forms. Some pathogenic trichomonads such as *Trichomonas vaginalis* do not form cysts and are directly transmitted from host to host. The termite flagellates are mainly transmitted by proctodeal feeding (Grassé, 1952; Honigberg, 1970).

9.4 Energy metabolism

Although much information is available on the biology, ecology and morphology of diverse amitochondriate flagellates, their metabolic biochemistry has been explored only in a handful of diplomonads (free-living *Hexamita* spp. (Biagini *et al.*, 1997b) and parasitic *Giardia lamblia*) and parabasalid (several parasites) species available in axenic cultures. Pelobionts remain essentially unexplored in this respect (Chapman-Andresen and Hamburger, 1981).

ATP production is dependent on glycolysis and also on amino acid catabolism (Coombs and Müller, 1995; North and Lockwood, 1995; Müller, 1998). In the absence of mitochondrial energy conservation, these pathways provide only a few moles of ATP for each substrate molecule catabolized, thus these organisms are rather profligate in their substrate utilization. Pyruvate is formed via classical glycolysis, and is further metabolized in the cytosol, or in type II amitochondriates also in the hydrogenosomes (Figure 9.9). In diplomonads the main metabolic end products are acetate, ethanol and alanine. The nature of end products is taxon-dependent in parabasalids; in *Trichomonas vaginalis* they are glycerol and lactate, formed in the cytosol, and acetate and H_2 formed in the hydrogenosomes. The proportion of the end products formed is affected by pO_2 of the medium (Ellis *et al.*, 1992; Paget *et al.*, 1993). Under anaerobic conditions the more reduced products, ethanol or lactate predominate, while the presence of ozone favours acetate production. The maintenance energy, that is, the amount of energy needed for all cellular functions save the increase in biomass, approaches 50 per cent in *Trichomonas* (ter Kuile, 1994), an exceptionally high value.

The amino acid arginine is also an important energy substrate in *Giardia* and in parabasalids, since they have the ATP-producing arginine dihydrolase pathway (North and Lockwood, 1995).

Besides the absence of mitochondrial energy conservation, the amitochondriate

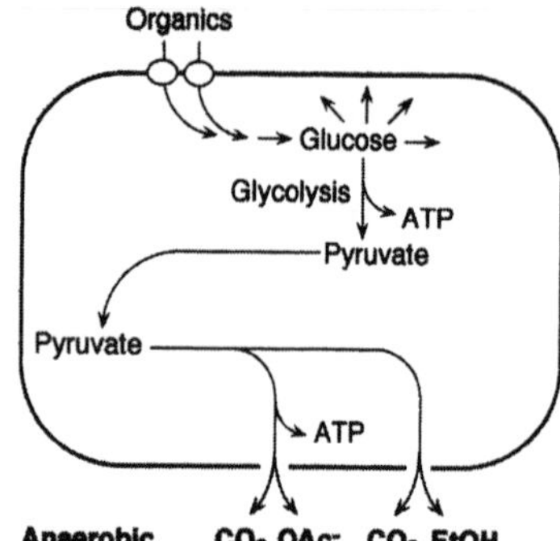

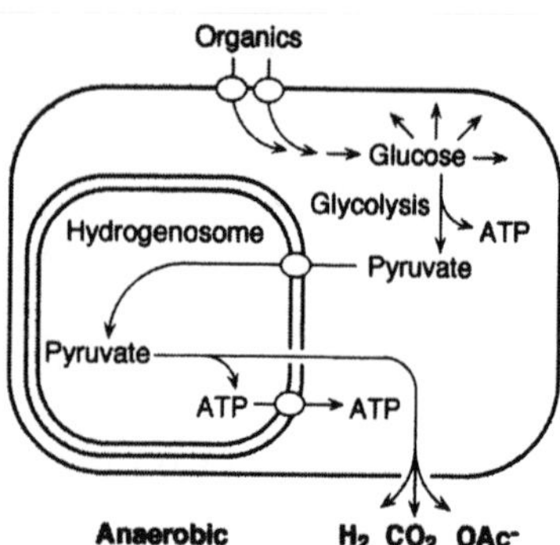

Figure 9.9 Schematic representation of carbohydrate catabolism in Type I (*Figure a*) and Type II (*Figure b*) amitochondriate flagellates: acetate (OAc-), ethanol (EtOH), adenosine triphosphate (ATP)

Source: Martin and Müller 1998, reproduced with the permission of the publisher

protists share a number enzymatic characteristics (Müller, 1998). Pyruvate is oxidized by an iron-sulphur enzyme, pyruvate: ferredoxine oxidoreductase (PFO) and not the mitochondrial type pyruvate dehydrogenase complex. This enzyme is linked to additional iron-sulphur proteins. Furthermore, inorganic pyrophosphate (PPi) replaces ATP as phosphoryl donor in the phosphorylation of glucose-6-phosphate by PPi-linked phosphofructokinase. This reaction enhances the meagre ATP yield of glucose fermentation (Mertens, 1993). Type I and type II amitochondriate metabolism is defined by the absence or presence of hydrogenosomes (Müller, 1998; Figure 9.9). In parabasalids, pyruvate oxidation takes place in this organelle, with the formation of molecular hydrogen catalyzed by Fe-hydrogenase (Müller, 1993). Such a reaction is unusual for eukaryotes. The fate of acetyl-CoA from pyruvate oxidation is also different in the two types of organisms. In *Giardia* it is converted to the end products, ethanol and acetate by unusual enzymes (Sánchez and Müller, 1996; Sánchez, 1998) found only in one other protist, *Entamoeba* and in various prokaryotes. End-product formation in parabasalids, both in hydrogenosome and cytosol, is catalyzed by enzymes that occur in mitochondriate eukaryotes as well.

Although amitochondriate protists clearly lack the energy-conserving function of mitochondria, this trait is probably not ancestral but represents secondary adaptations. Increasing evidence suggests that the ancestors of these organisms did encounter in their past the endosymbiotic event leading to the development of mitochondria. A number of genes of likely endosymbiotic origin, indicating a probable mitochondrial ancestry, have been detected in the nuclear genome of amitochondriate protists (see for example Katz, 1998; Müller, 1997a, 1997b; Rosenthal *et al.*, 1997), with several of these genes sharing a most recent common ancestor with their mitochondrial homologues. Losses during evolution are frequent, thus a disappearance of characteristic mitochondrial functions and components can be attributed to adaptations to life in anoxic environments (Fenchel and Finlay, 1995; Martin and Müller, 1998). Common features of amitochondriate and mitochondriate organisms probably derived from their common ancestor (Martin and Müller, 1998).

The origin of enzymes that differentiate beween the two groups of organisms is also a puzzling question (Müller, 1998). Some of these probably have been inherited from the common organellar ancestor that arose in the original endosymbiotic event (Martin and Müller, 1998), but others could have been acquired by the eukaryotic cell from other prokaryotes at various stages of their history (Brown *et al.*, 1998; Rosenthal *et al.*, 1997; Doolittle, 1998).

9.5 Mutual relationships of the major lineages

Ultrastructural studies have revealed a multiple of clearly distinguishable organizational types among protists, but in essence no synapomorphies that would permit a hierarchical arrangement of these. The result has been an increasing proliferation of independent major taxa (Patterson and Sogin, 1992) but no natural taxonomy. As was shown earlier, each of the major groups of amitochondriate protists discussed here could be recognized as having sufficiently well-defined morphology to assume their monophyletic status. Morphology, however, provided no clues to their relative taxonomic position, just as it did not for the rest of the groups. Information on macromolecular sequences has raised the hope that sequencing homologous macromolecules from a sufficient number of representatives of each group will give us a clear picture of large-scale phylogenetic relationships. The greatest stock has been put into the gene for the small subunit ribosomal RNA, with a spectacular number of protist species sequenced in the past decade and half (Sogin, 1991). More recently, protist databases for various proteins have also started expanding. In parallel with accumulation of data there has been a constant improvement in methods for sequence analysis and phylogenetic reconstruction. All this effort has provided a wealth of important and useful information, and has led to diverse phylogenetic inferences. However, in essence, the place of the amitochondriate flagellate groups on a global phylogenetic tree of the eukaryotes, and thus their relationship to other protist lineages remains unresolved (Philippe and Adoutte, 1998; Embley and Hirth, 1998; Philippe and Laurent, 1998; Katz, 1998).

REFERENCES

Amos, W. D., Grimstone, A. V., Rothschild, L. J. and Allen, R. D. (1979) Structure, protein composition and birefringence of the root fiber in the flagellate *Trichomonas*. *Journal of Cell Science*, **35**, 139–164.

Bauchop, T. (1977) Foregut fermentation. In *Microbial Ecology of the Gut* (eds R. T. J. Clarke and T. Bauchop), London: Academic Press, pp. 223–248.

Bernard, C., Simpson, A. G. B. and Patterson, D. J. (1997) An ultrastructure study of a free-living retortamonad, *Chilomastix cuspidata* (Larsen et Patterson, 1990) n. comb. (Retortamonadida, Protista). *European Journal of Protistology*, **33**, 254–265.

Benchimol, M., Almeida, J. C. A. and de Souza, W. (1996) Further studies on the organization of the hydrogenosome in *Tritrichomonas foetus*. *Tissue and Cell*, **28**, 287–299.

Berchtold, M. and König, H. (1995) Phylogenetic position of two uncultivated trichomonads

Pentatrichomonoides scroa and *Metadevescovina extranea* Kirby from the hindgut of the termite *Mastotermes darwiniensis* Frogatt. *Systematic and Applied Microbiology*, **18**, 567–573.

Biagini, G. A., Finlay, B. J. and Lloyd, D. (1997a) Evolution of hydrogenosomes. *FEMS Microbiology Letters*, **155**, 133–140.

Biagini, G. A., Suller, M. T. E., Finlay, B. J. and Lloyd, D. (1997b) Oxygen uptake and antioxidant responses of the free-living diplomonad *Hexamita* sp. *Journal of Eukaryotic Microbiology*, **44**, 447–453.

Bradley, P. J., Lahti, C. J., Plümper, E. and Johnson, P. J. (1997) Targeting and translocation of proteins into the hydrogenosome of the protist *Trichomonas*: similarities with mitochondrial protein import. *EMBO Journal*, **16**, 3484–3493.

Branke, J. M., Berchtold, M., Breunig, A. and König, H. (1996) 16S-like rDNA sequence and phylogenetic position of the diplomonad *Spironucleus muris* (Lavier 1936). *European Journal of Protistology*, **32**, 227–233.

Bricheux, G. and Brugerolle, G. (1997) Molecular cloning of actin genes in *Trichomonas vaginalis* and phylogeny inferred from actin sequences. *FEMS Microbiological Letters*, **153**, 205–213.

Broers, C. A. M., Meijers, H. H. M., Symens, J. C., Stumm, C. K., Vogels, G. D. and Brugerolle, G. (1993) Symbiotic association of *Psalteriomonas vulgaris* n. spec. with *Methanobacterium formicicum*. *European Journal of Protistology*, **29**, 98–105.

Brown, D. M., Upcroft, J. A., Edwards, M. R. and Upcroft, P. (1998) Anaerobic bacterial metabolism in the ancient eukaryote *Giardia duodenalis*. *International Journal for Parasitology*, **28**, 149–164.

Brugerolle, G. (1991a) Flagellar and cytoskeletal systems in amitochondrial flagellates: Archamoebae, Metamonada and Parabasala. *Protoplasma*, **164**, 70–90.

Brugerolle, G. (1991b) Cell organization in free-living amitochondriate heterotrophic flagellates. In *The Biology of Free-living heterotrophic Flagellates* (eds D. J. Patterson and J. Larsen), Oxford: Clarendon Press, pp. 133–148.

Brugerolle, G. and Lee, J. J. (2000) Phylum Parabasalia. In *An Illustrated Guide to the Protozoa* (eds J. J. Lee, G. F. Leedale, D. J. Patterson and P. C. Bradbury), *2nd edn*, Society of Protozoologists, in press.

Brugerolle, G., Breunig, A. and König, H. (1994) Ultrastructure study of *Pentatrichomonoides* sp. a trichomonad flagellate from *Mastotermes darwiniensis*. *European Journal of Protistology*, **30**, 372–378.

Brugerolle, G., Bricheux, G. and Coffe, G. (1996) Actin cytoskeleton demonstration in *Trichomonas vaginalis* and in other trichomonads. *Biology of the Cell*, **88**, 29–36.

Brugerolle, G. and König, H. (1997) Ultrastructure and organization of the cytoskeleton in *Oxymonas*, an intestinal flagellate of termites. *Journal of Eukaryotic Microbiology*, **44**, 305–313.

Brugerolle, G. and Patterson, D. J. (1997) Ultrastructure of *Trimastix convexa* Hollande, an amitochondriate anaerobic flagellate with a previously undescribed organization. *European Journal of Protistology*, **33**, 121–130.

Brugerolle, G. and Viscogliosi, E. (1994) Organization and composition of the striated roots supporting the Golgi apparatus, the so-called parabasal apparatus, in parabasalid flagellates. *Biology of the Cell*, **81**, 277–285.

Cavalier-Smith, T. (1993) Kingdom Protozoa and its 18 phyla. *Microbiological Reviews*, **57**, 953–994.

Cavalier-Smith T. and Chao, E. E. (1996) Molecular phylogeny of the free-living archezoan *Trepomonas agilis* and the nature of the first eukaryote. *Journal of Molecular Evolution*, **43**, 551–562.

Chapman-Andresen, C. and Hamburger, K. (1981) Respiratory studies on the giant amoeba *Pelomyxa palustris*. *Journal of Protozoology*, **28**, 433–440.

Cleveland, L. R., Hall, S. R., Sanders, E. P. and Collier, J. (1934) The wood-feeding roach *Cryptocercus*, its protozoa and the symbiosis between protozoa and roach. *Memoires of the American Academy of Sciences,* **17**, 155–342.

Coombs, G. H. and Müller, M. (1995) Energy metabolism in anaerobic protozoa. In *Biochemistry and Molecular Biology of Parasites* (eds J. J. Marr and M. Müller), London: Academic Press, pp. 33–47.

Dacks, J. B. and Redfield, R. J. (1998) Phylogenetic placement of *Trichonympha. Journal of Eukaryotic Microbiology*, **45**, 445–447.

Doolittle, W. F. (1998) You are what you eat: a gene transfer ratchet could account for bacterial genes in eukaryotic nuclear genomes. *Trends in Genetics*, **14**, 307–311.

Ellis, J. E., Cole, D. and Lloyd, D. (1992) Influence of oxygen on the fermentative metabolism of metronidazole-sensitive and resistant strains of *Trichomonas vaginalis. Molecular and Biochemical Parasitology*, **56**, 79–88.

Embley, T. M. and Hirt, R. P. (1998) Early branching eukaryotes? *Current Opinion in Genetics and Development*, **8**, 624–629.

Fenchel, T. and Finlay, B. J. (1995) Ecology and evolution in anoxic worlds. In *Ecology and Evolution* (eds R. M. May and P. Harvey), Oxford: Oxford University Press, pp. 276.

Fenchel, T., Bernard, C., Esteban, G., Finlay, B. J., Hansen, P. J. and Iversen, N. (1995) Microbial diversity and activity in a Danish fjord with anoxic deep water. *Ophelia*, **43**, 45–100.

Germot, A., Brugerolle, G. and Viscogliosi, E. (1996) The undulating membrane of trichomonads; the structure and immunolabelling of its cytoskeleton. *European Journal of Protistology*, **32**, 298–305.

Grassé, P. P. (1952) Classe des Zooflagellata ou Zoomastigina. In *Traité de Zoologie tome I (1)* (ed. P. P. Grassé), Paris: Masson, pp. 704–982.

Gunderson, J., Hinkle, G., Leipe, D., Morrison, H., Stickel, S. K., Odelson, D. A., Breznak, J. A., Nerad, T. A., Müller, M. and Sogin, M. (1995) Phylogeny of trichomonads inferred from small-subunit rRNA sequences. *Journal of Eukaryotic Microbiology*, **42**, 411–415.

Hashimoto, T., Nakamura, Y., Kamaishi T. and Hasegawa, M. (1997) Early evolution of eukaryotes inferred from protein phylogenies of translation elongation factors 1a and 2. *Archiv für Protistenkunde*, **148**, 287–295.

Hollande, A. and Carruette-Valentin, J. (1971) Les atractophores, l'induction du fuseau et la division cellulaire chez les Hypermastigines, étude infrastructurale et révision systématique des Trichonymphines et des Spirotrichonymphines. *Protistologica*, **7**, 3–100.

Honigberg, B. M. (1970) Protozoa associated with termites and their role in digestion. In *Biology of termites* (eds K. Krishna and F. M. Weesner), New York: Academic Press, pp. 1–36.

Honigberg, B. M. (1978) Trichomonads of veterinary importance. In *Parasitic Protozoa, vol. 2* (ed. J. P. Kreier), New York: Academic Press, pp. 163–273.

Katz, L. A. (1998) Changing the perspectives on the origin of eukaryotes. *Trends in Ecology and Evolution*, **13**, 493–497.

Keeling, P. J., Poulsen, N. and McFadden, G. I. (1998) Phylogenetic diversity of parabasalian symbionts from termites, including the phylogenetic position of *Pseudotrypanosoma* and *Trichonympha. Journal of Eukaryotic Microbiology*, **45**, 643–650.

Kulda, J. and Nohýnková, E. (1978) Flagellates of human intestine and of intestines of other species. In *Parasitic Protozoa, vol. 2* (ed. J. P. Kreier), New York: Academic Press, pp. 2–138.

Kulda, J. and Nohýnková, E. (1995) *Giardia* in humans and animals. In *Parasitic Protozoa, 2nd edn , vol. 10* (ed. J. P. Kreier), San Diego: Academic Press, pp. 225–422.

Lavette, A. (1970) Sur le genre *Projoenia* et les affinités des Joeniidae (Zooflagellés, Metamonadina). *Comptes Rendus de l'Académie des Sciences de Paris*, **270**, 1695–1698.

Leipe, D. D., Gunderson, J. H., Nerad, T. A. and Sogin, M. (1993) Small subunit ribosomal RNA of *Hexamita inflata* and the quest for the first branch in the eukaryotic tree. *Molecular and Biochemical Parasitology*, **59**, 41–48.

Lingle, L. W. and Salisbury, J. L. (1997) Centrin and the cytoskeleton of the protist *Holomastigotoides. Cell Motility and the Cytoskeleton*, **36**, 377–390.

Martin, W. F. and Müller, M. (1998) The hydrogen hypothesis of the first eukaryote. *Nature*, **392**, 37–41.

McBee, R. H. (1977) Fermentation in the hindgut. In *Microbial Ecology of the Gut* (eds R. T. J. Clarke and T. Bauchop), London: Academic Press, pp. 185–222.

Meng, T. C., Aley, S. B., Svärd, S. G., Smith, M. W., Huang, B., Kim T. C. and Gillin, F. D. (1996) Immunolocalization and sequence of caltractin/centrin from the early branching eukaryote *Giardia lamblia. Molecular and Biochemical Parasitology*, **79**, 103–108.

Moriya, S., Okhuma, M. and Kudo T. (1998) Phylogenetic position of symbiotic protist *Dinenympha exilis* in the hindgut of the termite *Reticulitermes speratus* inferred from the protein phylogeny of elongation factor 1a. *Gene*, **210**, 221–227.

Mertens, E. (1993) ATP versus pyrophosphate: Glycolysis revisited in parasitic protists. *Parasitology Today*, **9**, 122–126.

Müller, M. (1993) The hydrogenosome. *Journal of General Microbiology*, **139**, 2879–2889.

Müller, M. (1997a) Evolutionary origin of hydrogenosomes. *Parasitology Today*, **13**, 166–167.

Müller, M. (1997b) What are the microsporidia? *Parasitology Today*, **13**, 455–456.

Müller, M. (1998) Enzymes and compartmentation of core energy metabolism of anaerobic protists – a special case in eukaryotic evolution? In *Evolutionary Relationships Among Protozoa* (eds G. H. Coombs , K. Vickerman, M. A. Sleigh and A. Warren), Dordrecht: Kluwer, pp. 109–131.

Nohria, A., Alonso, R. A. and Peattie, D. A. (1992) Identification and characterization of g-giardin and the g-giardin gene from *Giardia lamblia. Molecular and Biochemical Parasitology*, **56**, 27–38.

North, M. J. and Lockwood, B. C. (1995) Amino acid and protein metabolism. In *Biochemistry and Molecular Biology of Parasites* (eds J. J. Marr and M. Müller), London: Academic Press, pp. 67–88.

Odelson, D. A. and Breznak, J. A (1985) Nutrition and growth characteristics of *Trichomitopsis termopsidis*, a cellulolytic protozoan from termites. *Applied and Environmental Microbiology*, **49**, 614–621.

Ohkuma, M., Ohtoko, K., Grunau, C., Moriya, S. and Kudo, T. (1998) Phylogenetic identification of the symbiotic hypermastigote *Trichonympha agilis* in the hindgut of the termite *Reticulitermes speratus* based on small-subunit rRNA sequence. *Journal of Eukaryotic Microbiology*, **45**, 439–444.

Paget, T. A., Kelly, M. L., Jarroll E. L., Lindmark, D. G. and Lloyd, D. (1993) The effect of oxygen on fermentation in *Giardia lamblia. Molecular and Biochemical Parasitology*, **57**, 65–72.

Patterson, D. J. and Sogin, M. L. (1992) Eukaryotic origins and protistan diversity. In *The Origins and Evolution of the Cell* (eds H. Hartman and K. Matsuno), Singapore: World Scientific, pp. 13–46.

Pecka, Z., Nohýnková, E. and Kulda, J. (1996) Ultrastructure of *Cochlosoma anatis* Kotlan, 1923 and taxonomic position of the family Cochlosomatidae (Parabasalidea: Trichomonadida). *European Journal of Protistology*, **32**, 190–201.

Philippe, H. and Adoutte, A. (1998) The molecular phylogeny of Eukaryota: solid facts and uncertainties. In *Evolutionary Relationships Among Protozoa* (eds G. H. Coombs, K. Vickerman, M. A. Sleigh and A. Warren), Dordrecht: Kluwer, pp. 25–56.

Philippe, H. and Laurent, J. (1998) How good are deep phylogenetic trees? *Current Opinion in Genetics and Development*, **8**, 616–623.

Rosenthal, B., Mai, Z., Caplivsksi, D., Ghosh, S., de la Vega, H., Graf, T. and Samuelson, J. (1997) Evidence for the bacterial origin of genes encoding fermentation enzymes of the amitochondriate protozoan parasite *Entamoeba histolytica. Journal of Bacteriology*, **179**, 3736–3747.

Rozario, C., Morin, L., Roger, A. J., Smith, M. W. and Müller, M. (1996) Primary structure and phylogenetic relationships of glyceraldehyde-3-phosphate dehydrogenase genes of free-living and parasitic diplomonad flagellates. *Journal of Eukaryotic Microbiology*, **43**, 330–340.

Sánchez, L. B. 1998. Aldehyde dehydrogenase(CoA-acetylating) and the mechanism of ethanol formation in the amitochondriate protist, *Giardia lamblia*. *Archives of Biochemistry and Biophysics*, **354**, 57–64.

Sánchez, L. B. and Müller, M. (1996) Purification and characterization of the acetate forming enzyme, acetyl-CoA synthetase (ADP-forming) from the amitochondriate protist, *Giardia lamblia*. *FEBS Letters*, **378**, 240–244.

Siddall, M. E., Hong, H. and Desser, S. S. (1992) Phylogenetic analysis of the Diplomonadida (Wenyon 1925, Brugerolle 1975): Evidence for heterochrony in protozoa and against *Giardia lamblia* as a 'missing link'. *Journal of Protozoology*, **39**, 361–367.

Silberman, J. D., Clark, C. G. and Sogin, M. L. (1996) *Dientamoeba fragilis* shares a recent common evolutionary history with the trichomonads. *Molecular and Biochemical Parasitology*, **76**, 311–314.

Simpson, A. G. B., Bernard, C., Fenchel, T. and Patterson, D. J. (1997) The organisation of *Mastigamoeba schizophrenia* n. sp.: more evidence of ultrastructural idiosynchrasy and simplicity in pelobiont protists. *European Journal of Protistology*, **33**, 87–98.

Simpson, G. B., Van Den Hoff, J., Bernard, C., Burton, H. R. and Patterson D. J. (1997) The ultrastructure and systematic position of the Euglenozoon *Postgaardi mariagerensis*, Fenchel *et al.*, *Archiv für Protistenkunde*, **147**, 213–225.

Sogin, M. L. (1991) Early evolution and the origin of eukaryotes. *Current Opinion in Genetics and Development*, **1**, 457–463.

Sterud, E. (1998) Ultrastructure of *Spironucleus torosa* Pointon et Morrison, 1990 (Diplomonadida: Hexamitidae), in cod *Gadus morhua* (L.) and saithe *Pollachius virens* (L.) from South-Eastern Norway. *European Journal of Protistology*, **34**, 69–77.

ter Kuile, B. H. (1994) Carbohydrate metabolism and physiology of the parasitic protist *Trchomonas vaginalis* studied in chemostasts. *Microbiology*, **140**, 2495–2502.

Thivend, P., Fonty, G., Jouany, J. P., Durand, M. and Gouet, P. (1985) Le fermenteur rumen. *Reproduction , Nutrition et Dévelopement*, **25**, 729–753.

Viscogliosi, E. and Brugerolle, G. (1993a) Cytoskeleton in trichomonads. I. Immunological and biochemical comparative study of costal proteins in the genus *Tritrichomonas*. *European Journal of Protistology*, **29**, 160–170.

Viscogliosi, E. and Brugerolle, G. (1993b) Cytoskeleton in trichomonads. II. Immunological and biochemical characterization of the preaxostylar fibres and undulating membrane in the genus *Tritrichomonas*. *European Journal of Protistology*, **29**, 381–389.

Viscogliosi, E. and Brugerolle, G. (1994a) Striated fibers in trichomonads: Costa proteins represent a new class of proteins forming striated roots. *Cell Motility and the Cytoskeleton*, **29**, 82–93.

Viscogliosi, E. and Brugerolle, G. (1994b) Cytoskeleton in trichomonads. III. Study of the morphogenesis during division by using monoclonal antibodies against cytoskeletal structures. *European Journal of Protistology*, **30**, 129–138.

Viscogliosi, E., Durieux, I., Delgado-Viscogliosi, P., Bayle, D. and Dive, D. (1996) Phylogenetic implication of iron-containing superoxide dismutase genes from trichomonad species. *Molecular and Biochemical Parasitology*, **80**, 209–214.

Viscogliosi, E. and Müller, M. (1998) Phylogenetic relationships of the glycolytic enzyme, glyceraldehyde-3-phospho dehydrogenase, from parabasalid flagellates. *Journal of Molecular Evolution*, **47**, 190–199.

Viscogliosi, E., Philippe, H., Baroin, A., Perasso R. and Brugerolle, G. (1993) Phylogeny of trichomonads based on partial sequences of large subunit rRNA and on cladistic analysis of morphological data. *Journal of Eukaryotic Microbiology*, **40**, 411–421

Weekers, P. H. H., Kleyn, J. and Vogels, G. D. (1997) Phylogenetic position of *Psalteriomonas lanterna* deduced from the SSUrDNA sequence. *Journal of Eukaryotic Microbiology*, **44**, 467–470.

Chapter 10

Adaptations to parasitism among flagellates

Keith Vickerman

ABSTRACT

Adaptations are usually regarded as the material products of natural selection acting on differential reproductive success. For a parasite, reproductive success depends not only on its survival and multiplication within a given host, but also on its ensuring transmission to a new host. Adaptations relating to transmission and niche selection within the host are discussed using *Giardia lamblia* and *Trypanosoma brucei* as examples of an intestinal parasite transmitted by cysts through the faeco-oral route and a blood parasite transmitted by vector, respectively.

The flagellum may function not only in important migratory movements within the host, but also in attachment mechanisms securing the parasite in a particular site. Studies on trypanosomatid parasites have provided some of the best illustrations of adaptation to parasitism in two different hosts. In *T. brucei*, the ability to switch patterns of energy metabolism adaptively with a change of host by repressing and activating the single mitochondrion has been investigated extensively, as has the parasite's ability to evade the mammalian host's immune response by repeatedly changing the glycoprotein antigen composing the parasite's surface coat. Subversion of host defences by manipulation of the activities of the macrophage host cell by the leishmanias is also discussed. These adaptations for survival depend upon preadaptations developed in the previous host. Some caution is necessary, however, in identifying adaptations to parasitism in the absence of knowledge of the free-living forebears of parasitic flagellates which may have been preadapted to a parasitic role.

Caution is also advised in discussing parasite adaptations when extrapolating from observations on *in vitro* or *in vivo* models to the parasite in nature. Most worrying of parasite adaptations are those that appear to have developed in parasites going nowhere, that is, dead-end stages of the life-cycle. It is possible that some of these adaptations may be the result of division of labour and the operation of kin selection within the parasite population, but some mysteries, such as the dead-end behaviour of the facultatively parasitic amoeboflagellate *Naegleria fowleri* in the human brain, are still in need of a plausible explanation in evolutionary terms.

10.1 Introduction

Many of the major taxonomic groups of flagellates contain organisms that live in partnership with another of an entirely different species (Table 10.1). Parasitism is a

Table 10.1 Occurrence of parasitic and free-living forms in flagellate taxa. Traditional classification based on Corliss (1994)

Phylum	*Order/family*	*Representative genera*	
		Free-living	*Parasitic*
Metamonada	Diplomonadida	*Hexamita**, *Trepomonas**	*Giardia*, *Spironucleus*
	Oxymonadida	none	*Saccinobaculus*, *Notila*
	Retortamonadida	*Chilomastix**	*Retortamonas*
Parabasala	Trichomonadida	*Pseudotrichomonas*	*Trichomonas*, *Histomonas*, *Dientamoeba*
	Hypermastigida	none	*Trichonympha*, *Barbulanympha*
Percolozoa	Schizopyrenida	*Percolomonas*, *Naegleria**	
Euglenozoa	Euglenida	*Euglena*, *Colacium**, *Peranema*, *Rhynchopus**	*Hegneria*# *Euglenomorpha*#
	Kinetoplastida		
	Bodonidae	*Bodo*, *Dimastigella**	*Ichthyobodo*, *Cryptobia* *Trypanoplasma*
	Trypanosomatidae	none	*Trypanosoma*, *Crithidia* *Phytomonas*, *Leishmania*
Dinozoa	Dinoflagellida	*Peridinium*, *Ceratium* *Gymnodinium*, *Oxyrrhis*	*Blastodinium*, *Symbiodinium*, *Amoebophrya*
	Syndiniida	none	*Syndinium*, *Hematodinium*

Notes:
* Genera with parasitic as well as free-living representatives
\# These genera are old and unconfirmed records. There are no recent records

way of life in which one partner, the parasite, is dependent for its survival on at least one gene of the other partner, the host. If genetic dependency operates in both directions we speak of mutualism rather than parasitism or more commonly, this side of the Atlantic, of symbiosis. In practice, the distinction is often a difficult one to draw. For example, the hindgut of the lower termites is packed with diverse flagellates, some of which digest cellulose and lignin for the host, but whether they all do is debatable (that is, some could be regarded as parasites), and whether the host is dependent upon all species is questionable.

In this chapter I shall discuss as adaptations those features of organization of flagellates that ostensibly make an efficient contribution to their ability to survive as parasites. In evolution, adaptations are commonly believed to arise in response to natural selection, so convincing demonstrations of adaptation should be geared to proof of increased fitness, that is, differential reproductive success. For parasites an important part of that reproductive success entails survival for transmission to new hosts. In identifying adaptations to parasitism, the parasitologist has to be wary of certain unjustified assumptions. For a start, parasitologists, like most biologists, tend to assume that, as a result of the corrective feedback of natural selection, life forms display the best available organization for their particular ecological niche. However, a supposed adaptation to parasitism may also have been a feature of the free-living ancestor, and we know less about free-living than about parasitic protozoa. In addition, parasitologists often study parasites as long-established laboratory-adapted strains and tend to assume that the parasite's behaviour in the laboratory parallels exactly its conduct in the natural host. Laboratory descendents of wild parasites have inevitably been subject to artificial selection, however, and may bear as much resemblance to their ancestor as a chihuahua does to a wolf. Illustrative examples of these pitfalls will be provided later.

Although it sounds like first-rank heresy, perhaps another assumption that might be questioned is that all adaptations are moulded by natural selection. This query is relevant to the practice of some parasitologists of constructing parasite life-cycles in which some stages show undoubted adaptations to survival in the host, but these apparently do not lead to infection of a further host, that is, they lead nowhere. How can we possibly explain such adaptations in terms of their selective value? I shall come back to this topic in my final discussion.

What sort of adaptations does a successful parasite need? First, it has to have adaptations that enable it to secure entry to a host. Second, it has to adapt to changing conditions within the host, in order to acquire nutrients for energy and growth, and to be able to select appropriate niches within the host. Third, it has to avoid host defence mechanisms in order to continue the infection. Last, it not only has to produce progeny, it also has to ensure their transmission to another host.

10.2 Adaptations for transmission

In order to travel from host to host, a parasite must adopt a form that protects it from the ravages of the outside world, or opt for a lifestyle that enables it to be transmitted directly from one host to another without having to face life outside.

Flagellates and other protozoa that live in the alimentary tract of animals usually pass from host to host in encysted form. The cyst wall is an adaptation that protects the parasite from desiccation and destruction during its journey from the faeces of one host to the food of the next. *Giardia lamblia* (Figure 10.1) of humans, is transmitted in this way, as are most of the amitochondriate gut parasites (Brugerolle and Müller, Chapter 9, this volume). Some amitochondriate parasites may be transmitted directly from one host to another, for example, *Trichomonas vaginalis* which is transmitted venereally in humans, and the flagellate symbionts of social wood-eating insects which are transmitted after moulting by direct consumption of the hindgut contents of a fellow host (proctodaeal feeding).

Adaptation to transmission by vector is, however, the most frequent solution to the problem of avoiding the hazards of the outside world: the parasite alternates between two different kinds of host, the feeding habits of the vector being exploited to arrange transmission. Such two-host life-cycles are characteristic of the trypanosomes and leishmanias which are the most important flagellate pathogens of humans and domestic animals. In the course of each life-cycle, the parasite has to adapt to a series of changing environments, usually in both hosts. In these kinetoplastid flagellates, changes in energy metabolism through cyclical activation and repression of the single mitochondrion with its prominent kinetoplast (mass of mitochondrial DNA) may play a vital part in such adaptation, as also may changes in the character of the parasite's surface.

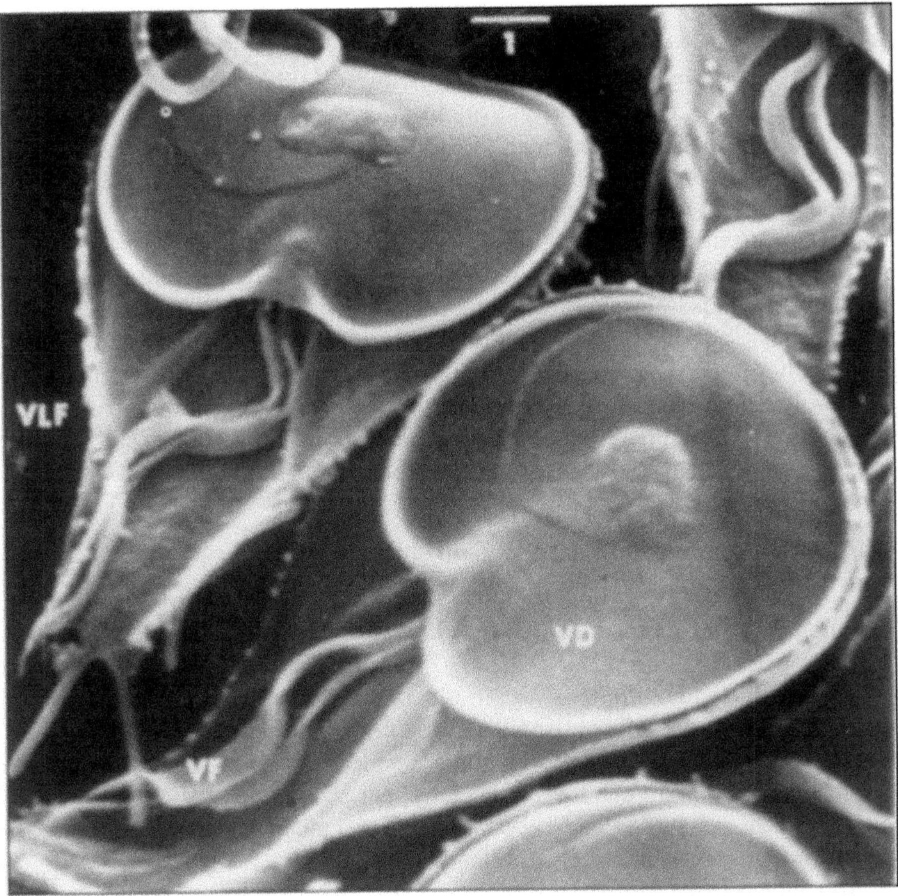

Figure 10.1 Scanning electron micrograph of trophozoites of *Giardia lamblia* in biopsy specimen from human jejunum. Ventral view of flagellates showing ventral adhesive disc (VD) and fringed ventrolateral flange (VLF). The edge of the helically-wound band of microtubules can be seen within the disc. The ventral flagella have vane-like expansions, and their beating in the ventral trough creates a suction pressure beneath the disc
X 10,000 (Scale bar = 1 μm)

Source: Micrograph: K. Vickerman

The sleeping sickness parasite, *Trypanosoma brucei*, multiplies in the blood and the body fluids of humans and other mammals, and is transmitted by the bloodsucking tsetse fly (*Glossina* spp.), undergoing multiplication and migration in the fly finally to invade the salivary glands to be injected as the infective metacyclic form when the fly takes a blood meal. *Trypanosoma cruzi*, causative agent of Chagas' disease, multiplies inside human/mammalian heart muscle (and other cells), then invades the bloodstream to be taken up by bloodsucking bugs (*Hemiptera*, for example *Rhodnius* and *Triatoma* spp.), and undergo multiplication in the midgut, producing metacyclic trypanosomes in the hindgut for deposition in faeces on the skin of the next host during feeding; the metacyclics invade the host's body via the skin. The various species of *Leishmania* multiply inside macrophages of skin (causing dermal leishmaniasis) or viscera (causing kala azar), and are taken up by the female sandfly (*Phlebotomus* or *Lutzomya* spp.) during a blood meal to multiply and undergo development in the midgut (more rarely hindgut), metacyclics swimming to the foregut so that they can enter the skin of a new host when the fly bites. The trypanosomatid flagellates have the broadest host range of any major group of protozoan parasites (Vickerman, 1994).

Another digenetic (two-host) life-cycle is exhibited by *Phytomonas* spp., trypanosomatids living in the latex and phloem of flowering plants and transmitted by sap-sucking hemipterans after invading their salivary glands (Camargo, 1999). Several trypanosomatids (*Leptomonas*, *Crithidia*, *Herpetomonas*, *Blastocrithidia* spp.), however, have monogenetic (one host) life cycles, parasitizing the insect gut, especially. Oddly, those gut trypanosomatids that parasitize terrestrial hosts are often transmitted by desiccation-resistant cyst-like forms that lack a cell wall; how such seemingly naked parasites survive between hosts is unknown (Reduth and Schaub, 1988).

10.2.1 Transmission and niche selection in a cyst-transmitted gut parasite: **Giardia lamblia**

The diplomonad *Giardia lamblia* (*G. intestinalis*) is the most commonly-reported intestinal parasitic infection of humans (Figure 10.1). Its trophic phase is precisely located in the jejunum (upper small intestine), where it undergoes binary fission while adhering firmly to the sides of the villi; it is absent from the villar tips. Transmission of the parasite occurs via the faeco-oral route, by means of quadri-nucleate cysts formed by detached parasites in the lower small intestine. *Giardia lamblia* is readily cultivated *in vitro*, where encystation and excystation can be induced at will (for comprehensive review and references see Kulda and Nohýnková, 1995).

Attachment in the jejunum is crucial for the parasite's full exploitation of its host's digestive activities. It is secured by a ventral adhesive disc which is supported by a gross spiral of microtubules reinforced by ribbons made up of giardin filaments (Brugerolle and Müller, Chapter 9 this volume). The host's enteric epithelium is constantly being renewed from dividing cells in the crypts; epithelial cells move up the villus to be shed into the gut lumen at its tip. The parasite rides this villar escalator, finding it necessary to detach before reaching the tip and reattach along the sides of a villus; the eight flagella execute clumsy swimming movements during transfer. The flagellate reattaches on a region of the villar epithelium where membrane-bound digestive enzymes (disaccharidases, amino and carboxy-peptidases) present maximum activity. The parasite is then in an ideal position to mop up products of digestion: glucose and amino acids. Beating of the flagella in the ventral groove (which emerges from beneath the adhesive disc; Figure 10.1) is believed to cause the suction force for adhesion of the parasite to the microvillar border, though contraction of myosin/actin-based cytoskeletal elements in the ventrolateral flange surrounding the disc has also been ascribed a role in attachment.

Giardia encysts by secreting a tough cyst wall containing fibrillar acidic leucine-rich proteins with galactosamine as a major sugar component. The events of excystation and encystation in the human intestine have been pieced together from observations on *in vitro*-cultivated parasites on the one hand, and our understanding of intestinal physiology on the other (Lujan *et al.*, 1998). Trophozoite emergence from the cyst in the small intestine is induced by stomach acid. Cyst production can be initiated *in vitro* in the presence of low concentrations of bile salts under slightly alkaline (pH 7.8) conditions. Trophozoites need an abundant supply of cholesterol, a condition best satisfied in the jejunum where cholesterol and other lipids are absorbed by the host. Detached trophozoites that travel further down the intestine encounter an environment poor in cholesterol which triggers differentation of the cyst wall secretory pathway. Cholesterol added to the lipid-deficient encystation medium blocks cyst production. The genes responsible for two of the cyst wall proteins have been characterized, but the relationship between reception of the stimulus (cholesterol depletion) and activation of these genes remains to be determined. The packaging and release of cyst wall proteins is associated with the *de novo* genesis of an identifiable Golgi apparatus in the trophozoite. Nuclear division (to produce four nuclei) and dismantling of the elaborate architecture of the adhesive disc precede maturation of the cyst as a water-resistant body ready to endure the harsh conditions of the outside world. Transmitting cysts are usually regarded as being a cryptobiotic phase in the life-cycle, but amazingly *G. lamblia* cysts retain active respiratory activity (for energy metabolism of *Giardia*, see Brugerolle and Müller, Chapter 9 this volume), a property that may account for their susceptibilty to desiccation.

Although many parasitic diplomonads appose their anterior ends to the microvillar border of the gut epithelium, none has aspired to the extraordinary development of the cytoskeleton found in the sucking disc of *Giardia* spp. in order to secure attachment. Yet a sucking disc of similar structure occurs in the trichomonadid *Cochlosoma anatis* of ducks (Pecka *et al.*, 1996), a remarkable example of evolutionary convergence.

10.2.2 Transmission and niche selection in vector-transmitted parasites: **Trypanosoma brucei** *and other digenetic trypanosomatids*

Trypanosoma brucei is a parasite of game animals and livestock causing *nagana* in the latter, but some genetic variants will infect humans, causing sleeping sickness. For contrast with the faeco-orally transmitted gut parasites such as *Giardia lamblia*, *T. brucei* will be used to illustrate adaptation to transmission by vector. A summary of the life-cycle is shown in Figure 10.2. Allusion to other trypanosomatids and comparison with other flagellate parasites will be made where appropriate, however.

The life-cycles of digenetic trypanosomatids are often characterized by extensive migrations in the vector and adaptation to a series of habitats in one host. For *T. brucei*, these habitats include blood, lymph, connective tissue and cerebrospinal fluids in the mammal, gut and salivary glands in *Glossina*. Striking adaptive changes occur in the surface membrane, mitochondrion and glycosomes, and in the occurrence of receptor-mediated endocytosis, as the trypanosome passes through its life-cycle (Vickerman, 1985): the preadaptation of the infective stages to the next host, be it insect or mammal, is particularly noticeable. These changes are summarized in Figure 10.2 and discussed in sections 10.3 and 10.4 of this chapter.

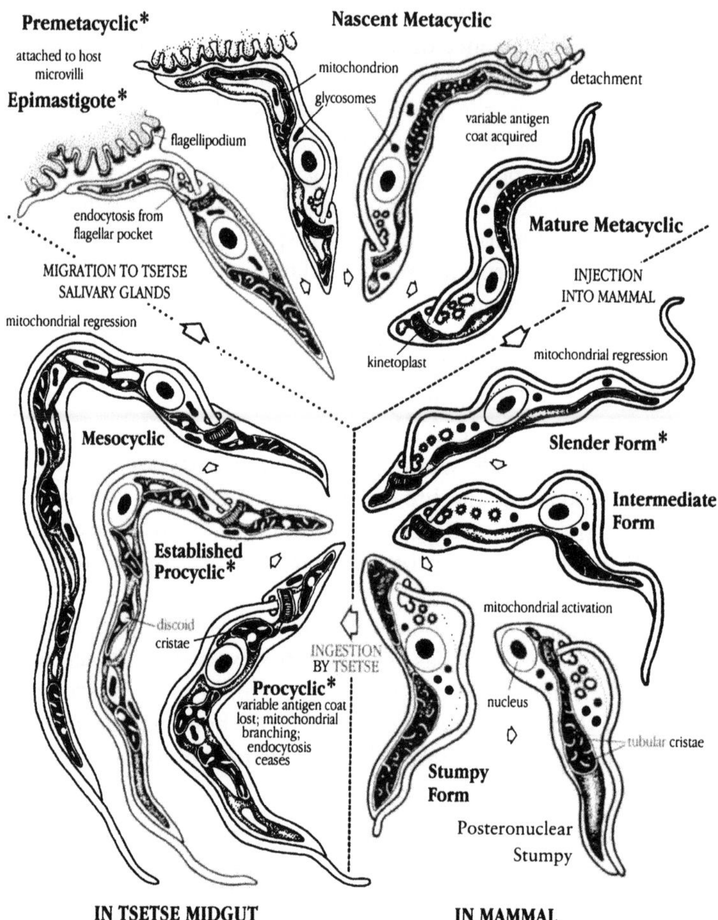

Figure 10.2 Schematic diagram of *Trypanosoma brucei* developmental cycle in mammal and tsetse fly, showing changes in cell surface, mitochondrion, glycosomes and receptor-mediated endocytosis, also in relative sizes of different stages. Stages possessing the variable antigen (VSG) coat lie to the right, procyclin-coated stages to the left of the midline. The mitochondrion is shown partly in section to show changes in cristae. Postero-nuclear stumpy forms are produced by some stocks only, and do not play an essential part in the life cycle.

* Stages in which division occurs.

In the wild, less than 10 per cent of fly infections will proceed as far as the salivary glands. Genetical studies suggest that a sexual process involving meiosis and syngamy occurs occasionally in the salivary glands. Mammal and most fly developmental stages are diploid, and which stages act as gametes is not known, though the epimastigote or premetacyclic seem likely choices. Migratory stages with varying morphology between the mesocyclic and epimastigote stages (Van Abeele *et al.*, 1999) have not yet been characterized ultrastructually.

Source: Based on Vickerman, 1985

For the moment, I shall concentrate on the significance of flagellar activity in such complex life-cycles.

In digenetic parasites, flagella may be important in migratory swimming movements and habitat selection within a particular host. Flagellar beating often provides the propulsive force for penetration of physical barriers during migration. *Trypanosoma brucei* is extracellular throughout its life-cycle but has to penetrate the chitinous peritrophic membrane surrounding the blood meal in the tsetse fly midgut, entering and leaving the ectotrophic space during its forward migration in the vector. In its mammalian host, penetration of blood vessel endothelium and choroid plexus epithelium are necessary for invasion of connective tissue and brain from the circulation. The marathon journey from tsetse midgut to salivary glands via the oesophagus, mouthparts and salivary ducts represents a major swimming feat on the part of the trypanosome (Van den Abeele *et al.*, 1999). Our continuing lack of understanding of the basis of the flagellum-based tropisms underlying such migratory activities is regrettable and represents a major challenge for future researchers.

Flagella may also play a vital part in enabling the parasite to remain in a suitable site within the host, by providing a mechanism for attachment to host surfaces. In particular, attachment appears to be a prerequisite for differentiation of the infective metacyclic stage in the vector (Vickerman and Tetley, 1990). Considerable modification of the flagellar membrane and associated cytoskeleton occurs at sites of attachment, with intraflagellar fibrils converging on hemidesomosome-like attachment complexes where the substratum is insect cuticle (for example *T. vivax* in mouthparts of tsetse, *T. cruzi* in rectum of *Triatoma*) or on punctate desomosome-like complexes where binding of gross expansions of the flagellar membrane ('flagellipodia') to living epithelium microvilli occurs, as in *T. brucei* in the tsetse salivary gland (Figure 10.2). There is recent evidence that attachment of the *T. cruzi* flagellum to *Triatoma* involves hydrophobic interaction with surface waxes, not lectin binding to the chitin beneath (Schmidt *et al.*, 1998), while the attachment to living membranes is to proteins which act as ligands for flagellar surface glycoconjugates, for example the lipophosphoglycan of *Leishmania major* binding to *Phlebotomus papatasi* midgut microvilli receptors (Dillon and Lane, 1999; see section 10.4.2). The additional flagellar cytoskeletal components include a 70kDa protein as yet uncharacterized (Kohl and Gull, 1998).

This development of the flagellum as an attachment organelle appears to be unique to the kinetoplastids and vital to completion of the life-cycle of many parasitic species, but whether the mechanism of attachment is the same in all cases, and whether attachment also occurs in free-living species, remain to be seen. More is known about the cytoskeletal proteins (Kohl and Gull, 1998) involved in attachment (see Figure 10.3) of flagellum to body to form the 'undulating membrane', a prominent feature of trypanosomes, and believed to be an adaptation of parasitic flagellates that gives additional thrust to the flagellum in moving through viscous media (for example blood, gut contents; Holwill and Taylor, Chapter 3 this volume).

Trypanosoma cruzi and the leishmanias forgo a functional flagellum while they are multiplying inside a host cell, though *T. cruzi* reacquires it before the resulting progeny burst forth again into the extracellular world. Continuous flagellar beating in *T. brucei* when it is being passively carried in the circulating blood, or when it is

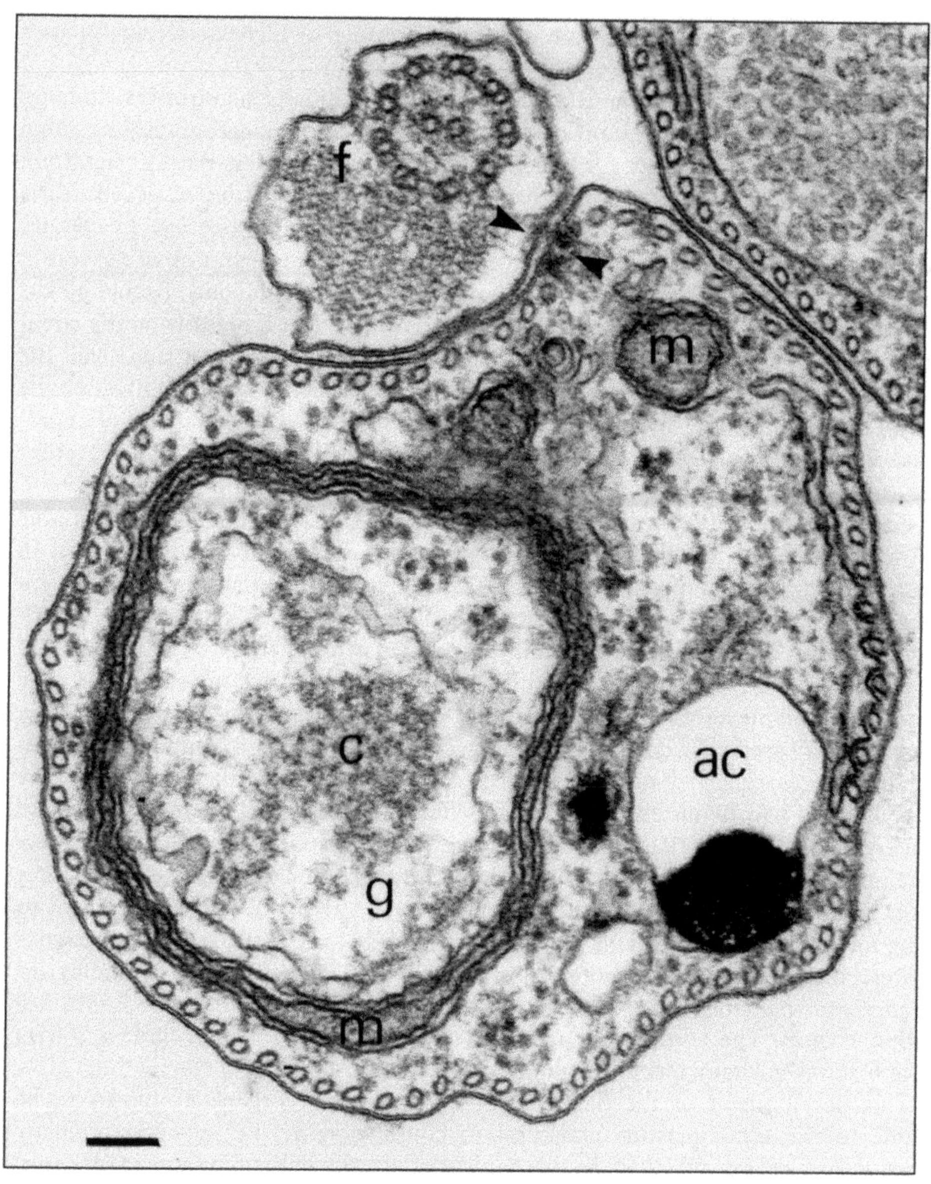

Figure 10.3 Transmission electron micrograph of anterior region of procyclic *Trypanosoma brucei* trypomastigote from established culture in glucose-containing medium. The mitochondrion (m) shows the branched form typical of the procyclic stage, while the glycosome (g) has the spherical form more typical of bloodstream trypomastigotes (see Figure 10.2) and contains a crystalloid (c). An acidicalcisome (ac) contains heavy polyphosphate deposits. The macular desmosome-like attachment between body and flagellum (f) is visible between arrowheads
X 10,000 (Scale bar = 0.1 μm)

Source: Micrograph: K. Vickerman

attached to the salivary epithelium of the vector, is less obviously adaptive, however. One possible explanation is that the beating flagellum exerts a rotary pump-like activity, circulating the contents of the flagellar pocket which is the principal (if not the only) site of endocytosis and exocytosis in this organism (Overath *et al.*, 1997). When flagellates are densely packed within host cavities – as in the *T. brucei* - packed fly salivary gland, in the larval mosquito gut teeming with *Crithidia*, or in palm phloem vessels choked with *Phytomonas* (Camargo, 1999), flagellar activity probably serves to circulate the surrounding medium so that all organisms have access to nutrients and oxygen. One of the most overpopulated habitats imaginable is the paunch of wood-eating termites; it contains a seething mass of motile flagellates (oxymonads, trichomonads, hypermastigids) and bacteria. They are not, however, moving around aimlessly, but through their flagellar activity are maintaining themselves in an environment optimal for redox potential and nutrient supply (Brune *et al.*, 1995).

10.3 Energy metabolism: mitochondrion and glycosome

One of the most spectacular examples of adaptation during the course of any parasite's life-cycle is the switch in energy substrate utilization by *Trypanosoma brucei* on passing from mammalian to insect host. This switch is associated with cyclical repression and activation of the single mitochondrion, and this cycle in turn is predicated upon an intact kinetoplast, the massed mitochondrial DNA characteristic of the entire order Kinetoplastida (for energy metabolism of amitochondriates see Brugerolle and Müller, Chapter 9 this volume).

Trypanosomes have no known storage carbohydrates and are largely dependent upon a continuing exogenous energy supply. The long slender multiplicative stage (LS) of *T. brucei* in the mammal obtains its energy from glycolysis which occurs in the glycosome, a special type of peroxisome, unique to kinetoplastids and absent from the related euglenids (Opperdoes *et al.*, 1998). Compartmentalization of the glycolytic chain enables the trypanosome to conduct this process more efficiently than other eukaryotic cells (Clayton and Michels, 1996). Pyruvate produced by glycolysis does not enter the mitochondrion, but is released into the blood. The unbranched mitochondrion is repressed and lacks cristae (Figure 10.2). The LS forms depend exclusively on substrate-level phosphorylation to generate ATP, yet their oxygen consumption is high. In most eukaryotic organisms, mitochondrial cytochrome oxidase catalyzes the interaction of the reducing equivalents produced during respiratory oxidation reactions with atmospheric oxygen, but in bloodstream *T. brucei* electron transfer to oxygen is carried out by a cyanide-insensitive glycerol-3-phosphate oxidase (GPO), and not by cytochromes. Glycerol-3-phosphate produced in the glycosome during the reoxidation of NADH is oxidized by GPO which is located in the mitochondrial membrane (Tielens and Van Hellemond, 1998).

In the non-multiplicative short stumpy (SS) form derived from the LS in each parasitaemic wave (section 10.4.1), the mitochondrion swells, acquires tubular cristae and the ability to utilize proline and succinate as energy sources (Figures 10.2, 10.4). This switch preadapts the SS to life in the tsetse midgut where a total switch

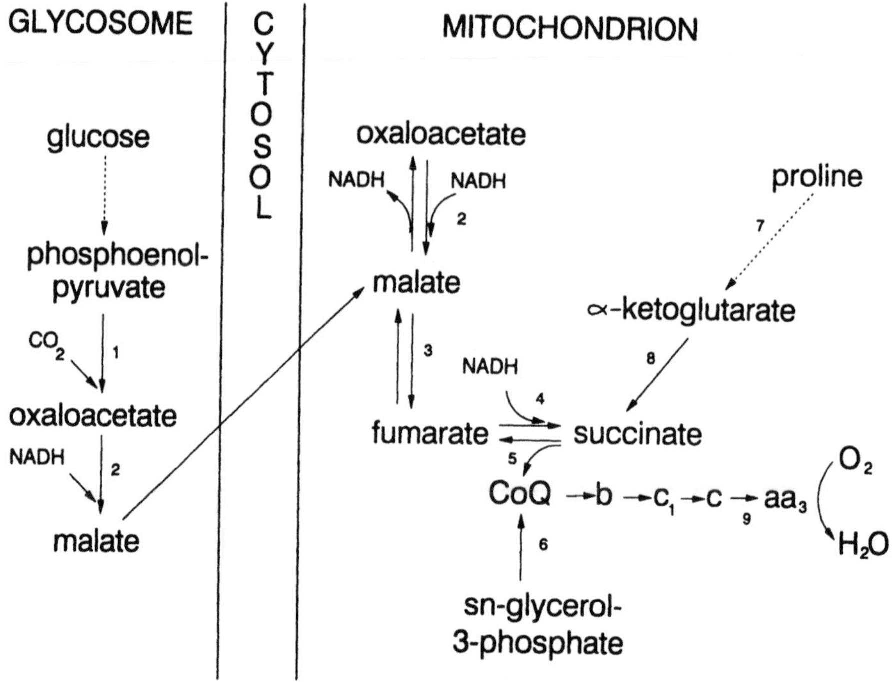

Figure 10.4 Pathways of glycolysis, proline oxidation and mitochondrial metabolism in *Trypanosoma brucei* procyclic trypomastigotes. Glucose breakdown takes place largely in the glycosome. Malate may be transformed into pyruvate by the malic enzyme in the cytosol (not shown).

Enzymes: 1, phosphoenolpyruvate carboxykinase; 2, malate dehydrogenase; 3, fumarase; 4, NADH-fumarate reductase; 5, succinate dehydrogenase; 6, glycerol-3-phosphate dehydrogenase; 7, proline oxidase; 8, alpha-ketoglutarate dehydrogenase; 9, cytochrome oxidase. The enzymes citrate synthase and isocitrate dehydrogenase appear not to be active, so the Krebs' cycle is incomplete

Source: Turrens, 1991

occurs from ultilization of glucose to utilization of proline as an energy source (Matthews, 1999). This switch is coincident with transformation to the procyclic (PC) trypomastigote, expansion of the mitochondrion into a network, and acquisition of plate-like cristae and a cytochrome chain for oxidative phosphorylation. Mitochondrial repression is reinitiated in the tsetse fly's salivary gland, where the multiplicative epimastigote (EPM) mitochondrion reverts to tubular cristae, and transformation of the EPM to the metacyclic stage is accompanied by the mitochondrial network becoming an unbranched structure again. The more repressed mitochondrion of the metacyclic stage may token preadaptation to life in the mammalian bloodstream (Vickerman, 1985). Admittedly nothing is known of the metabolism of the salivary gland stages as, unlike the PC stage, these are difficult to cultivate *in vitro*.

T. brucei utilizes its host's principal energy source, glucose in the mammalian bloodstream, and the amino acid proline (used as a source of energy for tsetse flight) after transformation to the PC form (Turrens, 1991). Much of our understanding of the metabolism of PC *T. brucei*, however, comes from parasites serially cultivated *in vitro*, rather than from the recently generated PCs. The possibility that 'domestication' has resulted from *in vitro* selection of features not present in natural procyclics must therefore be borne in mind. Such a feature may be glucose ultilization by procyclics, with selection provided by serial cultivation in glucose-containing media. Discussions of the energy metabolism of procyclics often portray glycolysis and entry of pyruvate into the mitochondrion for oxidative decarboxylation, via a fully active Krebs cycle, as proceeding alongside utilization of proline via a partial Krebs cycle following its conversion to a-ketoglutarate (Figure 10.4). But glucose is eliminated from the tsetse bloodmeal within fifteen minutes of ingestion (Vickerman, 1985), so this pathway can be of little importance in nature as opposed to the test tube. Even more mystifying is the report that the serially-cultured procyclics have a glucose transporter (THT2) encoded by a different gene from the bloodstream form transporter (THT1), and that the expression of these genes is life-cycle stage-dependent (Tetaud *et al.*, 1997).

In the majority of trypanosomatids investigated, glycolysis plays a less important role than in bloodstream *T. brucei*, and cyclical activation and repression of the mitochondrion associated with a complex two-host life-cycle is less marked, owing to the presence of a more or less well-developed cytochrome chain and modified Krebs' cycle at all times. Interestingly, however, culture promastigotes (latex stages?) of *Phytomonas* from the spurge *Euphorbia characias* exhibit a glycolytic energy metabolism similar to that of *T. brucei* bloodstream forms (Sanchez-Moreno *et al.*, 1992). Intracellular *Leishmania* amastigotes appear to be more dependent on the oxidation of fatty acids than on that of carbohydrates (Hart *et al.*, 1981), but whether this oxidation is associated with the glycosome rather than the mitochondrion is not yet known.

The glycosome appears to be a universal feature of kinetoplastids (Opperdoes *et al.*, 1998) but the glycosomes of free-living kinetoplastids have not been investigated yet, so their comparison with those of parasitic forms is not possible. Another membrane-bound organelle recently identified in trypanosomes and leishmanias is the acidocalcisome (Docampo and Moreno, 1999), involved in the regulation of Ca^{++} and pH homeostasis, possibly also in osmoregulation. It may prove significant in kinetoplastid energy metabolism, as it contains abundant pyrophosphate (Figure 10.3) which may act as an energy store. Similar organelles occur in the unrelated parasites *Toxoplasma* and *Pneumocystis,* but whether acidocalcisomes have evolved several times as an adaptation to parasitism, or are also present in free-living protists, is not yet known.

Much attention has been given to the significance of the kinetoplast (mitochondrial DNA) in trypanosome respiratory adaptation, especially since the discovery of RNA editing in these organisms. The impression has been created that both kinetoplast and RNA editing are adaptations to parasitism (Maslov *et al.*, 1994), but as both also occur in the free-living bodonid kinetoplastids, this notion would seem unlikely to be true. The kinetoplast (Shapiro and Englund, 1995) is a remarkable structure composed of packed circular DNA molecules. In the Trypanosomatidae the circles are concatenated to form a network which is physically connected to the

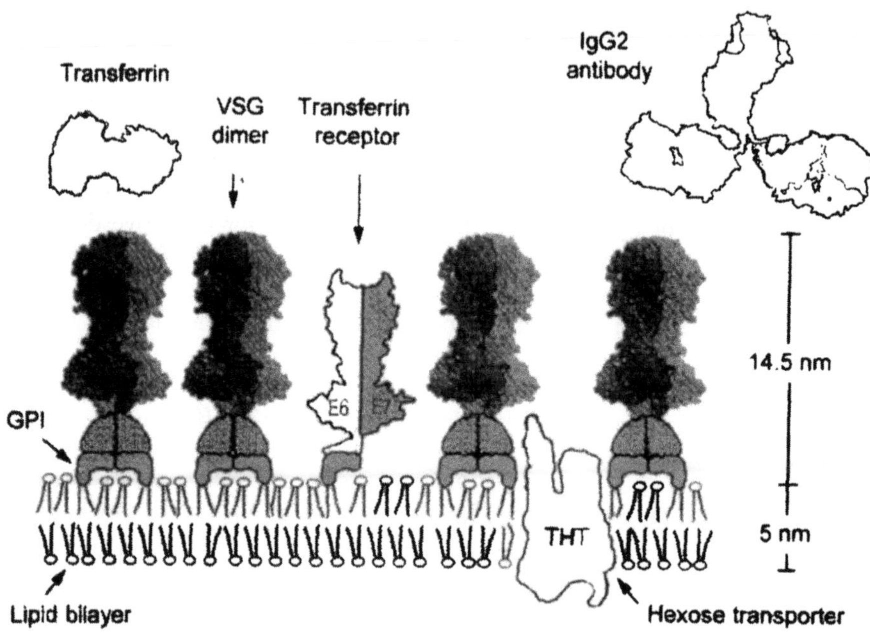

Figure 10.5 Schematic representation of the cell surface of bloodstream form *Trypanosoma brucei* showing the VSG coat as selective shield. The regularly spaced VSG dimers that form the surface coat have their C-terminal regions attached to glycosylphosphatidylinositol (GPI) anchors in the lipid bilayer. The dimensions of the VSG dimer are based on the crystal structure of the N-terminal domain of the protein; those of the anchor are not yet known. A transferrin receptor (involved in receptor-mediated endocytosis in the flagellar pocket membrane) is also a dimer made up of two monomers each of proteins encoded by two ESAGs (expression site-associated genes), E6 and E7. The ESAGs form part of a telomeric polycistronic transcription unit that includes the VSG gene. The trypanosome needs host transferrin as a source of iron. Membrane transporters (e.g. that for hexose sugars, depicted here) are shielded from host antibody by the VSG coat.

Source: After Borst and Fairlamb, 1998; based on Overath *et al.*, 1997

flagellar base via the mitochondrial membrane. During the cell cycle kinetoplast (kDNA) and nuclear DNA synthesis are synchronous. Division of the kDNA network into two daughter structures occurs after replication of the flagellar basal body and sprouting of a daughter flagellum. Division of the kinetoplast precedes division of the nucleus. The daughter basal bodies are associated not with the poles of the intranuclear mitotic spindle apparatus, but with the separating kinetoplasts (Kohl and Gull, 1998). The kDNA circles are of two kinds: a few large maxicircles (25–50 per kinetoplast, each 20–38 kb, depending on the species) and a large number (~5,000–27,000, each 0.46–2.5 kb) of minicircles. Maxicircles, which are similar to the mitochondrial DNAs of other eukaryotes, are homogenous in nucleotide sequence. They contain genes essential for mitochondrial activation: genes that encode mitochondrial ribosomal RNAs and subunits of some proteins involved in electron transport and ATP synthesis. When mechanically passaged repeatedly through mammals, maxicircles may suffer deletions which result in non-transmissibility by the tsetse fly, presumably through inability to activate the mitochondrion. Minicircles, on the other hand, are heterogeneous in nucleotide sequence, but evolve rapidly, and in a mechanically-passaged stock become homogenous in sequence.

For a long time, no better role could be found for the minicircles than holding the maxicircles together in a network. In recent years, however, it has become plain that the minicircles are transcribed into small RNA molecules (guide RNAs) that can potentially serve as templates for guiding insertions and deletions of uridine nucleotides in faulty maxicircle transcripts. This process of RNA editing (Alfonzo *et al.*, 1997; Sloof and Benne, 1997), correcting defects in the maxicircle genes for respiratory enzymes, represents an alternative means of handling genetic information quite unique to the kinetoplastids. There is evidence that editing is regulated to produce different mRNAs under different conditions, so that RNA editing can be viewed as a primitive way of changing the expression of genes

The bodonid kinetoplast is strikingly different. It may form a single multilayered mass close to the bases of the flagella, as in *Bodo,* or may take the form of dispersed kDNA masses within the mitochondrion, as in some *Cryptobia* spp. (Vickerman, 1990). It would appear that in both parasitic and free-living bodonids the kDNA does not form a catenated network, and may consist of circles of a single size class (Hajduk *et al.*, 1986), though in *Bodo saltans* minicircles are undoubtedly present (Blom *et al.*, 1998). The adaptive significance of the catenated kinetoplast network composed of two kinds of DNA circles in the all-parasitic trypanosomatids is dumbfoundingly intriguing! Further studies on the construction of bodonid kinetoplasts will be necessary before a meaningful comparison can be made, but stricter bipartitioning of the kDNA at cell division would seem to be afforded by the network of trypanosomatids through its close association with the flagellar basal bodies. In bodonids the uncatenated DNA is presumably farmed out in a more uneven manner.

RNA editing has now been demonstrated in the kinetoplast of the common free living kinetoplastid, *Bodo saltans* (Blom *et al.*, 1998), so it is not linked to a parasitic lifestyle. Indeed Cavalier-Smith (1997) has argued that RNA editing and glycosomes evolved in facultatively anaerobic free-living bodonids. Prolonged anaerobic glycolysis would allow the accumulation of harmful mutations in mitochondrial DNA, which would disadvantage these organisms on return to aerobic conditions unless they had by chance evolved RNA editing, and the origin of editing would be favoured in a multigenomic organelle, such as the kinetoplastid mitochondrion. Nothing is known about the metabolism of free-living bodonids, but their widespread ability to survive under anaerobic conditions (Bernard *et al.*, 1999) lends some support to this hypothesis.

10.4 The parasite surface and evasion of host defences

Parasite avoidance of host attack is of paramount importance in the evolutionary arms race between host and parasite. The molecular structure of the surface membrane which represents the host–parasite interface might be expected to make a vital contribution to this defence, and comparative studies on trypanosomatid parasites of mammals amply confirm this view. The surface of these parasites is dominated by glycoproteins and glycophospholipids that are rooted in the plasma membrane by glycosyl-phosphatidylinositol (GPI) anchors (Ferguson, 1997).

The parasitological significance of these glycoconjugates in the life-cycles of trypanosomatids is summarized in Table 10.2 (for details of chemical structure see Ferguson, 1997). Of particular importance for extracellular parasites is avoidance of the effects of activation of the mammalian complement system of serum proteins. Following such activation, some complement components can cause opsonization of parasites for phagocyte uptake, attraction of phagocytes to the site, or lysis of the parasite through insertion of the pore-forming complement attack complex in the surface membrane, resulting in disruption of ionic and osmotic control and in programmed cell death. Of particular importance to intracellular parasites is avoidance of oxygen radical-induced destruction. Two examples of defence strategies will be discussed: antigenic variation by the extracellular parasite *Trypanosoma brucei*, and subversion of the host cell's endocytotic and killing mechanisms by the intracellular leishmanias.

10.4.1 Evasion of the host's humoral response by African trypanosomes

One of the best-characterized adaptations of a parasite to survival in the face of host non-specific and specific defences is the antigenically-variable surface coat of *Trypanosoma brucei* and other African trypanosomes. The parasite avoids destruction by undergoing antigenic variation (for more detailed recent accounts and references see Cross, 1996; Barry, 1997; Borst *et al.*, 1997; Rudenko *et al.*, 1998). In its mammalian host the trypanosome population survives by sacrificing at regular intervals of a few days the majority of its members to the host's humoral immune response (Vickerman, 1989). This serial sacrifice is evident in graphs portraying the undulating parasitaemia of infected hosts. Each trypanosome carries on its surface a homogenous monomolecular layer of variant specific glycoprotein (VSG) which endows the parasite with a specific variable antigen type (VAT). Avoidance of the sacrificial act depends upon the trypanosome's ability to change the VSG in its surface coat and so undergo a change of VAT. In each bout of slaughter (parasitaemic remission) the dominant VATs are eliminated by the host's humoral (IgM) response, but the minority VATs, which have not yet provoked production of VAT-specific antibodies, are unaffected and continue to divide, replacing their dead comrades and giving rise to a recrudescence of parasitaemia. Remission and recrudescence occur repeatedly. At each remission, in an act of altruism, some members of the clone lay down their lives so that their fellows of identical (or near identical – see later) genotype can continue the line. This serial sacrifice prolongs the infection, in the case of the sleeping sickness trypanosomes (*T. brucei rhodesiense, T.b. gambiense*) to several months or even years. The chronic infection in the individual host increases the likelihood that that at some point a hungry tsetse fly will ingest part of the population and eventually transmit it to a new mammalian host.

Antigenic variation entails periodic replacement of the 12–15 nm-thick surface

Table 10.2 Parasitological significance of surface glycoconjugates in life cycles of trypanosomatids

Parasite	*Stage in life cycle*	*Surface glycoconjugate*	*Parasitological significance*
Trypanosoma brucei	Bloodstream TPM Metacyclic TPM	Variant-specific glycoprotein (VSG) Variant-specific glycoprotein	1 Macromolecular diffusion barrier 2 Antigenic variation
	Procyclic TPM	Procyclic acidic repetitive protein (procyclin)	Ligand for tsetse fly defence lectins
	EPM	Procyclic acidic repetitive protein	
Trypanosoma cruzi	Bloodstream TPM	GIPLs (Type 1) Large O-glycosylated mucins with a-galactose residues	1 Induce host macrophage TNF a and IL2, also high titre anti-galactose oligosaccharide antibodies 2. Major receptor sites for host-sialic acid transferred by sialidase: necessary for invasion of host cell
	AM	'Amastin'	
	EPM	GIPLs (Type 1) Small mucins	
	Metacyclic TPM	GIPLs (Type 1) Small mucins	GPI anchor contains ceramide: enables transfer of anchor to host cell and invasion
Leishmania major/L.donovani	AM	[No major glycoproteins] Lipophosphoglycan	 Inhibition of macrophage oxidative burst
	Procyclic PM	GIPLs (Type 2 mainly) Promastigote surface protease (gp63) Liphophosphoglycan	 Digestion of proteins in blood meal Adhesion to sandfly midgut-receptor
	Metacyclic PM	GIPLs (Type 2 mainly) Promastigote surface protease Lipophosphoglycan	Modulation of inducible NO synthase in macrophage Degradation of complement C3 to C3bi 1 Activation of complement to opsonize metacyclic for macrophage uptake 2 Inhibition of macrophage oxidative burst 3 Modulation of inducible NO synthase

Abbreviations: AM: amastigote; EPM: epimastigote; TPM: trypomastigote; GIPLs: glycoinositol phospholipids (Type 1 and Type 2 GIPLs differ in the structure of their glycan core); GPI: glycophosphatidylinositol; NO: nitric oxide; IL2: interleukin2; TNF: tumour necrosis factor (cytokines)

Source: Information from Ferguson, 1997; Alexander *et al.*, 1999; Dillon and Lane, 1999

coat which covers body and flagellum with one of different antigenic specificity. The coat is made up of ~10^7 copies of ~60kDa VSG arranged as a dense monolayer of homodimers on the parasite surface (Figure 10.5). In the non-immune host it acts as a macromolecular barrier, preventing insertion of the complement membrane attack complex, while allowing the free diffusion of small nutrient molecules to underlying trans-membrane transporter systems; it also acts as an antiphagocytic capsule in that, in the absence of opsonizing VAT-specific antibody, macrophages are not interested in the trypanosome. When the host produces a vigorous antibody response to a population of trypanosomes carrying a particular VSG, parasites belonging to that particular VAT are lysed (through activation of complement by the classical pathway) or opsonized so that they are engulfed and destroyed by macrophages of liver and spleen. In the meantime, however, some multiplicative slender forms have switched to expression of a different VSG coat. Within forty-eight hours of entering the vector the trypanosome has replaced its VSG with another glycoprotein altogether: procyclin; but later in the salivary glands of the fly, the parasites reacquire the VSG coat as they transform into metacyclic trypanosomes (Figure 10.2). This acquisition preadapts the metacyclic to life in the mammal.

Different VSGs owe their antigenic identity to extensive differences in primary sequence of the N-terminal region. The only conserved features in this sequence are cysteines that form disulphide bridges. Yet the VSGs appear to have a conserved tertiary structure which allows them all to produce a monolayer barrier that prevents host antibodies from recognizing the surface proteins. The main feature of this tertiary structure is two long alpha helices per monomer that are perpendicular to the cell surface and define the elongate shape of the VSG. Some non-VSG surface proteins such as the transferrin receptor (Figure 10.5) have a similar tertiary organization, suggesting that a master structure has evolved to allow a range of functions to be developed during the course of evolution of the African trypanosomes (Borst and Fairlamb, 1998).

New VATs are generated by a spontaneous switch in gene expression. The molecular basis of VAT switching is interesting because it illustrates how, contrary to the intuitive view of parasite evolution as progressive surrender of genes with increasing host dependence, the parasite's genome may be expanded substantially to ensure defeat of host attack (up to one eighth of the trypanosome's genes may encode VSGs). It also illustrates how clonal organisms may have mechanisms for generating raw material for natural selection to act upon, independent of recombination resulting from sexual processes.

A clone stock of *Trypanosoma brucei* has a repertoire of VATs based on about a thousand VSG genes. Only a small fraction of these (~25, M-VATs) are expressed in the infecting metacyclic population, which is therefore heterogeneous with respect to VAT. In order to be expressed, the VSG gene must come to lie in an expression site at the telomere (end) of a chromosome. Chromosomes of trypanosomes have been characterized by pulsed-field gel electrophoresis and similar techniques (Ersfeld *et al.*, 1999). Many VSG genes lie at the ends of chromosomes, and *T. brucei* has at least 100 mini-chromosomes (50–150 kb) whose sole function appears to be to carry telomeric VSG genes, in addition to thirty or more intermediate size (200–700 kb) and larger (over 2000 kb) chromosomes which may house internal VSG genes and carry M-VAT VSG genes at their telomeres. Only about twenty telomeres can act as VSG gene expression sites, however, usually in a mutually exclusive manner (Rudenko *et al.*, 1998).

For the expression of the majority of VSG genes, a silent 'basic copy' must produce an 'expression-linked copy' (ELC) of the gene which is transposed to a transcriptionally active expression site, where it displaces the resident gene, causing a switch in the VSG being transcribed and hence in VAT. What appears to happen is that recombination between ELC and resident gene occurs in homologous sequences of base pairs on each side of the VSG coding region. Sometimes partial gene conversion occurs when the expressed gene is only partly replaced by the incomplete ELC, through recombination occurring well within the coding region. A second method of VSG gene activation occurs when a telomeric gene is activated *in situ*, its particular telomere suddenly becoming an expression site as the previously active expression site is rendered inactive. The operational basis of this selective telomere activation is unknown, but when an ELC displaces a telomeric gene which has been activated *in situ,* VSG gene loss can occur and acquisition of new mosaic genes can take place through a process of partial gene conversion, as outlined earlier. The mechanism of antigenic variation therefore contains the seeds of VAT repertoire – and therefore of genotype – evolution.

Although we now have a fair understanding of the molecular basis of VAT switching, our view of the relationship between the switching event, at the level of the individual trypanosome, and the population fluxes of different trypanosome VATs in the blood, remains hazy. A host antibody does not induce switching, it merely acts as a selective agent. Natural selection should adjust the rate of VSG gene switching so that one new VAT is generated in each parasitaemic wave. A lower rate than this would lead to immediate resolution of the infection; a higher rate might be expected to lead to too many VATs growing and being responded to simultaneously, so that the VAT repertoire might quickly become exhausted. Early studies on the host's immune response suggested that the parasite presents antigenic stimulus to the host in a paced fashion. Antibodies to specific VATs arise in a time-dependent succession, as shown by agglutination reactions (which identify major VATs). In a given trypanosome stock, the major VATs appear to dominate the bloodstream in a loosely-defined sequence, but the sequence is semi-predictable, not a fixed linear series. At relapse, trypanosomes of a given VAT can give rise to several new VATs. What determines the major VAT sequence, and why one of several available minor VATs should be selected to become the next major sacrificial population, is still being debated.

One possible explanation of the major VAT succession and paced stimulation of the host's immune response comes from a study of VAT switching rates (Frank, 1999; Turner, 1999). Estimations of the frequency of VAT-switching in trypanosome populations after transmission through the tsetse fly vector suggests a higher figure ($0.97–2.2 \times 10^{-3}$ switches per cell per generation) than that previously obtained for syringe passaged infections ($10^{-5}–10^{-7}$ switches per cell per generation) and, contrary to expectations stated earlier, it now seems probable that most of the variable antigen genes are expressed as minority VATs early in the infection. Instability of VAT is a feature of trypanosome clones derived from metacyclic trypanosomes. High switching rates would have the obvious advantage of enabling the trypanosome population to establish itself in a partially immune host. The early generation of a great variety of VATs in the newly-infected host would ensure that the parasites evaded any immunity resulting from previous infection with cross-reacting trypanosome VAT repertoires, provided that they did not all stimulate an immune response. It is suggested that marked differences in switching rates between particular pairs of VATs might well account for the delay of growth of some VATs to antibody-inducing major VAT status until later in the infection (see Frank, 1999, for mathematical model) and the pattern of sequential dominance of VATs.

The selective advantage of high switching rates is, of course, that such rates could promote the rapid evolution of the VAT repertoire through the mechanisms discussed earlier, and such rapid change has been demonstrated in one *Trypanosoma brucei rhodesiense* VAT repertoire over a twenty-year period (Barry, 1997). The occurrence of multiple serodemes (VAT repertoires) within a trypanosome species, and the ability of such serodemes to evolve rapidly, make the prospect of devising a vaccine against African trypanosomiasis very bleak indeed.

The large number of minichromosomes would appear to be part of the parasite's adaptation to evade the host's immune response, as the minichromosomes act as a reservoir for a large number of telomeric VSG genes and are only found in the African trypanosomes (*T. brucei*, *T. congolense*, *T. vivax*) and their descendents (*T. evansi*, *T. equiperdum*). Although minichromosomes appear to be inherited faithfully by daughter trypanosomes in binary fission, until recently the way in which the 200 or so daughter minichromosomes are partitioned at cell division was beyond conjecture, because the minichromosomes lack centromeric sequences, and in electron micrographs only enough kinetochores could be observed to account for partitioning of the 'housekeeping' megachromosomes. Now Gull's group (Ersfeld *et al.*, 1999) have demonstrated that the daughter minichromosomes are segregated independently of the large chromosomes on the overlapping microtubules of the central mitotic spindle, probably several in a row, laterally stacked, each daughter being attached to a microtubule of complementary polarity. Although clearly part of the parasite's extraordinarily complex adaptation to evasion of the immune response, this method of segregating small DNA molecules may prove to be more widespread in eukaryotes. Trypanosomes have often pointed the way to novel, more widespread, biological phenomena.

Needless to say, antigenic variation is not the prerogative of African trypanosomes. Among other protists, such variation is known in *Plasmodium falciparum* and a few other malaria parasites, and, strangely enough, considering that it is not a blood parasite and its infection runs an acute course of one or two weeks, in *Giardia lamblia* (Nash, 1997). Its significance in this gut parasite is at present obscure.

10.4.2 Subversion of macrophage activity by leishmania parasites

While *Trypanosoma cruzi* commonly invades cardiac and other muscle cells, both *T. cruzi* and the leishmanias can survive and multiply inside the mammalian macrophage, a cell designed to destroy invading pathogens by phagocytozing and degrading them. In addition to their direct microbiocidal activities, macrophages are charged with orchestrating the acquired immune response by presenting microbial antigens to lymphocytes. In order to sustain an infection, then, parasites have to gain entry to the phagocytic cell, withstand or circumvent its killing and degradative functions, and subvert macrophage accessory cell activities in order to prevent the development of protective immunity. Both parasites multiply in the amastigote (unflagellated) form inside the host cell, but whereas leishmanias undergo their intracellular development within a parasitophorous vacuole which is part of the host cell's lysosomal system, *T. cruzi* escapes from the parasitophorous vacuole into the cytoplasmic matrix to multiply there, and thus avoids destructive agents in the vacuole. Also, unlike leishmanias, it reverts to the flagellate trypomastigote form in order to escape and invade another cell. Parasite surface components play a vital role in all these goings-on and in the host-parasite warfare that follows. For further information on *T. cruzi* see the spirited account of its cell invasion by Andrews (1995).

Here I shall try to show how *Leishmania* spp. have proved invaluable in investigating mammalian defences against intracellular parasites, and conversely how these parasites have exploited these defences to achieve their own ends (Alexander *et al.*, 1999; Bogdan and Rollinghoff, 1999).

Among the leishmanias, the early developmental stage (procyclic promastigote) in the gut of the sandfly (*Phlebotomus* spp.) vector expresses on its surface abundant lipophosphoglycan (LPG) and a metalloprotease, gp63. The LPG and gp63 molecules project through a turf-like glycocalyx of glycoinositolphospholipids (GIPLs) covering the lipid bilayer (Ferguson, 1997; see also Figure 10.6). These glycoconjugates illustrate well the adaptation of the parasite at the molecular level to life in both its hosts. Thus, gp63 is believed to protect the promastigote from digestive enzymes in the vector gut, while LPG facilitates adherence of the flagellum to midgut epithelium microvilli. In *L. donovani*, transformation from the multiplicative non mammal-infective procyclic to the non-dividing mammal-infective metacyclic stage involves a lengthening of the surface LPG molecules and upregulation of gp63 expression. These changes allow the metacyclics to withstand activation of the complement cascade when inoculated by the vector into the mammalian host. LPG activates complement and in the mammal procyclics are lysed as a result of insertion of the complement membrane attack complex (C5–C9) into the plasma membrane. In metacyclics, however, the thickened glycocalyx, produced by LPG elongation through an increased number of disaccharide repeat units, prevents insertion, and the complex is later shed from the surface.

The leishmania parasite actually exploits macrophage complement receptors to secure its binding to the macrophage surface, and then internalization. The protease activity of Gp63 cleaves C3b to C3b1, and these complementary components on the parasite's surface bind it to the macrophage's CR1 and CR3 receptors. Other macrophage receptors (for example mannose-fucose and fibronectin receptors) may also assist in parasite entry. Inside the macrophage, LPG and gp63 further protect the parasite from destruction by oxygen radicals and microbiocidal nitric oxide killing mechanisms. In gaining access to the host cell by the CR1 and CR3 receptors, the parasites fail to trigger the macrophage's respiratory burst that generates

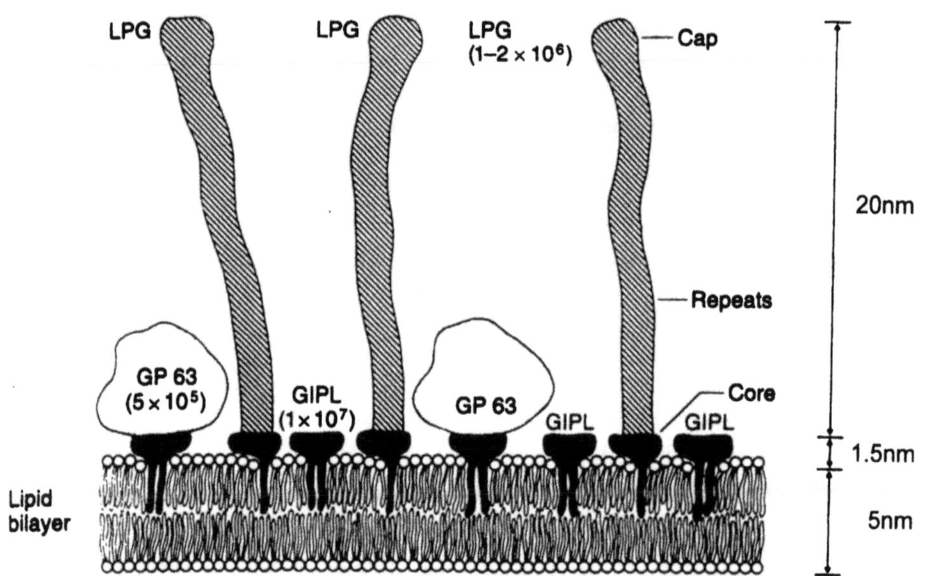

Figure 10.6 Schematic representation of the cell surface of *Leishmania* procyclic promastigote to show postulated relationship of major glycoconjugates. All have GPI anchors and include: the Zn^{++} metalloprotease with a molecular mass of 63kDA (gp63), lipophosphoglycan (LPG) a complex glycolipid, and glycoinositolphospholipids (GIPLs).

The LPG has a tetrapartite structure consisting of the membrane anchor, a phosphosaccharide core, phosphorylated oligosaccharide repeats and a terminal oligosaccharide cap structure. The abundance of these molecules per cell is indicated

Source: Based on Overath *et al.*, 1997

microbiocidal oxygen radicals. LPG transiently inhibits the fusion of acidified endosomes with the parasitophorous vacuole and scavenges any oxygen radicals generated. It also suppresses nitric oxide synthase (NOS) expression, and hence the production of nitric oxide. The gp63 protease protects the parasite from host lysosome-mediated cytolysis, and GIPLs also strongly inhibit NOS expression.

As the flagellated leishmania metacyclic transforms into the amastigote within the parasitophorous vacuole, phagosome-lysosome fusion occurs, and the parasites are able to survive in their acidic, hydrolase-rich environment. Transformation is accompanied by downregulation of LPG and gp63 expression, leaving the amastigote surface dominated by GIPLs. Amastigotes released from bursting host cells must therefore use mechanisms for entry into fresh host cells different from those of metacyclic promastigotes. One possibility is opsonization of amastigotes with antibody from the developing humoral immune response of the host and exploitation of the macrophage Fc (immunoglobulin) receptor to effect binding, but our understanding of amastigote adaptations for host cell entry is still limited.

Protective immunity to leishmania infections depends upon T cell-mediated responses rather than humoral (B cell-mediated) responses. In addition to adaptations for survival in the macrophage of the non-immune host, leishmania parasites may frustrate this cell-mediated immune response. Successful presentation of parasite antigens to T lymphocytes to induce an immune response is via Class II major histocompatibility molecules (MHC) on antigen-presenting cells such as macrophages. Such presentation induces expansion of interferon gamma-producing (CD4+ Th1 subset) lymphocytes, and interferon in turn induces NOS expression and nitric oxide production in the host macrophage. In *Leishmania amazonensis*-infected macrophages, Class II MHC molecules reaching the parasitophorous vacuole are endocytozed by amastigotes, and degraded both within the parasitophorous vacuole and in the parasite itself by cysteine proteases of both host and parasite origin. The induction of protective immunity against leishmaniasis is also dependent upon the production of the cytokine IL-12 by the infected macrophage which induces IFN production by Natural Killer cells and T cells. *Leishmania* metacyclic promastigotes, but not procyclic promastigotes, are powerful inhibitors of macrophage IL-12 production both *in vitro* and *in vivo*, though whether their amastigote successors can suppress IL-12 production is doubtful. LPG, present on metacyclics, but not on amastigotes, may be the regulatory molecule.

10.5 Parasites going nowhere: altruism and division of labour

The construction of life-cycles that include dead-end stages has always been regarded as suspect by orthodox Darwinians, for surely natural selection should not permit such extravagance. However, it is not inconceivable that something like kin selection – first invoked to account for evolution of sterile castes in the social insects – might operate in unicellular organisms to explain the existence of parasites that go nowhere (Vickerman, 1993).

Unicellular organisms were until relatively recently believed to lead somewhat selfish lives, each for himself, so to speak. Division of labour between cells, social cooperation, and some individuals laying down their lives so that others could continue the species, were not for them. But suddenly evidence is accumulating fast that this is not the case, and that many, if not all, individual protists may depend upon others for their survival (Christensen *et al.*, 1998). I have already mentioned (in section 4.1) the social interactions that occur between bloodstream trypanosomes during antigenic variation. A further opportunity to exhibit 'altruism' may occur in the vector's midgut, where trypanosome numbers appear to be controlled by some members of the population undergoing programmed cell death (Welburn and Maudlin, 1997). If confirmed, such population control measures may be looked upon as adaptive, as they avoid exhausting the host's resources, thus increasing the parasite population's chances of survival.

Division of labour, well exemplified among free-living protists by the cellular slime molds where some amoebae become reproductive spores while others develop into sterile stem cells in the fruiting body, may also occur among parasites. A possible example may be seen in the development of *Leishmania* in the sandfly. In addition to giving rise to metacyclic promastigotes that eventually infect the mammalian host, some of the attached procyclic promastigotes of the vector midgut may give rise to rounded paramastigotes attached to the culticularized foregut. These appear to be a parallel development to the metacyclics and do not appear to give rise to mammal-infective forms, so what is their function? A possible answer to this question comes from studies on what happens during transmission of leishmania parasites by the bite of the sandfly.

In order to infect the mammal, metacyclics have to be expelled from the sandfly gut through the food canal in the proboscis. But during biting, blood has to be ingested, pumped into the midgut through the action of the pharyngeal and cibarial pumps; this two-way movement occurs only in leishmania-infected sandflies, and the presence of the parasites themselves appears to induce this abnormal movement. Schlein *et al.* (1993) have suggested that chitinase secreted by the foregut-attached flagellates is responsible for inflicting damage by lysing the soft inner layer of the cuticle of the pumps. Closure of the cardiac valve is necessary to permit one-way flow of blood through the pumps; in infected flies the valve remains open, so that suction exerted by the pump acts in both directions. The result is that blood and midgut contents are ejected and reingested during feeding, but some ejected metacyclics are able to initiate infection of the mammalian host.

10.6 The great mystery of life

I began this discussion of adaptations to parasitism with a warning about the adaptationist-selectionist approach, and I shall end by returning to it. Certainly one of the great unanswered questions of biology (and therefore one of the great mysteries of life!) is whether all adaptations can be explained in terms of Darwinian selection. In this respect, some embarrassment to the 'Darwin rules' view of parasite evolution can be caused by the facultatively pathogenic amoebae such as *Entamoeba histolytica*, *Acanthamoeba castellanii* and *Naegleria fowleri*. As this is a volume on flagellates, I shall take as an example the amoeboflagellate *Naegleria fowleri*, causative agent of primary amoebic meningoencephalitis in humans (see John, 1998, for review of biology).

This amoeba is normally free-living in warm waters, preferring temperatures of 37–46°C and feeding on bacteria, but it can also invade the human body via the olfactory epithelium during swimming. It migrates along the axons of the olfactory neurons and the lymphatic spaces beside them to enter the meningeal spaces, adopting a cytophagous mode of feeding as it does so. The inflammatory reaction which the parasite induces in the meninges, and then in the brain itself, rapidly leads to the death of the host, usually in little more than a week. Once the amoeba has become invasive it ceases to produce the encysted stage which would enable it to endure adverse conditions outside a host. Unlike *Trypanosoma brucei*, once the parasite is in the brain it cannot leave (*T. brucei* can temporarily avoid host defences by exploiting the host's blood–brain barrier, but can re-enter the bloodstream). When the host dies, the amoeba dies with it, as post-mortem changes in the brain will quickly render it inviable. If the invasive amoeba is no longer capable of contributing to the infection of future hosts, how are we to account for its adaptations to life in the brain, as the entire parasitic phase would seem to be beyond the reach of natural selection? This phase can be compared with the post-reproductive phase in senescent multicellular animals, when deleterious genes may be freely expressed because thay cannot be selected against, neither can beneficial genes be selected for (Medawar, 1952).

Of course, one interpretation of the 'one-way ticket' invasive phase of *N. fowleri* is that the human host is not the natural one (that is, the host in which the parasite evolved), and that in the natural host infection does culminate in the production of viable transmissive cysts, whereas in the non-natural host the life-cycle is incomplete. Unfortunately the geographical distribution of the disease makes spotting a likely candidate host difficult. Is it possible that in such cases as the tissue-invading amoebae, preadaptation of the non-invasive form can reach such heights that opportunity is all that is required for these organisms to switch from one lifestyle to another?

REFERENCES

Alexander, J., Satoskav, A. R. and Russell, D. G. (1999) *Leishmania* spp.: models of intracellular parasitism. *Journal of Cell Science*, **112**, 2993–3002.

Alfonzo, J. D., Thiemann, O. and Simpson, L. (1997) The mechanism of U insertion/deletion RNA editing in kinetoplastid mitochondria. *Nucleic Acids Research*, **25**, 3751–3759.

Andrews, N. W. (1995) What we talk about when we talk about cell invasion: reflections on a trypanosome system. In *Molecular Approaches to Parasitology* (eds J. C. Boothroyd and R. Komuniecki), New York, Wiley-Liss, pp. 359–369.

Barry, J. D. (1997) The biology of antigenic variation in African trypanosomes. In *Trypanosomiasis and Leishmaniasis: Biology and Control* (eds G. Hide, J. C. Mottram, G. H. Coombs and P. H. Holmes), Wallingford: CAB International, pp. 89–107.

Bernard, C., Simpson, S. and Patterson, D. J. (1999) Some free-living flagellates (Protista) from anoxic habitats. *Ophelia,* (in press).

Blom, D., de Haan, A., van den Berg, M., Sloof, P., Jirku, M., Lukes, J. and Benne, R. (1998) RNA editing in the free-living bodonid *Bodo saltans. Nucleic Acids Research,* **26**, 1205–1213.

Bogdan, C. and Rollinghoff, M. (1999) How do protozoan parasites survive inside macrophages? *Parasitology Today,* **15**, 22–28.

Borst, P. and Fairlamb, A. H. (1998) Surface receptors and transporters of *Trypanosoma brucei. Annual Review of Microbiology,* **52**, 745–778.

Borst, P., Bitter, W., Blundell, P., Cross, M., McCulloch, R., Rudenko, G., Taylor, M. C. and Van Leeuwen, F. (1997) The expression sites for variant surface glycoproteins of *Trypanosoma brucei.* In *Trypanosomiasis and Leishmaniasis: Biology and Control* (eds G. Hide, J. C. Mottram, G. H. Coombs and P. H. Holmes),Wallingford: CAB International, pp. 109–131.

Brune, A., Emerson, D. and Breznak, J. A. (1995) The termite gut microflora as an oxygen sink: microelectrode determination of oxygen and pH gradients in guts of lower and higher termites. *Applied and Environmental Microbiology,* **61**, 2681–2687.

Camargo, E. P. (1999) *Phytomonas* and other trypanosomatid parasites of plants and fruit. *Advances in Parasitology,* **42**, 29–112.

Cavalier-Smith, T. (1997) Cell and genome evolution: facultative anaerobiosis, glycosomes and kinetoplast RNA editing. *Trends in Genetics,* **13**, 6–9.

Christensen, S. T., Leick, V., Rasmussen, L. and Wheatley, D. N. (1998) Signalling in unicellular eukaryotes. *International Review of Cytology,* **177**, 181–253.

Clayton, C. and Michels, P. (1996) Metabolic compartmentation in African trypanosomes. *Parasitology Today,* **12**, 465–471.

Corliss, J. O. (1994) An interim utilitarian ('user-friendly') hierarchical classification and characterisation of the protists. *Acta Protozoologica,* **33**, 1–51.

Cross, G. A. M. (1996) Antigenic variation in trypanosomes: secrets surface slowly. *BioEssays*, **18**, 283–291.

Dillon, R. J. and Lane, R. P. (1999) Detection of *Leishmania* lipophosphoglycan binding proteins in the gut of the sandfly vector. *Parasitology*, **118**, 27–32.

Docampo, R. and Moreno, S. N. J. (1999) Acidocalcisome: a novel calcium storage compartment in trypanosomatids and apicomplexan parasites. *Parasitology Today*, **15**, 443–448).

Ersfeld, K., Melville, S. E. and Gull, K. (1999) Nuclear and genome organization of *Trypanosoma brucei*. *Parasitology Today*, **15**, 58–63.

Ferguson, M. A. J. (1997) The surface glycoconjugates of trypanosomatid parasites. *Philosophical Transactions of the Royal Society of London Series B*, **352**, 1295–1302.

Frank, S. A. (1999) A model for sequential domination of antigenic variants in African trypanosome infections. *Proceedings of the Royal Society, Series B*, **266**, 1397–1401.

Hart, D. T, Vickerman, K and Coombs, G. H. (1981) Respiration of *Leishmania mexicana* amastigotes and promastigotes. *Molecular and Biochemical Parasitology*, **4**, 39–51.

Hajduk, S. L., Siqueira, A. M. and Vickerman, K. (1986) Kinetoplast DNA of *Bodo caudatus*. *Molecular and Cellular Biology*, **6**, 4372–4378.

John, D. T. (1998) Opportunistic amoebae. In *Topley and Wilson's Microbiology and Microbial Infections vol. 5* (eds F. E. G. Cox, J. P. Kreier and D. Wakelin.), London: Arnold, pp. 1179–1192.

Kohl, L. and Gull, K. (1998) Molecular architecture of the trypanosome cytoskeleton. *Molecular and Biochemical Parasitology*, **93**, 1–9.

Kulda, J, and Nohýnková, E. (1995) *Giardia* in humans and animals. In *Parasitic Protozoa, 2nd edn, vol. 10* (eds J. P. Kreier and J. R. Baker), San Diego: Academic Press, pp. 225–422.

Lujàn, H. D., Mowatt, M. R. and Nash, T. E. (1998) The molecular mechanisms of *Giardia* encystation. *Parasitology Today*, **14**, 446–450.

Maslov, D. A., Avila, H. A., Lake, J. A. and Simpson, L. (1994) Evolution of RNA editing in kinetoplastid protozoa. *Nature*, **365**, 345–348.

Matthews, K. R. (1999) Developments in the differentiation of *Trypanosoma brucei*. *Parasitology Today*, **15**, 76–80.

Medawar, P. B (1952) *An Unsolved Problem of Biology*. London: H. K. Lewis.

Nash, T. E. (1997) Antigenic variation in *Giardia lamblia* and the host's immune response. *Philosophical Transactions of the Royal Society, Series B*, **352**, 1369–1375.

Opperdoes, F., Michels, P. A. M., Adje, C. and Hannaert, V. (1998) Organelle and enzyme evolution in trypanosomatids. In *Evolutionary Relationships among Protozoa* (eds G. H. Coombs, K. Vickerman, M. A. Sleigh and A. Warren), Dordrecht: Kluwer Academic, pp. 229–253.

Overath, P., Stierhof, Y. D. and Wiese, M. (1997) Endocytosis and secretion in trypanosomatid parasites – tumultuous traffic in a pocket. *Trends in Cell Biology*, 7, 27–33.

Pecka, Z., Nohýnková, E. and Kulda, J. (1996) Ultrastructure of *Cochlosoma anatis* Kotlan 1923 and taxonomic position of the family Cochlosomatidae (Parabasalidea: Trichomonadida). *European Journal of Protistology*, **32**, 190–201.

Reduth, D. and Schaub, G. A. (1988) The ultrastructure of cysts of *Blastocrithidia triatomae* Cerisola *et al.* 1971 (Trypanosomatidae): a freeze-fracture study. *Parasitology Research*, **74**, 301–306.

Rudenko, G., Chaves, I., Dirks-Mulder, A. and Borst, P. (1998) Selection for activation of a new variant surface glycoprotein gene expression site in *Trypanosoma brucei* can result in deletion of the old one. *Molecular and Biochemical Parasitology*, **95**, 97–109.

Sanchez-Moreno, M., Lasztity, D., Coppens, I. and Opperdoes, F. (1992) Characterisation of carbohydrate metabolism and demonstration of glycosomes in a *Phytomonas* sp. isolated from *Euphorbia characia*. *Molecular and Biochemical Parasitology*, **54**, 185–200.

Schlein, Y., Jacobson, R. L. and Messer, G. (1993) *Leishmania* infections damage the feeding mechanisms of the sandfly vector and implement transmission by bite. *Proceedings of the National Academy of Sciences USA,* **89**, 9944–9948.

Schmidt, J., Kleffmann, T. and Schaub, G. A. (1998) Hydrophobic attachment of *Trypanosoma cruzi* to a superficial layer of the rectal cuticle in the bug *Triatoma infestans. Parasitology Research,* **84**, 527–536.

Seay, M. B., Heard, P. L. and Chaudhari, G. (1996) Surface Zn-protease as a molecule for defense of *Leishmania mexicana amazonensis* promastigotes against cytolysis inside macrophage phagolysosomes. *Infection and Immunity,* **64**, 5129–5137.

Shapiro, T. A. and Englund, P. T. (1995) The structure and replication of kinetoplast DNA. *Annual Review of Microbiology,* **49**, 117–143.

Sloof, P. and Benne, R. (1997) RNA editing in kinetoplastid parasites: what to do with U. *Trends in Microbiology,* **5**, 189–195.

Tetaud, E., Barrett, M. P., Bringaud, F. and Baltz, T. (1997) Kinetoplastid glucose transporters. *Biochemical Journal,* **325**, 569–580.

Tielens, A. G. M. and Van Hellemond, J. J. (1998) Differences in energy metabolism between Trypanosomatidae. *Parasitology Today,* **14**, 265–271.

Turner, C. M. R. (1999) Antigenic variation in *Trypanosoma brucei* infections: an holistic view. *Journal of Cell Science,* **112**, 3187–3192.

Turrens, J. F. (1991) Mitochondrial metabolism of African trypanosomes. In *Biochemical Protozoology* (eds G. Coombs and M. North), London: Taylor and Francis, pp.145–153.

Van den Abbeele, J., Claes, Y., Van Bockstaele, D., Le Ray, D. and Coosemans, M. (1999) *Trypanosoma brucei* spp.: development in the tsetse fly: characterisation of the post-mesocyclic stages in the foregut and proboscis. *Parasitology,* **118**, 469–478.

Vickerman, K. (1985) Developmental cycles and biology of pathogenic trypanosomes. *British Medical Bulletin,* **41**, 105–114.

Vickerman, K. (1989) Trypanosome sociology and antigenic variation. *Parasitology,* **99**, 537–547.

Vickerman, K. (1990) Phylum Zoomastigina: Class Kinetoplastida. In *Handbook of Protoctista* (eds. L. Margulis, J. O. Corliss, M. Melkonian and D. Chapman), Boston: Jones and Bartlett, pp. 200–210.

Vickerman, K. (1993) Natural selection and the life cycles of protozoan parasites. *Proceedings of the Zoological Society of Calcutta, Haldane Commemoration Volume*, pp. 41–42.

Vickerman, K. (1994) The evolutionary expansion of the trypanosomatid flagellates. *International Journal for Parasitology,* **24**, 1317–1331.

Vickerman, K. and Tetley, L. (1990) Flagellar surfaces of parasitic protozoa and their role in attachment. In *Ciliary and Flagellar Membranes* (ed. R. A. Bloodgood), New York: Plenum, pp. 267–303.

Welburn, S. and Maudlin, I. (1997) Control of *Trypanosoma brucei brucei* infections in tsetse *Glossina morsitans. Medical and Veterinary Entomology,* **11**, 286–289.

Chapter 11

Flagellates and the microbial loop

Johanna Laybourn-Parry and Jacqueline Parry

ABSTRACT

Flagellate Protozoa, particularly nanoflagellates, dominate the protozoan communities of the planktonic habit. They play a cardinal role in transferring carbon and in nutrient regeneration within the microbial plankton. They feed mainly on heterotrophic bacteria, but are capable of exploiting dissolved organic carbon and phototrophic bacteria, as well as some components of the phytoplankton. Many of the coloured photosynthetic nanoflagellates are capable of phagotrophy on bacteria, thereby supplementing their carbon budgets by mixotrophy. Heterotrophic flagellate abundances are controlled by bottom-up factors which influence the productivity of bacteria, but may also be subject to top-down control from grazers. Cladocera, ciliates and rotifers exploit them as food, but may also compete with them for bacterial food resources. Aggregates of particulate organic matter are a common feature of the planktonic environment. They form foci of high microbial activity, and many flagellates are adapted to living in the attached mode, achieving high densities on particles. In some stable aquatic water columns, for example meromictic lakes, flagellates form distinct strata in specialized communities which attract specific populations of predators.

11.1 Introduction

Flagellate Protozoa are ubiquitous in all aquatic environments and soils, from the equator to polar regions. They are usually, with few exceptions, the most numerous protozoan component of planktonic and benthic aquatic environments (Table 11.1). Soils have been less widely investigated, but flagellates are frequently an important element in these communities, where they are still effectively aquatic organisms, living in the moisture films around soil particles and in the interstices between particles (Cowling, 1994). The flagellates are a diverse group both morphologically and physiologically. They contain a wide range of nutritional types ranging from heterotrophs which consume bacteria, algae and other protozoa, through complex degrees of mixotrophy, to purely autotrophic species.

The potential importance of flagellate Protozoa in planktonic processes, such as the biogeochemical cycling of nutrients and carbon and the flow of energy, was recognized some decades ago (Pomeroy, 1974). The term 'microbial loop' was coined by Azam *et al.* (1983). The concept of the microbial loop placed bacteria,

Table 11.1 Abundances of heterotrophic nanoflagellates, their bacterial food supply and ciliates in a range of freshwater and marine ecosystems

Site	*Bacteria x 10^9 l^{-1}*	*Heterotrophic flagellates x 10^6 l^{-1}*	*Ciliates x 10^3 l^{-1}*	*Source*
Limfjorden (Denmark)	0.5–15.2	0.2–15.2	1.4–162	Anderson and Sørensen (1986)
Coastal off Georgia	—	0.3–6.3	—	Sherr *et al.* (1984)
Red Sea	52–88	0.6–1.2	—	Weisse (1989)
North Atlantic	0.16–0.52	0.78–0.86	—	Weisse and Scheffel-Möser (1991)
Inshore S. Ocean	0.2–0.8	1.6–4.2	—	Leakey *et al.* (1996)
North Sea	1.2–2.7	0.7–6.6	—	Neilsen and Richardson (1989)
Loch Ness (oligotrophic)	0.23–0.71	0.012–0.273	0.05–2.1	Laybourn-Parry *et al.* (1994)
Crooked Lake, Antarctica (ultra-oligotrophic)	0.12–0.45	0–0.051	0.010–0.5	Laybourn-Parry *et al.* (1992, 1995)
Bohemian Reservoir (mesotrophic)	max. 4.9	max. 1.4	max 66	Simek and Straskrabová (1992)
Lake Constance (eutrophic)	1.9–11.8	0.7–3.1	9–133	Weisse *et al.* (1990)

heterotrophic flagellates and other protozoans into the structure of the planktonic food web (Figure 11.1). The term, however, is rather misleading in the light of current knowledge. Recent investigations have demonstrated that the microbial component is effectively an additional series of components within trophic levels, and that they interact extensively with metazoans and some components of the phytoplankton. These interactions have been well illustrated by manipulation experiments (Reimann, 1985; Weiss and Scheffel-Möser, 1991) and modelling of the microbial food web (Blackburn *et al.*, 1997).

As the database on the ecology and physiology of flagellates has grown, the intricacies of their interactions with bacteria, other protozoan groups, metazoans and the pool of dissolved organic carbon (DOC) are being revealed. Some groups are still poorly researched, particularly the dinoflagellates. Traditionally thought of as phytoplankton, about half of the extant species are colourless and heterotrophic, and many of those possessing photosynthetic pigments are not entirely dependent on photosynthesis (Gaines and Elbrächter, 1987). The heterotrophic forms are frequently abundant, and rival ciliates in their ability to graze algae and bacteria (Hansen, 1991; Nakamura *et al.*, 1995).

Many phytoflagellates practice phagotrophy on bacteria and compete for bacterial food resources with other bacterivores. On occasions they may exert a greater grazing impact than heterotrophic nanoflagellates in freshwater sub-tropical lakes (Sanders and Porter, 1988) and in Antarctic meromictic lakes (Roberts and Laybourn-Parry, 1999). Some heterotrophic flagellates are able to exploit the DOC

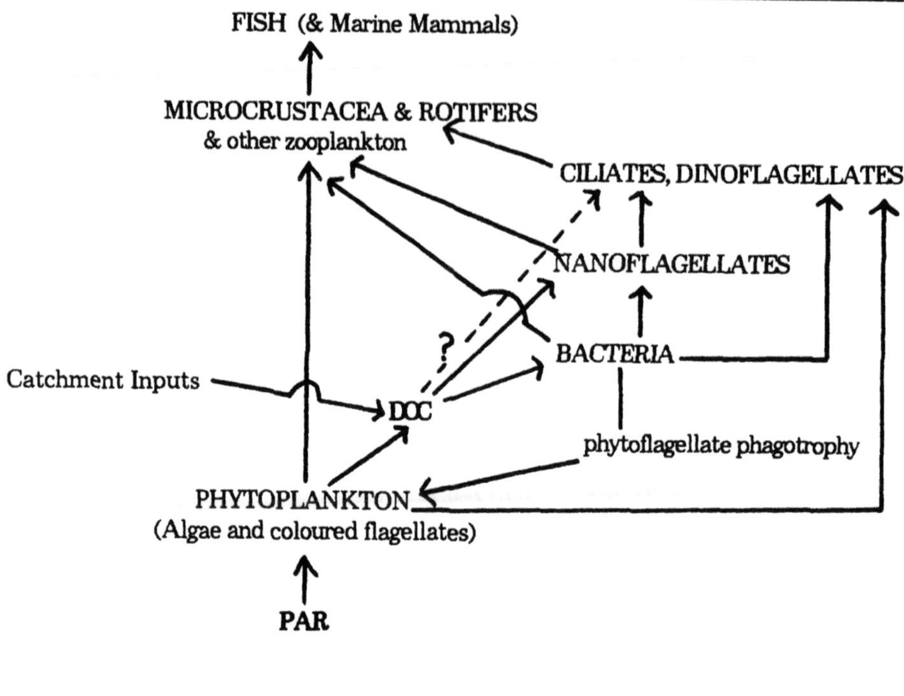

Figure 11.1 A simplified scheme of the microbial loop

pool directly, thereby short-circuiting the microbial loop and exploiting an energy source which has traditionally been thought of as only available to bacteria (Marchant and Scott, 1993).

11.2 Functional dynamics of the microbial loop

In the model most widely accepted, the pool of DOC, which is exploited by the bacterioplankton, is largely derived from the phytoplankton (Figure 11.1). The phytoplankton exude a portion of their photosynthate into the water as high molecular weight DOC, particularly towards the end of a bloom when nutrients become limiting. It is estimated that up to 60 per cent of primary production may be released in this way and is used by bacteria (Bell and Kuparinen 1984; Small *et al.*, 1989). Other sources of DOC include excretion by zooplankton, and leakage from phytoplankton cells during inefficient or 'sloppy' feeding by zooplankton (Eppley *et al.*, 1981). Viral induced lysis of bacterial cells may recycle significant amounts of bacterial carbon to the DOC pool (Bratbak *et al.*, 1990). Moreover, viruses form a food source for heterotrophic flagellates in marine waters, though they ingest them at lower rate than bacteria (González and Suttle, 1993). Bacterivore ciliates have also been shown to contribute to the pool, their activity shifts the dominance to lower molecular weight fractions (Taylor *et al.*, 1985). The importance of allochthonous carbon in freshwaters, and the role of the microbial plankton in transferring that carbon to higher trophic levels, have only recently been recognized. Many lakes at higher latitudes receive significant inputs of humic carbon derived from the catchment. This colours the water and inhibits photosynthesis. In Loch Ness, for example, the daily carbon demand of the bacterioplankton cannot be met by exudation from primary production (Laybourn-Parry *et al.,* 1994). In this lake and in Lake Pääjärvi (Finland), it is the protozooplankton, and in particular the heterotrophic nanoflagellates, that transfer allochthonous carbon from the bacterioplankton to zooplankton (Laybourn-Parry *et al.,* 1994; Kankalla *et al.,* 1996).

The importance of the microbial loop varies seasonally and is related in most lakes to the patterns of primary production, which in turn are controlled by light and nutrient availability. In the annual cycle the typical pattern for temperate clearwater lakes is an increase in the components of the microbial loop in spring (Simek and Straskrabová, 1992; Laybourn-Parry and Rogerson, 1993; Weisse *et al.,* 1990) (Figure 11.2 a). After the stratification of the water column in summer, the protozooplankton, which is usually dominated by heterotrophic flagellates, plays a crucial role in the recycling of nutrients (phosphorus and nitrogen) to the bacterio- and phytoplankton (Laybourn-Parry, 1992). In humic lakes, however, it is rainfall and catchment inputs which mediate patterns in microbial activity (Laybourn-Parry *et al.,* 1994) (Figure 11.2 b). Annual studies of microbial plankton in marine waters are limited for obvious logistic reasons, but inshore or estuarine studies indicate a strong seasonality in microbial activity (Leakey *et al.,* 1994, Laybourn-Parry *et al.,* 1992) (Figure 11.2 c).

In Lake Constance the response of the microbial plankton to the phytoplankton spring bloom was very pronounced. More than 50 per cent of the primary production was channelled through the microbial loop (Weisse *et al.,* 1990). During this phase the heterotrophic nanoflagellate carbon demand was 93 mg C m^2 day^{-1} compared with ciliate carbon demand of 368 mg C m^2 day^{-1} . During the bloom the flagellates grew rapidly, but they suffered heavy grazing pressure from ciliates. This predation pressure resulted in only 25 per cent of bacterial production being removed by flagellate grazing. In the Baltic Sea the spring carbon requirement of the heterotrophic flagellates was almost in balance with bacterial production (Kousa and Kivi, 1989). The increase in flagellate numbers commenced before the bacterial spring peak. In this case pico-phytoplankton were also exploited as a food resource during the entire spring bloom. In contrast, investigations in the North Sea have indicated that the microbial loop was relatively unimportant in carbon turnover during the spring bloom, and that the bulk of primary production was transferred to the benthos (Nielsen and Richardson, 1989). In summer, however, after stratification, the biomass of bacteria and their flagellate predators increased significantly, and it is during this time that the microbial loop played an important role in carbon cycling.

Community dynamics in aquatic systems are viewed as being controlled by either bottom-up (nutrients and light) or top-down processes (predation pressure). The complexities of bottom-up versus top-down control of flagellates are still poorly researched, but the data which are emerging from freshwaters suggest that bottom-up factors appear to be consistently more important than top-down factors in controlling their abundance (Gasol *et al.,* 1995). There are, however, some studies which indicate the reverse situation (Weisse, 1991; Carrias *et al.,* 1998). The dimension of time is important in such considerations, not only in relation to season

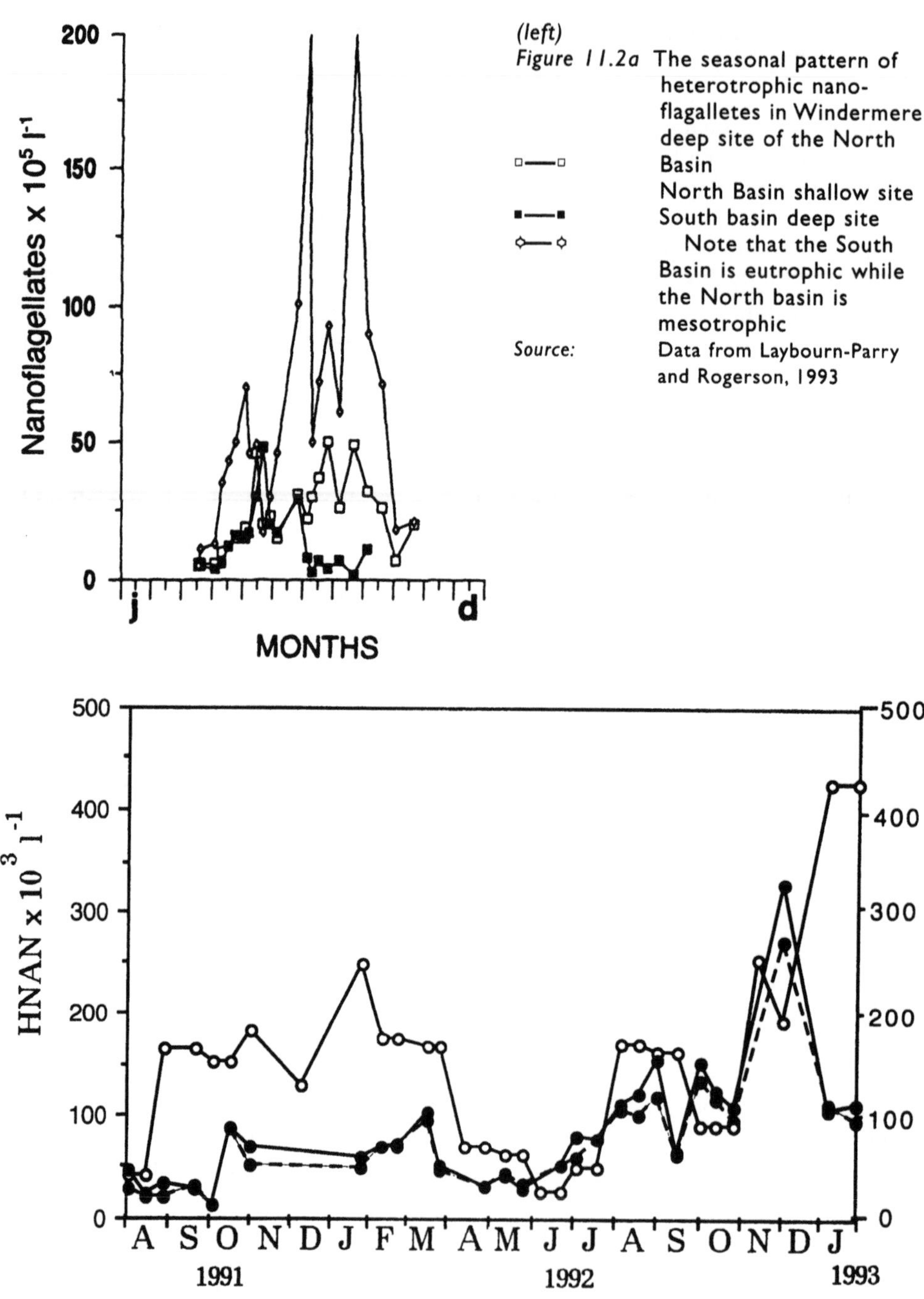

(left)
Figure 11.2a The seasonal pattern of heterotrophic nano-flagalletes in Windermere deep site of the North Basin (□—□); North Basin shallow site (■—■); South basin deep site (◇—◇). Note that the South Basin is eutrophic while the North basin is mesotrophic

Source: Data from Laybourn-Parry and Rogerson, 1993

Figure 11.2b The seasonal pattern of heterotrophic nanoflagellates in Loch Ness, a lake supported by allochthonous carbon inputs. Solid line mean values in the top 30m of the water column, dotted line mean values in the top 100m. Open circles mean monthly rainfall

Source: Data from Laybourn-Parry and Walton, 1998

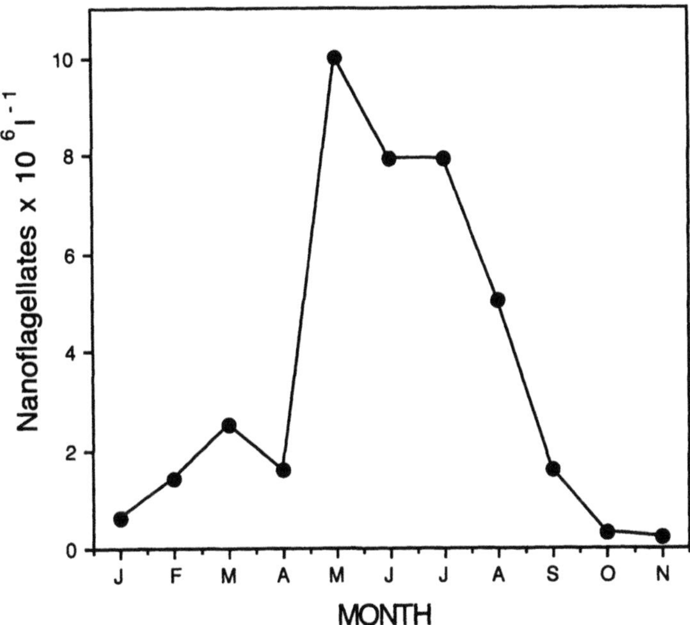

Figure 11.2c Seasonal patterns of heterotrophic nanoflagellate abundance in Loch Striven (a sea loch)

Source: Data from Laybourn-Parry *et al.*, 1992

but also in the frequency of sampling and the detail of the database. The elements of the microbial loop have short life-cycles and changes can occur over very short time scales. The predators of flagellates vary between marine and freshwaters, and this impacts on potential top-down control of their populations. Metazoan grazing of flagellates appears not to be a feature of marine plankton dynamics. Freshwater zooplankton contains several filter-feeding taxa (the Cladocera and rotifers) which are known to exert grazing pressure on flagellates. These organisms have few marine representatives. Marine crustacean zooplankton is dominated by copepods, which feed on larger particles (which may include ciliates). Calanoids do not impact significantly on small cells like flagellates (Burns and Schallenberg, 1996).

In a survey of sixteen Quebec lakes aimed at unravelling the complexities of bottom-up versus top-down control, Gasol *et al.* (1995) found a close correlation between phosphorus, chlorophyll and bacteria and heterotrophic flagellates (Figure 11.3). Bacteria, total phosphorus and chlorophyll *a* concentrations were the main factors which mediated the abundances of flagellates, particularly in spring. However, this varied among lakes depending on their trophic status. When the effect of resource availability was removed, the biomass of cladoceran predators of flagellates became an important determining factor in their abundance. Flagellate abundance declined, and the ratio of bacteria per flagellate increased, as cladocerans increased. In contrast, ciliate predators were only important in some lakes. The majority of these lakes were oligotrophic to mesotrophic. In contrast Carrias *et al.* (1998) concluded that free-swimming

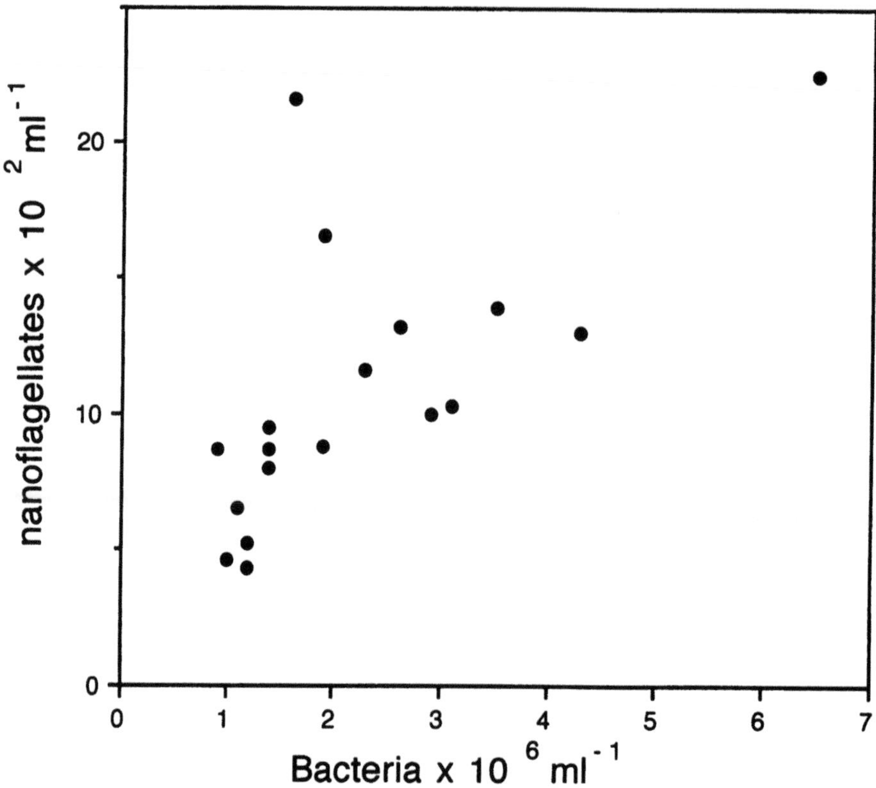

Figure 11.3 Heterotrophic nanoflagellate abundances in relation to bacterial abundances in sixteen Quebec lakes
Source: Data from Gasol *et al.*, 1995

flagellate abundance in oligotrophic Lake Pavin was strongly controlled by metazoan predation pressure.

The seasonal taxonomic make-up of the community of flagellates is also controlled by grazing in lakes (Weisse, 1991). The mean cell size of the community, and hence its taxonomic makeup, changed throughout the year in Lake Constance. In the spring the community was dominated by larger forms, but after stratification in the clearwater phase the mean cell size was reduced to one third. These differences were attributed to differing grazing impact by ciliates and metazoans (Cladocera and rotifers). Temporal changes in the taxonomic composition of the heterotrophic flagellate community is poorly documented. Rather, workers enumerate both phototrophic and heterotrophic flagellates, and may also categorize them into size classes. This relates to the difficulties of identification without detailed light, electron microscopy or laser confocal microscopy in studies aimed at studying protozooplankton dynamics rather than taxonomy.

The limited data indicates that freshwater communities are dominated by similar taxa across latitudes. *Monas* (*Spumella*) is common, as are choanoflagellates, bicosoecids and *Kathablepharis ovalis* Skuja (Carrick and Fahnenstiel, 1989; Sanders *et al.*, 1989; Bennett *et al.*, 1990; Carrias *et al.*, 1998). In the marine environment bicosoecids and choanoflagellates are also a dominant component of the nanoflagellate community (Anderson and Sørensen, 1986; Leakey *et al.*, 1996). Mixotrophic phototrophic nanoflagellates are common in some lakes, for example in sub-tropical Lake Oglethorpe the genera *Dinobryon* and *Ochromonas* exerted a significant grazing impact on bacteria (Sanders *et al.*, 1989). Meromictic Antarctic lakes have a high degree of mixotrophy. In the Dry Valley lakes of Southern Victoria Land cryptophytes dominate the flagellate community, outgrazing the heterotrophic forms during the summer (Roberts and Laybourn-Parry, 1999). In a saline meromictic lake (Ace Lake) in the Vestfold Hills, *Pyramimonas* dominates the autotrophic plankton and appears to indulge in mixotrophy, and possibly complete heterotrophy during the dark austral winter (Bell and Laybourn-Parry, 1999).

We tend to classify freshwater nanoflagellates into one of two categories in plankton dynamics investigations, either phototrophic or heterotrophic. However, there is clear evidence from studies where grazing studies have been undertaken in conjunction with seasonal changes in abundance and biomass, that the nutritional demarkation is not that clear-cut. Many species of coloured nanoflagellates, and indeed some of the larger flagellates (for example, dinoflagellates), have considerable nutritional versatility. They switch from phototrophy to mixotrophy in response to a range of environmental factors, and may periodically be major consumers of bacteria (see section 11.3).

11.3 Interactions between flagellates and dinoflagellates and their food resources

The heterotrophic nanoflagellates are largely bacterivores. Consequently there have been a number of attempts to relate or couple their abundances with those of their prey. Some investigators have found a positive correlation between the abundance of bacteria and heterotrophic nanoflagellates (Davis *et al.*, 1985; Berninger *et al.*, 1991). However, this relationship has been disputed and there is an increasing body of evidence to indicate that the correlation is tenuous or non-existent, for example Gasol and Vaqué (1993), who reviewed the literature, adding other data from their own unpublished observations. Bacterial abundances ranged from 10^5–10^{11} ml^{-1} in the oceans, rivers, lakes and sediments. The heterotrophic nanoflagellates ranged between 10^2-10^6 ml^{-1} across these various ecosystems. The relationship between flagellates and bacteria was not strong in any ecosystem, but it was statistically significant in lakes and rivers.

There are a number of factors which account for the apparent lack of coupling. First, heterotrophic nanoflagellates are not the only bacterivores in the plankton or sediments. There is plenty of evidence to show that phototrophic nanoflagellates also feed on bacteria, sometimes at high levels (such as Bennett *et al.*, 1990). Ciliates also graze bacteria (Christofferson *et al.*, 1990; Sherr and Sherr, 1987; Laybourn-Parry *et al.*, 1996). Rotifers and Cladocera may also exploit bacteria (Porter *et al.*, 1983; Carrias *et al.*, 1998). Bacterial biomass can be attacked by viruses, causing cell lysis and decrease in abundance (Procter and Fuhrman, 1990). There may also be significant top-down control on heterotrophic nanoflagellates which prevents them reaching abundances the bacterial food resources could potentially support (Gasol and Vaqué, 1993). Moreover, the impact and balance of these microbial and metazoan interactions with bacterial communities will change over short and longer time scales.

The grazing rates of flagellates vary in relation to a range of environmental factors, of which temperature is one of the most important (see Table 11.2). As one would anticipate, higher environmental temperatures result in higher grazing rates and impact on bacterial production. In low temperature environments, such as polar lakes and seas, grazing rates are low and only a relatively small proportion of bacterial production is predated. In warmer lakes and marine waters the uptake of bacteria can be high, and in some cases total daily bacterial production can be consumed by the flagellate community. The grazing rates shown in Table 11.2 are community rates and take no account of taxonomic makeup.

Studies on individual species have demonstrated that there are significant differences between species, even within close temperature ranges (Eccleston-Parry and Leadbeater, 1994). Consequently any seasonal change in the taxonomic makeup of a community may have a marked impact on the overall grazing pressure and the proportion of daily bacterial production removed. Clearance rates vary considerably between species, from a maximum of 0.1-203 nl cell^{-1} hour^{-1}. There is apparently no relationship between the size of the flagellate and uptake and clearance rates (Eccleston-Parry and Leadbeater, 1994). Bacterial type and size can also influence the clearance and grazing rates of nanoflagellates. For example, *Paraphysomonas imperforata* exhibits a wide range of clearance and uptake rates when fed on a spectrum of bacteria and algal species (Goldman and Dennett, 1990; Caron *et al.*, 1986; Eccleston-Parry and Leadbeater, 1994). It is evident that within a community of heterotrophic nanoflagellates, the observed grazing rates are determined by a complex array of factors including temperature, food type, the individual feeding capabilities of constituent flagellate species and probably also competition between species.

While heterotrophic dinoflagellates have attracted less attention than nanoflagellates, they are nonetheless often abundant in marine waters. In coastal Antarctic waters for example abundances of 46,500 cells l^{-1} have been recorded in summer (Archer *et al.*, 1996), in the Kattegat (Denmark) they corresponded to 13–77 per cent of phytoplankton biomass (Hansen, 1991) and in the Gulf of Gdansk (southern Baltic) they contributed between 73 and 76 per cent of heterotrophic protozooplankton annual biomass (Bralewska and Witek, 1995). They feed on ciliates, diatoms, nanoflagellates and other dinoflagellates, using a variety of mechanisms. At such high abundances they must play a significant role in carbon cycling within the marine microbial plankton.

In Antarctic waters nine taxa of dinoflagellates removed 4.8 per cent of autotrophic biomass day^{-1}, equivalent to 25 per cent of daily primary production (Archer *et al.*, 1996). Their clearance and grazing rates can be high. In *Oblea rotunda* clearance rates ranged between 80 and 700 nl individual^{-1} hour^{-1} (Strom and Buskey, 1993), whereas in some species they tend to be lower, for example *Oxyrrhis marina* where rates were between 14 and 21 nl individual^{-1} hour^{-1} (Goldman *et al.*, 1989).

A number of phytoflagellate genera in freshwaters are capable of feeding on bacteria, and more correctly should be categorized as mixotrophs. These include

Table 11.2 Grazing rates and impact of heterotrophic nanoflagellates in a range of aquatic ecosystems.

Site	*Grazing rate bacteria cell⁻¹ h⁻¹*	*% bacterial production removed day⁻¹*	*Temperature °C*	*Source*
Crooked Lake, Antarctica (hyper-oligotrophic)	max. 4.8	0.1–9.7	0.5–4	Laybourn-Parry *et al.* (1995)
Loch Ness, Scotland (oligotrophic)	0.4–1.0	max 27	5–15	Laybourn-Parry and Walton (1998)
Bohemian reservoir (mesotrophic)	15.8–24.5	10–64	—	Simek and Straskrabová (1992)
Lake Vechten, Netherlands (eutrophic)	168	max 100	10–20	Bloem and Bar-Gilissen (1989)
Resolute Passage High Arctic	3–64	—	<1	Laurion *et al.* (1995)
Antarctic coastal	0.06–8.23	<0.3–30.6	-0.4–1.4	Leakey *et al.* (1996)
Georgian Coastal (USA)	—	30–50	12–29	Sherr *et al.* (1984)
Red Sea	4.5–33	max 76	22 -27	Weisse (1989)

chrysomonads such as *Dinobryon, Ochromonas, Poterioochromonas* and *Chrysostephnosphaera,* members of the Prymnesiida such as *Chrysochromulina ericina* and *C. spinifera* and some cryptomonads such as *Cryptomonas* (Porter, 1988; Sanders, 1991). There is debate as to why these phytoflagellates graze bacteria. Some workers argue that it is a means of acquiring nutrients for photosynthesis during periods of limitation (Nygaard and Tobiesen, 1993), while others suggest that the prime purpose of phagotrophy is to obtain carbon to support growth (Bird and Kalff, 1986; Caron *et al.*, 1990; Jones *et al.*, 1993). A number of factors have been shown to affect the degree of phagotrophy, including light, dissolved organic carbon concentration and concentrations of potential food (Sanders *et al.*, 1990; Jones *et al.*, 1993).

Clearance rates of mixotrophic coloured nanoflagellates are comparable with many of those reported for heterotrophic nanoflagellates. The grazing rates reported for mixotrophic flagellates from temperate to sub-tropical temperatures ranged between 7 and 77 bacteria cell^{-1} hour^{-1} (Sanders and Porter, 1988; Bennett *et al.*, 1990; Nygaard and Tobiesen, 1993). In cold Antarctic lakes ingestion rates ranged between 0.2 and 3.6 bacteria cell^{-1} hour^{-1}, which was lower than that achieved by the heterotrophic flagellates in the same assemblage (4–6.7 bacteria cell^{-1} hour^{-1}) (Roberts and Laybourn-Parry, 1999). Loss of chlorophyll occurs with increased dependency on phagotrophy (Caron *et al.*, 1990; Sanders *et al.*, 1990).

Mixotrophic lacustrine nanoflagellates can, on occasions, outgraze heterotrophic nanoflagellates. In Lake Oglethorpe, a sub-tropical eutrophic lake, mixotrophic nanoflagellates equalled or exceeded the grazing impact of heterotrophic flagellates in the surface water during February (Bennett *et al.*, 1990). In Lakes Hoare and Fryxell in the McMurdo Dry Valleys of Antarctica, mixotrophic flagellates also had a greater grazing impact on the bacterioplankton, in this case throughout the entire austral summer (Roberts and Laybourn-Parry, 1999). Mixotrophic flagellates removed between 2 and 31 per cent of bacterial biomass day^{-1} in Lake Fryxell and 0.2 to 3.4 per cent in Lake Hoare. In contrast, heterotrophic flagellates grazed between 0.1 and 3.7 per cent and 0.7 and 2.3 per cent day^{-1} respectively. Data on mixotrophic dinoflagellate grazing rates is limited, but the indications are that they can have significant grazing capabilities. *Gymnodinium sanguineum* feeding on nanociliate populations in Chesapeake Bay had clearance rates of 0 to 5.8 μl cell^{-1} hour^{-1}, giving ingestion rates of 0.06 prey cell^{-1} hour^{-1}, which resulting in the daily removal of between 6 and 67 per cent of the standing stock of <20μm ciliates (Bockstahler and Coats, 1993).

It is evident that flagellate protozoa play a cardinal role in grazing bacterial, viral, algal and protozoan biomass, and even exploit the pool of DOC in aquatic ecosystems. The indications are that competitive interactions between different flagellate populations occur, and that at least some species are able to discriminate between prey types and size, and exercise selective feeding (Chrzanowski and Simek, 1990). Competition and feeding selectivity are two areas which need investigation.

11.4 Flagellate growth dynamics

The growth of heterotrophic flagellates is a function of a variety of factors among which food type, food concentration and temperature play the major role. Heterotrophic nanoflagellate specific growth rates, and hence doubling times, vary significantly. Laboratory-based studies give us an indication of potential maximum growth rates (see Table 11.3). The disparity in rates often seen for the same species under similar conditions is probably a reflection of genetic differences between populations.

In situ measurements in marine and freshwaters generally reveal much lower growth rates than those seen in laboratory populations (Table 11.3). These values are lower for several reasons. First, they are a measure of growth under fluctuating ambient temperatures, which are usually lower than those applied to laboratory studies. Second, they are based on the exploitation of natural bacterial assemblages, which may be available at sub-optimal levels, and may not all be palatable or able to sustain strong growth. Given this fact, it is likely that in the natural environment flagellates have to clear a larger volume of water to obtain sufficient food to sustain growth. Hence they have to expend more energy in feeding and have less energy available to partition into growth and reproduction. Competition for food between flagellate species and other bacterivores may also be reflected in lower *in situ* growth rates. Moreover, there may be locomotory behaviour related to avoidance of predation, which incurs additional energy expenditure in natural populations.

Heterotrophic dinoflagellates have comparable specific growth rates. *Oblea rotunda* achieved rates between 0. 019 and 0.027 h^{-1}. Differences were related to the food phytoplankton type (Strom and Buskey, 1993), and in another study a range of species had growth rates between 0.013 and 0.048 h^{-1} (Hansen, 1992). These rates are, however, significantly lower than their major protozoan competitors, the ciliates.

Table 11.3 Specific growth rates of cultured and natural assemblages of heterotrophic nanoflagellates

Species	*Max. specific growth rate h⁻¹*	*Temperature °C*	*Source*
Bodo designis	0.16	20	Eccleston-Parry and Leadbeater (1994)
Ciliophrys infusonium	0.045	20	"
Codosiga gracilis	0.052	20	"
Jakoba libera	0.036	20	"
Paraphysomonas imperforata	0.21	20	"
Diaphanoeca grandis	0.12	21	Anderson (1989)
Ochromonas sp.	0.016	18	Ammerman et *al.* (1994)
Stephanoeca diplocostata	0.076	18	Geider and Leadbeater (1988)
Stephanoeca diplocostata	0.031	20	Leadbeater and Davies (1984)
In situ natural community	0.06–0.027	15	Bjørnsen et *al.* (1988)
"	0.013–0.012	2.0–4.8	Laybourn-Parry et *al.* (1996)
"	0.043–0.093	5–15	Laybourn-Parry and Walton (1998)
"	0.003–0.033	4–15 approx.	Weisse (1991)
"	0.068–0.114	18–22 approx.	Bloem and Bar-Gilissen (1989)

11.5 The role of flagellates in nutrient regeneration

Protozoa play a crucial role in the recycling of nitrogen and phosphorus within the 'microbial loop'. This is because their main prey type, bacteria, act as nutrient sinks (Jürgens and Güde, 1990) and are consumers of nutrients rather than producers (Goldman *et al.,* 1985; LeCorre *et al.,* 1996). It is therefore the role of protozoa to remineralize organic nutrients (within their prey) to soluble inorganic nutrients, thereby providing picoplankton with a valuable source of nutrients in times of exogenous nutrient depletion (Jürgens and Güde, 1990). Dolan (1997) stated that the weight-specific excretion rate is inversely proportional to the size of the protist, thus implying that flagellates are the dominant remineralizers within the protozoan group. Estimates of flagellate weight specific regeneration rates vary from 2.8–148 μgN mg^{-1}DW h^{-1} (Andersson *et al.,* 1985; Goldman *et al.,* 1985; Caron and Goldman, 1990; Ferrier-Pagés and Rassoulzadegan, 1994; Suzuki *et al.,* 1996). In addition, microheterotrophs (1–15μm) are responsible for 64 per cent of the ammonium regeneration in the English Channel (LeCorre *et al.,* 1996) and could account for 50 per cent of the recycled nitrogen in oligotrophic Mediterranean waters (Ferrier-Pagés and Rassoulzadegan, 1994).

The major proportion of remineralized nutrients is of a specific form. Phosphorus is predominantly excreted as soluble reactive phosphorus (SRP) (Jürgens and Güde, 1990). Dissolved organic phosphorus (DOP) release can occur, but values vary considerably (Andersen *et al.,* 1986). There is no doubt that nitrogen is excreted mainly as ammonium (Goldman *et al.,* 1985), but other nitrogenous sources are also excreted such as urea (Goldman *et al.,* 1985), and amino acids (Andersson *et al.,* 1985), though in much smaller amounts. Secondary uptake of these remineralized nutrients by prey can potentially complicate laboratory grazing experiments. Ferrier-Pagés and

Rassoulzadegan (1994) estimated that about 2.8×10^{-12} μg-atoms N $cell^{-1}$ $hour^{-1}$ can be taken up by secondary uptake, but this can be avoided in experiments by using heat-killed bacteria as the prey (Ferrier-Pagés and Rassoulzadegan, 1994) or by using a bacteriostatic antibiotic (Eccleston-Parry and Leadbeater, 1995). In the field, bacterial growth coincides with protistan growth and excretion (Snyder and Hoch, 1996), as bacteria compete more successfully than phytoplankton for nutrients at low concentrations, particularly phorphorus (Currie and Kalff, 1984).

Few experiments have examined the role of specific flagellates on regeneration processes; of these there has been a bias towards nitrogen remineralization. A feature common to all experiments is that nutrient regeneration is most intense at the beginning of the exponential growth phase in batch cultures (Ferrier-Pagés and Rassoulzadegan, 1994; Eccleston-Parry and Leadbeater, 1995; Suzuki *et al.*, 1996), since flagellate growth is exponential and nutrient excretion is linear. Variability in specific growth rates (and hence regeneration rates) can be overcome by using continuous cultures (Bloem *et al.*, 1988; Jurgens and Gude, 1990; Eccleston-Parry and Leadbeater, 1995) or mechanistic models (Suzuki *et al.*, 1996), but such methodologies have been rarely applied.

Calculated regeneration efficiencies (Re) of phosphorus by heterotrophic flagellates feeding on bacteria have been found to vary within the extremely large range of 4–91 per cent (Andersson *et al.*, 1985; Andersen *et al.*, 1986; Jürgens and Güde, 1990; Eccleston-Parry and Leadbeater, 1995). A smaller range of 5–68 per cent has been found when using phytoplankton as prey (Goldman and Caron, 1985; Goldman *et al.*, 1985; Andersen *et al.*, 1986). Calculated regeneration efficiencies of nitrogen by heterotrophic flagellates feeding on phytoplankton prey can range from 7 to 70 per cent (Goldman and Caron, 1985; Goldman *et al.*, 1985, 1987), whereas a range of 13 to 41 per cent (Andersson *et al.*, 1985; Goldman *et al.*, 1985) has been found when bacteria are the prey. Higher values have been recorded (Ferrier-Pagés and Rassoulzadegan, 1994; Eccleston-Parry and Leadbeater, 1995), but on average, estimates of 30–50 per cent are most common (Caron and Goldman, 1990). Suzuki *et al.* (1996) have suggested that Re values of 50 per cent indicate one trophic step beginning with the prey of interest, 60 per cent indicates two to three trophic steps and 90 per cent indicates five to six trophic steps, and that this supports the concept of multistep foodwebs in the <20μm fractions (Wikner and Hagström, 1988).

Although variation in Re values during the flagellate growth cycle is common, very little interspecific variation in values between different flagellate species has been found (Andersson *et al.*, 1985; Goldman and Caron, 1985; Goldman *et al.*, 1985; Andersen *et al.*, 1986; Jürgens and Güde, 1990), possibly due to the choice of species used in these experiments, those with relatively high specific growth rates. Eccleston-Parry and Leadbeater (1995) did find interspecific variation in P regeneration efficiencies between flagellates with 'fast' and 'slower' growth rates with *Paraphysomonas imperforata* and *Bodo designis* (μ = 0.2 and 0.14 h^{-1}) releasing significantly more bacterial-bound phosphorus than *Jakoba libera* and *Stephanoeca diplocostata* (μ = 0.029 and 0.026 h^{-1}). However, no interspecies differences for the regeneration of nitrogen were found. *P. imperforata* is known to have the ability to conserve nitrogen (Goldman *et al.*, 1985), and Ferrier-Pagés and Rassoulzadegan (1994) found that *Pseudobodo* sp. also seems to conserve nitrogen during the exponential phase of growth. Eccleston-Parry and Leadbeater (1995) speculated that the possible reason for interspecies differences in P regeneration capacities was related to their life-cycle strategies/growth rates of the protozoa, and a recent study by Dolan (1997) suggests that, in the case of orthophosphate, excretion rates are indeed predictable based on flagellate size and growth rate, but ammonium excretion is unrelated to growth rate. The specific growth rates of flagellates have been found to be independent of the nutritional status of the prey (Goldman and Caron, 1985; Eccleston-Parry and Leadbeater, 1995); however, they can affect the quantity and quality of the nutrients excreted by the flagellates (Jürgens and Güde, 1990).

Nakano (1994) found that the ammonium regeneration rate of flagellates was closely related to the ingested bacterial nitrogen, while its phosphate regeneration rate depended on bacterial carbon:phosphorus ratios. Eccleston-Parry and Leadbeater (1995), on the other hand, found no correlation between the nitrogen and P content of prey and the level remineralized using phosphorus-rich and N-rich bacterial states, and suggested that the carbon:nitrogen and carbon:phosphorus ratios of the flagellates themselves influenced the level of nutrients regenerated. In fact, recent studies on zooplankton grazers (cladocerans and copepods) have shown that the level of P regeneration depends on the phosphorus content of the consumer (Lyche *et al.*, 1997). Published data on the carbon:nitrogen:phosphorus ratios of flagellates are limited. Currently, estimates of flagellate carbon:nitrogen ratios range from 2.5 to 7 (Bloem *et al.*, 1988; Nakano, 1994; Eccleston-Parry and Leadbeater, 1995) while carbon : phosphorus ratios range from 18 to 109 (Bloem *et al.*, 1988; Nakano, 1994; Eccleston-Parry and Leadbeater, 1995).

With regard to the quality of nutrients regenerated by flagellates, no significant relationship has been found between the N:P ratio of the nutrients excreted and that of the bacteria ingested (Nakano, 1994; Eccleston-Parry and Leadbeater, 1995). Rapidly growing protozoa appear to excrete orthophosphate (SRP) relative to ammonia well in excess of the Redfield ratio of 16:1, that is, nitrogen : phosphorus ratios in the range from 2:1 to 8:1 (Dolan, 1997), which suggests that protozoa may be more important in the regeneration of orthophosphate than ammonia. However, this may be spurious if one considers that the dominant excretory compound, SRP, is entirely comprised of orthophosphate (Dolan, 1997).

Information on the role of mixotrophic flagellates in nutrient regeneration processes is limited, even though they are able to show maximum excretion rates similar to like-sized heterotrophic protists (Dolan, 1997). However, very basic differences in patterns of nutrient regeneration by a mixotroph, *Ochromonas* sp. and a pure heterotroph, *Spumella* sp., have been recorded (Rothhaupt, 1997). The heterotroph, as expected, released SRP and ammonium when feeding on bacteria. Regeneration by the mixotroph depended on light availability and bacterial density, so that in continuous darkness, SRP and ammonium were released, but in the light when the mixotroph was growing phototrophically, these minerals were taken up by the mixotroph and not released during the nocturnal dark phase. However, if bacterial densities were high in the light, the mixotroph fed heterotrophically and released SRP and ammonium. It must be stressed that this behaviour is not to be expected in all mixotrophs, as there is a whole spectrum of mixotrophic behaviour which would ultimately lead to a spectrum of remineralization behaviours. Even so, the role of mixotrophic flagellates as nutrient remineralizers does warrant more attention, particularly since their abundance and the level of bacterivory appear to increase when mineral nutrients are limiting (Arenovski *et al.,* 1995), which coincides with the fact that the importance of microbial consortia in recycling processes is likely to increase as systems approach oligotrophy (Harrison, 1980). Since their abundance is tightly coupled to heterotrophic flagellates (Pick and Caron, 1987), their role as nutrient remineralizers may be more important than they have previously been given credit for.

11.6 Aggregates and microbial consortia

A close examination of a water sample from any aquatic environment will reveal particles of organic matter. These aggregates of particulate organic matter (POC) are usually formed by bacterial growth and aggregation (Sheldon *et al.*, 1967; Biddanda, 1985). Experimental investigations have shown that the aggregates are formed by bacterial utilization of DOC. Free-floating and attached bacteria exude extracellular POC in the form of slime, capsular material and appendages, producing extensive networks of extracellular carbon around particles, for example dead diatom frustules and inorganic particles, thereby creating flocs or aggregates (Biddanda, 1985). Once formed, the flocs rapidly develop associated flagellate populations which may be attached (for example choanoflagellates) or associated with the aggregate for feeding purposes. Other protozoa including amoebae and ciliates feeding on bacteria and flagellates join the community, producing a microbial consortium.

Aggregates support higher bacterial concentrations than the surrounding water and achieve the largest proportion of bacterial productivity in marine and freshwater plankton (Bell and Albright, 1981; Kirchman and Mitchell, 1982; Rogerson and Laybourn-Parry, 1992a; Grossart and Simon, 1993). The attached bacteria are larger than those in the surrounding water, and this, together with their active metabolic state, probably renders them attractive prey to flagellate and ciliate predators. Flagellates apparently select larger bacteria in preference to smaller forms (Chrzanowski and Simek, 1990). There appear to be taxonomic differences between attached and unattached bacteria. Based on rRNA analysis in marine waters, specific taxonomic groups appeared to favour attachment to particles (Delong *et al.*, 1993).

Flagellate abundances in microbial consortia are also much higher than in the surrounding water (Caron, 1991). This is consistent across freshwater to marine habitats. In the Clyde estuary 80 per cent of the heterotrophic flagellate community was either attached to or associated with particles, and 44 per cent of the phototrophic flagellates were attached (Rogerson and Laybourn-Parry, 1992b). In oceanic waters flagellates constituted 2–25 per cent of microbial biomass, the proportion varying with distance offshore and depth (Taylor *et al.*, 1986).

Detailed taxonomic analysis of marine particles extending from the surface to the ocean floor revealed forty different species of zooflagellate (Patterson *et al.,* 1993). They included a range of choanoflagellate species, bicosoecids and *Bodo* species among others. Most of the species described were small and bacterivorous. The ability to feed on particles and to exploit the attached niche varies among species. Some flagellates can graze very effectively on attached bacteria, while others, such as

Monas and *Cryptobia*, are unable to do so (Caron, 1987). There are distinct changes in species composition on particles, with depth in the marine water column. As depth increased both species diversity and abundance decreased. Species which survive at depth must be barotolerant. Sensitivity to increased pressure does vary between species (Turley and Carstens, 1991) and thus one sees species successions in relation to depth as particles undergo sedimentation.

Microbial consortia associated with particles or aggregates are an integral part of the complex microbial community dynamics operating in the water columns of the sea, lakes and estuaries. The formation and dissolution of aggregates is a dynamic process, changing both temporally and spatially within a three-dimensional environment. Dead or senescent phytoplankton cells frequently provide the nucleus of microbial aggregates. As they are transformed into aggregates they go through a series of events. First there is growth of bacteria, followed by further bacterial colonization, the aggregation of detritus, the colonization and growth of protozoa, and their predation of bacteria which eventually leads to the disruption of the aggregate (Biddanda and Pomeroy, 1988). The grazing activities of flagellates and other protozoa have the effect of disrupting the integrity of the aggregate as they remove bacteria and associated attachment structures. This process has been likened to the breakup of forest litter by detritivores.

In nutrient-rich waters, like those of estuaries and eutrophic lakes, aggregates are effectively a specialized niche in an otherwise rich environment. However, in oligotrophic lakes and the open ocean, the aggregate and its surrounding microzone offers a rich micro-environment in depaurate waters, allowing higher levels of productivity than would be possible by organisms freely suspended in the water. The microbial consortium is surrounded by a nutrient-rich microzone (P, N and DOC) as a result of exudation of carbon from POC and phototrophic flagellates and excretion of phosphorus and nitrogen by flagellates and other protozoa.

11.7 Spatial distribution patterns

In most situations, wind-driven turbulence in the euphotic zone results in a fairly homogenous distribution of flagellates and other protozoa, because they are unable to override the effects of water movements. In lakes, seasonal thermal stratification, or permanent chemical stratification in meromictic lakes, creates conditions which may favour the development of discrete layers of flagellates and their predators. Deep maxima of phytoflagellates associated with the metalimnion (boundary layer) have been reported from a variety of lakes. These are made up of chrysophytes (Pick *et al.*, 1984), dinoflagellates (Gálvez *et al.*, 1988), *Pyramimonas* (Bell and Laybourn-Parry, 1999) and cryptophytes (Gervais, 1998; Roberts and Laybourn-Parry, 1999). The phytoflagellates effectively live in a shaded environment where light levels are extremely reduced relative to surface levels of radiation. They effect a trade-off between living in the shade and other advantages, including the avoidance of predation and access to nutrients.

Deep layer phytoflagellates sustain growth by low-light-adapted photosynthesis (Gervais, 1998). In Antarctic perennially ice-covered lakes where light levels are severely attenuated, and even in summer may be only 1–2 per cent of surface levels, deep layer phytoflagellates exhibit remarkably high levels of photosynthetic efficiency (Vincent, 1981). Cryptophytes are able to migrate into the anoxic sulphide-rich lower water, enabling them to pick up nutrients and avoid ciliate predation (Gasol *et al.,* 1993; Gervais, 1998). In temperate lakes these phytoflagellates sustain growth solely by photosynthesis; there is no evidence of phagotrophy on bacteria.

Deep maxima of phytoflagellates attract specialized ciliate predators. In both temperate and Antarctic lakes large populations of specific ciliate species are found in association with the phytoflagellates. In Lake Cisó high numbers of the ciliates *Coleps* and *Prorodon* occur with the phytoflagellates, while in Lake Schlachtensee the dominant predator was *Coleps* (Pedrós-Alió *et al.,* 1995; Gervais, 1998). A dense population of *Plagiocampa* exploits the cryptophytes in Lake Fryxell (Roberts and Laybourn-Parry, 1999). In all cases the phytoflagellates can be clearly seen in various stages of digestion inside their predators under epifluorescence microscopy. Neither *Coleps* or *Prorodon* can follow their prey into the anoxic hypolimnion, but *Holophyra* is found periodically in lower anoxic waters. The grazing impact of these ciliate predators is not large. In Lake Cisó only 2-25 per cent of the phytoflagellates suffer predation, and in Lake Fryxell 0.02–20.5 per cent.

11.8 Future directions

There are a series of aspects of flagellate ecophysiology which demand further research. In terms of feeding biology we need to know much more about their ability to select different food types and sizes of food, and the mechanisms by which they do this. There is clearly a continuum between complete heterotrophy and complete autotrophy. Many of the coloured flagellates are capable of mixotrophy, either using phagotrophy or by taking up organic compounds. The relative dependence on photosynthesis versus heterotrophy in mixotrophic species and the factors which mediate the balance are still poorly understood in the natural environment. Mixotrophic species can exert very significant grazing pressure on bacterioplankton in some lakes and are clearly an important element in the transfer of carbon to higher trophic levels. Most studies of mixotrophic flagellates relate to the freshwater environment. More attention needs to be addressed to marine species. Are mixotrophic flagellates only significant in freshwaters?

While we do have information on the growth rates and doubling times of flagellates and their rates of energy ingestion, we have very few data for their growth efficiencies. A general figure of 40–60 per cent gross conversion or growth efficiency (Fenchel, 1982; Caron *et al.,* 1986) is widely used when constructing models of microbial community dynamics. It is dangerous to assume that a figure that is derived from laboratory cultures at relatively high temperatures can be applied with any accuracy to natural communities. Where conversion efficiencies have been derived under a field food and temperature regimen, they are usually significantly lower than 40 per cent. For example in Lake Biwa gross growth efficiency ranged between 11 and 53 per cent (Nagata, 1988) and in Crooked Lake (Antarctica) it was around 24 per cent (Laybourn-Parry *et al.,* 1995). We do need more basic ecophysiological data which can be used to construct more accurate dynamic models.

The taxonomy of communities during seasonal successions has received scant attention. We do have reasonably good data on patterns of abundance, but these usually lack taxonomic analysis. Flagellate taxonomy is time-consuming, but where it has been done there are clear successions, and often the changed population makeup causes significant changes in grazing pressure, because of taxon specific differences in grazing rates. The development of molecular probes and a wider use of laser confocal microscopy, which is less time-consuming than electron microscopy but provides comparable levels of magnification, may go some way to resolving this deficiency.

REFERENCES

Ammerman, J. W., Fuhrman, J. A., Hagström, Å. and Azam, F. (1984) Bacterioplankton growth in seawater. *Marine Ecology Progress Series*, **18**, 31–39.

Andersen, O. K., Goldman, J. C., Caron D. A. and Dennett, M. R (1986) Nutrient cycling in a microflagellate food chain. III. Phosphorus dynamics. *Marine Ecology Progress Series*, **31**, 47–55.

Andersen, P. (1989) Functional biology of the choanoflagellate *Diaphanoeca grandis* Ellis. *Marine Microbial Food Webs*, **3**, 35–50.

Andersen, P. and Sørensen, H. M. (1986) Population dynamics and trophic coupling in pelagic microorganisms in eutrophic coastal waters. *Marine Ecology Progress Series*, **33**, 99–109.

Andersson, A., Lee, C., Azam, F. and Hagström, A. (1985) Release of amino acids and inorganic nutrients by heterotrophic marine microflagellates. *Marine Ecology Progress Series*, **23**, 99–106.

Archer, S. D., Leakey, R. J. G., Burkill, P. H. and Sleigh, M. A. (1996) Microbial dynamics in coastal waters of east Antarctica: herbivory by heterotrophic dinoflagellates. *Marine Ecology Progress Series*, **139**, 239–255.

Arenovski, A. L., Lin Lim, E. and Caron, D. A. (1995) Mixotrophic nanoplankton in oligotrophic surface waters of the Sargasso Sea may employ phagotrophy to obtain major nutrients. *Journal of Plankton Research*, **17**, 801–820.

Azam, F., Fenchel, T., Field, J. G., Gray, J. S., Meyer-Reil, R. A. and Thingstad, F. (1983) The ecological role of water column microbes in the sea. *Marine Ecology Progress Series*, **10**, 257–263.

Bell, C. R. and Albright, L. J. (1981) Attached and free-floating bacteria in the Fraser River estuary, British Columbia, Canada. *Marine Ecology Progress Series*, **6**, 317–327.

Bell, E. and Laybourn-Parry, J. (1999) Annual plankton dynamics in an Antarctic saline lake. *Freshwater Biology*, **41**, 507–519.

Bell, R. T. and Kuparinen, J. (1984) Assessing phytoplankton and bacterioplankton production during early spring in Lake Erken, Sweden. *Applied and Environmental Microbiology*, **48**, 1221–1230.

Bennett, S. J., Sanders, R. W. and Porter, K. G. (1990) Heterotrophic, autotrophic and mixotrophic nanoflagellates: seasonal abundances and bacterivory in a eutrophic lake. *Limnology and Oceanography*, **35**, 1821–1832.

Berninger, U-G., Finlay, B. J. and Kuupp-Leinkki, P. (1991) Protozoan control of bacterial abundances in freshwater. *Limnology and Oceanography*, **36**, 139–147.

Biddanda, B. A. (1985) Microbial synthesis of macroparticulate matter. *Marine Ecology Progress Series*, **20**, 242–251.

Biddanda, B. A. and Pomeroy, L. R. (1988) Microbial aggregation and degradation of phytoplankton-derived detritus in seawater. I. Microbial succession. *Marine Ecology Progress Series*, **42**, 79–88.

Bird, B. F. and Kalff, J. (1986) Bacterial grazing by planktonic lake algae. *Science*, **231**, 493–495.

Bjørnsen, P. K., Riemann, B., Horsted, S. J., Nielsen, T. G. and Pock-Sten, J. (1988) Trophic interactions between heterotrophic flagellates and bacterioplankton in manipulated sea water enclosures. *Limnology and Oceanography*, **33**, 409–420.

Blackburn, N., Azam, F. and Hagström, Å. (1997) Spatially explicit simulations of a microbial food web. *Limnology and Oceanography*, **42**, 613–622.

Bloem, J. and Bär-Gilissen, M-J. B. (1989) Bacterial activity and protozoan grazing potential in a stratified lake. *Limnology and Oceanography*, **34**, 297–309.

Bloem, J., Starink, M., Bär-Gilissen, M-J. B. and Cappenberg, T. E. (1988) Protozoan grazing, bacterial activity and mineralization in two-stage continuous culture. *Applied and Environnmental Microbiology*, **54**, 3113–3121.

Bockstahler, K. R. and Coats, D. W. (1993) Grazing of the mixotrophic dinoflagellate *Gymnodinium sanguineum* on ciliate populations in Chesapeake Bay. *Marine Biology*, **116**, 477–487.

Bralewska, J. M. and Witek, Z. (1995) Heterotrophic dinoflagellates in the ecosystem of the Gulf of Gdansk. *Marine Ecology Progress Series*, **117**, 241–248.

Bratbak, G., Heldal, M., Norland, S. and Thingstad, T. F. (1990) Viruses as partners in spring bloom microbial trophodynamics. *Applied and Environmental Microbiology*, **56**, 1400–1405.

Burns, C. W. and Schallenburg, M. (1996) Relative impacts of copepods, cladocerans and nutrients on the microbial food web of a mesotrophic lake. *Journal of Plankton Research*, **15**, 683–714.

Caron, D. A. (1987) Grazing of attached bacteria by heterotrophic microflagellates. *Microbial Ecology*, **13**, 203–218.

Caron, D. A. (1991) Heterotrophic flagellates associated with sedimenting detritus. In *The Biology of Heterotrophic Flagellates* (eds. D. J. Patterson and J. Larsen), Clarendon Press, Oxford, pp. 77–92.

Caron, D. A. and Goldman, J. C (1990) Protozoan nutrient regeneration. In *Ecology of Marine Protozoa* (ed. G. M. Cupriulo), Oxford: Oxford University Press (1992), pp. 283–306.

Caron, D. A., Goldman, J. C. and Dennett, M. R. (1986) Effect of temperature on growth, respiration and nutrient regeneration by an omnivorous microflagellate. *Marine Ecology Progress Series*, **24**, 243–254.

Caron, D. A., Porter, K. G. and Sanders, R. W. (1990) Carbon, nitrogen and phosphorus budgets for the mixotrophic phytoflagellate *Poterioochromonas malhamensis* (Chrysophyseae) during bacterial ingestion. *Limnology and Oceanography*, **35**, 433–443.

Carrias, J-F., Amblard, C., Quiblier-Lloberas, C. and Bourdier, G. (1998) Seasonal dynamics of free and attached heterotrophic nanoflagellates in an oligomesotrophic lake. *Freshwater Biology*, **39**, 91–102.

Carrick, H. J. and Fahnenstiel, G. L. (1989) Biomass, size structure, and composition of phototrophic and heterotrophic nanoflagellate communities in Lakes Huron and Michigan. *Canadian Journal of Fisheries and Aquatic Sciences*, **46**, 1922–1928.

Christoffersen, K., Riemann, B., Hansen, L. R., Klysner, A. and Sørensen, H. B. (1990) Qualitative importance of the microbial loop and plankton community structure in a eutrophic lake during a bloom of cyanobacteria. *Microbial Ecology*, **20**, 253–272.

Chrzanowski, T. H. and Simek, K. (1990) Prey-size selection of freshwater flagellated protozoa. *Limnology and Oceanography*, **35**, 1429–1436.

Cowling, A. J. (1994) Protozoan distribution and adaptation. In *Soil Protozoa* (ed. J. F. Darbyshire), Wallingford: CAB International, pp. 5–42.

Currie, D. J. and Kalff, J. (1984) The relative importance of bacterioplankton and phytoplankton in phosphorus uptake in freshwater. *Limnology and Oceanography*, **29**, 311–321.

Davis, P. G., Caron, D. A., Johnson, P. W. and Sieburth, J. McN. (1985) Phototrophic and apochlorotic components of picoplankton and nanoplankton in the North Atlantic: geographic, vertical, seasonal and diel distributions. *Marine Ecology Progress Series*, **21**, 15–26.

Delong, E. F., Franks, D. G. and Alldredge, A. L. (1993) Phylogenetic diversity of aggregate-attached vs free-living marine bacterial assemblages. *Limnology and Oceanography*, **38**, 924–934.

Dolan, J. R. (1997) Phosphorus and ammonia excretion by planktonic protists. *Marine Geology*, **139**, 109–122.

Eccleston-Parry, J. D. and Leadbeater, B. S. C. (1994) A comparison of the growth kinetics of six marine heterotrophic nanoflagellates fed with one bacterial species. *Marine Ecology Progress Series*, **105**, 167–177.

Eccleston-Parry, J. D. and Leadbeater, B. S. C. (1995) Regeneration of phosphorus and nitrogen by four species of heterotrophic nanoflagellates feeding on three nutritional states of a single bacterial strain. *Applied and Environmental Microbiology*, **61**, 1033–1038.

Eppley, R. W., Horrigan, S. G., Fuhrmann, J. A. *et al.* (1981) Origins of dissolved organic matter in southern California coastal waters: experiment on the role of zooplankton. *Marine Ecology Progress Series*, **6**, 149–159.

Fenchel, T. (1982) Ecology of heterotrophic microflagellates. II. Bioenergetics and growth. *Marine Ecology Progress Series*, **8**, 225–231.

Ferrier-Pagés, C. and Rassoulzadegan, F. (1994) N-remineralization in planktonic protozoa. *Limnology and Oceanography*, **39**, 411–418.

Gaines, G. and Elbrächter, M. (1987) Heterotrophic nutrition. In *The Biology of the Dinoflagellates* (ed. F. R. J. Taylor), Oxford: Blackwell Scientific, pp. 224–268.

Gálvez, J. A., Niell, F. X. and Lucena, J. (1988) Description and mechanism of formation of a deep chlorophyll maximum due to *Ceratium hirundinella* (O. F. Müller) Bergh. *Archiv für Hydrobiologie*, **112**, 143–155.

Gasol, J. M., García-Cantizano, J., Massana, R., Guerrero, R. and Pedrós-Alió, C. (1993) Physiological ecology of a metalimnetic *Cryptomonas* population: relationships to light, sulfide and nutrients. *Journal of Plankton Research*, **15**, 255–275.

Gasol, J. M. and Vaqué, D. (1993) Lack of coupling between heterotrophic nanoflagellates and bacteria: a general phenomenon across aquatic systems? *Limnology and Oceanography*, **38**, 657–665.

Gasol, J. M., Simons, A. M. and Kalff, J. (1995) Patterns in the top-down versus bottom-up regulation of heterotrophic nanoflagellates in temperate lakes. *Journal of Plankton Research*, **17**, 1879–1903.

Geider, R. J. and Leadbeater, B. S. C. (1988) Kinetics and energetics of growth of the marine choanoflagellate *Stephanoeca diplocostata. Marine Ecology Progress Series*, **47**, 169–177.

Gervais, F. (1998) Ecology of cryptophytes coexisting near a freshwater chemocline. *Freshwater Biology*, **39**, 61–78.

Goldman, J. C. and Caron, D. A. (1985) Experimental studies on an omnivorous microflagellate: implications for grazing and nutrient regeneration in the marine microbial food chain. *Deep Sea Research*, **32**, 899–915.

Goldman, J. C., Caron, D. A., Andersen, O. K. and Dennett, M. R. (1985) Nutrient cycling in a microflagellate food chain. I. Nitrogen dynamics. *Marine Ecology Progress Series*, **24**, 231–242.

Goldman, J. C., Caron, D. A. and Dennett, M. R. (1987) Nutrient cycling in a microflagellate food chain. IV. Phytoplankton–microflagellate interactions. *Marine Ecology Progress Series*, **38**, 75–87.

Goldman, J. C., Dennett, M. R. and Gordin, H. (1989) Dynamics of herbivorous grazing by the heterotrophic dinoflagellate *Oxyrrhis marina. Journal of Plankton Research*, **11**, 391–407.

Goldman, J. C. and Dennett, M. R. (1990) Dynamics of prey selection by an omnivorous flagellate. *Marine Ecology Progress Series*, **59**, 183–194.

González, J. M. and Suttle, C. A. (1993) Grazing by marine nanoflagellates on viruses and virus-sized particles: ingestion and digestion. *Marine Ecology Progress Series*, **94**, 1–10.

Grossart, H-P. and Simon, M. (1993) Limnetic macroscopic organic aggregates (lake snow): occurrence, characteristics, and microbial dynamics in Lake Constance. *Limnology and Oceanography*, **38**, 532–546.

Hansen, P. J. (1991) Quantitative importance and trophic role of heterotrophic dinoflagellates in a coastal pelagic food web. *Marine Ecology Progress Series*, **73**, 253– 261.

Hansen, P. J. (1992) Prey size selection, feeding rates and growth dynamics of heterotrophic dinoflagellates with special emphasis on *Gyrodinium spirale*. *Marine Biology*, **114**, 327–334.

Harrison, W. G. (1980) Nutrient regeneration and primary production in the sea. In *Primary Productivity in the Sea*, Brookhaven Symposium on Biology 31, Plenum, pp. 433–460.

Jones, H. L. J., Leadbeater, B. S. C. and Green, J. C. (1993) Mixotrophy in marine species of *Chrysochromulina* (Prymnesiophyceae): ingestion and digestion of a small green flagellate. *Journal of the Marine Biological Association of the UK*, 73, 283–296.

Jürgens, K. and H. Güde. 1990. Incorporation and release of phosphorus by planktonic bacteria and phagotrophic flagellates. *Marine Ecology Progress Series*, **59**, 271–284.

Kankaala, P., Arvola, L., Tulonen, T. and Ojala, A. (1996) Carbon budget for the pelagic food web of the euphotic zone in a boreal lake (Lake Pääjärvi). *Canadian Journal of Fisheries and Aquatic Science*, **53**, 1663–1674.

Keller, M. D., Shapiro, L. P., Haugen, E. M., Cucci, T. L., Sherr, E. B. and Sherr, B. F. (1994) Phagotrophy of fluorescently labelled bacteria by an oceanic phytoplankter. *Microbial Ecology*, **28**, 39–52.

Kirchman, D. and Mitchell, R. (1982) Contribution of particle-bound bacteria to total micro-heterotrophic activity in five ponds and two marshes. *Applied and Environmental Microbiology*, **43**, 200–209.

Kousa, H. and Kivi, K. (1989) Bacteria and heterotrophic flagellates in the pelagic carbon cycle in the northern Baltic Sea. *Marine Ecology Progress Series*, **53**, 93–100.

Laurion, I., Demers, S. and Vézina, A. F. (1995) The microbial food web associated with ice algal assemblage: biomass and bacterivory of nanoflagellate protozoans in Resolute Passage (High Canadian Arctic). *Marine Ecology Progress Series*, **120**, 77–87.

Laybourn-Parry, J. (1992) *Protozoan Plankton Ecology*. London: Chapman and Hall.

Laybourn-Parry, J. and Rogerson, A. (1993) Seasonal patterns of protozooplankton in Lake Windermere, England. *Archiv für Hydrobiologie*, **129**, 25–43.

Laybourn-Parry, J., Rogerson, A. and Crawford, D. W. (1992) Temporal patterns of protozooplankton abundance in the Clyde and Loch Striven. *Estuarine, Coastal and Shelf Science*, **35**, 533–543.

Laybourn-Parry, J., Walton, M., Young, J., Jones, R. I. and Shine, A. (1994) Protozooplankton and bacterioplankton in a large oligotrophic lake – Loch Ness, Scotland. *Journal of Plankton Research*, **16**, 1655–1670.

Laybourn-Parry, J., Ellis-Evans, J. C. and Bayliss, P. (1995) The dynamics of heterotrophic nanoflagellates and bacterioplankton in a large ultra-oligotrophic Antarctic lake. *Journal of Plankton Research*, **17**, 1835–1850.

Laybourn-Parry, J., Ellis-Evans, J. C. and Butler, H. (1996) Microbial dynamics during the summer ice-loss phase in maritime Antarctic lakes. *Journal of Plankton Research*, **18**, 495–511.

Laybourn-Parry, J. and Walton, M. (1998) Seasonal heterotrophic flagellate and bacterial plankton dynamics in a large oligotrophic lake – Loch Ness, Scotland. *Freshwater Biology*, **39**, 1–8.

Leadbeater, B. S. C. and Davies, M. E. (1984) Developmental studies on the loricate choanoflagellate *Stephanoeca diplocostata* Ellis. III. Growth and turnover of silica, preliminary observations. *Journal of Experimental Marine Biology and Ecology*, **81**, 251–268.

Leakey, R. J. G., Fenton, N. and Clarke, A. (1994) The annual cycle of planktonic ciliates in nearshore waters at Signy Island, Antarctica. *Journal of Plankton Research*, **16**, 841–856.

Leakey, R. J. G., Archer, S. D. and Grey, J. (1996) Microbial dynamics in coastal waters of East Antarctica: bacterial production and nanoflagellate bacterivory. *Marine Ecology Progress Series*, **142**, 3–17.

LeCorre, P., Wafar, M., Lhelguen, S. and Maguer, J. F. (1996) Ammonium assimilation and regeneration by size-fractionated plankton in permanently well-mixed temperate waters. *Journal of Plankton Research*, **18**, 355–370.

Lynche, A., Andersen, T., Christoffersen, K., Hessen, D., Berger Hansen, P. H. and Klyssner, A. (1997) Zooplankton as sources and sinks for phosphorous in mesocosms in a P-limited lake. In *Verhandlungen Proceedings Travaux 26th Congress, São Paulo* (eds. W. D. Williams and A. Sladeckova), Stuttgart, p. 355.

Marchant, H. J. and Scott, F. J. (1993) Uptake of sub-micrometre particles and dissolved organic matter by Antarctic choanoflagellates. *Marine Ecology Progress Series*, **92**, 59–64.

Nagata, T. (1988) The microflagellate-picoplankton food linkage in the water column of Lake Biwa. *Limnology and Oceanography*, **33**, 504–517.

Nakamura, Y., Shin-ya, S. and Hiromi, J. (1995) Growth and grazing of a naked heterotrophic dinoflagellate, *Gyrodinium dominans. Aquatic Microbial Ecology*, **9**, 157–164.

Nakano, S. (1994) Carbon-nitrogen-phosphorus ratios and nutrient regeneration of a heterotrophic flagellate fed on bacteria with different elemental ratios. *Achiv für Hydrobiologie*, **129**, 257–271.

Nielsen, T. G. and Richardson, K. (1989) Food chain structure of the North Sea plankton communties; seasonal variations of the role of the microbial loop. *Marine Ecology Progress Series*, **56**, 75–87.

Nygaard, K. and Tobiesen, A. (1993) Bacterivory in algae: a survival strategy during nutrient limitation. *Limnology and Oceanography*, **38**, 273–279.

Patterson, D. J., Nygaard, K., Steinberg, G. and Turley, C. M. (1993) Heterotrophic flagellates and other protists associated with oceanic detritus throughout the water column in the mid North Atlantic. *Journal of the Marine Biological Association of the UK*, **73**, 67–95.

Pedrós-Alió, C., Massana, R., Latasa, M., Gracía-Cantizano, J. and Gasol, J. M. (1995) Predation by ciliates on a metalimnetic *Cryptomonas* population: feeding rates, impact and effects of vertical migration. *Journal of Plankton Research*, **17**, 2131–2154.

Pick, F. R., Nalewajko, C. and Lean, D. R. S. (1984) The origin of a metalimnetic chrysophyte peak. *Limnology and Oceanography*, **29**, 125–134.

Pick, F. R. and Caron, D. A. (1987) Picoplankton and nanoplankton biomass in Lake Ontario: relative contibution of phototrophic and heterotrophic communities. *Canadian Journal of Fisheries and Aquatic Science*, **44**, 2164–2172.

Pomeroy, L. R. (1974) The ocean's food web, a changing paradigm. *Bioscience*, **9**, 499–504.

Porter, K. G. (1988) Phagotropic phytoflagellates in microbial food webs. *Hydrobiologia*, **149**, 89–97.

Porter, K. G., Feig, Y. S. and Vetter, E. F. (1983) Morphology, flow regimes and filtering rates of *Daphnia, Ceriodaphnia* and *Bosmina* fed on natural bacteria. *Oecologia* (Berlin), **58**, 156–163.

Procter, L. M. and Fuhrman, J. A. (1990) Viral mortality of marine bacteria and cyanobacteria. *Nature*, **343**, 60–62.

Riemann, B. (1985) Potential influence of fish predation and zooplankton grazing on natural populations of freshwater bacteria. *Applied and Environmental Microbiology*, **50**, 187–193.

Roberts, E. and Laybourn-Parry, J. (1999) Mixotrophic chrysophytes and their predators in the Dry Valley lakes of Antarctica. *Freshwater Biology*, **41**, 737–746.

Rogerson, A. and Laybourn-Parry, J. (1992a) Bacterioplankton abundance and production in the Clyde, Scotland. *Archiv für Hydrobiologie*, **126**, 1–14.

Rogerson, A. and Laybourn-Parry, J. (1992b) Aggregate dwelling protozooplankton communities in estuaries. *Archiv für Hydrobiologie*, **125**, 411–422.

Rothhaupt, K. O. (1997) Nutrient turnover by freshwater bacterivorous flagellates: differences between a heterotrophic and a mixotrophic chrysophyte. *Aquatic Microbial Ecology*, **12**, 65–70.

Sanders, R. W. (1991) Trophic strategies among heterotrophic flagellates. In *The Biology of Heterotrophic Flagellates* (eds. D. J. Patterson and J. Larsen), Oxford: Clarendon Press, pp. 21–38.

Sanders, R. W. and Porter, K. G. (1988) Phagotrophic phytoflagellates. *Advances in Microbial Ecology*, **10**, 167–192.

Sanders, R. W., Porter, K. G., Bennett, S. J. and Debaise, A. E. (1989) Seasonal patterns of bacterivory by flagellates, ciliates, rotifers, and cladocerans in a freshwater planktonic community. *Limnology and Oceanography*, **34**, 673–687.

Sanders, R. W., Porter, K. G. and Caron, D. A. (1990) Relationship between phototrophy and phagotrophy in the mixotrophic chrysophyte *Poterioochromonas malhamensis*. *Microbial Ecology*, **19**, 97–109.

Sheldon, R. W., Evelyn, T. P. T. and Parsons, T. R. (1967) On the occurrence and formation of small particles in seawater. *Limnology and Oceanography*, **12**, 367–375.

Sherr, B. F., Sherr, E. B. and Newell, S. Y. (1984) Abundance and productivity of heterotrophic nanoplankton in Georgia coastal waters. *Journal of Plankton Research*, **6**, 195–202.

Sherr, E. B. and Sherr, B. F. (1987) High rates of consumption of bacteria by pelagic ciliates. *Nature*, **325**, 710–711.

Simek, K. and Straskrabová, V. (1992) Bacterioplankton production and protozoan bacterivory in a mesotrophic reservoir. *Journal of Plankton Research*, **14**, 773–787.

Small, L. F., Landry, M. R., Eppley, R. W., Azam, F. and Carlucci, A. F. (1989) Role of plankton in the carbon and nitrogen budgets of Santa Monica Basin, California. *Marine Ecology Progress Series*, **56**, 57–74.

Snyder, R. A. and Hoch, M. P. (1996) Consequences of protist-stimulated bacterial production for estimating protozoan growth efficiencies. *Hydrobiologia*, **341**, 113–123.

Strom, S. L. and Buskey, E. J. (1993) Feeding, growth and behaviour of the thecate heterotrophic dinoflagellate *Oblea rotunda*. *Limnology and Oceanography*, **38**, 956–977.

Suzuki, M. T., Sherr, E. B. and Sherr, B. F. (1996) Estimation of ammonium regeneration efficiencies associated with bactivory in pelagic food webs via a N-15 tracer method. *Journal of Plankton Research*, **17**, 411–428.

Taylor, G. T., Iturriaga, R. and Sullivan, C. W. (1985) Interactions of bacterivorous grazers and heterotrophic bacteria with dissolved organic matter. *Marine Ecology Progress Series*, **23**, 129–141.

Taylor, G. T., Karl, D. M. and Pace, M. L. (1986) Impact of bacteria and zooflagellates on the composition of sinking particles: an *in situ* experiment. *Marine Ecology Progress Series*, **29**, 141–155.

Turley, C. M. and Carstens, M. (1991) Pressure tolerance of oceanic flagellates: implications for remineralization of organic matter. *Deep Sea Research*, **38**, 403–413.

Vincent, W. F. (1981) Production strategies in Antarctic inland waters: phytoplankton ecophysiology in a permanently ice-covered lake. *Ecology*, **62**, 1215–1224.

Weisse, T. (1989) The microbial loop in the Red Sea: dynamics of pelagic bacteria and heterotrophic nanoflagellates. *Marine Ecology Progress Series*, **55**, 241–250.

Weisse, T. (1991) The annual cycle of heterotrophic freshwater nanoflagellates: role of bottom-up versus top-down control. *Journal of Plankton Research*, **13**, 167–185.

Weisse, T., Müller, H., Pinto-Coelho, R. M., Schweizer, A., Springmann, D. and Baldringer, G. (1990) Response of the microbial loop to the phytoplankton in a large prealpine lake. *Limnology and Oceanography*, 35, 781–793.

Weisse, T. and Scheffel-Möser, U. (1991) Uncoupling the microbial loop: growth and grazing loss rates of bacteria and heterotrophic nanoflagellates in the North Atlantic. *Marine Ecology Progress Series*, 71, 195–205.

Wilkner, J. and Hagström, A. (1988) Evidence for a tightly coupled nanoplanktonic predator–prey link regulating the bacterivores in the marine environment. *Marine Ecology Progress Series*, 50, 137–145.

Chapter 12

Functional diversity of heterotrophic flagellates in aquatic ecosystems

Hartmut Arndt, Désirée Dietrich, Brigitte Auer, Ernst-Josef Cleven, Tom Gräfenhan, Markus Weitere and Alexander P. Mylnikov

ABSTRACT

There is a lack of taxonomic resolution due to methodological problems in most ecological reports despite the significant contribution of heterotrophic flagellates (HF) to the carbon cycle of most aquatic ecosystems. The determination of HF species on a quantitative level is difficult, especially for most athecate and aloricate taxa. The dominant taxonomic groups among heterotrophic nano- and microflagellate communities within different marine, brackish and limnetic pelagic communities (heterokont taxa, dinoflagellates, choanoflagellates, kathablepharids) and benthic communities (euglenids, bodonids, thaumatomastigids, apusomonads) seems to be surprisingly similar. HF among Protista *incertae sedis*, often neglected in ecological studies, were abundant in all investigated habitats. The taxonomic variety of HF reflects the large diversity of functions of HF such as predominant bacterivory, herbivory, carnivory, detritivory and omnivory, respectively. Typical benthic HF can contribute significantly to pelagic HF communities especially in limnetic and marine coastal waters. High tolerances to changes in salinity give rise to the assumption that several species are able to live in both marine and freshwater habitats. The functional diversity of HF is discussed with respect to the feeding ecology, life strategies, tolerances to extreme abiotic and biotic conditions and distribution patterns. Considering the strong predation pressure by metazoans and protists on HF communities, many morphological and behavioural features of HF may be explained as predator avoidance mechanisms.

12.1 Introduction

This chapter will concentrate on free-living obligate heterotrophic flagellates (HF), which are abundant or otherwise important components of aquatic ecosystems. For a long time ecologists have dealt with flagellates as they do with bacteria, seeing them as bright spots under the epifluorescence microscope. HF were mainly considered as one trophic group: bacterivores. Nevertheless, treating heterotrophic nanoflagellates as a black box was a very important step in the understanding of heterotrophic flagellates as one major component in aquatic food webs, which is able to transfer a significant amount of bacterial production to higher trophic levels (Fenchel, 1982b; Azam *et al.*, 1983; Berninger *et al.*, 1991). A major feature of the microbial loop (Azam *et al.*, 1983) is that a significant portion of phytoplankton primary production is excreted in the form of dissolved organic matter which then provides a substrate for bacterial growth, thereby creating a food source for protozoans (mainly HF). Protozoan production (HF and ciliates) enters the traditional food web via grazing by metazoans (see Laybourn-Parry and Parry, Chapter 11 this volume). The concept of the microbial loop has subsequently been revised in several important respects. For instance, quantitative studies have indicated that herbivory by nano- and microflagellates plays an important role within the carbon flux (see for example Sherr and Sherr, 1994); bacterial communities are not only grazed by protozoans but are also structured by protistan grazers (for example Güde, 1979; Turley *et al.*, 1986; Jürgens *et al.*, 1997a); mixotrophy is of potential importance in many different aquatic ecosystems and among various groups of protists (for example Sanders, 1991a; Caron and Finlay, 1994); predation by metazoans on protists can be very selective (Sanders and Wickham, 1993; Arndt, 1993a) but predation may also enhance flagellates via indirect effects (for example, support of bacterial growth: Hahn *et al.*, 1999; Arndt *et al.*, 1992).

In contrast to these recent findings, our knowledge of the ecology and quantitative composition of heterotrophic flagellates is still poor. Following the pioneering work of Fenchel (for review see Fenchel, 1986a), the autecology of only about a dozen – mainly bacterivorous – species has been studied in detail (for example Caron *et al.*, 1990; Goldman and Caron, 1985; Jürgens, 1992; Jones and Rees, 1994; Geider and Leadbeater, 1988; Eccleston-Parry and Leadbeater, 1994a). This is in contrast to the large number of known HF species and to the high species diversity (up to more than 100) recorded by taxonomists from the intensively investigated sites (for example Patterson *et al.*, 1989; Vørs, 1992; Vørs *et al.*, 1995; Patterson and Simpson, 1996; Tong *et al.*, 1998).

Patterson, Larsen and co-workers' *The Biology of Free-Living Heterotrophic Flagellates* (Patterson and Larsen, 1991) was a first attempt to combine the quantitative data of field work and the taxonomic knowledge of specialists. Ten years ago, however, quantitative data were generally restricted to counts of 'HNF' (heterotrophic nanoflagellates). Recently a more detailed database has become available on the quantitative composition of heterotrophic flagellates of several pelagic communities from marine, brackish and fresh waters. The idea of this overview is to summarize the available information and to make a first attempt to compare the major flagellate fauna in pelagic and benthic habitats at marine and limnetic sites. Against this background, the ecology of different systematic groups and their specific adaptations to environmental conditions will be compared, and their functions in different ecosystems will be discussed.

12.2 Methodological problems of quantitative studies

The number of publications presenting quantitative data for pelagic heterotrophic flagellates in aquatic habitats is steadily increasing (Sanders *et al.*, 1992; Gasol and Vaqué, 1993). One important problem is, however, that most of the quantitative data are based on counts of fixed samples analysed by means of epifluorescence microscopy. One disadvantage of this procedure is that the presence of one or more flagella on a cell may be difficult to resolve. Thus, epifluorescence nanofauna counts may also include, for example, small naked amoebae, yeasts, zoospores belonging to a range of organisms, nanociliates (though these are mostly distinguishable), disrupted cells from very different eucaryotes and maybe even large bacteria with a large nucleomorph (Patterson and Larsen, 1991; Arndt, 1993b). Tests involving the counting of HF with epifluorescence microscopy gave reproducible results when carried out on laboratory cultures (Caron, 1983; Bloem *et al.*, 1986). As far as we are aware, diverse HF communities from field samples have never been tested adequately to determine the level of accuracy.

A second problem of epifluorescence microscopy technique is the significant non-uniform shrinkage of fixed flagellates, making estimates of biovolume difficult to calculate. Choi and Stoecker (1989) found that, upon fixation, *Paraphysomonas* could shrink to 38 per cent of its original volume. A third fundamental problem is that in general only about fifty to seventy flagellates are counted per membrane filter, and therefore forms that contribute only 1–2 per cent to total abundance are easily overlooked. Though they are much less abundant than nanoflagellates, the contribution of large flagellates to total flagellate biomass in very different ecosystems can be significant (Sherr and Sherr, 1994; Arndt and Mathes, 1991). Large heterotrophic flagellates (LHF, ^{3}15µm) are often disrupted by routine fixatives (for example, some chrysomonads and many representatives of Protista *incertae sedis*) and thus difficult to quantify.

Most of the above mentioned problems also apply to the study of benthic HF which is additionally complex due to the masking of animals by sediment particles. According to the review given by Alongi (1991), nearly all recent estimates of benthic HF densities suffer from methodological problems. Extraction methods, such as sea ice or fixatives, are very selective and underestimate especially nanoflagellates. Even with countings using epifluorescence microscopy (Bak and Nieuwland, 1989), and combined density gradient centrifugation (Alongi, 1990; Starink *et al.*, 1994), the taxonomic composition of the community could not be adequately studied.

As an alternative method, several authors proposed a live-counting technique analysing small droplets by means of light microscopy (for example Massana and Güde, 1991; Gasol, 1993; Arndt and Mathes, 1991). One important advantage of this method is that morphological and behavioural features hidden in fixed samples can be used to differentiate the major taxonomic groups of HF, sometimes even to species level. Biovolumes can be estimated from video-micrographs. Two important disadvantages of live-counting are that samples have to be analysed within a short period after sampling, and that it is a time-consuming procedure. Thus, the counting of parallel samples is limited. We have good experiences of counting pelagic flagellates (undiluted samples) and benthic flagellates (diluted by a factor of 5– >20 with filtered water) in droplets of 5–20µl in a miniaturized version of a Sedgewick-Rafter chamber with a height of about 0.2 mm. Samples are analysed not longer than one to two hours after sampling by means of a phase contrast microscope (Zeiss Axioskop) equipped with video-enhancement: 20x, 40x objectives are used for quantitative counts. The concentration of flagellates is adjusted so that counting of a chamber can be completed within a few minutes (this is especially critical for pelagic samples). The use of a 63x LD-objective or water immersion objectives (63x and 100x) with a long working distance is helpful for a species determination in relatively undisturbed samples.

A reasonable compromise should be a combination of live-counts, best in combination with a cultivation of dominant flagellates for a safe determination, and epifluorescence counts. These seem to be indispensable prerequisites for determination of the taxonomic composition of heterotrophic flagellates in field samples and for a reliable estimate of HF abundances and biovolumes, respectively. Most of the data sets discussed here have made use of live-counts, at least as an additional method. Recent advances in molecular biological methodology (for example, development of specific probes) may serve as elegant methods for future studies to determine field abundances of distinct HF species in fixed samples (for example Lim *et al.*, 1999; Tong and Sleigh, pers. comm.). Molecular methods may be especially helpful for the determination of common species that are very difficult to distinguish by morphological features (for example, several chrysomonads: Bruchmüller, 1998).

12.3 Composition of flagellate communities

The possible evolutionary relationships between various groups of flagellates are still incompletely understood (Karpov, Chapter 16 and Cavalier-Smith, Chapter 17, this volume). Major problems for ecologists include: first, that there is no comprehensive key available for the identification of HF species, and second, that autecological studies on flagellates have been carried out for only a restricted number of species or strains. In order to characterize the possible functional role of organisms in an ecosystem, ecologists must rely on the autecological data for a few, hopefully, representative, species, which can be combined with the knowledge on the distribution and quantitative composition of communities.

12.3.1 Pelagic flagellates

The abundances of *nanoflagellates (HNF)* in different pelagic habitats can vary from about 20 to >20,000 HNF per ml (mostly between 100 and 10,000 HNF/ml) and are mostly related to the abundance of bacteria (for review, see Sanders *et al.*, 1992). Despite this considerable variability in abundance, the available information on the taxonomic composition of HF from pelagic communities indicates a surprising conformity regarding the dominant taxonomic groups (Figure 12.1 A, B). On annual average about 20–50 per cent of HNF biomass is formed by heterokont taxa (mainly chrysomonads and bicosoecids). Not only in marine but also in freshwaters, choanoflagellates contribute another significant part of about 5–40 per cent of the biomass. Kathablepharids appear to be a very important group in most lakes, accounting for about 10–>25 per cent of average HNF biomass. If *Leucocryptos* is included within the kathablepharids, then they may reach a similar importance in marine waters (Vørs *et al.*, 1995). Kinetoplastids always occur in plankton communities but are generally of reduced quantitative importance (1–8 per cent of annual mean HNF biomass). Five other groups – small dinoflagellates, thaumatomastigids, apusomonads, colourless cryptomonads and euglenids – are also generally present in plankton communities, but commonly form only minor parts of the HNF biomass.

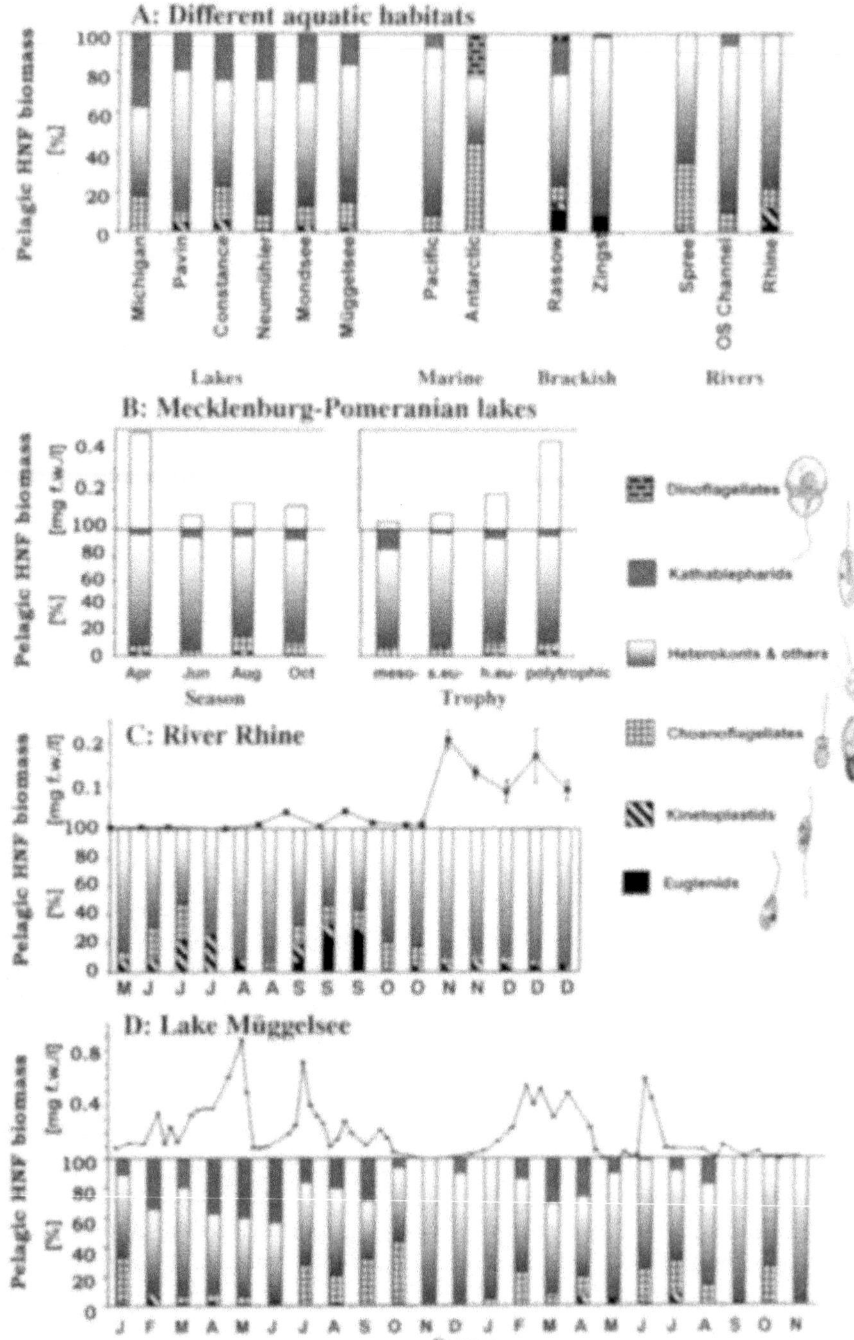
A: Different aquatic habitats
Pelagic HNF biomass [%]
Michigan
Pavin
Constance
Neumühler
Mondsee
Müggelsee
Pacific
Antarctic
Rassow
Zingst
Spree
OS Channel
Rhine
Lakes
Marine
Brackish
Rivers
B: Mecklenburg-Pomeranian lakes
Pelagic HNF biomass [mg f.w./l] [%]
Apr
Jun
Aug
Oct
Season
meso-
s.eu-
h.eu-
polytrophic
Trophy
Dinoflagellates
Kathablepharids
Heterokonts & others
Choanoflagellates
Kinetoplastids
Euglenids
C: River Rhine
Pelagic HNF biomass [mg f.w./l] [%]
M J J J A A S S S O O N N D D D
D: Lake Müggelsee
Pelagic HNF biomass [mg f.w./l] [%]
J F M A M J J A S O N D J F M A M J J A S O N
Season

There is always a significant part of HNF that cannot be assigned to one of the major groups (Protista *incertae sedis*) or that cannot be identified (for reasons of comparison both of these flagellate groups were included among 'heterokonts & others' in Figure 12.1).

Despite the similarities in the general composition of the investigated plankton communities, there are also several specific features. Several groups of marine flagellates, such as ebriids, silicoflagellates, acanthoecid choanoflagellates, and many genera of dinoflagellates, do not occur in lake plankton. Acanthoecid choanoflagellates, with their siliceous loricae, are a well-studied component of marine and brackish waters (for example Buck and Garrison, 1988; Leadbeater, 1974; Kuuppo, 1994). Aloricate HNF have only recently been studied in more detail. They might generally be much more abundant than choanoflagellates (Kuuppo, 1994; Vørs *et al.*, 1995).

In most plankton communities, considerable seasonal changes in the abundance, biomass and composition of HNF have been observed (for example, Carrick and Fahnenstiel, 1989; Mathes and Arndt, 1995; Carrias *et al.*, 1998; Weisse and Müller, 1998). In samples from the River Rhine and Lake Müggelsee (Figure 12.1 C, D) the biomass changed by a factor of 100 in the course of a year. HNF dynamics in temperate regions are generally characterized by maxima in spring, due to increased food supply, and minima in early summer, due to intensive grazing pressure by metazoans (for example, Weisse, 1991; Cleven, 1995). The flagellate dynamics of the Rhine were correlated more to the water discharge than to seasonal events. High run-offs cause a release of HF from predation pressure by benthic filter-feeders (Weitere *et al.*, unpublished).

The biomass of *large heterotrophic flagellates* (>15–200μm, LHF) seems to be as important as that of nanoflagellates (Figure 12.2 A). This fact was stressed among

(left)
Figure 12.1 Comparison of the composition of heterotrophic nanoflagellate (HNF) communities from very different pelagic environments (percentage of mean biomass in mg fresh weight/litre)

A Different pelagic habitats (seasonal mean values except for the marine data): Lake Michigan (USA): Carrick and Fahnenstiel (1989) (cryptomonads were considered as kathablepharids: Carrick, pers. comm.); L. Pavin (France): Carrias (1996); L. Constance (Germany): Springmann (unpubl.); L. Neumühler See (Germany): Mathes and Arndt (1995); L. Mondsee (Austria): Salbrechter and Arndt (1994); L. Müggelsee (Berlin, Germany): Arndt (1994); Equatorial Pacific: Vørs *et al.* (1995); Antarctica: Hewes *et al.* (1990; mean of two cruises); Rassower Strom and Zingster Strom (shallow Baltic coastal waters, Germany, 8 and 5 PSU): Arndt (unpubl.); River Spree (Leipsch/Alt-Schadow) and Oder-Spree-Channel (Fürstenwalde, Germany): Arndt (unpubl.); R. Rhine (Cologne, Germany): Weitere (unpubl.)

B Mecklenburg-Pomeranian lakes (Germany): 55 lakes of different trophy sampled four times per year (1996 and 1997; Auer and Arndt, unpubl.). Data are summarized for different seasons and different degrees of trophy.

C River Rhine at Cologne (Germany), seasonal changes of biomass and composition of HNF (1998; Weitere, unpubl.).

D Lake Müggelsee (Berlin, Germany), seasonal changes of biomass and composition of HNF during two successive years (1989 and 1990; Arndt, 1994 and unpubl.)

limnologists decades ago (for example Sandon, 1932; Nauwerck, 1963). However, its quantitative importance in connection with the microbial food web has only recently been considered (for example Sherr and Sherr, 1989; Arndt and Mathes, 1991). The annual mean size distribution of total flagellate abundance and biovolume (using live-counting technique) from Lake Müggelsee (Figure 12.2 B) indicates that large HF are much less abundant than small forms, but contribute significantly to HF biomass. With regard to lake communities, the relative contribution of nano- and microflagellate biomass varies depending on the trophic status of lakes, with a tendency for microflagellate biomass to dominate under hypertrophic conditions (Figure 12.2 D; Auer, unpublished). Colourless dinoflagellates probably form the major part of the biomass of microflagellates in most marine and freshwaters (Smetacek, 1981; Lessard and Swift, 1985; Arndt and Mathes, 1991). In coastal marine waters, ebriids can occasionally form a substantial part of LHF biomass (for example Smetacek, 1981). In freshwaters, large chrysomonads (for example *Spumella*) are, besides dinoflagellates (for example *Gymnodinium helveticum*), important LHF, seasonally varying in their contribution (Figure 12.2 C). In some lakes (Mischke, 1994) and in brackish waters (Arndt, unpubl.) large members of the Protista *incertae sedis* can occur in significant abundance.

There are major seasonal changes in the contribution of the different size groups of HF in lakes and in coastal waters. The largest HF are generally recorded during the spring bloom of phytoplankton, whereas the smallest forms dominate under conditions of high metazoan grazing pressure in summer, or when food concentrations are low during winter (Figure 12.2 C, D; Smetacek, 1981; Weisse, 1991). At times of high metazoan grazing pressure, LHF often disappear from pelagic communities (Auer *et al.*, unpublished).

Quantitative changes in the vertical distribution of HF are well documented for marine and freshwaters. Generally, maximum numbers occur at sites of highest food concentrations, mostly at the surface (Fenchel, 1986a). Not much is known about the vertical changes in the taxonomic composition. In the mesotrophic Lake Mondsee, chrysomonads and kathablepharids did not change significantly regarding their percentage contribution, whereas kinetoplastids, choanoflagellates, bicosoecids and dinoflagellates decreased at greater depths (Salbrechter and Arndt, 1994).

(right)
Figure 12.2 Comparison of the composition of large heterotrophic flagellate (LHF, >15µm) communities in different pelagic environments (percentage of mean biomass in mg fresh weight/litre)

A Contribution of LHF to total flagellate biomass. Lake Valencia (Venezuela) and Lake Lanao (Philippines): Lewis (1985); River Danube at Budapest (Hungary): Arndt and Mathes (1991); data from all other sites from references as in Figure 12.1 A.

B Size distribution of heterotrophic flagellates regarding biomass and abundance for Lake Müggelsee (1989; Berlin, Germany): Arndt (1994).

C Seasonal changes of LHF and metazoan biomass and LHF contribution to total flagellate biomass in Lake Müggelsee (1989 and 1990; Arndt *et al.*, 1993 and Arndt and Mathes, 1991)

D Size distribution of HF communities in relation to season and trophy and metazooplankton biomass in fifty-five Mecklenburg-Pomeranian lakes (Auer *et al.*, unpubl.)

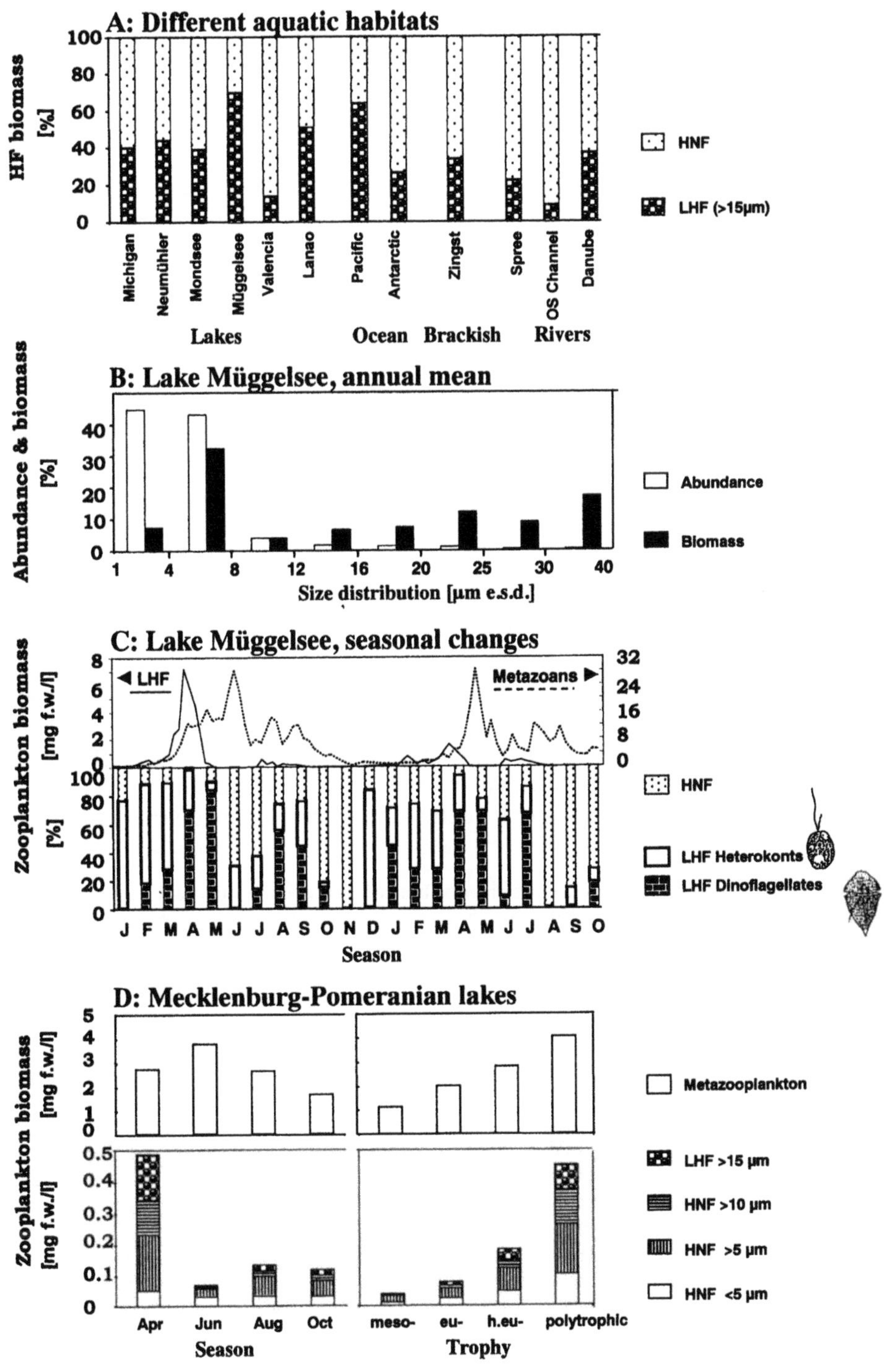
A: Different aquatic habitats
HF biomass [%]
HNF
LHF (>15µm)
Michigan
Neumühler
Mondsee
Müggelsee
Valencia
Lanao
Pacific
Antarctic
Zingst
Spree
OS Channel
Danube
Lakes
Ocean
Brackish
Rivers
B: Lake Müggelsee, annual mean
Abundance & biomass [%]
Abundance
Biomass
Size distribution [µm e.s.d.]
C: Lake Müggelsee, seasonal changes
Zooplankton biomass [mg f.w./l]
LHF
Metazoans
[%]
HNF
LHF Heterokonts
LHF Dinoflagellates
Season
D: Mecklenburg-Pomeranian lakes
Zooplankton biomass [mg f.w./l]
Metazooplankton
LHF >15 µm
HNF >10 µm
HNF >5 µm
HNF <5 µm
Apr
Jun
Aug
Oct
Season
meso-
eu-
h.eu-
polytrophic
Trophy

12.3.2 Benthic flagellates

The taxonomic diversity of benthic flagellate communities has been summarized by Patterson *et al.* (1989). Recently numerous marine sites have been studied, and many new species, among them many common ones, have been added to our knowledge (for example Vørs, 1992; Patterson and Simpson, 1996; Tong *et al.*, 1998). Regarding the quantitative importance of benthic flagellates, owing to methodological problems (see section 12.2) we are only just beginning to be able to compare different habitats. Abundances which have been reported range from below 100 to several millions HF per ml sediment (Gasol, 1993). HF abundance seems to be positively correlated to bacterial abundance and grain size; however, available data sets are still contradictory (Alongi, 1991; Hondeveld *et al.*, 1994). Quantitative data on the taxonomic composition are very sparse and often restricted to a few taxonomic groups (for example Bark, 1981; Baldock *et al.*, 1983).

Most of the data on the structure of benthic flagellate communities presented in Figure 12.3 A come from our own recent studies. They give in most cases only a sporadic picture of the community structure in the aerobic surface layer (seasonal averages only from the Baltic at Hiddensee and from Antarctica). Generally, euglenids are most important regarding their biomass contribution (20–85 per cent of HF biomass; for Ladberger Mühlenbach, Figure 12.3 C, they were included among the 'others' category), followed by bodonids (5–20 per cent of HF biomass). Both euglenozoan groups are generally also the most diverse groups (for example Patterson *et al.*, 1989). Though only recently considered, thaumatomastigids and apusomonads are typical components of benthos communities, and contribute together about 1–20 per cent of the HF biomass. Colourless chrysomonads may be abundant at certain sites, but contribute in general not more than 30 per cent to HF biomass, mostly much less. Seawater ice extractions reveal a high diversity of benthic

(right)

Figure 12.3 Comparison of the composition of benthic heterotrophic flagellate (HF) communities from very different environments (percentage of mean biomass in mg fresh weight/cm³ sediment)

A Different benthic sites (seasonal mean values except for the lake, the marine sites and River Rhine; the water depth at the sampling sites is indicated on the abscissa): L. Speldrop (Lower Rhine, Germany): Arndt *et al.* (unpubl.); Antarctic site in the Potter Cove, King George Island: Dietrich (unpubl.); Aegean Sea, Sporades Basin at 1,250m depth: Arndt and Hausmann (unpubl.); Hiddensee (Fährinsel; southern Baltic, Germany): Dietrich (unpubl.); Rassower Strom and Zingster Strom (southern Baltic, Germany; 8 and 5 PSU): Arndt (unpubl.); Antarctic melt water stream, King George Island: Dietrich (unpubl.); Ladberger Mühlenbach (Westfalia, Germany): Cleven (unpubl.); River Rhine at Cologne (Germany): Altmann and Arndt (unpubl.).

B Hiddensee (Fährinsel; southern Baltic, Germany), seasonal changes of biomass and composition of HF during 1996/97 (Dietrich and Arndt, unpubl.).

C Ladberger Mühlenbach (Westfalia, Germany), seasonal changes of biomass and composition of HF during 1997/98 (Cleven, unpubl.)

D Vertical changes of HF biomass and composition at two brackish sites of the southern Baltic (cf. 12.3 A; mean of five samplings): Arndt (unpubl.).

E Vertical changes of HF biomass and composition in the brook Ladberger Mühlenbach (Westfalia, Germany; May 1997): Cleven (unpubl.)

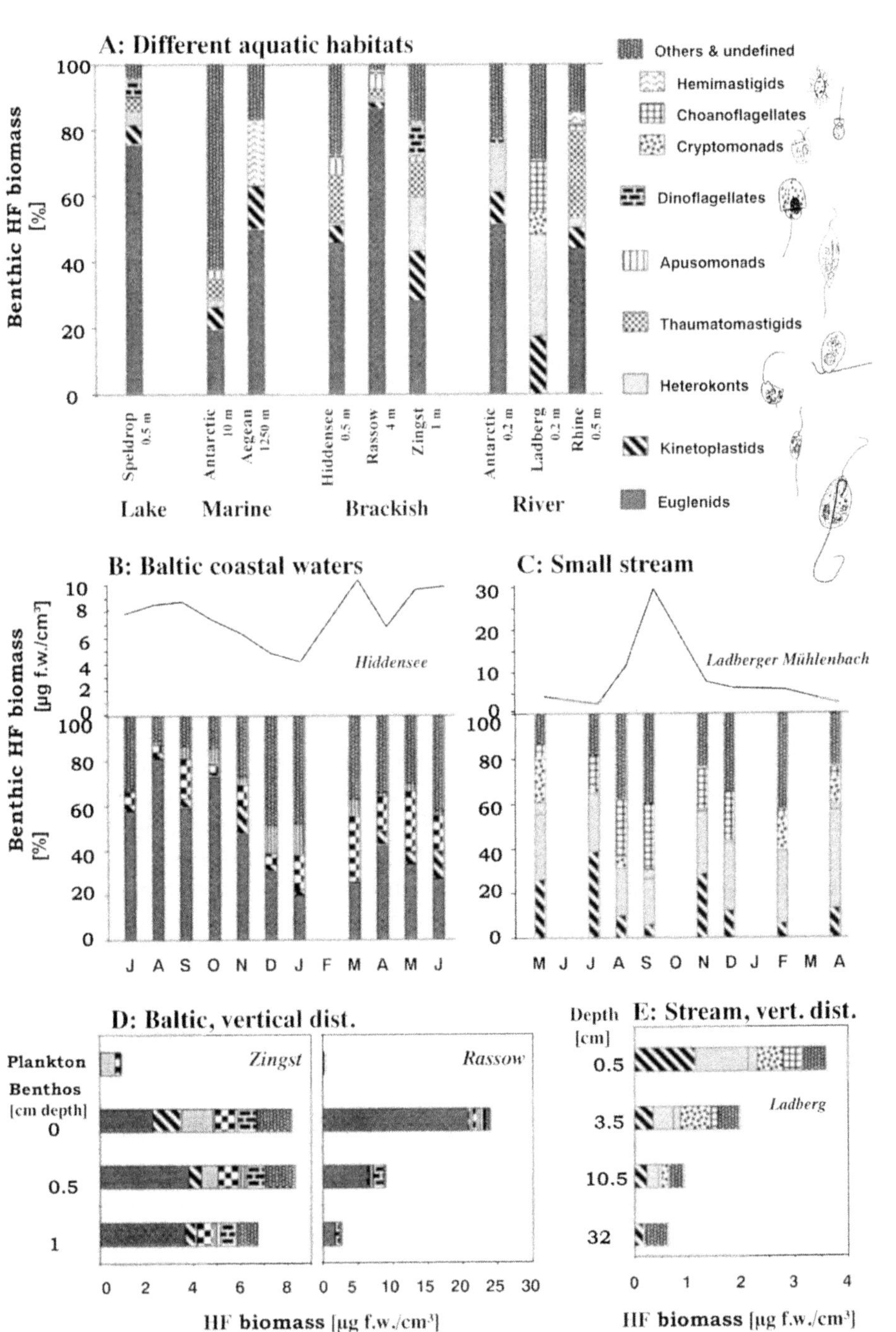
A: Different aquatic habitats
Benthic HF biomass [%]
Speldrop 0.5 m
Antarctic 10 m
Aegean 1250 m
Hiddensee 0.5 m
Rassow 4 m
Zingst 1 m
Antarctic 0.2 m
Ladberg 0.2 m
Rhine 0.5 m
Lake
Marine
Brackish
River
Others & undefined
Hemimastigids
Choanoflagellates
Cryptomonads
Dinoflagellates
Apusomonads
Thaumatomastigids
Heterokonts
Kinetoplastids
Euglenids
B: Baltic coastal waters
Benthic HF biomass [µg f.w./cm³]
Hiddensee
Benthic HF biomass [%]
J A S O N D J F M A M J
C: Small stream
Ladberger Mühlenbach
M J J A S O N D J F M A
D: Baltic, vertical dist.
Zingst
Rassow
Plankton
Benthos [cm depth]
0
0.5
1
HF biomass [µg f.w./cm³]
Depth [cm]
E: Stream, vert. dist.
Ladberg
0.5
3.5
10.5
32
HF biomass [µg f.w./cm³]

dinoflagellates from marine sites (Hoppenrath, pers. comm.). It seems from our data that their contribution to total biomass of HF is up to 20 per cent. Occasionally of importance are colourless cryptomonads, choanoflagellates, cercomonads, bicosoecids, pedinellids and even hemimastigids. Several genera (for example *Percolomonas*, *Ancyromonas*) and other members of Protista *incertae sedis* are also temporarily abundant, but should not contribute significantly to average HF biomass. There is an obvious variability in the contribution of the major HF groups at different sites, which is certainly related to the quality of sediments and accompanying factors (Patterson *et al.*, 1989; Alongi, 1991). Our knowledge regarding the specific requirements of the different taxa and the specific top-down and bottom-up effects, however, is very limited at present.

Two seasonal patterns of HF dynamics from a brackish site and a small stream are presented in Figure 12.3 (B, C), indicating that absolute biomass and the relative contribution of HF groups changes throughout the year. Bark (1981) found a seasonal succession, from bodonids and euglenids in early summer to a predominance of diplomonads in autumn, in a small highly eutrophic lake. The typical coupling of phytoplankton spring bloom and HF known from the pelagial is not generally recorded from benthic HF communities, where summer and also winter peaks have been reported (Bark, 1981; Bak and Nieuwland, 1989; Hondeveld *et al.*, 1994; Starink *et al.*, 1996). In benthic systems seasonal changes might be masked by the long-term storage of organic material in the sediment, which gives rise to a relatively continuous bacterial production.

All of these insights are restricted to studies of the aerobic surface layer. The vertical distribution of HF abundances is mainly related to the chemical properties of the different sediment layers, and highest numbers are generally found in the upper layers of sediment (for review see Alongi, 1991). Again, little is known regarding the taxonomic composition of flagellates at different depths. In all three vertical profiles presented in Figure 12.3 (D, E), biomass decreases with depth, and the relative contribution of euglenids (in Figure 12.3 E included among the 'others' category) and bodonids increases. In highly reducing conditions deeper in the sediment, the community structure changes significantly towards diplomonads, Protista *incertae sedis*, percolozoans, and several undefined forms; however, euglenids were still present (Arndt unpubl.). Amitochondriate flagellates like diplomonads (for example *Trepomonas*, *Hexamita*) and archamoebae (for example *Pelomyxa*, *Mastigella*) can predominate under anaerobic conditions (for review see Fenchel and Finlay, 1995; Brugerolle, Chapter 9 this volume). Even anaerobic communities can be relatively diverse. Mylnikov (1978) recorded about thirty species of archamoebids, trichomonads and diplomonads, as well as several species typical for aerobic sites, from the sediments of the Ivanjkovski Reservoir in Russia.

12.3.3 Bentho-pelagic coupling of flagellate communities

From a variety of studies regarding the structure of heterotrophic nanoflagellate communities, it seems that several species are well adapted to be wanderers between the two environments, that of the pelagial and that of the benthal. HF can populate the pelagial as a result of resuspension of the sediment surface layer, by rafting of biofilms or by active migration. Some forms, such as thaumatomastigids (for example Shirkina, 1987), dinoflagellates (for example Burkholder and Glasgow, 1997) and cercomonads (*Massisteria*: Patterson and Fenchel, 1990), can rapidly change between amoeboid and flagellated forms. HF are among the smallest organisms which can temporarily find suitable conditions to graze and reproduce even on aggregates (size 50 to > 3000 μm) in the pelagial. Their density is close to that of water, and they can live suspended for days or even longer, enough to reproduce several times before reaching the sediment surface, or to form cysts when conditions deteriorate. Aggregates with an organic matrix from different origins suspended in the pelagial have been found to be important constituents of oceanic (Alldredge and Silver, 1988), estuarine (Zimmermann and Kausch, 1996) and fresh waters (Grossart and Simon, 1993), and are known to be 'hot spots' of microbial activity. Aggregates can serve as micro-habitats showing ten to a thousandfold higher HF concentrations than the surrounding water (for review see Caron, 1991). Aggregate-dwelling flagellates in the pelagial have been reported especially from estuarine waters (Rogerson and Laybourn-Parry, 1992; Zimmermann and Kausch, 1996; Garstecki *et al.*, unpublished results). However, even in the pelagial of the open ocean at least cysts of bentho-pelagic species occur. Cultures from oceanic sites have revealed typical benthic genera (for example *Bodo*, *Amastigomonas*, *Cercomonas*, *Caecitellus*: Patterson *et al.*, 1993; Vørs *et al.*, 1995; Arndt and Hausmann, unpublished).

Figure 12.4 gives a quantitative example for the relative contribution of HNF (data calculated per area) separated for first, typical pelagic species (free-swimming HF and HF which are only occasionally found in benthic samples), second, typical benthic species found during the time of sampling in the pelagial (for example several bodonids, thaumatomastigids, apusomonads and cercomonads) and third, for flagellates from the benthal of shallow coastal waters around the Island Rugia (Baltic, Germany). In three out of ten occasions benthic species in the pelagial contributed at least one quarter to all flagellates calculated per area. These flagellates form a bentho-pelagic flagellate community, living either in the upper 1–3 mm of the sediment (fluff) or on particles in the pelagial. Those flagellates which are typically adapted to move between sand grains, such as large euglenids, flattened dinoflagellates and several amoeboid forms, only rarely occur in the pelagial, and avoid probable resuspension by moving deeper into the sediment.

12.4 Comparative ecology of abundant flagellate groups

12.4.1 Feeding ecology

The feeding behaviour of HF can differ between the various groups of flagellates. However, the size spectrum of particles consumed is much larger than previously assumed (for reviews see Sanders, 1991b; Radek and Hausmann, 1994; Sleigh, Chapter 8 this volume). It can range from 0.2 μm (choanoflagellates) up to >50 μm (some dinoflagellates), thus deviating strongly from the original assumption that HF are primarily bacterivorous. It is known that several nanoflagellates from marine and freshwaters also feed on nanophytoplankton (for example, Sherr *et al.*, 1991; Cleven,

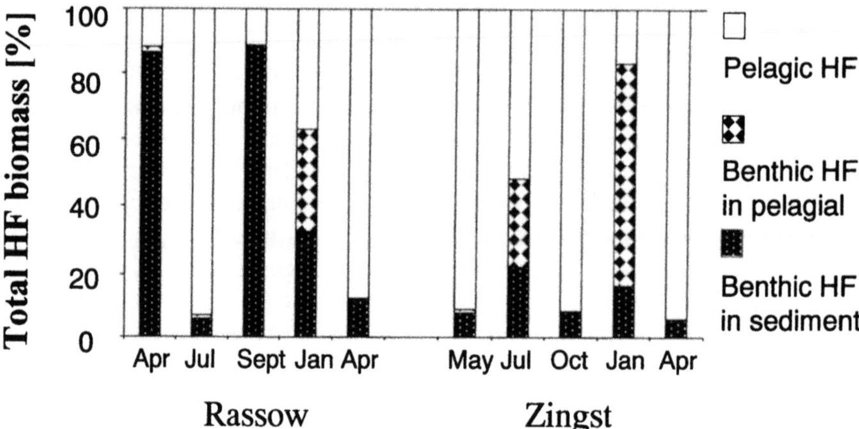

Figure 12.4 Bentho-pelagic coupling of heterotrophic flagellate populations at two brackish water sites of the southern Baltic (Rassower Strom, 4m depth, 8 PSU; Kirr-Bucht at Zingst, 1m depth; 5 PSU; Arndt, unpubl.). The relative contribution of HF biomass to total HF (data calculated per area) was separated for typical pelagic species, typical benthic species found during the time of sampling in the pelagial, and for those flagellates which were found in the benthal at the time of sampling. Each column is based on data from four layers (three parallel samples of each of the pelagial (at 0.5m depth) and three sediment layers: upper fluff layer of about 3mm thickness, 0–5mm sediment, 5–10mm sediment). In accordance with our experience in the study area, it was assumed that there were no significant vertical differences in the pelagial and that, due to oxygen deficiency, HF biomass in sediment layers below 10mm depth is the same as that for the 5–10mm layer

1995), and may also consume other flagellates, ciliates and even small metazoans (Sanders, 1991b). An extreme example of feeding on larger particles is the piscivory by the dinoflagellate *Pfiesteria piscicida* (Burkholder and Glasgow, 1997). The knowledge of food vacuole contents was used by several authors to relate abundances of HF to different feeding types or feeding guilds (for example Pratt and Cairns, 1985). Figure 12.5 indicates the size spectrum of food particles for pelagic and benthic communities. It seems that the contribution of exclusive bacterivores in pelagic habitats is only about one tenth in limnetic sites (mainly choanoflagellates, bicosoecids and kinetoplastids). In marine and brackish sites the percentage can be significantly larger. In benthic habitats about one quarter of HF biomass consists of bacterivores (mainly kinetoplastids, bicosoecids, some Protista *incertae sedis* and cryptomonads).

In the pelagial, the majority of all HF are most probably omnivores feeding on different trophic levels, except for a few specialists known mainly among dinoflagellates (see for example Schnepf and Elbrächter, 1992). Most dinoflagellates, chrysomonads, some Protista *incertae sedis* and kathablepharids are known to feed on bacteria, as well as on algae and most probably also on heterotrophic protists. Figure 12.6 illustrates this important – though often neglected – feature of many HF. Both bacterivory (feeding on *Aerobacter*) and carnivory (predation on a bacterivorous *Cercomonas*) occurs in the same strain of a small *Paraphysomonas* (diameter

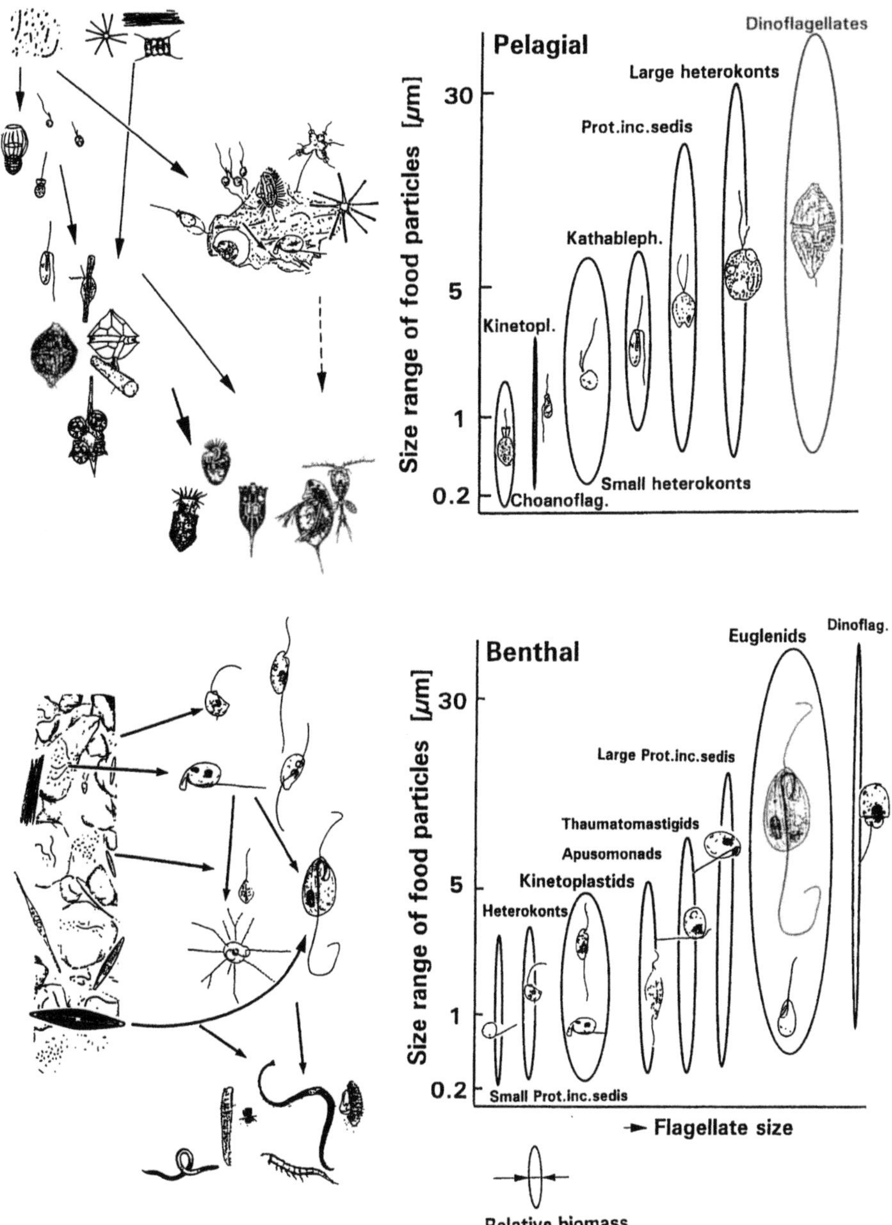

Figure 12.5 Right panels: food size spectrum of important pelagic and benthic flagellate taxa in relation to their mean body size. The relative importance regarding their biomass contribution is indicated by the width of the symbols. Data summarized in Figures 12.1 to 12.3 served as a database.

Left panels: schematic drawing of major fluxes of matter through a typical pelagic and benthic HF community. The species drawn are only representatives of functional groups and might be different in marine, brackish and limnetic sites

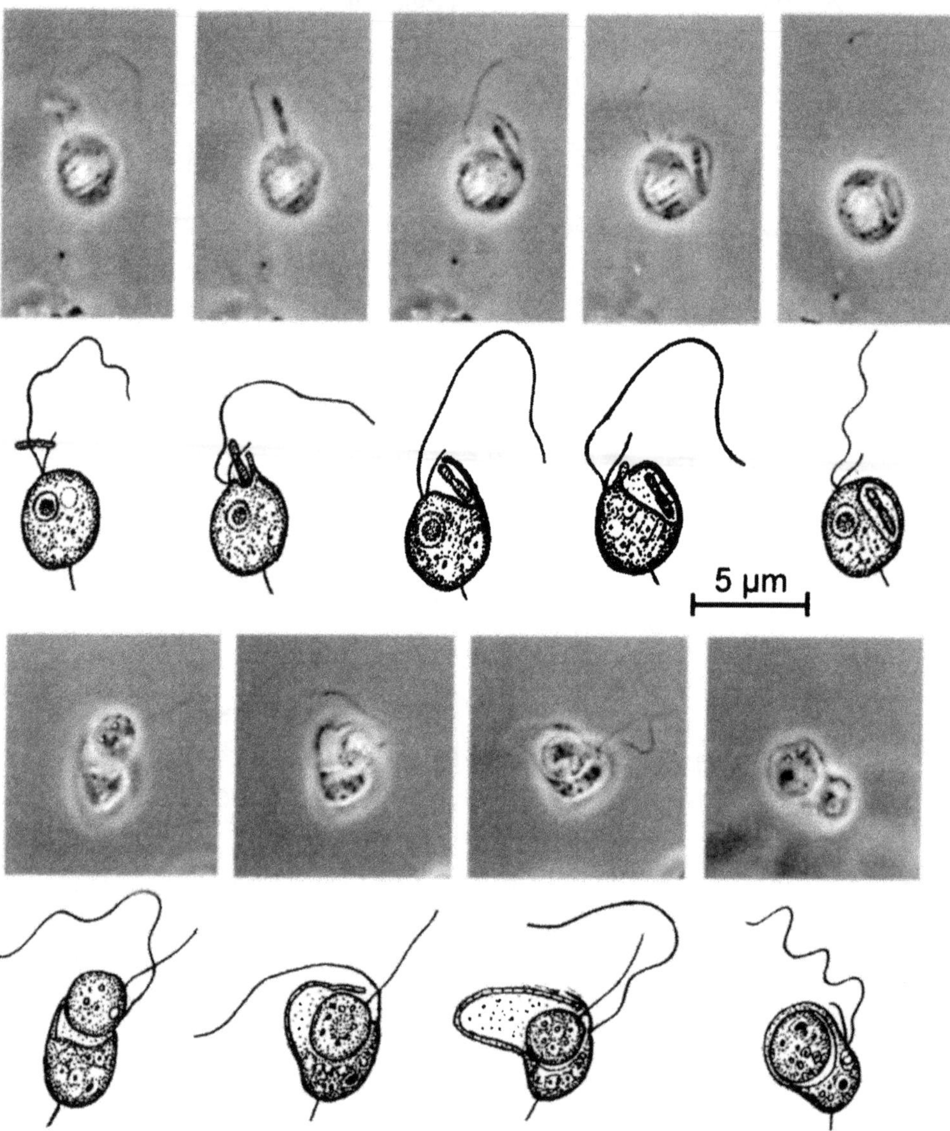

Figure 12.6 Documentation of omnivory by a small *Paraphysomonas sp.* (diameter 3µm) feeding on a bacterium (*Aerobacter*) and another heterotrophic flagellate (*Cercomonas*). Food particles driven by the filter current towards the cell surface are handled by flagella and within 2–4 seconds a long pseudopodium is formed. With the help of the short flagellum even agile flagellates are pushed into the pocket formed by the rapidly extending pseudopodium. The food vacuole is closed after a period of about 5–10 (bacteria) up to 30 (flagellates) seconds following the first contact with the prey. Upon closing the food vacuole, flagellates are completely immobilized and their membrane is disrupted, probably by enzymatic and/or osmotic processes

Source: Arndt *et al.*, unpublished

3-4µm). Large flagellates like *Gymnodinium helveticum* and *Ceratium hirundinella* were found to feed on a wide food spectrum such as bacteria, detritus, algae and other protists (Sandon, 1932; Arndt, unpubl.).

The ability of most flagellates to take up particles often larger than their original body volume by engulfment of the whole prey, aided by pseudopod-like structures (chrysomonads, some dinoflagellates like *Gymnodinium*), a cytostome (for example kathablepharids), or by other mechanisms (for example pallium-feeding and myzocytosis outside the theca of dinoflagellates (Schnepf and Elbrächter, 1992)) seems to be an important requirement for survival at low food concentrations in the pelagial. These flagellates may have two modes of feeding: grazing of large food particles upon occasional contacts, and the uptake of small food items, transported towards the cell by a filter current created by the flagella. This feeding behaviour seems to be an adaptation to the life in the diluted environment of the pelagial similar to the feeding behaviour of calanoid copepods, which switch between filter feeding of small and grasping of large food particles (for example Vanderploeg and Paffenhöfer, 1985). Besides bacterivory, HF can reach the same importance as algivores, as can ciliates in marine (Lessard and Swift, 1985; Sherr and Sherr, 1994) and in limnetic pelagic communities (Arndt *et al.*, 1993).

Little is known regarding the relative importance of large food particles (for example algae) and bacteria for different large flagellates. But according to energetical considerations (Fenchel, 1986b), the importance of interception feeding on small bacteria should decline with the size of the flagellate. This would mean that the flagellate groups included in Figure 12.5, arranged according to their mean size, might also represent feeding niches. Choanoflagellates, as filter-feeders, are known to be able to feed on the smallest prokaryotes up to a size of about 2–3µm. Bodonids, associated with marine or lake snow, feed on particles not much larger than about 5µm, which they grasp from the surface of particles. Small chrysomonads (generally 3–8µm) feed on particles from 0.2µm up to their own size, by direct interception feeding aided by pseudopod formation at the base of the flagella (Figure 12.6). Regarding the bacterivory of nanoflagellates, it is well-known that they select for large bacteria up to a certain limit (for example Simek and Chrzanowski, 1992). Kathablepharids consume bacteria and even cryptomonads (Cleven, 1995) with the help of extrusomes and a cytopharynx. Chrysomonads among LHF with an average size between 15 and 25µm diameter can engulf food items up to 50µm in length (Arndt, unpubl.); the same is known for ebriids and dinoflagellates (Smetacek, 1981). But even much larger items such as colonies of centric diatoms can be consumed (for example Jacobson and Anderson, 1986).

Feeding strategies of pelagic HF are clearly separated by the degree of contact to the substrate: first, free-swimming forms (for example colourless kathablepharids, dinoflagellates, large chrysomonads and some thaumatomastigids), second, forms which are loosely and temporarily attached by protoplasmic threads (small chrysomonads, some choanoflagellates, pedinellids) or flagellum (for example bodonids, thaumatomastigids, apusomonadids, bicosoecids) and third, attached forms (for example some loricate bicosoecids and choanoflagellates). At least half of the biomass of total HF is composed of flagellates living in more or less close contact to sestonic particles, such as algae, colonies of algae, lake or marine snow and so on (the second and third groups mentioned). Most of these flagellates create filter currents and feed mostly on small food items (0.2–5µm). Due to hydrodynamic forces, attachment has been considered an important mechanism for an efficient particle concentration (Fenchel, 1986b). Another, though generally minor, part of pelagic HF biomass, is due to flagellates crawling over surfaces of detrital particles, where bacteria are significantly concentrated compared with the surrounding water. In contrast to these attached HF, free-swimming flagellates seem to feed preferably on relatively large food particles such as algae and other protists.

According to our present knowledge regarding the composition of benthic communities (aerobic surface layer), there should be a separation of feeding types similar to that found in pelagic communities (Figure 12.5). Among typical bacterivores are suspension feeders grazing on bacteria of the pore water (bicosoecids, *Bodo saltans*, choanoflagellates, pedinellids) and forms that grasp more or less attached bacteria (most bodonids, small euglenids, several representatives of apusomonads, cercomonads, and Protista *incertae sedis*). In the benthos, abundance of bacteria and algae is generally up to three orders of magnitude higher than in pelagic sites, making grasping abundant small particles (for example bacteria) as well as large particles (diatoms, other protists) an efficient way of nutrition. Benthic flagellates often possess specialized feeding organelles (see section 12.4.1). As in pelagic environments, there are many flagellates that feed preferentially on large particles. Many euglenids and dinoflagellates are known to feed on large diatoms and other protists. In addition, there are predatory forms such as *Metopion* and *Metromonas* which seem to feed preferably on other protists. Comparable again to pelagic waters, most forms probably have a relatively wide food spectrum (Sanders, 1991b; Figure 12.5). It seems that predominant bacterivores and herbivores are, as in pelagic systems, of similar importance (Epstein, 1997; Dietrich and Arndt, in press). There are only a few detailed studies on the feeding of benthic flagellates. The dinoflagellate *Oxyrrhis marina* consumes bacteria, algae and heterotrophic flagellates (for example bodonids, bicosoecids, chrysomonads; Premke and Arndt, unpubl.) as well as dissolved nutrients (Sanders, 1991b). There are differences in the mobility of the HF in the sediment which should affect the feeding niche. The behaviour ranges between fast movement between the sand grains (for example some bodonids and euglenids) and ambush predation (for example *Massisteria*).

Not only the size but also the concentration of food particles is an important factor influencing the co-occurrence of organisms. Data in literature about the incipient limiting concentration of food particles for HF are very variable for similar species, and range between 10^4 and 10^7 bacteria per ml (Eccleston-Parry and Leadbeater, 1994b). LHF which depend on the frequency of contacts with large food items (for example algae) significantly increase in biomass and in their relative contribution to total HF biomass with increases in lake trophy (Mathes and Arndt, 1994; Figure 12.2 D). Choanoflagellates, which are known to be very effective filterers, should have advantages over other flagellates when food concentrations are low. The relative contribution of choanoflagellates to HNF biomass increased from about 5 per cent in hypertrophic lakes to about 11 per cent in mesotrophic lakes (Auer, unpublished; Figure 12.1 A). Bodonids crawling on particles in the pelagial showed the opposite trend, 1 per cent in mesotrophic and about 5 per cent in hypertrophic lakes. From our present knowledge, however, it seems that generally food concentration has a major influence on absolute numbers but only a minor effect on the composition of HF assemblages regarding the importance of taxonomic groups (Figure 12.1 A).

Another aspect of feeding niches is the different strategies adopted to survive periods of starvation. Such strategies comprise rapid encystment and excystment, changes in food sources used, and dramatic changes in metabolic rates and other physiological and cytological changes (for review see Fenchel, 1986a). Osmotrophy, though known to occur under conditions of cultivation in a great variety of organisms (Sanders, 1991b), seems to play a minor role in pelagic systems where concentrations of easily degradable DOC are very low. In the pore water of benthic systems, however, DOC concentrations can be several orders of magnitude higher compared with the pelagial. Some benthic species seem to be typical osmotrophs (for example *Astasia* and *Chilomonas*: Pringsheim, 1963). This aspect requires more attention, especially for deeper sediment layers.

12.4.2 Tolerances to abiotic factors

The tolerance ranges of flagellate species or taxonomic groups to abiotic conditions are still a neglected topic in protistology. There is only anecdotal knowledge available even about major abiotic factors such as salinity, temperature and oxygen.

A special phenomenon of HF is the observation of several, at least morphologically, identical species in marine and freshwaters (for example Mylnikov and Zhgarev, 1984; Larsen and Patterson, 1990). This gives rise to the assumption that flagellates have a wide range of tolerance with regard to salinity. This hypothesis is supported by our studies on the tolerances of single clones of ten freshwater and eight marine strains (Figure 12.7). At least some species occur under both marine and freshwater conditions. Species lists of very different sites often include *Bodo saltans, B. designis* and *Rhynchomonas nasuta* as well as *Percolomonas cosmopolitus,* which indicates their general tolerance. Our results also indicate that there are several species that have a more or less narrow range of tolerance, and that salinity may well be a limit for the distribution of several freshwater species (Mylnikov, 1983). Occurrences at extremely high salinities compared with the surrounding water have been reported from sea ice biota (for example Ikävalko and Thomsen, 1997).

Temperature is another important factor for the distribution of organisms. It is well-known that growth rates of flagellates increase with a rise of temperature (for example Fenchel, 1986a; Choi and Peters, 1992). Most determinations of HF growth rates have been carried out at temperatures of 10–20 °C. However, it seems that HF are well adapted to also live and reproduce at very low temperatures. For the pelagial, Choi and Peters (1992) determined growth rates of *Paraphysomonas* strains of 0.5–0.8 d^{-1}at –1.5°C. Recent studies of growth rates at about 0°C of benthic HF revealed values of 0.1 d^{-1} (total HF community) and 1 d^{-1} (small heterokonts) for temperate brackish waters, and values of 0.3 d^{-1} (total HF community) and 0.7 d^{-1} (small heterokonts) for Antarctic freshwaters (Dietrich and Arndt, in press).

A third important factor for flagellates, especially benthic HF, is tolerance of low oxygen concentrations. The drastic vertical changes in the structure of HF communities indicate that tolerances differ strongly between different species. Under anoxic

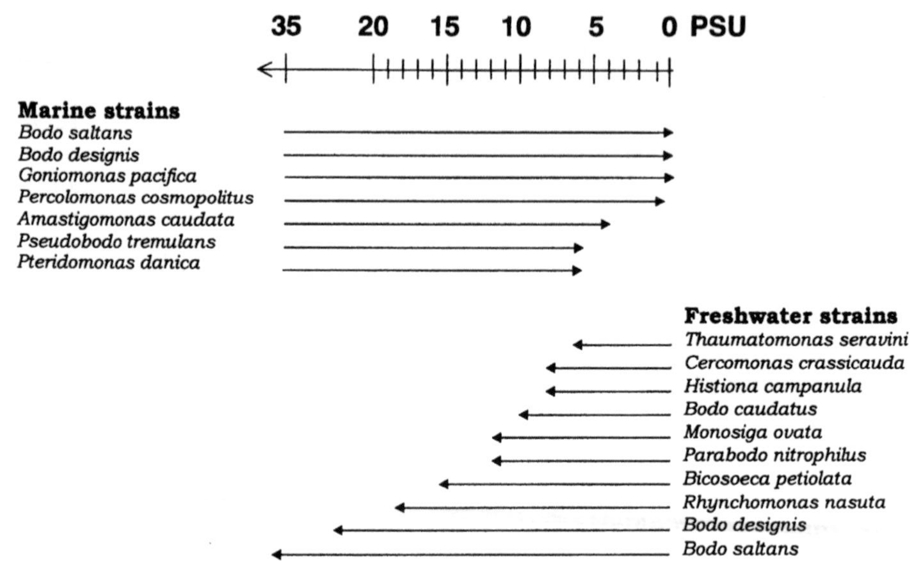

Figure 12.7 Survival of HF strains, isolated from different marine and freshwater sites, at different salinities. Marine strains were transferred in steps from 35, 30, 25, 20, 15, 10, 8, 6, 4, 2, to 0 PSU every day; freshwater strains were transferred in steps of 1 PSU every day, except for *Rhynchomonas* and *B. saltans*, which were transferred in steps of 2 PSU every day. Flagellates were fed by *Aerobacter aerogenes* ad libidum, as they were fed in maintenance cultures. All experiments were done in five parallel petri dishes for both, experimental and control sets

Source: Mylnikov and Gräfenhan, unpublished

conditions, typically amitochondriate flagellates appear in field samples, mainly diplomonads, archamoebids and trichomonads. It seems that several normally aerobic forms can also reproduce, or at least survive, in anaerobic conditions by physiological adaptations (for a review, see Fenchel and Finlay, 1995).

Pressure is assumed to limit the distribution of flagellates in the deep sea (Turley and Carstens, 1991). However, high numbers of HF have also been reported from the deep sea sediments (Alongi, 1991). Live counts from a depth of 1200m in the Mediterranean revealed comparable compositions of flagellate communities to those in shallow waters (Arndt and Hausmann, unpubl.; Figure 12.1 A).

12.4.3 Spatial distribution patterns and life strategies

Horizontal distribution patterns of heterotrophic HF species in the macro- and microscale are not well documented. The examples given in Figure 12.8 indicate a

(right)
Figure 12.8 Distribution patterns of heterotrophic flagellates

A Horizontal (3m depth) and vertical distribution of the heterotrophic dinoflagellate *Gymnodinium helveticum* in the pelagial of Lake Mondsee, Austria, during May 1993 (Salbrechter, in Arndt, 1994)

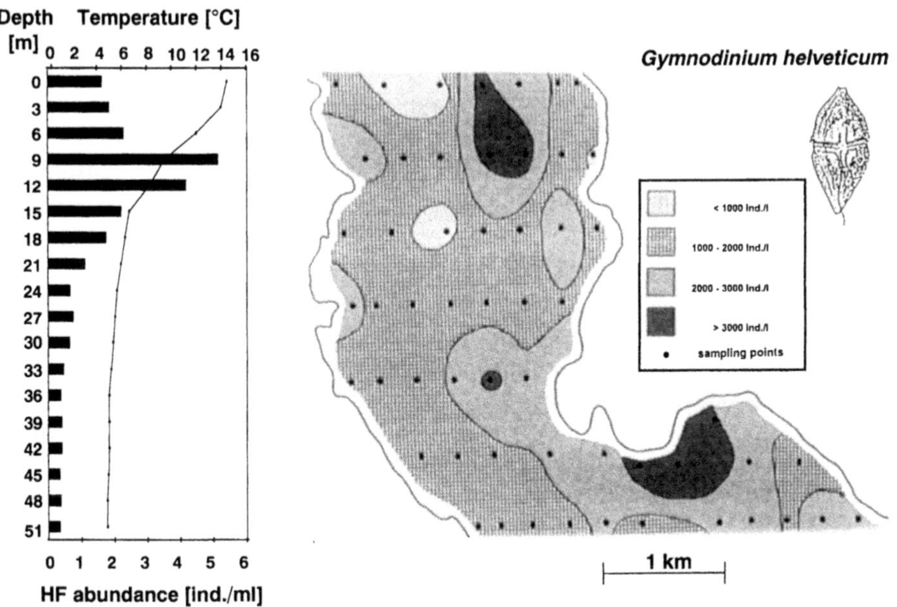

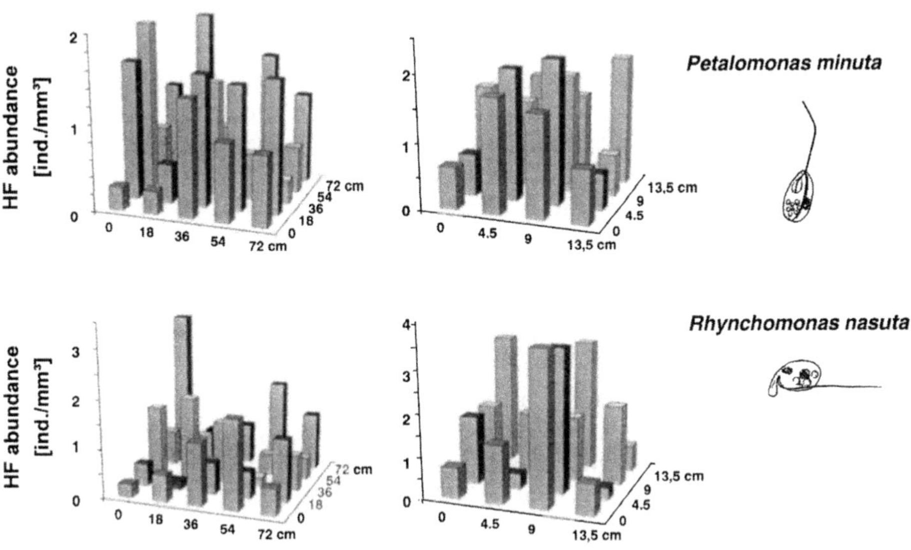

B Patterns of the horizontal microdistribution of two benthic nanoflagellates, the euglenid *Petalomonas minuta* and the bodonid *Rhynchomonas nasuta*, in the upper 0–3mm sediment layer at Kloster, Hiddensee (southern Baltic; horizontal distances of samplings 18 and 4.5 cm, resp.; Gräfenhan, Heinrichs and Arndt, unpubl.)

several-fold change in abundance in the vertical and horizontal distribution of the heterotrophic dinoflagellate *Gymnodinium helveticum*. This pattern for a pelagic HF species is similar to those known for phototrophic protists and metazooplankters. Patchy distribution in the micro-scale also seems to be an important feature of benthic HF populations (Figure 12.8 B). Variations in abundance were of a similar order of magnitude as is known for benthic ciliates and meiofauna organisms (Arlt, 1973). The causes for the horizontal distribution patterns observed for pelagic and benthic HF are not well understood, but are most probably a combined effect of several different factors such as active migration, passive transport, food concentration and reproduction, and predation pressure.

Life conditions in small habitats, for example suspended detritus flocs and sand grains, can change very rapidly. An important part of HF life strategies comprises the ability to form cysts. Among the most abundant forms, at least chrysomonads, choanoflagellates, bodonids, dinoflagellates, thaumatomastigids and cercomonads are able to survive unfavourable conditions in the form of cysts. Several species do not form cysts in cultures, but this does not necessarily mean that they are not able to form cysts under field conditions. In many cases the specific environmental cues are probably missing under conditions of cultivation. The speed of encystment and excystment seems to be important for survival and competition of species, but comparative data are rare.

A special property of HF is their ability to grow at relatively high rates, as high as the average growth rate of bacteria in field populations, enabling them to control bacteria dynamics. Specific growth rates are very much dependent on food conditions. Under favourable conditions in cultures, many different HNF species can grow with maximum rates (for example, Fenchel, 1982a; Eccleston-Parry and Leadbeater, 1994a). Thus growth rates reported from literature are very variable. There are only a few clear trends: doubling times of large flagellates are known to range between less than a day and several days (Carrick *et al.*, 1992; Falkenhayn and Lessard, 1992; Hansen *et al.*, 1997), whereas some nanoflagellates may divide within an interval of 3–4 hours. The succession of flagellates in planktonic or benthic samples generally starts with a dominance of heterokonts, mainly chrysomonads. In benthic samples there are also bicosoecids and pedinellids, indicating this group to be an r-strategist compared with other flagellate groups. To survive periods of starvation, HF have been reported to lower significantly their metabolic rates (Fenchel, 1986a).

12.4.4 Mortality of flagellates owing to predation

In pelagic communities predation plays a major role for the regulation of HF abundances. Field studies on the succession of seasonal events in the pelagial revealed that high abundances of metazooplankters mostly correspond with low abundances of flagellates. Laboratory and a few field experiments have shown that ciliates, rotifers, cladocerans and copepods (Arndt, 1993a; Sanders and Wickham, 1993; Cleven, 1996; Jürgens *et al.*, 1997b) can act as voracious predators of heterotrophic flagellates. There are only a few quantitative estimates of the different loss factors acting under field conditions (for example Weisse, 1991; Arndt and Nixdorf, 1991). Figure 12.9 gives an example for the seasonal changes regarding the quantitative impact of different predators on nano- and microflagellates in a hypertrophic lake.

Generally the major part of flagellate production is consumed by predators. From early summer until autumn, losses were mainly owing to mesozooplankton, particularly cladocerans, which caused a drastic decline of HF biomass especially of slowly reproducing large forms (for data of biomass and composition of HF see Figures 12.1 D and 12.2 C). In winter and spring, however, a major part of HF mortality was owing to other heterotrophic protists. Interestingly, spring mortality of HNF was mainly owing to other HF in the size fraction 5–28μm (*Kathablepharis*, large chrysomonads, *Gymnodinium*) and to a lesser extent to small oligotrichs and prostomatid ciliates. This result supports the importance of omnivory among HF (see sections 12.4.1 and 12.4.6). Losses of large flagellates during spring were mainly owing to ciliates. As is indicated in Figure 12.9, other losses in addition to predation cannot be neglected. Such losses could include encystment or lysis owing to starvation or other unfavourable conditions, and mortality owing to viral infections or parasites. According to our estimate, these losses can account for more than fifty per cent of daily production at certain times, and thus offer an interesting topic for future studies.

Even less information is available on the fate of benthic flagellate production. Ciliates have been shown to be potential predators of benthic flagellates (for example, Fenchel, 1969; Epstein *et al.*, 1992). Recent size-fractionation experiments

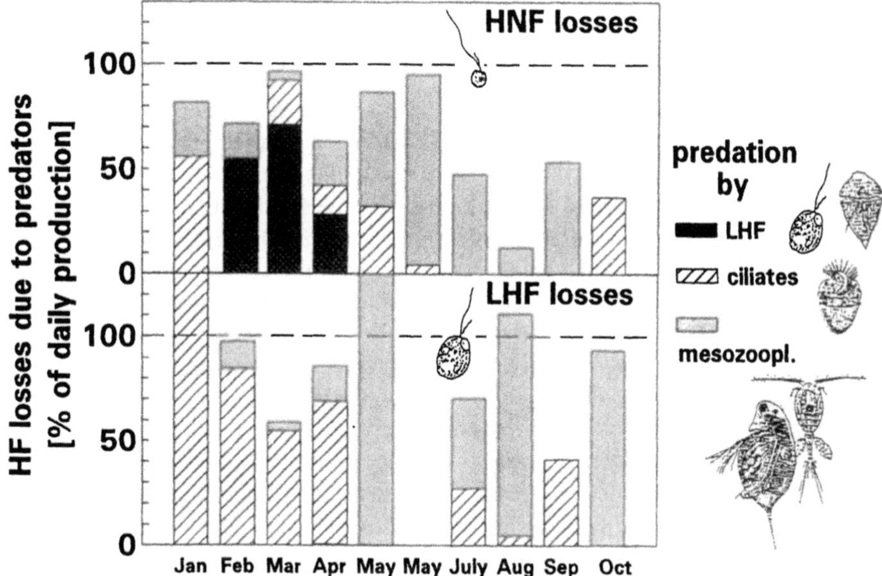

Figure 12.9 Seasonal changes in the losses of the production of heterotrophic nanoflagellates (HNF; <15μm) and large flagellates (LHF; >15μm) due to the presence of different predators.

Results of size fractionation experiments (Lake Müggelsee, Berlin; cf. Arndt and Nixdorf, 1991; Arndt, 1994) using two parallel 1-litre bottles with lake water filtered through sieves of mesh sizes of either 44μm (without metazoans), 20/28μm (without large ciliates) or 5/10μm (without small ciliates and large flagellates), incubated for one day *in situ*. HF production was estimated from population growth rate in the course of the experiment in the fraction 5/10μm (HNF) or 20/28 μm (LHF) in the surface layer of a Baltic muddy sand showed that between 0 and 120 per cent of daily HF production can be lost through predation by ciliates and meiofauna (Dietrich and Arndt, in press). There are at least indications that in benthic systems too, predation can be an important factor for the occurrence of flagellates.

A consumption of a significant part of HF production by predators would strongly select for avoidance strategies. On the basis of the data outlined, several behavioural and morphological features of HF might well be interpreted today as adaptations to avoid predation. First, there are morphological adaptations, such as the possession of loricae (for example many bicosoecids, choanoflagellates), thecae with appendices (dinoflagellates), scales (chrysomonads and thaumatomastigids), and an increase in size due to the formation of colonies (for example several heterokonts, choanoflagellates and spongomonads). Second, there are chemical defences such as toxin production and extrusomes (for example dinoflagellates). Third, there are behavioural adaptations such as attachment to large particles (for example chrysomonads, bicosoecids and choanoflagellates), retraction of protoplasmic threads (for example pedinellids and chrysomonads), vertical migration (for example dinoflagellates) and swimming patterns (for example dinoflagellates and kathablepharids). Fourth, there are adaptations of life strategies such as high reproduction rates (for example chrysomonads and pedinellids) and survival of parts of the population in the sediment in cysts.

Since, in general, predation by protists is underestimated, several of these defence mechanisms have not been interpreted as such. For instance, scales and loricae may not be effective to avoid predation by metazoans but can probably reduce the impact by other protists. Association or attachment of HF to large particles might reduce their vulnerability to protozoan suspension feeders and to a certain extent also to metazoan filter feeders. In the sediment, masking by particles can play a certain role. It may be that vertical migration into the anoxic zone can protect some species such as euglenids from aerobic predators. Motionless forms (for example *Massisteria*) might also be difficult for predators to detect. Species which seems to be not very much protected from predation often have very high reproduction rates (for example *Spumella*) or survive periods with strong predation pressure as cysts in the sediment.

Figure 12.5 (lower panels) may serve as a summary to illustrate the diverse functions of HF, acting as bacterivores, algivores, detritivores, carnivores, and probably mainly as omnivores, in pelagic as well as in benthic systems. Since flagellate ecology is a new field, many questions require further intensive studies. Detailed autecological studies are available for only very few species (even now the determination of species on a quantitative level is difficult for many athecate and aloricate forms). The systematic position of several ecologically important groups of flagellates is not yet clear. It seems that several important species have yet to be described.

ACKNOWLEDGEMENTS

We are thankful for financial support by the German Research Foundation (DFG, Bonn, Germany; H.A.), by the Federal Ministry for Education and Science (BMBF, Bonn, Germany; H.A.), by the Deutsche Bundesstiftung Umwelt (Osnabrück, Germany; D.D., M.W.) and by the Regional Environmental Council of Mecklenburg-Vorpommern (LAUN, Schwerin, Germany, H.A.). We are grateful for many discussions with, and comments by, several colleagues. Special thanks are due to Stephen Wickham and Klaus Jürgens for helpful comments. Last but not least we are very thankful to the two editors of the volume, Barry Leadbeater and John Green, for help, patience and hospitality and much more.

REFERENCES

Alldredge, A. L. and Silver, M. W. (1988) Characteristics, dynamics and significance of marine snow. *Progress in Oceanography*, **20**, 41–82.

Alongi, D. M. (1990) Abundances of benthic microfauna in relation to outwelling of mangrove detritus in a tropical coastal region. *Marine Ecology Progress Series*, **63**, 53–63.

Alongi, D. M. (1991) Flagellates of benthic communities: their characteristics and methods of study. In *The Biology of Free-Living Heterotrophic Flagellates* (eds D. J. Patterson and J. Larsen), Oxford: Clarendon Press, pp. 57–75.

Arlt, G. (1973) Vertical and horizontal microdistribution of the meiofauna in the Greifswalder Bodden. *Oikos*, **Suppl. 15**, 105–111.

Arndt, H. (1993a) Rotifers as predators on components of the microbial web (bacteria, heterotrophic flagellates, ciliates) – a review. *Hydrobiologia*, **255/256**, 231–246.

Arndt, H. (1993b) A critical review of the importance of rhizopods (naked and testate amoebae) and actinopods (heliozoa) in lake plankton. *Marine Microbial Food Webs*, 7, 3–29.

Arndt, H. (1994) Protozoen als wesentliche Komponente pelagischer Ökosysteme von Seen. *Kataloge des Ö. Landesmuseums,* **NF 71**, 111–147.

Arndt, H. and Mathes, J. (1991) Large heterotrophic flagellates form a significant part of protozooplankton biomass in lakes and rivers. *Ophelia*, **33**, 225–234.

Arndt, H., Güde, H., Macek, M. and Rothhaupt, K. O. (1992) Chemostats used to model the microbial food web: evidence for the feedback effect of herbivorous metazoans. *Archiv für Hydrobiologie, Beihefte Ergebnisse der Limnologie*, **37**, 187–194.

Arndt, H., Krocker, M., Nixdorf, B. and Köhler, A. (1993) Long-term annual and seasonal changes of meta- and protozooplankton in Lake Müggelsee (Berlin): effects of eutrophication, grazing activities, and the impact of predation. *Internationale Revue der gesamten Hydrobiologie*, **78**, 379–402.

Arndt, H. and Nixdorf, B. (1991) Spring clear-water phase in a eutrophic lake: control by herbivorous zooplankton enhanced by grazing on components of the microbial web. *Verhandlungen der Internationalen Vereinigung für Theoretische und Angewandte Limnologie*, **24**, 879–883.

Azam, F., Fenchel, T., Field, J. G., Gray, J. S., Meyer-Reil, L-A. and Thingstad, F. (1983) The ecological role of water-column microbes in the sea. *Marine Ecology Progress Series*, **10**, 257–263.

Bak, R. P. M. and Nieuwland, G. (1989) Seasonal fluctuations in benthic protozoan populations at different depths in marine sediments. *Netherlands Journal of Sea Research*, **24**, 37–44.

Baldock, B. M., Baker, J. H. and Sleigh, M. A. (1983) Abundance and productivity of protozoa in chalk streams. *Holarctic Ecology*, **6**, 238–246.

Bark, A. W. (1981) The temporal and spatial distribution of planktonic and benthic protozoan communities in a small productive lake. *Hydrobiologia*, **85**, 239–255.

Berninger, U-G., Caron, D. A., Sanders, R. W. and Finlay, B. J. (1991) Heterotrophic flagellates of planktonic communities, their characteristics and methods of study. In *The Biology of Free-Living Heterotrophic Flagellates* (eds D. J. Patterson and J. Larsen), Oxford: Clarendon Press, pp. 39–56.

Bloem, J., Bär-Gilissen, M-J. B. and Cappenberg, T. E. (1986) Fixation, counting, and manipulation of heterotrophic nanoflagellates. *Applied and Environmental Microbiology*, **52**, 1266–1272.

Bruchmüller, I. (1998) Molekularbiologische Charakterisierung und phylogenetische Einordnung heterotropher Nanoflagellaten und prostomatider Ciliaten des Süsswassers. Dissertation, Mathematisch-Naturwissenschaftliche Fakultät, Christian-Albrechts-University of Kiel.

Buck, K. R. and Garrison, D. L. (1988) Distribution and abundance of choanoflagellates (Acanthoecidae) across the ice-edge zone in the Weddell Sea, Antarctica. *Marine Biology*, **98**, 263–269.

Burkholder, J. M. and Glasgow, H. B., Jr. (1997) *Pfiesteria piscicida* and other *Pfiesteria*-like dinoflagellates: behavior, impacts, and environmental controls. *Limnology and Oceanography*, **42**, 1052–1075.

Caron, D. A. (1983) Technique for enumeration of heterotrophic and phototrophic nanoplankton, using epifluorescence microscopy, and comparison with other procedures. *Applied and Environmental Microbiology*, **46**, 491–498.

Caron, D. A. (1991) Heterotrophic flagellates associated with sedimenting detritus. In *The Biology of Free-Living Heterotrophic Flagellates* (eds D. J. Patterson and J. Larsen), Oxford: Clarendon Press, pp. 77–92.

Caron, D. A. and Finlay, B. J. (1994) Protozoan links in food webs. In *Progress in Protozoology* (eds K. Hausmann and N. Hülsmann), Stuttgart: Gustav Fischer Verlag, pp. 125–130.

Caron, D. A., Goldman, J. C. and Dennett, M. R. (1990) Carbon utilization by the omnivorous flagellate *Paraphysomonas imperforata*. *Limnology and Oceanography*, **35**, 192–201.

Carrias, J-F., Amblard, C., Quiblier-Lloberas, C. and Bourdier, G. (1998) Seasonal dynamics of free and attached heterotrophic nanoflagellates in an oligotrophic lake. *Freshwater Biology*, **39**, 91–101.

Carrias, J-F. (1996) La boucle microbienne en milieu lacustre: structure et fonctionnement des communautés picoplanctoniques et de protistes flagellés et ciliés. Doctoral thesis, Universite Blaise Pascal, Clerment-Ferrand.

Carrick, H. J. and Fahnenstiel, G. L. (1989) Biomass, size distribution, and composition of phototrophic and heterotrophic nanoflagellate communities in Lakes Horon and Michigan. *Canadian Journal of Fisheries and Aquatic Sciences*, **46**, 1922–1928.

Carrick, H. J., Fahnenstiel, G. L. and Taylor, W. D. (1992) Growth and production of planktonic protozoa in Lake Michigan: *in situ* versus *in vitro* comparisons and importance to food web dynamics. *Limnology and Oceanography*, **37**, 1221–1235.

Choi, J. W. and Peters, F. (1992) Effects of temperature on two psychrophilic ecotypes of a heterotrophic nanoflagellate, *Paraphysomonas imperforata*. *Applied and Environmental Microbiology*, **58**, 593–599.

Choi, J. W. and Stoecker, D. K. (1989) Effects of fixation on cell volume of marine planktonic protozoa. *Applied and Environmental Microbiology*, **55**, 1761–1765.

Cleven, E-J. (1995) Grazing-Kontrolle der Bakterien- und Flagellaten (HNF) – Produktion durch ausgewähltes Protozooplankton im Bodensee. Dissertation, Fakultät für Biologie, University of Konstanz.

Cleven, E-J. (1996) Indirectly fluorescently labelled flagellates (IFLF): a tool to estimate the predation on free-living heterotrophic flagellates. *Journal of Plankton Research*, **18**, 429–442.

Dietrich, D. and Arndt, H. (2000) Biomass partitioning of benthic microbes in a Baltic inlet: relationships between bacteria, algae, heterotrophic flagellates and ciliates. *Marine Biology*, in press.

Eccleston-Parry, J. D. and Leadbeater, B. S. C. (1994a) A comparison of the growth kinetics of six marine heterotrophic nanoflagellates fed with one bacterial species. *Marine Ecology Progress Series*, **105**, 167–177.

Eccleston-Parry, J. D. and Leadbeater, B. S. C. (1994b) The effect of long-term low bacterial density on the growth kinetics of three marine heterotrophic nanoflagellates. *Journal of experimental marine Biology and Ecology*, **177**, 219–233.

Epstein, S. S., Burkovsky, I. V. and Shiaris, M. P. (1992) Ciliate grazing on bacteria, flagellates, and microalgae in a temperate zone sandy tidal flat: ingestion rates and food niche partitioning. *Journal of Experimental Marine Biology and Ecology*, **165**, 103–123.

Epstein, S. S. (1997) Microbial food webs in marine sediments. II. Seasonal changes in trophic interactions in a sandy tidal flat community. *Microbial Ecology*, **34**, 199–209.

Falkenhayn, C. J. and Lessard, E. J. (1992) Growth rates of heterotrophic dinoflagellates in the north Atlantic. *ASLO Abstracts*, 1992.

Fenchel, T. (1969) The ecology of marine microbenthos. IV. Structure and function of the benthic ecosystem, its chemical and physical factors and the microfauna communities with special reference to the ciliated protozoa. *Ophelia*, **6**, 1–182.

Fenchel, T. (1982a) Ecology of heterotrophic microflagellates. III. Adaptations to heterogeneous environments. *Marine Ecology Progress Series*, **9**, 25–33.

Fenchel, T. (1982b) Ecology of heterotrophic microglagellates. IV. Quantitative occurrence and importance as bacterial consumers. *Marine Ecology Progress Series*, **9**, 35–42.

Fenchel, T. (1986a) The ecology of heterotrophic flagellates. In *Advances in Microbial Ecology, vol. 9* (ed. K. C. Marshall), New York: Plenum Press, pp. 57–97.

Fenchel, T. (1986b) Protozoan filter feeding. *Progress in Protistology*, **1**, 65–113.

Fenchel, T. and Finlay, B. J. (1995) *Ecology and Evolution in Anoxic Worlds*. Oxford: Oxford University Press.

Gasol, J. M. (1993) Benthic flagellates and ciliates in fine freshwater sediments: calibration of a live counting procedure and estimation of their abundances. *Microbial Ecology*, **25**, 247–262.

Gasol, J. M. and Vaqué, D. (1993) Lack of coupling between heterotrophic nanoflagellates and bacteria: a general phenomenon across aquatic systems? *Limnology and Oceanography*, **38**, 657–665.

Geider, R. J. and Leadbeater, B. S. C. (1988) Kinetics and energetics of growth of the marine choanoflagellate *Stephanoeca diplocostata*. *Marine Ecology Progress Series*, **47**, 169–177.

Goldman, J. C. and Caron, D. A. (1985) Experimental studies on an omnivorous microflagellate: implications for grazing and nutrient regeneration in the marine microbial food chain. *Deep-Sea Research*, **32**, 899–915.

Grossart, H-P. and Simon, M. (1993) Limnetic macroscopic organic aggregates (lake snow): occurrence, characteristics, and microbial dynamics in Lake Constance. *Limnology and Oceanography*, **38**, 532–564.

Güde, H. (1979) Grazing by protozoa as selection factor for activated sludge bacteria. *Microbial Ecology*, **5**, 225–237.

Hahn, M. W., Moore, E. R. B. and Höfle, M. G. (1999) Bacterial filament formation, a defense mechanism against flagellate grazing, is growth rate controlled in bacteria of different phyla. *Applied and Environmental Microbiology*, **65**, 25–35.

Hansen, P. J., Björnsen, P. K. and Hansen, B. W. (1997) Zooplankton grazing and growth: scaling within the 2–2,000-µm body size range. *Limnology and Oceanography*, **42**, 687–704.

Hewes, C. D., Sakshaug, E., Reid, F. M. H. and Holm-Hansen, O. (1990) Microbial autotrophic and heterotrophic eucaryotes in Antarctic waters: relationships between biomass and chlorophyll, adenosine triphosphate and particulate organic carbon. *Marine Ecology Progress Series*, **63**, 27–35.

Hondeveld, B. J. M., Nieuwland, G., van Duyl, F. C. and Bak, R. P. M. (1994) Temporal and spatial variations in heterotrophic nanoflagellate abundance in North Sea sediments. *Marine Ecology Progress Series*, **109**, 235–243.

Ikävalko, J. and Thomsen, H. A. (1997) The Baltic Sea ice biota (March 1994): a study of the protistan community. *European Journal of Protistology*, **33**, 229–243.

Jacobson, D. M. and Anderson, D. M. (1986) Thecate heterotrophic dinoflagellates: feeding behavior and mechanisms. *Journal of Phycology*, **22**, 249–258.

Jones, R. I. and Rees, S. (1994) Characteristics of particle uptake by the phagotrophic phytoflagellate, *Dinobryon divergens*. *Marine Microbial Food Webs*, **8**, 97–110.

Jürgens, K. (1992) Is there plenty of food for heterotrophic flagellates in eutrophic waters? *Archiv für Hydrobiologie, Beihefte Ergebnisse der Limnologie*, **37**, 195–205.

Jürgens, K., Arndt, H. and Zimmermann, H. (1997a) Impact of metazoan and protozoan grazers on bacterial biomass distribution in microcosm experiments. *Aquatic Microbial Ecology*, **12**, 131–138.

Jürgens, K., Wickham, S. A., Rothhaupt, K. O. and Santer, B. (1997b) Feeding rates of macro- and microzooplankton on heterotrophic nanoflagellates. *Limnology and Oceanography*, **41**, 1833–1839.

Kuuppo, P. (1994) Annual variation in the abundance and size of heterotrophic nanoflagellates on the SW coast of Finland, the Baltic Sea. *Journal of Plankton Research*, **16**, 1525–1542.

Larsen, J. and Patterson, D. J. (1990) Some flagellates (Protista) from tropical marine sediments. *Journal of Natural History*, **24**, 801–937.

Leadbeater, B. S. C. (1974) Ultrastructural observations on nanoplankton collected from the coast of Jugoslavia and the Bay of Algiers. *Journal of the Marine Biological Association of the United Kingdom*, **54**, 179–196.

Lessard, E. J. and Swift, E. (1985) Species-specific grazing rates of heterotrophic dinoflagellates in oceanic waters, measured with a dual-label radioisotope technique. *Marine Biology*, **87**, 289–296.

Lewis, W. M. (1985) Protozoan abundances in the plankton of two tropical lakes. *Archiv für Hydrobiologie*, **104**, 337–343.

Lim, E. L., Dennett, M. R. and Caron, D. A. (1999) The ecology of *Paraphysomonas imperforata* based on studies employing oligonucleotide probe identification in coastal water samples and enrichment cultures. *Limnology and Oceanography*, **44**, 37–51.

Massana, R. and Güde, H. (1991) Comparison between three methods for determining flagellate abundance in natural waters. *Ophelia*, **33**, 197–203.

Mathes, J. and Arndt, H. (1994) Biomass and composition of protozooplankton in relation to lake trophy in north German lakes. *Marine Microbial Food Webs*, **8**, 357–375.

Mathes, J. and Arndt, H. (1995) Annual cycle of protozooplankton (ciliates, flagellates and sarcodines) in relation to phyto- and metazooplankton in Lake Neumühler See (Mecklenburg, Germany). *Archiv für Hydrobiologie*, **134**, 337–358.

Mischke, U. (1994) Influence of food quality and quantity on ingestion and growth rates of three omnivorous heterotrophic flagellates. *Marine Microbial Food Webs*, **8**, 125–143.

Mylnikov, A. P. (1978) Colourless flagellates in the benthos of Ivanjkovski Reservoir (Zoomastigophorea Calkins, Protozoa) (in Russian). *Biologiya Vnutrennych Vod. Informaticionny Bulleten*, **39**, 13–18.

Mylnikov, A. P. (1983) The adaptation of freshwater zooflagellates to increased salinity (in Russian). *Biologiya Vnutrennych Vod. Informaticionny Bulleten*, **61**, 21–24.

Mylnikov, A. P. and Zhgarev, N. A. (1984) The flagellates in the littoral of the Barents Sea and freshwater lakes (in Russian). *Biologiya Vnutrennych Vod. Informaticionny Bulleten*, **63**, 54–57.

Nauwerck, A. (1963) Die Beziehungen zwischen Zooplankton und Phytoplankton im See Erken. *Symbolae Botanicae Upsalienses*, **17**, 1–163.

Patterson, D. J. and Fenchel, T. (1990) *Massisteria marina* Larsen and Patterson 1990, a widespread and abundant bacterivorous protist associated with marine detritus. *Marine Ecology Progress Series*, **62**, 11–19.

Patterson, D. J. and Larsen, J. (eds) (1991) *The Biology of Free-Living Heterotrophic Flagellates*. Oxford: Systematics Association, Clarendon Press.

Patterson, D. J. and Simpson, A. G. B. (1996) Heterotrophic flagellates from coastal marine and hypersaline sediments in Western Australia. *European Journal of Protistology*, **32**, 1–24.

Patterson, D. J., Larsen, J. and Corliss, J. O. (1989) The ecology of heterotrophic flagellates and ciliates living in marine sediments. *Progress in Protistology*, **3**, 185–277.

Patterson, D. J., Nygaard, K., Steinberg, G. and Turley, C. M. (1993) Heterotrophic flagellates and other protists associated with oceanic detritus throughout the water column in the mid North Atlantic. *Journal of the Marine Biological Association of the United Kingdom*, **73**, 67–95.

Pratt, J. R. and Cairns, J. (1985) Functional groups in the protozoa: roles in differing ecosystems. *Journal of Protozoology*, **32**, 415–423.

Pringsheim, E. G. (1963) *Farblose Algen*. Jena: Gustav Fischer.

Radek, R. and Hausmann, K. (1994) Endocytosis, digestion, and defecation in flagellates. *Acta Protozoologica*, **33**, 127–147.

Rogerson A. and Laybourn-Parry, J. (1992) Aggregate dwelling protozooplankton communities in estuaries. *Archiv für Hydrobiologie*, **125**, 411–422.

Salbrechter, M. and Arndt, H. (1994) The annual cycle of protozooplankton in the mesotrophic, alpine Lake Mondsee (Austria). *Marine Microbial Food Webs*, **8**, 217–234.

Sanders, R. W. (1991a) Mixotrophic protists in marine and freshwater ecosystems. *Journal of Protozoology*, **38**, 76–81.

Sanders, R. W. (1991b) Trophic strategies among heterotrophic flagellates. In *The Biology of Free-Living Heterotrophic Flagellates* (eds D. J. Patterson and J. Larsen), Oxford: Clarendon Press, pp. 21–38.

Sanders, R. W. and Wickham, S. A. (1993) Planktonic protozoa and metazoa: predation, food quality and population control. *Marine Microbial Food Webs*, **7**, 197–223.

Sanders, R. W., Caron, D. A. and Berninger, U-G. (1992) Relationship between bacteria and heterotrophic nanoplankton in marine and freshwaters: an inter-ecosystem comparison. *Marine Ecology Progress Series*, **86**, 1–14.

Sandon, H. (1932) *The Food of Protozoa*. Cairo: Egyptian University, Publications of the Faculty of Science no. 1.

Schnepf, E. and Elbrächter, M. (1992) Nutritional strategies in dinoflagellates: a review with emphasis on cell biological aspects. *European Journal of Protistology*, **28**, 3–24.

Sherr, B. F. and Sherr, E. B. (1989) Trophic impacts of phagotrophic Protozoa in pelagic foodwebs. In *Recent Advances in Microbial Ecology* (eds T. Hattori, Y. Ishida, Y. Maruyama, R. Y. Morita and A. Uchida), Tokyo: Japan Scientific Societies Press, pp. 388–393.

Sherr, E. B. and Sherr, B. F. (1994) Bacterivory and herbivory: key roles of phagotrophic protists in pelagic food webs. *Microbial Ecology*, **28**, 223–235.

Sherr, E. B., Sherr, B. F. and McDaniel, J. (1991) Clearance rates of <6μm fluorescently labeled algae (FLA) by estuarine protozoa: potential grazing impact of flagellates and ciliates. *Marine Ecology Progress Series*, **69**, 81–92.

Shirkina, N. I. (1987) The morphology and life cycle of *Thaumatomonas lauterborni* De Saedeleer (Mastigophora, Diesing) (in Russian). In *Fauna and Biology of Freshwater Organisms*, Leningrad (St Petersburg): Nauka, pp. 87–107.

Simek, K. and Chrzanowski, T. H. (1992) Direct and indirect evidence of size-selective grazing on pelagic bacteria by freshwater nanoflagellates. *Applied and Environmental Microbiology*, **58**, 3715–3720.

Smetacek, V. (1981) The annual cycle of protozooplankton in the Kiel Bight. *Marine Biology*, **63**, 1–11.

Starink, M., Bär-Gilissen, M-J., Bak, R. P. M. and Cappenberg, T. E. (1994) Quantitative centrifugation to extract benthic protozoa from freshwater sediments. *Applied and Environmental Microbiology*, **60**, 167–173.

Starink, M., Bär-Gilissen, M-J., Bak, R. P. M. and Cappenberg, T. E. (1996) Bacterivory by heterotrophic nanoflagellates and bacterial production in sediments of a freshwater littoral system. *Limnology and Oceanography*, **41**, 62–69.

Tong, S. M., Nygaard, K., Bernard, C., Vørs, N. and Patterson, D. J. (1998) Heterotrophic flagellates from the water column in Port Jackson, Sydney, Australia. *European Journal of Protistology*, **34**, 162–194.

Turley, C. M. and Carstens, M. (1991) Pressure tolerance of oceanic flagellates: implications for remineralization of organic matter. *Deep-Sea Research*, **38**, 403–413.

Turley, C. M., Newell, R. C. and Robins, D. B. (1986) Survival strategies of two small marine ciliates and their role in regulating bacterial community structure under experimental conditions. *Marine Ecology Progress Series*, **33**, 59–70.

Vanderploeg, H. A. and Paffenhöfer, G-A. (1985) Modes of algal capture by the freshwater copepod *Diaptomus sicilis* and their relation to food-size selection. *Limnology and Oceanography*, 30, 871–885.

Vørs, N. (1992) Heterotrophic amoebae, flagellates and heliozoa from the Tvärminne area, Gulf of Finland, in 1988–1990. *Ophelia*, **36**, 1–109.

Vørs, N., Buck, K. R., Chavez, F. P., Eikrem, W., Hansen, L. E., Østergaard, J. B. and Thomsen, H. A. (1995) Nanoplankton of the equatorial Pacific with emphasis on the heterotrophic protists. *Deep-Sea Research II*, **42**, 585–602.

Weisse, T. (1991) The annual cycle of heterotrophic freshwater nanoflagellates: role of bottom-up versus top-down control. *Journal of Plankton Research*, **13**, 167–185.

Weisse, T. and Müller, H. (1998) Planktonic protozoa and the microbial food web of Lake Constance. *Archiv für Hydrobiologie, Special Issues Advances in Limnology*, **53**, 223–254.

Zimmermann, H. and Kausch, H. (1996) Microaggregates in the Elbe Estuary: structure and colonization during spring. *Archiv für Hydrobiologie, Special Issues Advances in Limnology*, **48**, 85–92.

Chapter 13

Geographic distribution and diversity of free-living heterotrophic flagellates

D. J. Patterson and W. J. Lee

ABSTRACT

As yet, no consensus has emerged on the geographic distribution of free-living heterotrophic flagellates or on the overall diversity of the group. A survey of the literature on the distribution of flagellates from marine zones reveals that more than half the species have been reported from a single location. This suggests that these organisms are often endemic. In a series of original surveys of flagellates, about 40 per cent of the 350 species were reported from a single location, again suggestive of endemism. When the communities from these surveys are compared using the clustering algorithm in PRIMER, there is, in contrast, no evidence of endemism because the communities from geographic regions do not cluster together. Communities cluster on the type of habitat from which they were drawn, and this suggests that geographic location may play no part in the make-up of communities. The conflict of insights from information on species and information on communities creates uncertainty over the geographical distribution of flagellates. This is probably because the actual distribution of these organisms is obscured by factors extrinsic to their distribution, principally issues relating to under-reporting and to arguable species concepts. Our interpretation of available data is constrained by the morphological species concepts. Our interpretation is that there are not many species of heterotrophic flagellates (perhaps no more than 3,000) and most have a cosmopolitan distribution. We believe that there are assemblages of flagellates with distinctive taxonomic compositions, but that we are unable to describe these more precisely until we reduce the impact of factors which are external to the biology of the organisms but which influence our understanding of the structure of communities of heterotrophic flagellates.

13.1 Introduction: how diversity is linked to geography

The significance attached to microbial communities in aquatic ecosystems (Azam *et al.*, 1982) led to a resurgence of interest in the diversity, taxonomy, and distribution of free-living flagellates. Despite the stunning abundance of free-living flagellates in natural habitats (Berninger *et al.*, 1991), and despite a contemporary review of taxonomy to the level of genus (Patterson and Larsen, 1991), there is no recent synthesis in respect of the species-level diversity within this group.

Corliss (1982) reviewed the numbers of species in various protist groups, and concluded that there were fewer than 150,000 species of protists, and that of these only 200 or so were assignable to the heterotrophic flagellates (excluding dinoflagellates(4,200 species) and haptophytes (1,500 species)). May (1988, 1990) assessed the number of species of all organisms, and concluded that there was a relationship between size and species diversity from a size range from about 1 cm to 10 m. He showed that the known numbers of microbial species were several orders of magnitude fewer than were suggested by his analysis of larger species. He was unable to establish if this was a reflection of the true situation, or if it reflected the lack of taxonomic attention. John and Maggs (1997), considering diversity among the algae, believe that these organisms have been under-reported by at least one order of magnitude. Assessment of the microbial diversity in natural habitats by molecular means (Bowers *et al.*, 1998; DeLong *et al.*, 1993; Embley and Stackebrandt, 1994; Giovannoni *et al.*, 1990; Ward *et al.*, 1990) would also suggest that the microbial diversity has been significantly under-reported. However, from the perspective of ciliates, an alternative view has been emerging that these protozoa have a ubiquitous distribution, and that the number of species is not particularly great (for example Finlay, 1998; Finlay *et al.*, 1996a, 1996b, 1998).

The global diversity of species of heterotrophic flagellates is a function of local species diversity, the perceptible niches, and the size (and therefore number) of the spatial and temporal patches occupied. We therefore need to understand (*inter alia*) the size of the geographic patches occupied by the species of flagellate, if we wish to improve our understanding of their overall diversity. There is as yet little discussion of the biogeography of these organisms.

This account summarizes the results of a survey of the literature that includes information on the distribution of marine flagellates, and an analysis (Lee and Patterson, 1998) of original surveys of flagellate diversity in a variety of habitats world-wide.

13.2 Intrinsic and extrinsic factors

The geographic distribution of organisms is determined by their evolutionary history, by their physiological preferences, and by dispersal. The first two factors favour a patchy distribution, and the latter favours cosmopolitanism. We refer to these three factors as '*intrinsic*'. However, our perception of the geographic distribution of organisms can be distorted by '*extrinsic*' factors. Extrinsic factors are those which are not inherent to the biology of the organisms (Silva, 1984). In the study of the distribution of heterotrophic flagellates, extrinsic factors include methodological artefacts such as sampling or the interpretation of data (for example, species concepts). Extrinsic factors which might have these kinds of effects include the following.

13.2.1 Under-sampling

Under-sampling applies when the number of species collected from a site and made available to be reported is (significantly) less than the number of species present. It can be caused by a failure completely to sample sub-habitats, to apply selective or inadequate extraction techniques which favour selected taxa, or to sample volumes insufficient to reflect adequately the diversity over time, or to include sparsely distributed species.

Example It is likely that flagellates, like ciliates, are distributed in accordance with the redox gradients within sediments (Bernard *et al.*, 1999). Sampling techniques used in the study of benthic flagellates (Ekebom *et al.*, 1996) favour the aerobios and fail to report species which, although they occur in sediments, prefer anaerobic conditions (Bernard *et al.*, 1999).

Consequences Under-sampling may cause similar communities to be regarded as dissimilar, and is most likely to favour suggestions of endemism.

13.2.2 Under-reporting

Even if diversity is well sampled, only a small number of species present may be reported. Sometimes, only those species which are most abundant may be reported, with many species being present as cysts or other forms which act as a 'seedbank' (Finlay and Esteban, 1998) going unrecorded. Taxonomic preferences of the worker (see later) may also lead to many taxa being ignored (that is, under-reported).

Example Finlay (1998) has shown that snap-shot surveys of ciliates might only reveal one-tenth of the diversity present, the rest becoming evident presumably only as conditions became appropriate.

Consequences If under-reporting favours common species, the consequence will be an erroneous generalization of cosmopolitanism. This might happen when handling procedures favour bacterivorous species. Under-reporting which excludes species more randomly is likely to lead to a false sense of endemism.

13.2.3 Lack of replication

The problems alluded to earlier (and indeed some later) can include systemic errors, or can fail to reveal variation (and diversity) among samples. In the absence of any effective replication protocol, the significance of these issues is undetermined.

Example We have conducted replicated surveys of marine benthic communities at One Tree Island (Ekebom *et al.*, 1996) and at Cape Tribulation (Figure 13.6). The process requires maturation of samples over a period of between twenty-four and seventy-two hours, and in both cases, one of four replicates generated a community very different to the remainder. Differences between replicates may exceed that between sites.

Consequences The lack of replication is most likely to lead to erroneous conclusions of the distinctiveness of communities studied.

13.2.4 Taxonomic selectivity

Individual taxonomists usually favour certain taxa and do not report on all members of a community. There is, for example, a tradition of some groups of workers (such as Thomsen or Takahashi and co-workers) to prefer to study community diversity through untrastructural examination of whole mounts (Takahashi, 1987; Thomsen, 1978); or to focus on the water column (Vørs, 1992). Others, such as ourselves, favour sediment dwelling organisms studied live (Larsen and Patterson, 1990). Catalogues of species generated by different workers may not therefore contain the same species.

Example Lee and Patterson (1998) compared species lists developed by the same group of workers from many sites around the world. They included a survey of a similar site by an independent group of workers. The species lists generated by different workers from similar locations did not group together, indicating that the similarities among communities can be obscured by an extrinsic factor, most likely the taxonomic preferences of workers concerned.

Consequences Selectivity will lead to arguments of cosmopolitanism when studies of the same workers are compared, and to arguments favouring endemism when studies by different workers are compared.

13.2.5 Lumpers and splitters

Different taxonomists take different approaches about whether small differences in appearance should form the basis for new species or not. Finlay (1997) and Finlay *et al.* (1996a) emphasize the contribution of periodic taxonomic revisions which typically reduce the number of nominal species. It is our view that this process serves merely to resolve ambiguities in the application of a prevailing species concept in a rather arbitrary fashion.

Example Molina and Nerad (1991) recognized seven species of *Amastigomonas* (a splitters' perspective), although the lumpers' perspective is that only four have a discrete identity. *Pteridomonas danica* and *Actinomonas mirabilis* can only be distinguished confidently by reference to ultrastructural characteristics (Larsen and Patterson, 1990). This requirement for inaccessible technologies is characteristic of 'splitting'. We have noted that with the progression of our series of studies (compare Larsen and Patterson, 1990 with Lee and Patterson, 1999), the criteria for admission to species in some genera (such as *Notosolenus*, *Peranema*, *Bodo*, and *Petalomonas*) are becoming more relaxed and, with the expansion of the criteria, more restrictive (splitting) concepts are changing to more 'lumping' concepts.

Consequences When lists are compared, those produced by lumpers or involving broader species concepts will contain fewer species and will have greater overlap, and will favour interpretations of cosmopolitanism, whereas the use of lists generated by splitters is more likely to lead to conclusions of endemism.

13.2.6 Differing nomenclatural practices

There are circumstances in which an organism with an agreed identity may be referred to by different names. Several groups of flagellates contain heterotrophs and autotrophs. Some workers apply the International Code of Zoological Nomenclature (ICZN) while others apply the code for botanical nomenclature (ICBN). Because the codes are different, the same species may have more than one 'correct' name. Taxa in this situation have been referred to as ambiregnal (Patterson and Larsen, 1992).

Examples *Bodo saltans* (Figure 13.7g) is a distinctive kinetoplastid flagellate which attaches to the substrate by means of its recurrent flagellum and moves with a kicking motion. This is very distinctive, such that its identity is not in doubt. The same organism is referred to as *Pleuromonas jaculans* in the Russian literature (Hänel, 1979). Larsen and Patterson (1990) included a variety of taxa which bore than more than one name. *Dinema* and *Peranema* are generic names for euglenids which are accepted by the ICZN, but have to be rejected under the regulations of the ICBN. The species in these genera, *inter alia*, carry two names (for example *Dinema validum* is *Dinematomonas valida*).

Consequences Comparisons of accounts of species for which different names have been used are likely to cause incorrect conclusions of dissimilarity, and in the context of geography will favour incorrect conclusions of endemism. This can only be fixed by using a standardized nomenclature.

13.2.7 Uncertain identities

The widespread lack of type material adds to the difficulties of ensuring consistent identifications by a sparsely populated community of workers. The lack of information on 'within-species variation' and 'between-species variation' creates difficulties in establishing the correct identities of taxa.

Example Ekelund and Patterson (1998) and Lee and Patterson (2000) present tables which indicate the difficulties in confidently identifying taxa assigned to *Rhynchobodo* and *Protaspis*, respectively. *Lentomonas applanata* (*Entosiphon applanatum* of Preisig, 1979) may well be the same as *Ploeotia corrugata* of Larsen and Patterson, 1990. Yet the reporting of a critical diagnostic character (a protrusible siphon), without documentation, means that the correct generic assignment is uncertain (Ekebom *et al.*, 1996; Farmer and Triemer, 1994).

Consequences Uncertain identities may result in more than one name being used for the same species (or indeed more than one species being referred to by the same name). Problems in establishing clear identities of taxa can be alleviated, but not completely dispelled, by the use of uninterpreted records (photography, video, and so on). We think that it is more likely that differences in identification will exacerbate differences among species lists and will favour a sense of endemism.

13.2.8 Taxonomic instability

With the passage of time, taxonomists may divide species, merge them, or move them to different genera. These are normal changes in taxonomy.

Example *Percolomonas membranifera* (Larsen and Patterson, 1990) is now recognized as two species (*Carpediemonas bialata* and *C. membranifera*), *Bodo parvulus* is now *Caecitellus parvulus*, the current concept of *Peranema trichophorum* almost certainly contains several species.

Consequences Failure to apply nomenclatural corrections to the earlier usage will result in the same organism being reported under different names, and will favour incorrect conclusions of endemism.

13.2.9 Nomenclatural instability

The names of taxa change when synonymies are established. Nominal taxa with a short nomenclatural life will seem to have a restricted distribution unless the full taxonomic history is developed and appropriate nomenclature applied retrospectively.

Example Larsen and Patterson (1990) produced the first contemporary broad-ranging study of marine benthic flagellates, and their study includes over eighty new names. Within a decade, subsequent surveys have led to the elimination or change in use of the following: *Thecamonas* (now *Amastigomonas*) and three species in this genus (*T. filosa*, *T. mutabilis* and *T. trahens*) have new names (*A. filosa*, *A. mutabilis* and *A. debruynei* respectively), *Bodo cephaloporus* is now *Ancyromonas sigmoides*, *Percolomonas membranifera* is now *Carpediemonas membranifera*, and recognized as two species (the second is as yet unnamed), and *Bodo parvulus* is now *Caecitellus parvulus*. *Cryptobia libera* and *Heteromita ovata* are now *Jakoba libera* and *Cercomonas ovata*, respectively. *Peranema fusiforme* and *P. macrostoma* are now *Jenningsia fusiforme* and *J. macrostoma*, respectively. *Leucocryptos remigera* is now *Kathablepharis remigera*. *Percolomonas cuspidata* and *Pseudobodo minuta* are now *Chilomastix cuspidata* and *Cafeteria minuta*, respectively.

Consequences Nomenclatural instability will create an inappropriate sense of endemism, whether it involves division or amalgamation of taxa. It is often not possible to resolve the problems which arise, and this may well invalidate the use of much of the older literature.

13.2.10 Species concepts

Mayden (1997) holds the view that there are over twenty different concepts of species. Even without interrogating each concept, there is evidently an arbitrariness as to what is a species. The use of species concepts which embrace a greater diversity of organisms is more likely to lead to a sense of cosmopolitanism; narrow concepts may lead to a sense of endemism. At this time, the alpha taxonomy of heterotrophic flagellates is based mostly on a morphospecies concept. However, more discriminating criteria can be applied to some groups if they have scales or other excrescences which can be examined by electron-microscopy. Similarly, the application of molecular criteria in other protists suggests that traditional morphospecies offer a facade behind which hide a much greater number of physiological or molecular species (Cairns, 1993; John and Maggs, 1997).

Consequences The application of more discriminating species concepts will favour conclusions of endemism.

13.2.11 Subjective analysis of data

If the mechanism by which statements about endemism are developed from the data on distribution is not specified, then the appropriateness of the conclusions must be questioned. The case for endemism is usually based on the observation of a taxon from a single location, frequently on a single occasion. We do not regard such as presenting a case for endemism, until at least a variety of extrinsic factors has been addressed. We are of the view that it is the identification of a taxon in two or more non-contiguous regions which is evidence against endemism.

Example The case for endemism – driven largely by autecological considerations – is most strongly presented by Tyler (for example Tyler, 1996). The interpretation of occasional observations as bearing on biogeography has been applied to heterotrophic flagellates by, for example, Foissner *et al.* (1988).

Consequences The most likely incorrect assumption is that occasional records are indicators of endemism.

13.3 The literature

A survey has been carried out of over 300 publications which refer to marine flagellates in one or more of twenty-three marine zones (Figure 13.1) (Patterson and Lee, unpublished). This has produced information on almost 600 nominal species and 200 nominal genera of free-living heterotrophic flagellates, excluding dinoflagellates and haptophytes. The identities of taxa have not been confirmed because the calibre of documentation is insufficient to do this consistently, or indeed at all in some cases. Nomenclatural changes have not been made, as this process is contingent on confirming the identities of taxa. This exercise is therefore a purely bibliographic exercise.

The resulting curves (Figures 13.2a, b) reveal that almost half of the species and 40 per cent of the genera have only ever been reported from one zone. Fewer than ten species or genera have been observed in ten or more zones. On face value, this suggests that most taxa have a restricted distribution. However, species have not been reported equally from all twenty-three sites. More species have been reported from certain geographic sites: namely, and in rank order, Europe, Australia, North America, and the polar regions (Figure 13.3). The largest number of taxa have been reported from Europe. As most publications also relate to this geographic region, it

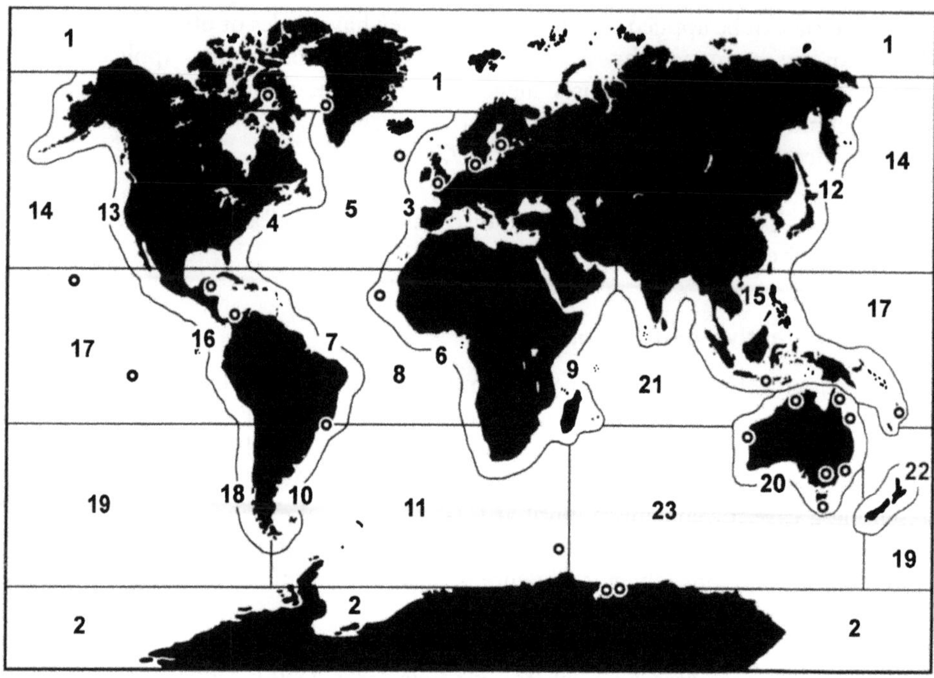

Figure 13.1 Map showing zones used in compiling geographic information from the literature, and showing sites used in the analysis presented as Figure 13.6.
Zones:

1 Arctic, but includes one site 64° N in W Greenland
2 Antarctica
3 Europe
4 NE America
5 North Atlantic
6 Intertropical W Africa
7 Caribbean
8 Equatorial Atlantic
9 S and E Africa
10 SE America
11 South Atlantic
12 NE Asia
13 NW America
14 North Pacific
15 Indonesia
16 Intertropical W. America
17 Equatorial Pacific
18 SW America
19 S Pacific
20 Australia
21 Indian Ocean
22 New Zealand
23 Southern Ocean

is possible that the absence of records from other locations reflects the absence of studies. The relationship between number of species for given sites and the number of papers relating to that site is positive and broadly linear.

Despite the strong suggestion of endemism within the literature, it is quite likely this insight could be flawed by a variety of extrinsic factors, the most substantial of which is likely to be under-sampling and under-reporting of taxa from many locations. This, and the inability to address issues of the correct identities and nomenclatures of taxa, will favour erroneous conclusions of endemism. The problems of identities will remain unresolved for many taxa, as those accounts in the literature which lack discriminating and uninterpreted records are of little value in assessing endemism.

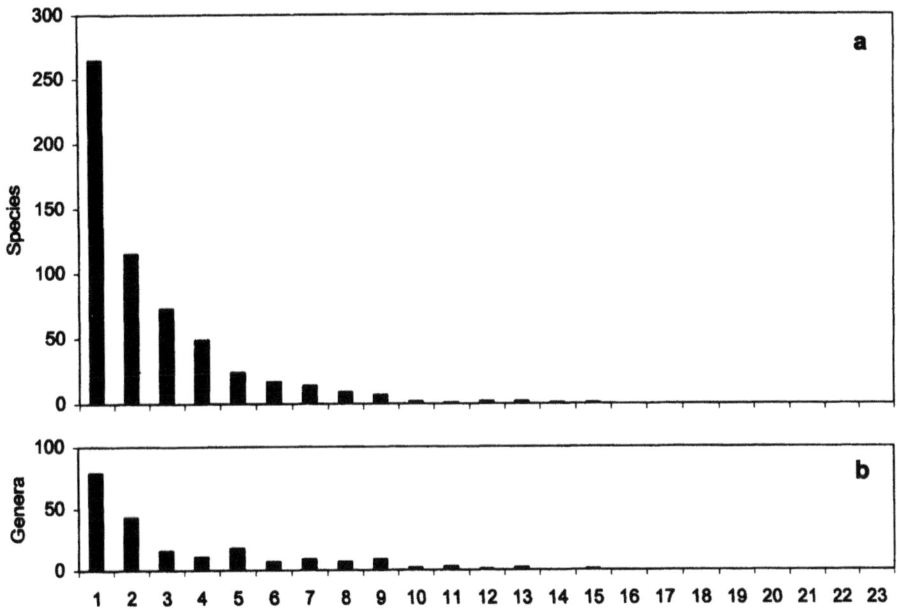

Figure 13.2 Summary of data from the literature on the distribution of heterotrophic flagellates: the number of regions (maximum 23) from which species (a) and genera (b) have been reported

13.4 Original surveys

The only series of recent studies which has sought to clarify the large-scale patchiness of free-living heterotrophic flagellates is that summarized elsewhere (Lee and Patterson, 1998). We have assembled species lists from over thirty sites (Figure 13.1). The coverage includes north and southern hemispheres, equatorial, tropical, temperate and polar locations, freshwater, marine and terrestrial (Ekelund and Patterson, 1998) plus some 'extreme' habitats such as hypersaline and anaerobic sites. Our studies have involved a small number of taxonomists, and have used standardized sampling and recording procedures. We rely heavily on photographic and video records to provide uninterpreted and archivable reference material. This approach minimizes many of the distortions caused by extrinsic factors, by either reducing them or standardizing their impact. It also standardizes issues relating to identity, taxonomy and nomenclature. We believe that the extent of under-reporting is likely to be similar for most studies. We have analysed the data using explicit algorithms. We apply a consistent morphospecies concept. We have yet been unable to establish if this species concept is appropriate to the task in hand (see later); and issues of under-sampling have not been addressed.

The frequency with which species and genera have been recorded in these surveys follows the same pattern as in the literature in presenting (roughly) hollow curves (Figure 13.4). Many species have been observed once, and increasingly fewer species recorded more frequently. This may be interpreted as indicating endemism. The same result would be obtained if species are being under-reported, either because they are

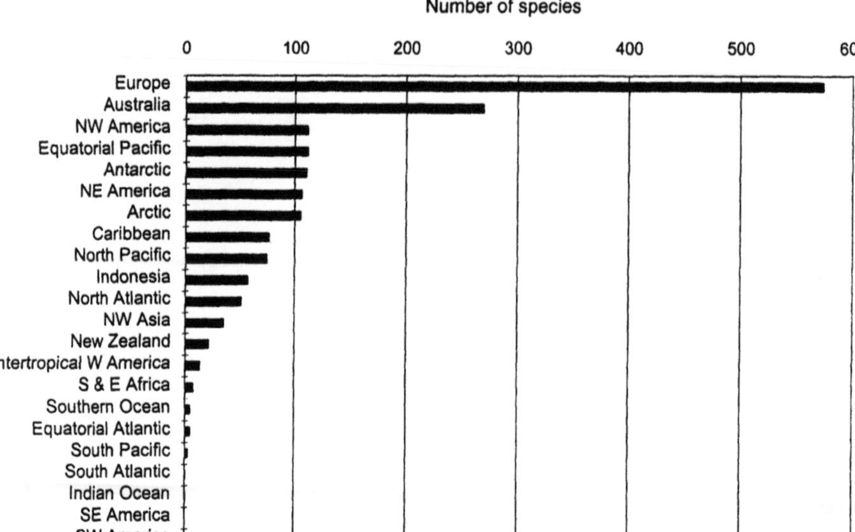

Figure 13.3 Summary of data from the literature on the distribution of heterotrophic flagellates: the number of species reported from each of the 23 geographic zones used

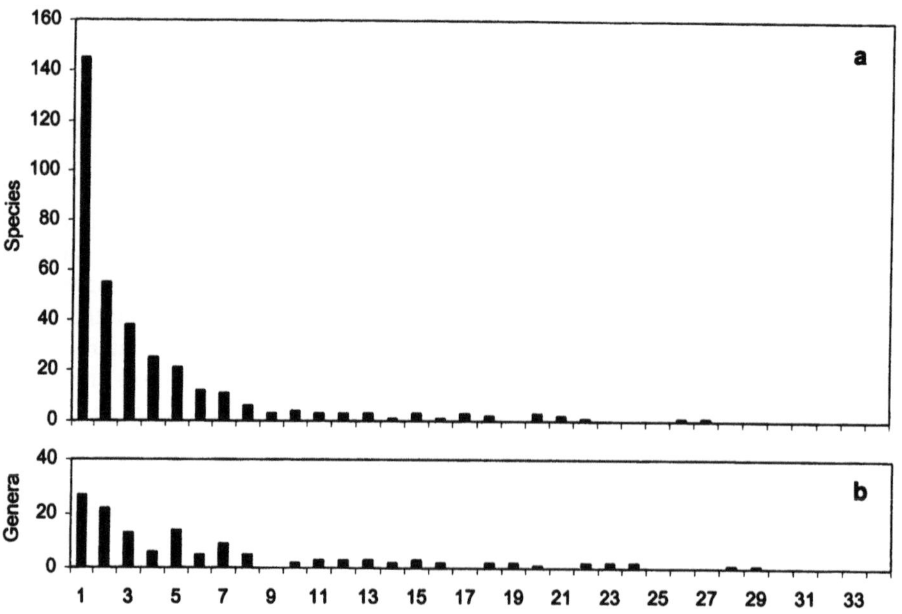

Figure 13.4 Summary of the frequency with which taxa have been reported in surveys (maximum being 34): (a) data on species; (b) data on genera

sparsely distributed and have been under-sampled, or are common but have not been adequately documented.

A relatively intensive survey involving twenty-two sampling events has been carried out at Botany Bay in Sydney, Australia (Lee and Patterson, 2000). From these surveys we conclude at least that previous studies are flawed by under-reporting. Of the ninety species recorded, only fifteen were encountered once. One-third of the species have been reported on over two-thirds or more of the sampling occasions (Figure 13.5). Unlike the data summarized in Figures 13.2 and 13.4, this indicates that at least some species are locally common and recurrent. We interpret other species-based syntheses as being marred by either under-reporting owing to logistic constraints and limits to the time committed to those surveys, or quite possibly under-sampling. Finlay and Esteban (1998) have argued that many species of ciliates, while occurring at a location, may be present in very low numbers or are encysted. Such species will be overlooked if the location is under-sampled. We note that as under-sampling and under-reporting are corrected, the case for cosmopolitanism will be strengthened.

The patterns of similarity of these communities has been assessed (Lee and Patterson, 1998) using the cluster algorithm within the PRIMER package (Clarke, 1993). The dendrogram based on presence/absence data for species has been updated and is presented in Figure 13.6. From this we draw several conclusions.

First, there is no evidence from this aspect of the study which suggests that there is a geographic determinant of community composition. Communities from Australian sites do not cluster together. Communities from several adjacent sites (Cape Tribulation 1–3) are not more similar to each other than they are to communities from more distant sites.

Second, communities from similar habitats do cluster together. The extreme habitats segregate from the remainder; thereafter, benthic communities can be shown to resemble each other and differ from water column communities, which also resemble each other. This is consistent with the view (Fenchel, 1987; Patterson *et al.*,

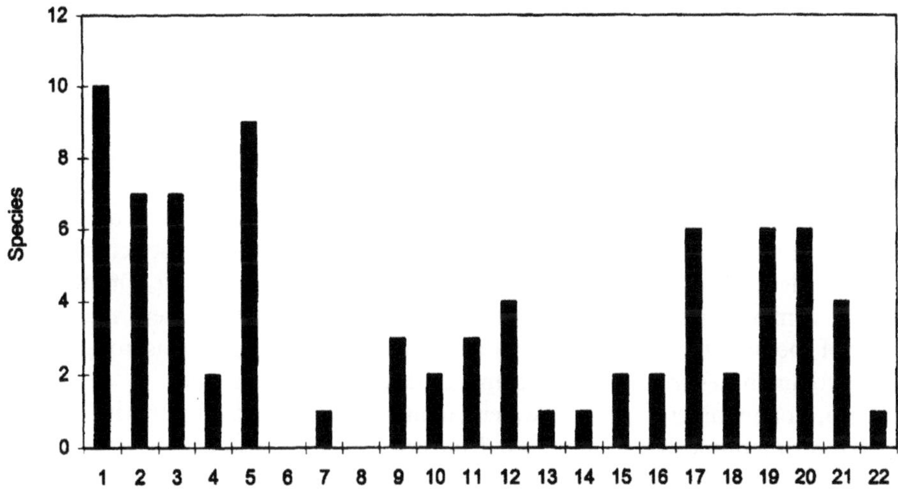

Figure 13.5 Frequency with which species have been reported in 22 surveys of Botany Bay

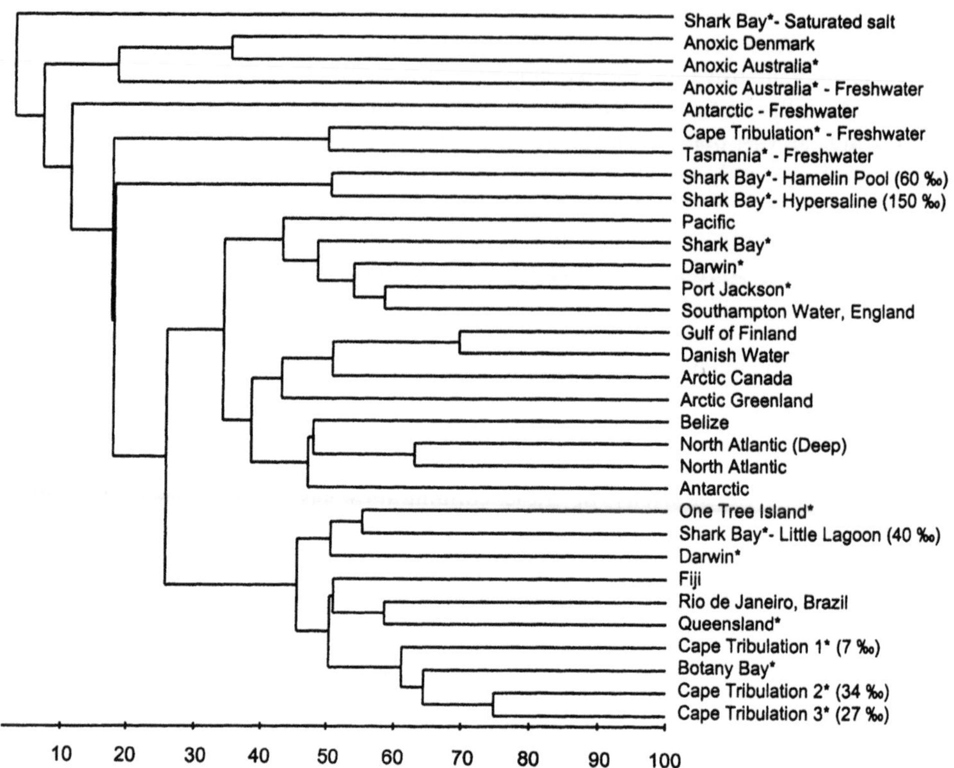

Figure 13.6 Dendrogram of similarities among communities surveyed. Bray-Curtis similarities based on presence/absence data, analysis by PRIMER. For authors and details of sites see Lee and Patterson (1998). Australian sites are marked (*). This diagram includes new surveys conducted at three sites at Cape Tribulation, Queensland (Australia) by one of us (W. J. Lee)

1989) that the distribution of protozoa is largely determined by ecophysiological factors such as their interaction with oxygen, redox potential, illumination, temperature, and so on. We recorded 189 species from the water column and 154 species from benthic sites; fifty- six have been reported from both. Benthic and water column communities are comparably diverse, each with about ninety genera (benthic: ninety-three; water column: ninety-nine). Figure 13.6 includes information on 138 genera. A comparison of the genera which make up the water column and benthic communities reveals that the parts of those communities which contribute to their distinctiveness are choanoflagellates in water column communities and euglenids in benthic communities.

Third, communities from extreme environments have relatively few species, the number of species which occur both in them and in 'normal' habitats is not great. As a result, communities from these sites appear distinctive in this study.

Fourth, communities from the water column separate broadly on the basis of temperature or latitude, but communities from sediments in warm regions (for example tropic or subtropic) do not cluster together. One warm water component (Belize) clusters with communities from cold water, but the species list from that site may have been distorted because this material was transported from Belize to Copenhagen for analysis. This suggests that there may be non-geographic factors with a very large spatial scale which might be reflected in patterns of distribution. Reference to these distributions as being 'biogeographical' would, in our view, be misleading.

Fifth, marine communities are distinct from freshwater communities, but about 10 per cent of the species can be found in freshwater and marine conditions. Indeed, some species are remarkably widespread. *Rhynchomonas nasuta* (Figure 13.7) has been reported from the northern and southern hemispheres, from the Atlantic and Pacific oceans; from polar, temperate, tropical and equatorial sites; from coastal sediments and open oceans; from surface waters and at depths up to 2 km; from freshwater, marine and terrestrial habitats; under complete (guaranteed) anoxia; and in hypersaline environments with five times the normal levels of salt. The most widespread species are shown in Figure 13.7, and are mostly small bacterivorous species.

The present study provides no evidence that geography determines the composition of communities of heterotrophic flagellates. There are several possible reasons for this. The first two are intrinsic causes.

The first is the argument by Finlay and Fenchel for ciliates (for example Fenchel, 1993; Finlay *et al.,* 1996a; Finlay, 1998). They suggest that because free-living protozoa occur in very high numbers, and because marine habitats are contiguous, there are few isolating mechanisms which might lead to sufficient isolation of populations to permit the emergence of new species. They also point out that dispersal is high and likely to prevent isolation of populations. The less intensively studied freshwater habitats may behave differently and have the potential to produce different insights.

A second reason may be that flagellates have a cosmopolitan distribution because they evolved early in eukaryotic history (Patterson, 1994). Almost certainly they have had an evolutionary history extending over 2 billion years and preceding any divisions of the world caused by movements of crustal plates. Over this time scale, the processes of dispersal will tend to obliterate traces of processes of speciation.

Turning to extrinsic factors, the differences between Figures 13.4 and 13.5 suggest that under-reporting influences the outcome of our studies, and impairs our capacity to reveal endemism. However, if we analyse communities with genus level data only, we continue to observe clustering of communities on the basis of the type of habitat (Lee and Patterson, 1998). The influence of under-sampling may be expected to be diminished when genera are used. The survival of the same pattern suggests that under-reporting, although it occurs (Figure 13.5), is not critical in determining our insights.

Finally, it may be that our concept of species is not sufficiently discriminatory properly to reveal species, and therefore will fail to reveal a geographically patchy distribution of the 'real' species, as has been argued to be the case with diatoms (Mann and Droop, 1996). The work reviewed in this survey calls upon a morphological species concept, in which a species is distinguished because there are one or more discrete and consistently recognizable differences between organisms assigned to that species, and other organisms. In the absence of reported sexuality for many of these species, we believe this is the only operational concept available to us. Cairns (1993) has suggested that morphological species may be facades behind which lurk

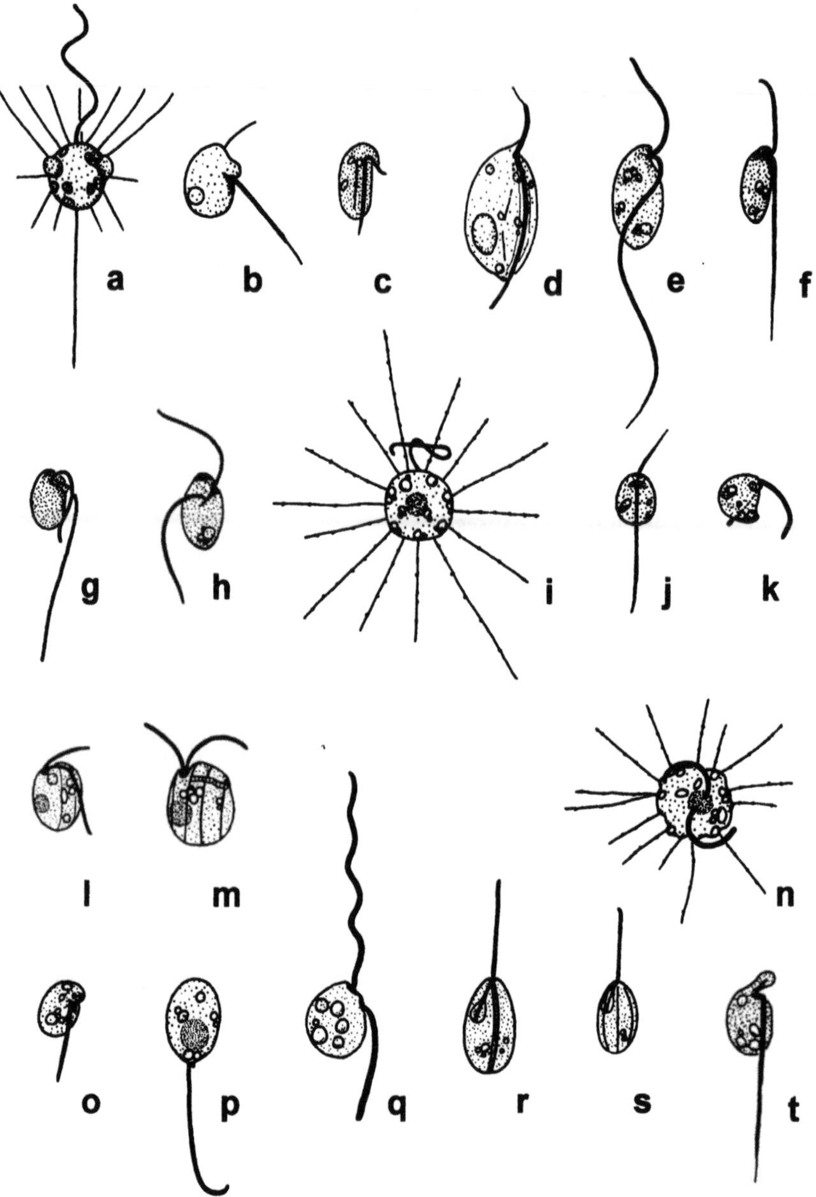

Figure 13.7 The twenty most commonly reported species of heterotrophic flagellates:

a *Actinomonas mirabilis/Pteridomonas danica*
b *Ancyromonas sigmoides*
c *Amastigomonas debruynei*
d *Amstigomonas mutabilis*
e *Bodo designis*
f *Bodo saliens*
g *Bodo saltans*
h *Bordnamonas tropicana*
i *Ciliophrys infusionum*
j *Caecitellus parvulus*
k *Cafeteria roenbergensis*
l *Goniomonas amphinema*
m *Goniomonas pacifica*
n *Massisteria marina*
o *Metopion fluens*
p *Metromonas simplex*
q *Pseudobodo tremulans*
r *Petalomonas minuta*
s *Petalomonas poosilla*
t *Rhynchomonas nasuta*

Scale bar = 5 μm

a rich diversity of taxonomical discrete entities. Molecular studies on protists strongly suggest that the Cairns' perspective is correct (Beam and Himes, 1987; Blackburn and Tyler, 1987; Jerome and Lynn, 1996; John and Maggs, 1997; Medlin *et al.*, 1991; Nanney and McCoy, 1976; Robinson *et al.*, 1992; Schoilin *et al.*, 1995; van Oppen *et al.*, 1996; Zakrys *et al.*, 1996). Taxa with rigid structures visible by electron-microscopy (such as choanoflagellates, *Paraphysomonas*, or coccolithophorids) have more species than would be discriminated by light microscopy (Patterson and Larsen, 1991).

There remain considerable opportunities to clarify what constitute taxonomic entities among the flagellates. Discriminating technologies (such as molecular technologies) and the use of cultures have a role to play in this. However, the primary issue is not to demonstrate the self-evident truth that concepts based on light microscopy are less discriminating than molecular concepts, but to establish which concept is the most appropriate for discussing particular issues relating to the distribution of flagellates.

Those taxa which have been studied by more observers are more speciose than comparable groups which have been studied less intensively. This point is illustrated with the euglenids, the autotrophic species of which have been subjected to considerably more scrutiny than have heterotrophic species. The number of species in autotrophic genera greatly exceeds those in heterotrophic sister genera (*Euglena* cf. *Astasia*), and the number of species per genus of autotrophs (typically hundreds) exceeds the number of species per genus of heterotroph (typically tens) (Huber-Pestalozzi, 1955). The application of morphological criteria, of criteria only accessible by light-microscopy, and the absence of a large community of workers are all likely to lead to non-discriminatory species concepts and a sense of cosmopolitanism.

13.5 Not all communities are the same

The distribution of marine dinoflagellates and freshwater algae (Kristiansen, 1996; Taylor, 1987) is generally confounded by species concepts and environmentally induced variation in the identity of taxa (Mann and Droop, 1996). None the less, there is plenty of evidence that there are assemblages, the composition of which is determined by environmental factors. The distributions of some species of well- studied groups correlate with environmental factors. The species composition of planktonic dinoflagellates differs in cold and warm waters (Taylor, 1976; Dodge and Marshall, 1994), and that of planktonic choanoflagellates from nearshore often differs from that of offshore (Thomsen *et al.*, 1997). Ciliate species are used as bioindicators in river, lake and waste water by hydrobiologists (Foissner and Berger, 1996). In this study, arguably, the two major groupings of water column communities may be separating on the basis of temperature regime. Yet, communities from sediments in the warm latitudes do not cluster together, and we are unable to reveal a warm-water assemblage of species in those habitats. None the less, there are suggestions that there are different assemblages of benthic free-living heterotrophic flagellates. Communities from extreme habitats (anoxic, hypersaline, deep oceanic) are impoverished in species diversity but quite distinctive (Figure 13.6). In surveys of three adjacent sites at Cape Tribulation, the communities from sites with normal salinity were more similar to a site 3,000 kilometres distant, than to an adjacent site with lowered salinity.

The structure of the communities is also determined by the recent history of the habitat (Finlay and Esteban, 1998), but our surveys have not been sufficiently intense to allow us to comment on this aspect. At this time, we are unable to characterize the number and nature of assemblages in benthic habitats.

13.6 The current model

In reviewing information on the free-living heterotrophic flagellates, we (Lee and Patterson, 1998) concluded that two general insights had emerged. First is the lack of strong evidence for endemism, and second is the fact that the number of (morpho)species in this group is considerably fewer than might be expected from the diversity of bacteria in natural ecosystems, or from the abundance of individuals of flagellates in aquatic ecosystems, or from the fact that over half of the types of eukaryotic organisation include, or are exclusively composed from, the adaptive group called flagellates (Patterson, 1999).

We argued for the retention of a model in which there is not a great diversity of morpho-species of heterotrophic flagellates, probably no more than 3,000 in total (dinoflagellates and haptophytes excluded), and that most of these species have a world-wide distribution. This is similar to the suggested numbers of morphospecies of ciliates (Finlay, 1998; Finlay *et al.*, 1996a, 1996b) and of marine phytoplankton (Sournia *et al.*, 1991). We believe that only two of the extrinsic factors discussed earlier have a significant impact on our perception of the distribution of flagellates. The first is that of under-reporting, and if corrected would only serve to enhance the sense of cosmopolitanism. The second is the current operational species concept which promotes a sense of cosmopolitanism. If more discriminating species concepts are shown to circumscribe taxa with discrete ecological identities and are therefore more appropriate, then it may transpire that some or many species have a restricted distribution which cannot be explained by autecological factors.

ACKNOWLEDGEMENTS

This paper pursues an idea developed by Naja Vørs and has been supported in large part by the Australian Biological Resources Study.

REFERENCES

Azam, F., Fenchel, T., Field, J. G., Gray, J. S., Meyer-Reil, L. A. and Thingstad, F. (1983) The ecological role of water-column microbes in the sea. *Marine Ecology Progress Series*, **10**, 257–63.

Beam, C. A. and Himes, M. (1987) Electrophoretic characterization of the *Crypthecodinium cohnii* (Dinophyceae) species complex. *Journal of Protozoology*, **34**, 204–217

Bernard, C., Simpson, A. G. B. and Patterson, D. J. (2000) Some free-living flagellates from anoxic habitats. *Ophelia*, in press.

Berninger, U-G., Caron, D. A., Saunders, R. W. and Finlay, B. J. (1991) Heterotrophic flagellates of planktonic communities, their characteristics and methods of study. In *The Biology of Free-Living Heterotrophic Flagellates* (eds D. J. Patterson and J. Larsen), Systematics

Association Special Volume 45, Oxford: Clarendon Press, pp. 39–56.

Blackburn, S. I. and Tyler, P. A. (1987) On the nature of eclectic species – a tiered approach to genetic compatibility in the desmid *Micrasterias thomasiana. British Phycological Journal*, 22, 277–298.

Bowers, N., Kroll, T. T. and Pratt, J. R. (1998) Diversity and geographic distribution of riboprints from three cosmopolitan species of *Colpoda* Müller (Ciliophora: Colpodea). *European Journal of Protistology*, **34**, 341–347.

Cairns J. J. (1993) Can microbial species with a cosmopolitan distribution become extinct? *Speculations in Science and Technology*, **16**, 69–73.

Clarke, K. R. (1993) Non-parametric multivariate analyses of changes in community structure. *Australian Journal of Ecology*, **18**, 117–143.

Corliss, J. O. (1982) Numbers of species comprising the phyletic groups assignable to the kingdom Protista. *Journal of Protozoology*, **29**, 499.

DeLong, E. P., Franks, D. G., Alldredge, A. L. (1993) Phylogenetic diversity of aggregate attached vs. free-living marine bacterial assemblages. *Limnology and Oceanography*, **38**, 924–934.

Dodge, J. D. and Marshall, H. G. (1994) Biogeographic analysis of the armored planktonic dinoflagellate *Ceratium* in the North Atlantic and adjacent seas. *Journal of Phycology*, **30**, 905–922.

Ekebom, J., Patterson, D. J. and Vørs, N. (1996) Heterotrophic flagellates from coral reef sediments (Great Barrier Reef, Australia). *Archiv für Protistenkunde*, **146**, 251–272.

Ekelund, F. and Patterson, D. J. (1998) Some heterotrophic flagellates from a cultivated garden soil in Australia. *Archiv für Protistenkunde*, **148**, 461–478.

Embley, T. M. and Stackebrandt, E. (1994) The use of 16s ribosomal RNA sequences in microbial ecology. In *Molecular Approaches to Environmental Microbiology* (eds P. W. Pickup, J. R. Saunders and G. A. Codds), Oxford: Chapman and Hall, pp. 39–62.

Farmer, M. and Treimer, R. E. (1994) An ultrastructural study of *Lentomonas applanatum* (Preisig) n. g. (Euglenida). *Journal of Eukaryotic Microbiology*, **41**, 112–119.

Fenchel, T. (1987) *Ecology of Protozoa: The Biology of Free-Living Phagotrophic Protists*. Madison: Science Tech.

Fenchel, T. (1993) There are more small than large species? *Oikos*, **68**, 375–378.

Finlay, B. J. (1997) The global diversity of protozoa and other small species. *Australian Biologist*, **10**, 131–150.

Finlay, B. J. (1998) The global diversity of protozoa and other small species. *International Journal of Parasitology*, **28**, 29–48.

Finlay, B. J., Corliss, J. O., Esteban, G. and Fenchel, T. (1996a) Biodiversity at the microbial level: the number of free-living ciliates in the biosphere. *Quarterly Review of Biology*, **71**, 221–237.

Finlay, B. J. and Esteban, G. F. (1998) Planktonic ciliates species diversity as an integral component of ecosystem function in a freshwater pond. *Protist*, **149**, 155–165.

Finlay, B. J., Esteban, G. F. and Fenchel, T. (1996b) Global diversity and body size. *Nature*, **383**, 132–133.

Finlay, B. J., Esteban, G. F. and Fenchel, T. (1998) Protozoan diversity: converging estimates of the global number of free-living ciliates species. *Protist*, **149**, 29–37.

Foissner, W. and Berger, H. (1996) A user-friendly guide to the ciliates (Protozoa, Ciliophora) commonly used by hydrobiologists as bioindicators in rivers, lakes, and waste waters, with notes on their ecology. *Freshwater Biology*, **35**, 375–482.

Foissner, W., Blatterer, H. and Foissner, I. (1988) The Hemimastigophora (*Hemimastix amphikineta* nov. gen., nov. spec.), a new protistan phylum from Gondwanian soils. *European Journal of Protistology*, **23**, 361–383.

Giovannoni, S. J., Britschgi, T. B., Moyer, C. L. and Field, K. G. (1990) Genetic diversity in Sargasso Sea bacterioplankton. *Nature*, **345**, 60–62.

Hänel, K. (1979) Systematik und Ökologie der farblosen Flagellaten des Abwassers. *Archiv für Protistenkunde*, **121**, 73–137.

Huber-Pestalozzi, G. (1955) Das Phytoplankton des Süsswassers. 4. Euglenophyceen. Stuttgart: Fischer.

Jerome, C. A. and Lynn, D. H. (1996) Identifying and distinguishing sibling species in the *Tetrahymena pyriformis* complex (Ciliophora, Oligohymenophora) using PCR/RFLP analysis of nuclear ribosomal DNA. *Journal of Eukaryotic Microbiology*, **43**, 492–497.

John, D. M. and Maggs, C. A. (1997) Species problems in eukaryotic algae: a modern perspective. In *Species: the Units of Biodiversity* (eds M. F. Claridge, H. A. Dawah and M. R. Wilson), London: Chapman and Hall, pp. 83–104.

Kristiansen, J. (1996) *Biogeography of Freshwater Algae*. Dordrecht: Kluwer Academic.

Larsen, J. and Patterson, D. J. (1990) Some flagellates (Protista) from tropical marine sediments. *Journal of Natural History*, **24**, 801–937.

Lee, W. J. and Patterson, D. J. (1998) Diversity and geographic distribution of free-living heterotrophic flagellates – analysis by PRIMER. *Protist*, **149**, 229–243.

Lee, W. J. and Patterson, D. J. (2000) Heterotrophic flagellates (Protista) from marine sediments of Botany Bay, Australia. *Journal of Natural History*, in press.

Mann, D. G. and Droop, S. J. M. (1996) Biodiversity, biogeography and conservation of diatoms. *Hydrobiologia*, **336**, 19–32.

May, R. M. (1988) How many species are there on earth? *Science*, **241**, 1441–1449.

May, R. M. (1990) How many species? *Philosophical Transactions of the Royal Society of London, Series B*, **330**, 293–304.

Mayden, R. L. (1997) A hierarchy of species concepts: the denouement in the saga of species problem. In *Species: the Units of Biodiversity* (eds M. F. Claridge, H. A. Dawah and M. R. Wilson), London: Chapman and Hall, pp. 381–424.

Medlin, L. K., Elwood, H. J., Stickel, S. and Sogin, M. L. (1991) Morphological and genetic variation within the diatom *Skeletonema costatum* (Bacillariophyta): evidence for a new species, *Skeletonema pseudocostatum*. *Journal of Phycology*, **27**, 514–524.

Molina, F. I. and Nerad, T. A. (1991) Ultrastructure of *Amastigomonas bermudensis* ATCC 50234 sp. nov. – a new heterotrophic marine flagellate. *European Journal of Protistology*, **27**, 386–396.

Nanney, D. L. and McCoy, J. W. (1976) Characterization of the species in the *Tetrahymena pyriformis* complex. *Transactions of the American Microscopical Society*, **95**, 664–682.

Patterson, D. J. (1999) The diversity of eukaryotes. *American Naturalist*, **154**, Sup., 96–124.

Patterson, D. J. and Larsen, J. (1991) *The Biology of Free-Living Heterotrophic Flagellates*, Systematics Association Special Volume 45. Oxford: Clarendon Press.

Patterson, D. J. and Larsen, J. (1992) A perspective on protistan nomenclature. *Journal of Protozoology*, **39**, 125–131.

Patterson, D. J., Larsen, J., and Corliss J. O. (1989) The ecology of heterotrophic flagellates and ciliates living in marine sediments. *Progress in Protistology*, **3**, 185–277.

Preisig, H. (1979) Zwei neue Vertreter der farblosen Euglenophyta. *Schweiz. Z. Hydrol.*, **41**, 155–160.

Robinson, B. S., Christy, P., Hayes, S. J. and Dobson, P. J. (1992) Discontinuous genetic variation among mesophilic *Naegleria* isolates: further evidence that *N. gruberi* is not a single species. *Journal of Protozoology*, **39**, 702–712.

Scholin, C. A., Hallegraeff, G. M. and Anderson, D. M. (1995) Molecular evolution of the *Alexandrium tamarense* 'species complex' (Dinophyceae): dispersal in the North American and West Pacific regions. *Phycologia*, **34**, 472–485.

Silva, P. C. (1984) The role of extrinsic factors in the past and future of green algal systematics. In *Systematics of the Green Algae* (eds D. E. G. Irvine and D. M. John), Systematics Association Special Volume 27, pp. 419–433.

Simpson, A. G. B., van den Hoff, J., Bernard, C., Burton, H. and Patterson, D. J. (1997) The ultrastructure and affinities of *Postgaardi mariagerensis* Fenchel *et al.*, an unusual free-living euglenozoon. *Archiv für Protistenkunde*, **147**, 213–225.

Sournia, A., Chrétiennot-Dinet, M-J. and Ricard, M. (1991) Marine phytoplankton: how many species in the world ocean? *Journal of Plankton Research*, **13**, 1093–1099.

Taylor, F. J. R. (1976) Dinoflagellates from the International Indian Ocean Expedition. *Bibliotheca Botanica*, **132**, 1–234.

Taylor, F. J. R. (1987) *The Biology of Dinoflagellates*. Oxford: Blackwell.

Takahashi, E. (1987) Loricate and scale bearing protists from Lützow-Holm Bay, Antarctica. II. Four marine species of *Paraphysomonas* (Chrysophyceae) including two new species from the fast-ice covered coastal areas. *Japanese Journal of Phycology*, **35**, 155–166.

Thomsen, H. A. (1978) Nanoplankton from the Gulf of Elat (= Gulf of Aquaba) with particular emphasis on choanoflagellates. *Israel Journal of Zoology*, **27**, 34–44.

Thomsen, H. A., Garrison, D. L. and Kosman, C. (1997) Choanoflagellates (Acanthoecidae, Choanoflagellida) from the Weddell Sea, Antarctica, taxonomy and community structure with particular emphasis on the ice biota; with preliminary remarks on choanoflagellates from Arctic Sea Ice (Northeast Water Polynya, Greenland). *Archiv für Protistenkunde*, **148**, 77–114.

Tyler, P. A. (1996) Endemism in freshwater algae. *Hydrobiologia*, **336**, 127–135.

van Oppen, M. J. N., Klerk, H., de Graaf, M., Stam, W. T. and Olsen, J. L. (1996) Hidden diversity in the marine algae: some examples of genetic variation below the species level. *Journal of the Marine Biological Association of the UK*, **76**, 239–242.

Vørs, N. (1992) Heterotrophic amoebae, flagellates and heliozoa from the Tvärminne area, Gulf of Finland, in 1988–1990. *Ophelia*, **36**, 1–109.

Ward, D. M., Weller, R. and Bateson, M. M. (1990) 16S rRNA sequences reveal numerous uncultured microorganisms in a natural community. *Nature*, **345**, 63–65.

Zakrys, B., Kucharski, R. and Moraczewski, I. (1996) Genetic and morphological variability among clones of *Euglena pisciformis* based on RAPD and biometric analysis. *Archiv für Hydrobiologie Supplement*, **114**, 1–21.

Chapter 14

Cosmopolitan haptophyte flagellates and their genetic links

L. K. Medlin, M. Lange, B. Edvardsen and A. Larsen

ABSTRACT

Certain cosmopolitan haptophyte species appear to have a more restricted distribution when examined with molecular markers than previously reported, if morphological markers alone are used to delimit the species and their distribution. These data suggest that cryptic species may be very common among taxa reported as cosmopolitan. The genetic relatedness of strains below the species level reflects their geographic origin. In some instances molecular markers, coupled with flow cytometric determinations of genome size, have indicated where species limits may be drawn in the absence of strong morphological markers separating the taxa. Future morphological investigations may identify features that can be used to separate the taxa that may have been overlooked before or not considered phylogenetically informative. Moreover, flow cytometric studies indicate that at least in the Haptophyta, many species may be linked in a heteromorphic life-cycle, further complicating our interpretation of their distribution in space and time. Our studies have demonstrated that a shift from a morphological to a phylogenetic species concept might alter the taxonomy of certain haptophytes, and also raises the question whether any of these species may have a cosmopolitan distribution.

14.1 Introduction

Flagellates are common in both the micro and nano fractions of the plankton, as well as being present in many benthic habitats. They exhibit a limited range of morphological markers that make their identification difficult to specialists untrained in taxonomy, and often require electron microscopy for definitive answers to species designations. A survey of some of the most common photosynthetic flagellates indicates that they can be assigned to a limited number of unrelated algal classes, but can be collectively referred to as microalgae (Margulis *et al.*, 1990). Primary distinguishing features for these different phytoflagellate classes, apart from their pigmentation, are their flagellar arrangement/apparatus and general body shape/covering. Within each class there are remarkable similarities in body shape, which undoubtedly reflect selection pressures on the group. Because of their limited morphological markers and small size, many of the more easily recognized species have been reported as cosmopolitan.

Although some species may indeed be cosmopolitan, such reports may be more commonly a result of insufficient data to resolve taxonomic differences. Mistaken cosmopolitanism can seriously underestimate biodiversity and our understanding of how species distributions change in time and space. Genetic studies from the marine ecosystem indicate that many marine taxa may be more genetically diverse than terrestrial or freshwater counterparts (van Oppen *et al.*, 1996; Gray, 1997). Cryptic species are also common and widespread across many taxonomic groups (Andersen *et al.*, 1998; Knowlton, 1993). These observations would support hypotheses that similar shapes evolve as solutions to similar problems in aquatic environments, and that when a fitness of form is achieved, molecular and morphological rates of evolution become widely disparate (Knowlton, 1993; Nanney *et al.*, 1998). An alternative approach has been taken by Tilman (1982) who maintains that as species optimize their ability to compete for non-limiting nutrients to become dominant (cosmopolitan) members of the ecosystem, they minimize phenotypic heterogeneity.

Life-cycle strategies of the flagellates, especially the phytoflagellates, are often neglected aspects of their biology and other factors affecting biodiversity estimates. Heteromorphic life-cycles are common features of the haptophytes (Billard, 1994). Often different life-cycle stages have been recognized as different species (Thomsen *et al.*, 1991). Different environmental conditions may favour one life-cycle stage over another (Davidson and Marchant, 1992; Edvardsen and Vaulot, 1996). Without a good knowledge of life-cycle strategies for these minute organisms, we may overestimate their biodiversity, and limit our knowledge of their distribution and the factors inducing the change from one stage into another.

Although our knowledge of phytoflagellate diversity and population structure has been hampered by their small size and paucity of morphological markers, and the difficulty of obtaining samples for long-term seasonal studies in open ocean environments, many more phytoflagellates are being brought into culture, often through the use of flow cytometry, so that we have greater opportunities to study their life histories and their genetic relatedness. The advent of molecular biological techniques has greatly enhanced our ability to analyse phytoplankton populations. Hypotheses regarding cosmopolitanism and species concepts can be tested, as molecular markers are developed to examine diversity at and below the species level.

A variety of molecular techniques can be employed to dissect the intricate problems surrounding genetic relatedness of organisms at all taxonomic levels. At higher levels, slower evolving genomic regions are commonly used, for example coding regions, such as the ribosomal RNA genes (small subunit or SSU, large subunit or LSU) and the gene encoding the large subunit of RUBISCO (ribulose 1,5-bisphosphate carboxylase). At lower taxonomic levels, non-coding spacer regions may be more appropriate to detect cryptic or sibling species (such as the internal transcribed spacer region (ITS) within the ribosomal gene cluster, the spacer separating the large and small subunits of RUBISCO (only in the chlorophyll *c* and the red algae but not in the dinoflagellates, see Palmer, 1995), the spacer between the pet B and D genes (Urbach, 1994), the spacer between the trnT and the trnF genes (Taberlet *et al.*, 1991), and the introns in the calmodulin genes (Côrte-Real *et al.*, 1994). There can be multiple copies of the calmodulin gene, and each copy may have several alleles: both cases can result in introns of different length.

In general, molecular analysis of multiple isolates of a single species tends to show that the isolates are related by geographic origin (Scholin *et al.*, 1994; Bakker *et al.*, 1995; Medlin *et al.*, 1996; Larsen and Medlin, 1997), but sometimes the geographic groupings uncover polyphyletic or paraphyletic taxa (Scholin *et al.*, 1994; Bakker *et al.*, 1995; van Oppen *et al.*, 1996; Medlin *et al.*, 1997), which suggest that cryptic species are probably present. At the population level, genetic variation can be measured by allozymes or by fingerprinting techniques such as restriction fragment length polymorphisms (RFLPs, AFLPs), randomly amplified polymorphic DNA (RAPDs) and variable numbers of tandem repeats (VNTRs), to produce a molecular marker unique for an individual.

The focus of this chapter is on the genetic diversity among selected species of the Prymnesiophyceae (Haptophyceae). These microalgal cells possess two equal to subequal, smooth flagella, with a third structure termed the haptonema placed between the flagella. Many genera are covered by organic scales, which can be mineralized (for example, in the coccolithophorids). We will concentrate on three major haptophyte genera that have cosmopolitan species and that are involved in harmful algal events.

14.2 *Phaeocystis*

Phaeocystis Lagerheim is a ubiquitous member of the marine phytoplankton, occurring from polar to tropical regions. It has a complex polymorphic life-cycle with both colonial and flagellate cells: each stage can reach bloom proportions to dominate marine ecosystems, although colonial stages are predominant in nutrient-rich waters. *Phaeocystis* plays a significant role in the production and accumulation of large amounts of dissolved organic compounds (DOC) (Lancelot *et al.*, 1987); in the global carbon cycle, especially in nutrient rich coastal and polar regions (Lancelot *et al.*, 1987; Smith *et al.*, 1991); and also contributes to the global sulphur cycle through the release of substantial amounts of dimethylsulphonio-propionate (DMSP), the precursor of dimethylsulphide (DMS) (Keller *et al.*, 1989; Liss *et al.*, 1994). Although DMSP primarily functions internally as an osmoregulatory solute in marine algae (Iverson *et al.*, 1989; Kirst *et al.*, 1991), as a cryoprotectant in ice algae (Kirst *et al.*, 1991) and as a methyl donor (Ishida, 1968), when it is released from the cells, it has a major impact on climate regulation (Charlson *et al.*, 1987). *Phaeocystis* may play another important ecological role, with its production of UV-B absorbing compounds (Davidson and Marchant, 1992).

Most life-cycle studies of *Phaeocystis* have been performed on *P. globosa* Scherffel (see review by Rousseau *et al.*, 1994). Free-living flagellate and non-flagellate cells alternate with a mucilaginous colony stage containing non-motile coccoid cells (Baumann *et al.*, 1994; Kornmann, 1955; Rousseau *et al.*, 1994). There is strong evidence that the microzoospores are haploid and may be involved in sexual reproduction: the non-motile free-living cells, macrozoospores and colonial cells are diploid (Vaulot *et al.*, 1994). However, the pathway and the conditions transforming one cell type into another are not understood.

Phaeocystis has caused many taxonomic problems because of its polymorphic life-cycle and extreme morphological variability. A cold-water form known as *Phaeocystis pouchetii* occurs in the northern hemisphere, and forms lobed colonies with cells arranged in packages of four. *Phaeocystis globosa*, a warm water form, forms spherical colonies with homogeneously distributed cells within the gelatinous colony-matrix. Most workers separated the two forms as distinct species until Kornmann (1955) concluded that *P. globosa* cell types were juvenile *P. pouchetii*. Others have also concluded that colony morphology is an unreliable specific feature (Kashkin, 1963; Chang, 1983). Instead, ultrastructural features of the flagellates are commonly used for species identification. Sournia (1988) reviewed the diagnostic features of *Phaeocystis* and recognized only two of its nine species (known at that time) as valid: the colony forming *P. pouchetii* (Hariot) Lagerheim, with *P. globosa* Scherffel as a later synonym and *P. scrobiculata* Moestrup known only from the flagellate stage. Sournia recommended marine ecologists to report colonial stages as *P. pouchetii* or *Phaeocystis* sp. to avoid confusion. Because of the world-wide occurrence of the colonial *Phaeocystis*, Baumann *et al.*, (1994) expressed doubt that a single species could acclimatize to temperatures ranging from -2°C to > 20°C. Reinvestigation of colony shapes in both juvenile and older stages of *P. pouchetii* and *P. globosa*, as well as studies on their temperature and light requirements, suggested their separation into species (Baumann *et al.*, 1994).

A third colonial species from Antarctic waters was recognized by Baumann *et al.* (1994). *Phaeocystis antarctica* Karsten was described at the turn of the century (Karsten, 1905) and had a colony morphology similar to *P. globosa* (Larsen and Moestrup, 1989), whereas temperature tolerances were similar to those of *P. pouchetii*. The separation of *P. globosa* and *P. antarctica* as distinct species is further supported by different pigment spectra (Buma *et al.*, 1991) and differences in DNA content (Vaulot *et al.*, 1994). Recently two new species of *Phaeocystis* from the Mediterranean Sea have been discovered, using both ultrastructural features of the flagellates and sequence data from their small subunit ribosomal RNA gene (Zingone *et al.*, 1999). One isolate forms loose cell aggregates without a definite shape and a marked external envelope, whereas the other isolate has only been observed as flagellate stages (Zingone *et al.*, 1999).

We have compared sequence data from the small subunit ribosomal RNA gene from fifteen colonial and three unicellular strains of *Phaeocystis* to resolve the issue of the number of colony-forming *Phaeocystis* species and the cosmopolitan distribution of its species, and to determine the relationship of *Phaeocystis* to other prymnesiophyte genera.

The Haptophyceae can be divided into two subclasses, the Pavlovophycidae and the Prymnesiophycidae, a division which is well supported by both molecular and morphological evidence. Among the subclass Prymnesiophycidae, the clade containing *Phaeocystis* is monophyletic with high bootstrap support and is sister taxon to all remaining species within the Prymnesiophycidae (Figure 14.1). Our analysis supports the separation of the three colony-forming species: *P. globosa, P. antarctica* and *P. pouchetii* (Figure 14.1). The absolute number of nucleotide differences separating the colony-forming strains of *Phaeocystis* are comparable to species differences within the protozoan *Tetrahymena* (zero to thirty-three) (Sogin *et al.*, 1986), the diatom *Skeletonema* (eleven) (Medlin *et al.*, 1991) and the eustigmatophyte *Nannochloropsis* (one to thirty-eight) (Andersen *et al.*, 1998), whereas more than fifty nucleotide differences separate the two new Mediterranean species from the other *Phaeocystis* species.

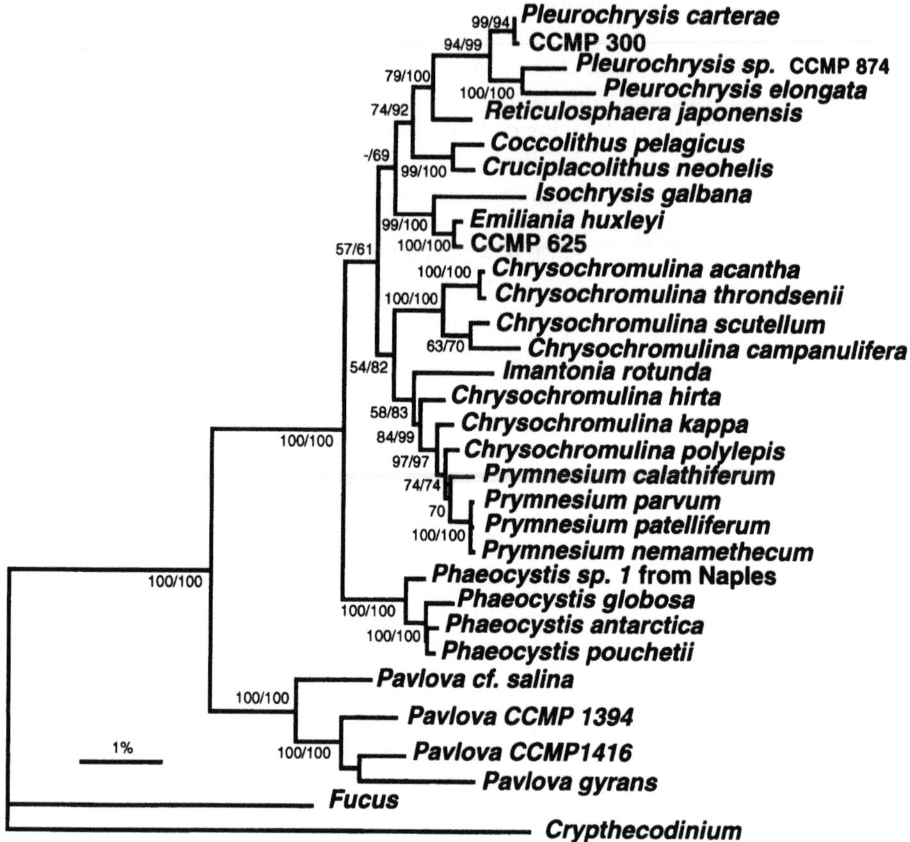

Figure 14.1 Phylogenetic reconstruction of the evolution of the Haptophyta inferred from a maximum likelihood analysis using 18S-rDNA sequence data. The scale bar corresponds to 1 base change per 100 nucleotide positions. Bootstrap values (100 replicates) from a maximum parsimony/neighbour-joining analysis are placed to the left of the nodes that they represent

A more detailed examination of the genus using multiple strains of each species revealed two species complexes (Figure 14.2). A cold-water complex contained two species: *P. pouchetii*, occurring in the Arctic Ocean and the northern part of the Atlantic, and *P. antarctica* from the Antarctic. The warm-water complex contained strains of *P. globosa*. Within both *P. globosa* and *P. antarctica* isolates, intraspecific sequence variation in the SSU rRNA gene ranged from 0 to 5 nucleotides (Medlin *et al.*, 1994; Lange, 1997). The separation of the three colony-forming *Phaeocystis* species is supported by differences in DNA content, pigment composition and colony morphology (Medlin *et al.*, 1994; Vaulot *et al.*, 1994). Although *P. globosa* strains cannot be distinguished into species using SSU rDNA sequence data, Vaulot *et al.* (1994) found three different genome sizes within *P. globosa* isolates as measured by flow cytometry. These findings suggest the occurrence of cryptic species within the *P. globosa* complex.

To obtain better resolution among our isolates, we explored the use of spacer regions to separate closely related *Phaeocystis* species and populations. The ITS1

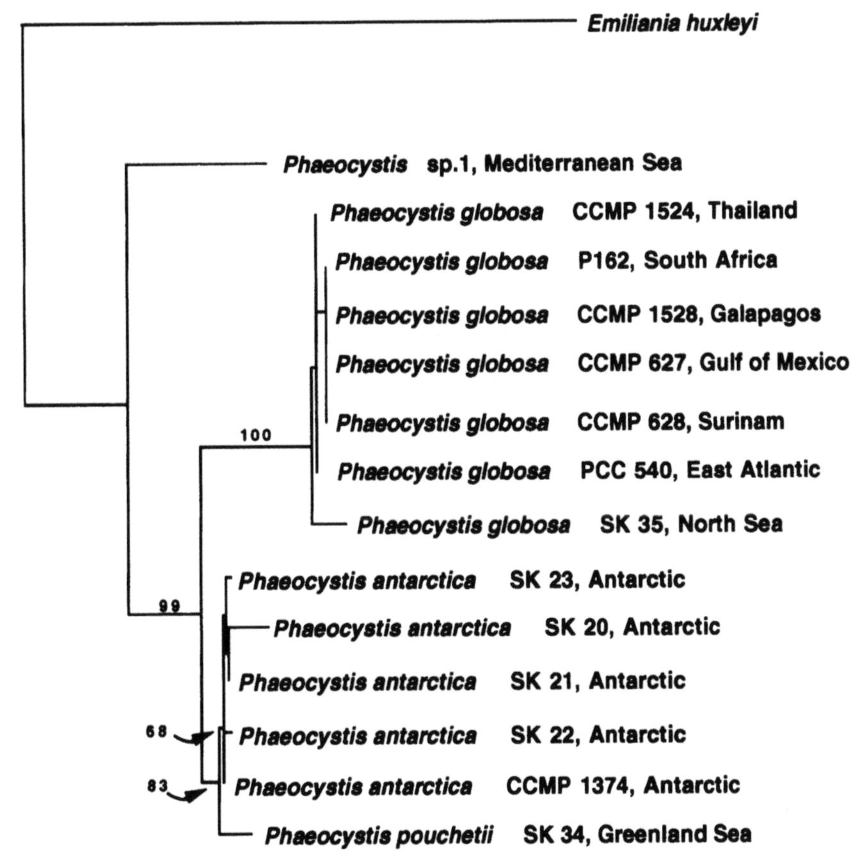

Figure 14.2 Neighbour-joining tree of *Phaeocystis* species/strains and *Emiliania huxleyi* based on 18S-rDNA distances. The scale bar below the tree corresponds to 1 base change per 100 nucleotide positions. Bootstrap values (100 replicates) are placed at the nodes that they represent

region of the ribosomal gene cluster was well suited to resolve not only phylogenetic relationships among both closely related *Phaeocystis* species, but also population level differences within *P. antarctica*. Similar sizes of ITS1 amplification products were found by agarose gel electrophoresis from all *Phaeocystis* strains (data not shown). Only a selection were sequenced because of the difficulty in obtaining clean sequence from this region, which may have been caused by the high G–C content and the strong folding of the secondary structure of this region (Lange, 1997). These included nine *P. antarctica* strains collected from different locations around Antarctica, one *P. pouchetii* strain from the Greenland Sea, one *P. globosa* strain from the North Sea, two tropical *P. globosa* strains from Surinam and Palau, and *Phaeocystis* sp. 1 from the Mediterranean (Figure 14.3). Only those regions of ITS1

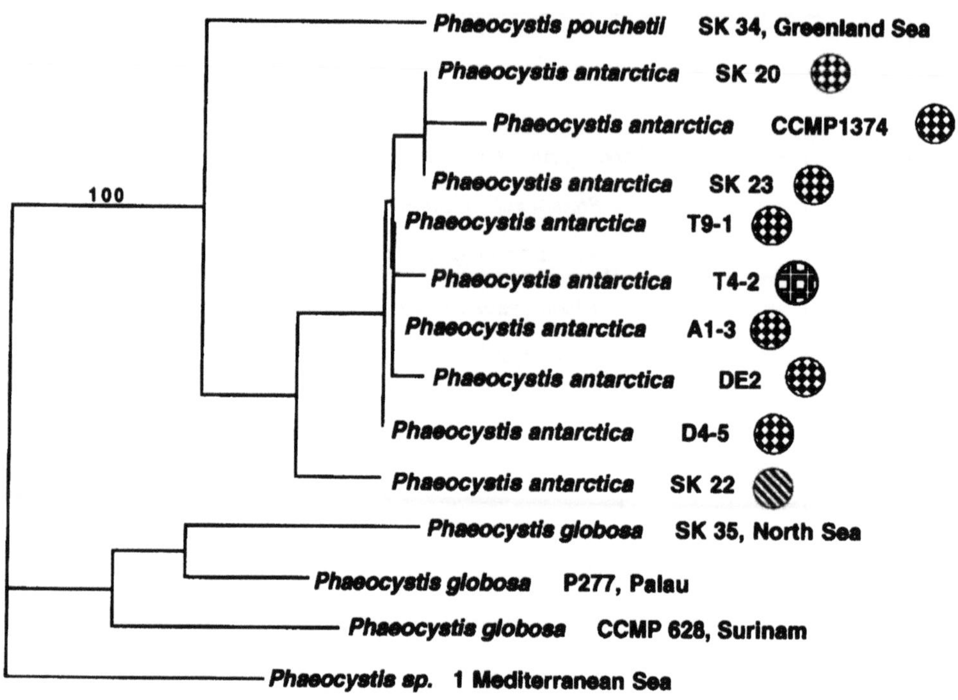

Figure 14.3 Neighbour-joining tree based on distances calculated from ITS1 sequence data from *Phaeocystis globosa* (strains SK 35, CCMP 628, P277), *Phaeocystis pouchetii* (SK 34), *Phaeocystis antarctica* (strains SK 20, SK 22, SK 23, CCMP 1374, DE2, T9-1, D4-5, A1-3, T4-2) and *Phaeocystis* sp. 1 from the Mediterranean. The distance bar below the tree corresponds to 1 base change per 100 nucleotides. Bootstrap values (100 replicates) are placed at the nodes that they represent. Only the cold water clade is well-supported, further indicating the likelihood that multiple species are found within the morphotype assigned to *Phaeocystis globosa*. The patterned circles represent different haplotypes whose origin can be found in Figure 14.4

that were well aligned among all *Phaeocystis* spp. were used for phylogenetic analysis.

The unicellular *Phaeocystis* sp. from the Mediterranean Sea was used as an outgroup to examine relationships among the cold-water species complex containing *P. antarctica* and *P. pouchetii* and *P. globosa*. *Phaeocystis antarctica* and *P. pouchetii* were related with high bootstrap support for all strains assignable to *P. antarctica*. Large sequence variations were found in this region among *P. globosa* strains from different geographical regions. Low bootstrap support for the relationships among *P. globosa* strains support the uniqueness of the strains in this species complex. One strain of *P. antarctica* (SK 22), isolated within the Antarctic Counter Current (ACC) was clearly related to *P. pouchetii*. The remaining eight *P. antarctica* strains isolated from the water masses defined by the Antarctic continental boundary currents (Figure 14.4) showed a high degree of similarity. Nevertheless, molecular analyses could separate *P. antarctica* strains into at least three different populations. Three strains from Prydz Bay (D4-5, T9-1, A1-3) differed only in a few base substitutions or deletions/insertions events from strains DE2 and CCMP 1374, and from the two identical *Phaeocystis* strains from the Weddell Sea (SK20; SK23). Strain T4-2 contained more insertions/deletions than the above mentioned cluster of *P. antarctica* strains.

Sequence data from the SSU rRNA gene have been used to explain the present day distributions of *Phaeocystis* species and to reconstruct their biogeographical history (Medlin *et al.*, 1994; Lange, 1997). These phylogenetic analyses suggest that the unicellular Mediterranean species diverged prior to the separation of the three colony-forming species. The warm-water species, *P. globosa,* diverged prior to the separation of the two cold-water forms, *P. pouchetii* and *P. antarctica.* Morphological features of the flagellate stages of *P. antarctica* and *P. pouchetii* are nearly identical, although their biogeographical distribution does not overlap. The morphological data and the SSU rDNA sequence data strongly suggest that both cold-water species evolved from a recent common warm-water ancestor. *P. antarctica* retained the colony morphology of the warm-water ancestor, whereas the colony morphology of *P. pouchetii* diverged.

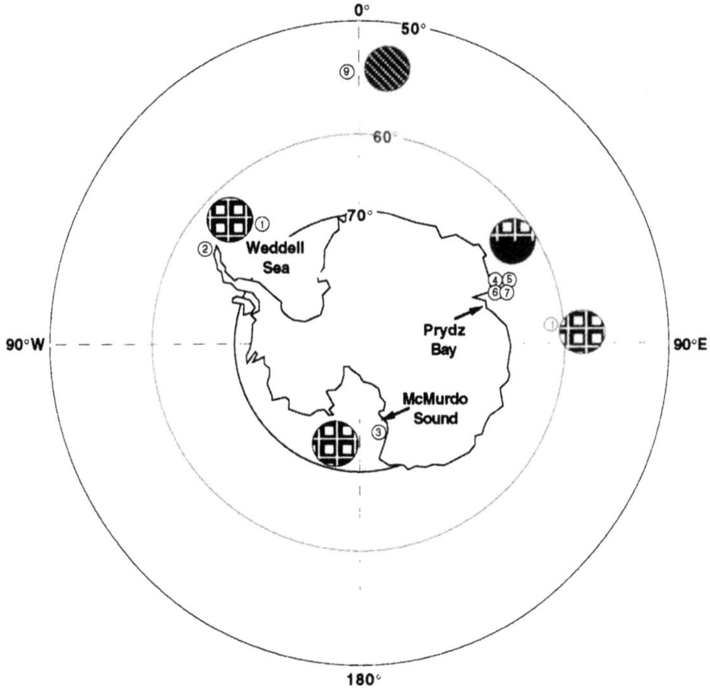

Figure 14.4 Locations of *Phaeocystis antarctica* strains from which ITS1 sequences were analysed (encircled numbers). 1 = A1–3, 2 = D4–5, 3 = DE2, 4 = T4–2, 5 = T9–1, 6 = SK 22, 7 = SK 20, 8 = SK 23, 9 = CCMP 1374. The locations of different haplotypes is indicated by the different patterns of the circles

A model that involves both vicariant and dispersal events has been invoked to explain the divergence of both polar species (Medlin *et al.*, 1994). Based on the assumption that 1 per cent base differences correspond to a divergence time of 25 million years as found for higher plants (Ochman and Wilson, 1987), Dasycladales (Olsen *et al.*, 1994) and diatoms (Kooistra and Medlin, 1996), the divergence times of *Phaeocystis* species were correlated with temperature changes in Antarctic surface water since the Cretaceous, and with important earth historical climatic and tectonic events (Medlin *et al.*, 1994; Lange, 1997). Following the interpretation of Medlin *et al.* (1994, 1995), *Phaeocystis* had a world-wide distribution during the warm Eocene period; the ancestor of *P. antarctica* was separated from warm-water *Phaeocystis* ancestral populations as a result of a vicariant event (the Drake passage opened and the ACC system was formed), and evolved as the first *Phaeocystis* cold-water species. The introduction of ancestors of *P. pouchetii* into the Arctic Sea was interpreted as a dispersal event, resulting from the opening of the Svalbard-Greenland Sea, and the migration of an ancestral *P. antarctica* populations from the southern to the northern hemisphere during colder times 15–10 million years ago. Such dispersal event from the south to the north during colder climate periods have allowed many populations to survive the crossing of equatorial waters (Crame, 1993).

We have further shown that populations of *P. antarctica* within the continental boundary water masses in the Antarctic appear to be well-mixed. Currents around the Antarctic continent keep the waters well mixed and effectively act as a barrier to population structure. Strain SK22 isolated within the ACC was clearly different from those populations in the continental boundary water masses, and was more closely related to *P. pouchetii* than it was to the other *Phaeocystis* strains isolated from the Antarctic continental water masses. Such data provide additional evidence to support the hypothesis of dispersal events from south to north to form disjunct bipolar populations, which evolved into separate species.

14.3 *Chrysochromulina*

All *Chrysochromulina* species described to date, except for four from freshwater lakes, are marine. They are a ubiquitous component of the marine nanoplankton in all seas, and may form a large part of the nanoplanktonic biomass in both coastal and oceanic waters (Estep *et al.*, 1984; Estep and MacIntyre, 1989; Thomsen *et al.*, 1994). Most often many *Chrysochromulina* species co-occur in the same water-mass, and normally in low cell concentrations (10^3–10^5 cells l^{-1}). In coastal waters adjacent to the North Atlantic, however, *Chrysochromulina* species have formed several harmful algal blooms (Edvardsen and Paasche, 1998).

At present, about fifty *Chrysochromulina* species have been described; forty-seven were listed by Jordan and Green (1994), to which can be added *C. papillata* Gao *et al.*, *C. quadrikonta* Kawachi et Inouye, *C. throndsenii* Eikrem, *C. scutellum* Eikrem et Moestrup, and *C. fragaria* Eikrem et Edvardsen. More than twenty species/forms await formal description (W. Eikrem, pers. comm.) and the true number of species may exceed 100 (Thomsen *et al.*, 1994). The species vary in body shape and dimensions, the cell length varying from 2–31 µm (Moestrup and Thomsen, 1995). They possess two, usually homodynamic, smooth flagella and a haptonema. The length of the haptonema also varies greatly, being in the range 3–180 µm. The cells are covered by two or more types of organic scales, whose morphology is the main character for distinguishing species, and which can usually be seen only in the electron microscope. The length of the haptonema relative to either the flagellar length or the cell diameter, the cell size and form, and the swimming pattern, are additional characters used in species determination that can be seen under the light microscope.

Information on haptophyte life-cycles is fragmentary, and sexual reproduction has been established only for a small number of species (Billard, 1994). In general, haptophyte life-cycles embrace an alternation between motile and non-motile stages. Non-motile, naked, amoeboid cells were observed in old cultures of many *Chrysochromulina* species (Parke *et al.*, 1955, 1956, 1958, 1959). In *C. polylepis*, however, two motile cell types have been described (Paasche *et al.*, 1990; Edvardsen and Paasche, 1992) that may be stages in a haplo-diploid life-cycle (Edvardsen and Vaulot, 1996). Had they been observed singly, it is highly likely that they would have been described as separate species.

As early as 1955, Parke and co-workers recognized the wide morphological variation within the genus *Chrysochromulina*, but at that time they did not want to erect genera on characters that could not be seen in the light microscope. Genetic analyses (Inouye, 1997; Medlin *et al.*, 1997) of *Chrysochromulina* species and related genera support morphological data (Birkhead and Pienaar, 1995) suggesting that *Chrysochromulina* is not a natural grouping.

Analyses of the nuclear-encoded SSU rRNA gene indicate that *Chrysochromulina* is paraphyletic and can be divided into two clades (Figure 14.1; Simon *et al.*, 1997; Medlin *et al.*, 1997; Edvardsen *et al.*, unpubl.). Based on SSU rDNA data, some species (*C. polylepis* Manton et Parke, *C. kappa* Parke et Manton, *C. hirta* Manton) are more closely related to *Prymnesium* species than to other *Chrysochromulina* species and comprise clade 1 (Figure 14.1). The *Chrysochromulina* and *Prymnesium* species in clade 1 have a spherical or oblong cell shape, and usually a haptonema that is shorter than their flagella. Many species in clade 1 have a compound root associated with flagellar root R1, a feature commonly found in coccolithophorids. Cultures of *P. parvum*, *P. patelliferum*, *P. nemamethecum* and *C. polylepis* were found to be toxic (review by Edvardsen and Paasche, 1998), and they all fall into clade 1.

All other *Chrysochromulina* species sequenced to date (*C. acantha* Leadbeater et Manton, *C. campanulifera* Manton et Leadbeater, *C. scutellum* Eikrem et Moestrup, and *C. throndsenii* Eikrem fall into the second clade (clade 2). These *Chrysochromulina* species have a saddle-shaped cell, and some have cup-shaped scales. The haptonema is longer or equal in length to the flagella. The species of clade 2 have a simple flagellar root R1 containing few microtubuli. The *Chrysochromulina* species of clade 2 were found to be non-toxic to *Artemia* (Crustacea) nauplii. These molecular and morphological data indicate that a revision of the taxonomy of *Chrysochromulina* is needed.

With over 100 base substitutions separating the most distantly related *Chrysochromulina* spp., the SSU rRNA gene is well suited to resolve species level differences in this genus. In addition, analyses of the first internal transcribed spacer

(ITS1) rDNA from eleven different *Chrysochromulina* species showed that this region could be used to resolve genetic relatedness at and below the species level in *Chrysochromulina* (Edvardsen and Medlin, 1998; Edvardsen, unpubl.) Interspecific variation in the ITS1 regions was very high, with about 50 per cent divergence between closely related species. Interspecific length variation in ITS1 was so great that it prevented reliable sequence alignment (Figure 14.5).

Because of the uncertainty of the identity and function of the alternate and authentic cell types found in *C. polylepis*, more detailed analyses of multiple strains of *C. polylepis* have been conducted. Flow cytometric analyses of the DNA-content per cell indicated that the authentic cells are haploid and the alternate cells are either haploid or diploid (Edvardsen and Vaulot, 1996). Edvardsen and Vaulot suggested that the authentic and alternate cell types of *C. polylepis* form parts of a haplo-diploid life-cycle where the haploid cells possibly may function as gametes and the diploid stage possibly may be the result of syngamy (Figure 14.6). The genome size and ploidy levels were examined in seventeen different *Chrysochromulina* species (Edvardsen, unpubl.). Two ploidy levels, assumed to represent haploid and diploid cells, were also found in *C. ericina, C. hirta and C. kappa* in addition to *C. polylepis*. These species all fall into clade 1 in the SSU gene tree together with *C. polylepis* and *Prymnesium* spp.

Nucleotide sequences from SSU rDNA and ITS1 rDNA from several strains of C.

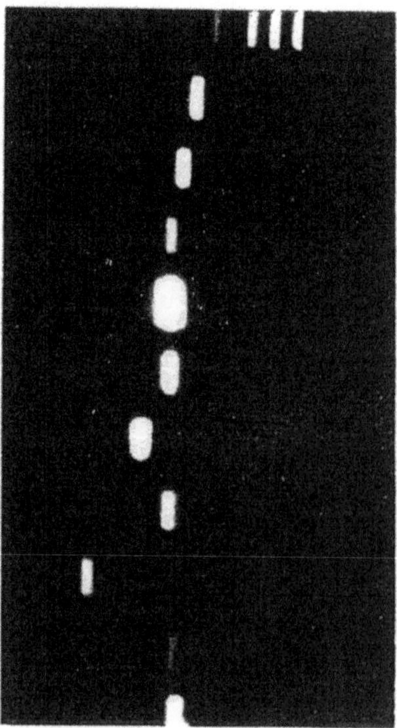

Figure 14.5 Agarose gel with ITS1 fragments from *Chrysochromulina* species, obtained by PCR using primers ITS1+ITS2. I DNA digested with PHIX144 and HaeIII were used as size markers

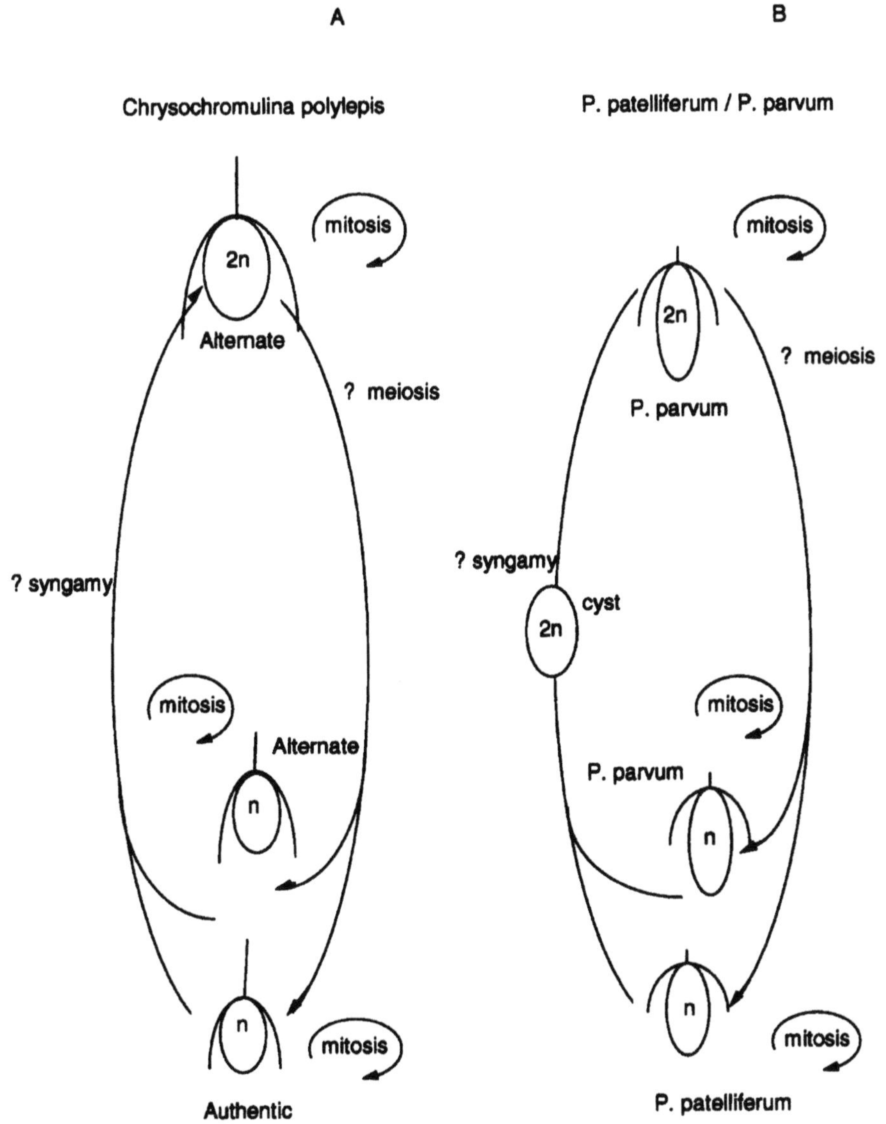

Figure 14.6 Proposed life-cycle linking (A) the authentic and alternate cell types of *C. polylepis* and (B) *P. parvum* to *P. patelliferum*

Sources: A modified from Edvardsen and Vaulot, 1996; B redrawn from Larsen and Edvardsen, 1998; reproduction courtesy of *Phycologia*

polylepis and from other *Chrysochromulina* species were compared to test further if the two cell types of *C. polylepis* belonged to the same species. Both cell types of *C. polylepis* strains had identical SSU rDNA. With the ITS1 region, both cell types from the same locality had identical sequence, whereas another strain, *C.* cf. *polylepis* from Britain, was sufficiently different from other *C. polylepis* strains to warrant a separate species or variety designation (Edvardsen and Medlin, 1998).

Another non-coding region, an intron in the nuclear-encoded calmodulin gene CaM-1, was used to compare different isolates of *C. polylepis* (Edvardsen and Medlin, 1998). Similar fragment lengths for six different clones of *C. polylepis* from South Norway were obtained from PCR amplification of a region including this intron (Figure 14.7). Different band patterns were obtained from *C. polylepis* isolated from Sweden and *C.* cf. *polylepis* from Britain. Because strain B152j of *C. polylepis* is diploid, we have interpreted the single band present in this strain as representing the homozygous condition for the intron, but if multiple copies of the calmodulin gene are present, then the third introns are all of the same length. Such might be the case for CCMP 287, which is haploid, but displays two different band lengths. However, this region has not been sequenced and therefore our interpretation of the banding patterns is speculative. This molecular marker appears somewhat better suited for studies at the population level in *Chrysochromulina* species because it can resolve geographically related clusters of strains.

All of these molecular markers have proven valuable in identifying or verifying a link between morphologically different, alternating forms of *Chrysochromulina* and in segregating populations by geographic or taxonomic affinities.

14.4 *Prymnesium*

Representatives of the genus *Prymnesium* Massart ex Conrad have been recorded from freshwater, brackish waters, low-salinity inshore waters and high-salinity offshore localities (for example Carter, 1937; Valkanov, 1964; Starmach, 1968; Green *et al.*, 1982; Kaartvedt *et al.*, 1991; Throndsen, 1969, 1983; Billard, 1983; Chang, 1985; Pienaar and Birkhead, 1994). Of the ten recorded *Prymnesium* species, some are of doubtful status, but the well-documented *Prymnesium calathiferum* Chang et Ryan, *Prymnesium parvum* Carter and *Prymnesium patelliferum* Green, Hibberd et al. are reported as toxic (Moestrup, 1994; Edvardsen and Paasche, 1998). *Prymnesium parvum* and *P. patelliferum* have been responsible for fish kills world-wide (Moestrup, 1994; Edvardsen and Paasche, 1998). Such harmful events have occurred mainly in brackish water localities (Edvardsen and Paasche, 1998). One harmful bloom of *P. calathiferum*, from a marine locality, is reported (Chang, 1985).

The genus *Prymnesium* is characterized by an oval cell shape, short haptonema and two heterodynamic flagella (Jordan and Green, 1994; Pienaar and Birkhead, 1994). The majority of these are described on the basis of the organic scale morphology surrounding the cell surface, but seen only with electron microscopy. However, four of the ten species, *P. czosnowskii* Starmach, *P. gladiociliatum* (Büttner) Jordan et Green, *P. minutum* Carter and *P. saltans* Massart ex Conrad, were described prior to the advent of electron microscopy and should be re-examined (Jordan and Green, 1994). Of these, *P. saltans* has been reported as toxic (Conrad, 1941).

Alternating life-cycle stages in *Prymnesium* include the occurrence of cysts (Carter, 1937; Conrad, 1941; Pienaar, 1980; Green *et al.*, 1982; Guo *et al.*, 1996), and a change between flagellate and non-motile cells (Padan *et al.*, 1967). The latter was observed when different types of growth media were provided (liquid and solid). Occurrences of cysts with walls composed of layers of scales (Pienaar, 1980) have been observed in *P. parvum* and *P. minutum* (Carter, 1937), *P. saltans* (Conrad,

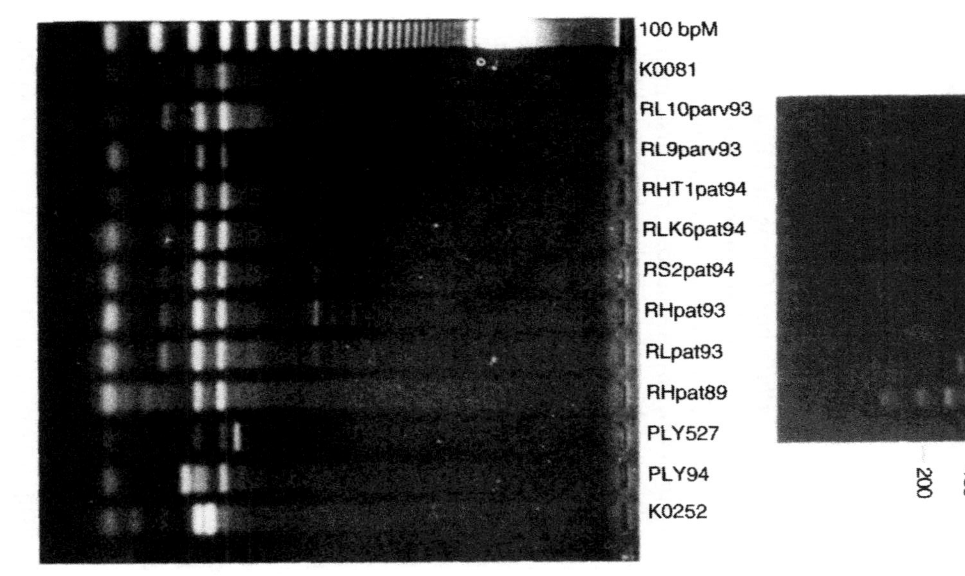

Figure 14.7 Agarose gel showing CaM-1band patterns obtained (A) from various isolates of *Chrysochromulina polyepis* using primers CAD 2 and CAD 3 (Edvardsen and Medlin, 1998, reproduction courtesy of *Phycologia*) and (B) from *Prymnesium* isolates using primers CAD 1, CAD 2 and CAD 3, in a nested PCR reaction (Côrte-Réal *et al.*, 1995) (Larsen and Medlin, 1997, reproduction courtesy of *Journal of Phycology*). 100 base pair ladder (Pharmacia) and l DNA digested with HindIII and EcoRI were used as size markers. Strain origin is as follows:
(A) *Chrysochromulina polylepis* from Norway = A, B11, B152J, B1511, K, S; from Sweden = CCMP 287; from the English Channel = PLY 200; (B) *Prymnesium parvum* from Norway = K0081, RL10parv93, RL9parv93; *P. patelliferum* from Norway = RLT1pat94, RLK6pat94, RS2pat94, RHpat93, RLpat93; *Prymnesium parvum* from England = PLY94; *P. patelliferum* from the English Channel = PLY527; *P. patelliferum* from Australia = K0252

1941) and *P. patelliferum* (Green *et al.*, 1982). Neither non-motile cells nor cysts have traditionally been coupled to any kind of sexual cycle in *Prymnesium*.

Prymnesium parvum and *P. patelliferum* are undoubtedly the most widely recorded species in the genus. They are, however, identical when examined in LM, but are defined as different species using electron microscopy on the basis of minor differences in the morphology of the organic scales surrounding their cell surfaces (Green *et al.*, 1982). In *P. parvum* the organic scales have a radial pattern on their proximal face and concentric rings on their distal face, whereas both scale-faces of both layers have a radial pattern in *P. patelliferum*. In addition, the scales of *P. patelliferum* have a central thickening, which cannot be found in *P. parvum* scales. The rims of the outer scales in *P. patelliferum* are broad and upright, whereas the rim of the scales in the inner layer is broad and inflexed in *P. parvum*. The rim of the inner *P. patelliferum* scales and the outer *P. parvum* scales is narrow and inflexed (Green *et al.*, 1982). Investigations of the flagellar root system (Green and Hori, 1994), autecology and toxicity (Larsen *et al.*, 1993; Larsen and Bryant, 1998) have not revealed further differences between *P. parvum* and *P. patelliferum*.

The SSU rDNA region of *P. parvum* and *P. patelliferum* differs by only two base pairs (Figure 14.1). In contrast, twenty-seven nucleotide differences were found between these two species and *Prymnesium calathiferum* (Chang and Ryan, 1985), which has an outer layer of far more elaborat scales than either *P. parvum* or *P. patelliferum*. In order to obtain more information about genetic variation within and between *P. parvum* and *P. patelliferum*, we sequenced the first internal transcribed spacer region (ITS1) of several strains, isolated from different geographical locations (Larsen and Medlin, 1997). Our analyses showed that the various strains were related by geographical origin, rather than by species affiliation (Figure 14.8; Larsen and Medlin, 1997). These results were supported by length variations in banding patterns produced by PCR of an intron within a calmodulin gene (Figure 14.7; Larsen and Medlin, 1997). The two to three major PCR products between 280 and 450 bp were assumed to be the major fragment lengths corresponding to the third intron in the calmodulin gene. The first nine lanes with identical banding patterns were obtained from *P. parvum* and *P. patelliferum* isolated from Norway and Denmark. Because *P. patelliferum* from Norway is haploid (Larsen and Edvardsen, 1998), we have assumed that there are multiple copies for calmodulin gene, which have resulted in introns of different lengths. The last three lanes show different banding patterns, which were also presumed to represent the multiple copies of the intron. These isolates of *P. parvum* and *P. patelliferum* originate outside of Scandinavia. *P. parvum* and *P. patelliferum* from one locality are genetically so closely related that they might be considered as one instead of two species.

Using a morphological species concept, *P. parvum* and *P. patelliferum* are regarded as two species. However, growth and toxicity experiments (Larsen *et al.*, 1993; Larsen and Bryant, 1998) as well as genetic analyses (Larsen and Medlin, 1997), indicated that the morphological characters used to separate the two species probably do not reflect an ancestor–descendant lineage. Using a phylogenetic species concept, *P. parvum* and *P. patelliferum* must therefore be regarded as one species.

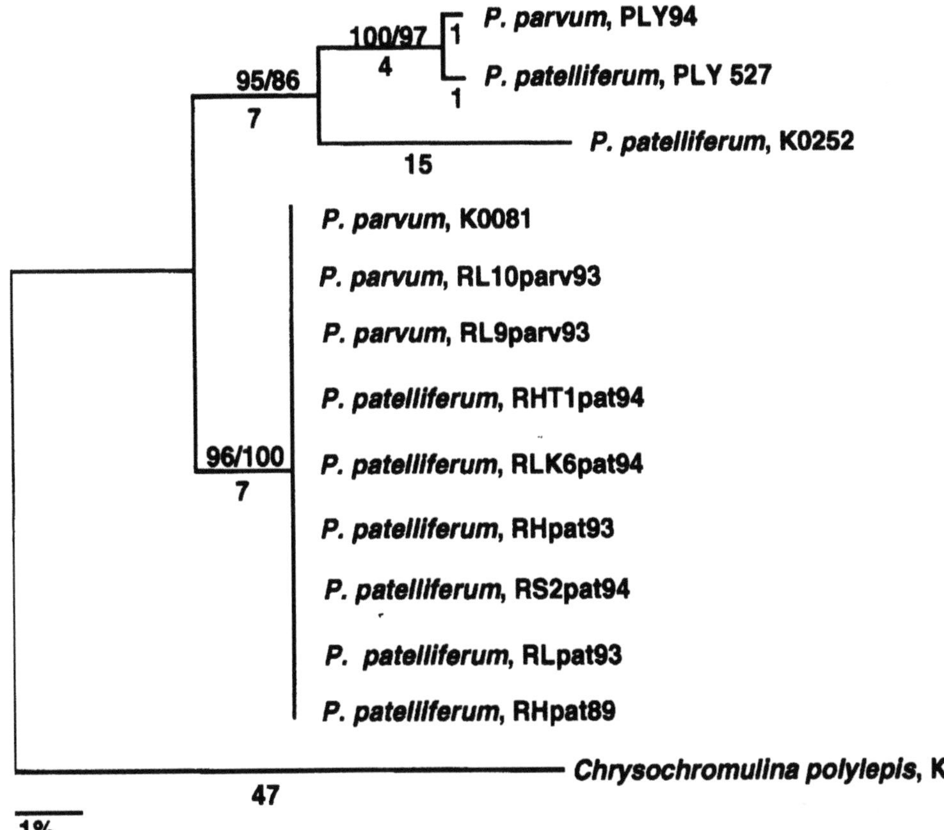

Figure 14.8 Phylogenetic analysis of relationships among four strains of *Prymnesium parvum* and eight strains of *P. patelliferum* deduced from ITS1-sequences. Figures above the internal nodes are bootstrap values based on a neighbour-joining analysis (left) and a maximum parsimony analysis (right). The absolute number of nucleotide differences between the strains are shown below the internal nodes. An isolate of *Chrysochromulina polylepis* is used as an outgroup. The distance corresponding to 1 change per 100 nucleotide positions is placed below the distance tree

Source: redrawn from Larsen and Medlin (1997), reproduction courtesy *Journal of Phycology*

This demonstrates that a shift from a morphological to a phylogenetic/genetic species concept might alter the taxonomy of these haptophytes and may be applicable to many other 'species pairs' linked in life-cycles.

The results of these genetic studies showed that *P. parvum* and *P. patelliferum* are apparently genetically identical in the regions of the genome investigated, and triggered speculations as to whether the two 'species', or morphotypes, could possibly be linked in a heteromorphic life-cycle. Thus, the two 'species' might represent different generations rather than being two separate gene pools (Larsen and Medlin, 1997). Flow cytometric analysis of several *P. parvum* and *P. patelliferum* strains revealed two major ploidy levels in the two 'species'. *Prymnesium patelliferum* cells were always haploid, whereas cells of *P. parvum* could be either haploid or diploid (Larsen and Edvardsen, 1998). Therefore a hypothetical haplo-diploid life-cycle that includes three flagellate scaly forms has been proposed (Figure 14.6).

Genetic as well as flow cytometric analyses of additional species within the genus might prove to give additional insight as to whether differences in scale morphology are suitable characters for defining species within *Prymnesium* or within the Haptophyta as a general rule.

ACKNOWLEDGEMENTS

This research was supported in part by grants to LKM from the DFG (SM 22/5-1) and the BMBF (TEPS 03F0161). The technical assistance of Ms U. Wellbrock is gratefully acknowledged. This is contribution number 1567 from the Alfred-Wegener-Institute.

REFERENCES

Andersen, R. A., Brett, R. W., Potter, D., and Sexton, J. P. (1998) Phylogeny of the Eustigmatophyceae based upon 18S rDNA, with emphasis on *Nannochloropsis*. *Protist*, **1**, 61–74.

Bakker, F. T., Olsen, J. L. and Stam, W. T. (1995) Evolution of nuclear rDNA ITS sequences in the *Cladophora albida/sericea* clade (Chlorophyta). *Journal of Molecular Evolution*, **40**, 640–651.

Baumann, M. E. M., Lancelot, C., Brandini, F. P., Sakshaug, E. and John, D. M. (1994) The taxonomic identity of the cosmopolitan prymnesiophyte *Phaeocystis*, a morphological and ecophysiological approach. *Journal of Marine Systems*, **5**, 23–39.

Billard, C. (1983) *Prymnesium zebrinum* sp. nov. et *P. annuliferum* sp. nov., deux nouvelles espèces apparentées à *P. parvum* Carter (Prymnesiophyceae). *Phycologia*, **22**, 141–151.

Billard, C. (1994) Life cycles. In *The Haptophyte Algae* (eds J. C. Green and B. S. C. Leadbeater),Systematics Association Special Volume 51, Oxford: Clarendon Press, pp. 167–186.

Birkhead, M. and Pienaar, R. N. (1995) The flagellar apparatus of *Chrysochromulina* sp. (Prymnesiophyceae). *Journal of Phycology*, **31**, 96–108.

Buma, A. G. J., Bano, N., Veldhuis, M. J. W. and Kraay, G. W. (1991) Comparison of the pigmentation of two strains of the prymnesiophyte *Phaeocystis* sp. *Netherlands Journal of Sea Research*, **27**, 173–182.

Carter, N. (1937) New or interesting algae from brackish water. *Archivs für Protistenkunde*, **90**, 1–68.

Chang, F. H. (1983) The mucilage producing *Phaeocystis pouchetii* (Prymnesiophyceae) cultured from the 1981 'Tasman Bay slime'. *New Zealand Journal of Marine and Freshwater Research*, **17**, 165–168.

Chang, F. H. (1985) Preliminary toxicity test of *Prymnesium calathiferum* n. sp. isolated from New Zealand. In *Toxic Dinoflagellates* (eds D. M. Anderson, A. W. White and D. Baden), New York: Elsevier, pp. 109–112.

Chang, F. H. and Ryan, K. G. (1985) *Prymnesium calathiferum* sp. nov. (Prymnesiophyceae), a new species isolated from Northland, New Zealand. *Phycologia*, **24**, 191–198.

Charlson, R. J., Lovelock, J. E., Andreae, M. O. and Warren, S. G. (1987) Oceanic phytoplankton, atmospheric sulphur, cloud albedo and climate. *Nature*, **326**, 655–661.

Conrad, W. (1941) Sur les Chrysomonadines à trois fouets. Aperçu synoptique. *Bulletin du Musée royal d'Histoire naturelle de Belgique*, **17**, 1–16.

Côrte-Real, H. B. S. M., Dixon, D. R., and Holland, P. W. H. (1994) Intron-targeted PCR, a new approach to survey neutral DNA polymorphism in bivalve populations. *Marine Biology*, **120**, 407–413.

Crame, J. A. (1993) Latitudinal range fluctuations in the marine realm through geological times. *TREE*, **8**, 161–166.

Davidson, A. T. and Marchant, H. J. (1992) The biology and ecology of *Phaeocystis* (Prymnesiophyceae). In *Progress in Phycological Research* (eds F. E. Round and D. J. Chapman), vol. 8, pp. 1–40.

Edvardsen, B. and Medlin, L. K. (1998) Genetic analysis of authentic and alternate forms of *Chrysochromulina polylepis* (Haptophyta). *Phycologia*, 37, 275–283.

Edvardsen, B. and Paasche, E. (1992) Two motile stages of *Chrysochromulina polylepis* (Prymnesiophyceae), morphology, growth and toxicity. *Journal of Phycology*, **28**, 104–114.

Edvardsen, B. and Paasche, E. (1998) Bloom dynamics and physiology of *Prymnesium* and *Chrysochromulina*. In *Physiological Ecology of Harmful Algal Blooms* (eds D. M. Anderson, A. D. Cembella and G. M. Hallegraeff), Heidelberg: Springer Verlag, pp. 193–208.

Edvardsen, B. and Vaulot, D. (1996) Ploidy analysis of the two motile forms of *Chrysochromulina polylepis* (Prymnesiophyceae). *Journal of Phycology*, **32**, 94–102.

Estep, K. W., Davis, P. G., Hargraves, P. E. and Sieburth, J. M. (1984) Chloroplast containing microflagellates in natural populations of north Atlantic nanoplankton, their identification and distribution; including a description of five new species of *Chrysochromulina* (Prymnesiophyceae). *Protistologica*, **20**, 613–634.

Estep, K. W. and MacIntyre, F. (1989) Taxonomy, lifecycle, distribution and dasmotrophy of *Chrysochromulina*, a theory accounting for scales, haptonema, muciferous bodies and toxicity. *Marine Ecology Progress Series*, **57**, 11–21.

Gray, J. S. (1997) *Marine Biodiversity: Patterns, Threats and Conservation Needs*. GEAMP Reports and Studies no. 62. London: International Maritime Organization.

Green, J. C., Hibberd, D. J. and Pienaar, R. N. (1982) The taxonomy of *Prymnesium* (Prymnesiophyceae) including a description of a new cosmopolitan species, *P. patellifera* sp. nov., and further observations on *P. parvum* N. Carter. *British Phycological Journal*, **17**, 363–382.

Green, J. C. and Hori, T. (1994) Flagella and flagellar roots. In *The Haptophyte Algae* (eds J. C. Green and B. S. C. Leadbeater), Systematics Association Special Volume 51, Oxford: Clarendon Press, pp. 91–109.

Guo, M., Harrison, P. J., and Taylor, F. J. R. (1996) Fish kills related to *Prymnesium parvum* N. Carter (Haptophyta) in the People's Republic of China. *Journal of Applied Phycology*, **8**, 111–117.

Inouye, I. (1997) Systematics of haptophyte algae in Asia-Pacific waters. *Algae (The Korean Journal of Phycology)*, **12**, 247–261.

Ishida, Y. (1968) Physiological studies on evolution of dimethyl sulfide from unicellular marine algae. *Memoires of the College of Agriculture Kyoto University*, **94**, 47–82.

Iverson, R. L., Nearhoof, F. L. and Andreae, M. O. (1989) Production of dimethylsulfonium propionate and dimethylsulfide by phytoplankton in estuarine and coastal waters. *Limnology and Oceanography*, **34**, 53–67.

Jordan, R. W and Green, J. C. (1994) A check-list of the extant Haptophyta of the world. *Journal of the Marine Biological Association of the UK*, 74, 149–174.

Kaartvedt, S., Johnsen, T. M., Aksnes, D. L., Lie, U. and Svendsen, H. (1991) Occurrence of the toxic phytoflagellate *Prymnesium parvum* and associated fish mortality in a Norwegian fjord system. *Canadian Journal of Fisheries and Aquatic Sciences*, **48**, 2316–2323.

Karsten, G. (1905) Das Phytoplankton des Antarktischen Meeres nach dem Material der Deutschen Tiefsee-Expedition 1898–1899. *Wissenschaftliche Ergebnisse der Deutschen Tiefsee-Expedition auf dem Dampfer 'Valdivia' 1898–1899*, Band II, Teil 2, 1–136.

Kashkin, N. I. (1963) Data on the ecology of *Phaeocystis pouchetii* (Hariot) Lagerheim, 1893 (Chrysophyceae). II. Habitat and specification of biogeographical characteristics. *Okeonologia*, 3, 697–705 (in Russian).

Keller, M. D., Bellows, W. K. and Guillard, R. R. L. (1989) Dimethyl sulphide production in marine phytoplankton: an additional impact of unusual blooms. In *Novel Phytoplankton Blooms Causes and Impacts of Recurrent Brown Tides and Other Unusual Blooms* (eds E. M. Cosper and E. J. Carpenter), Berlin: Springer, pp. 101–115.

Kirst, G. O., Thiel, C., Wolff, H., Nothnagel, J., Wanzek, M. and Ulmke, R. (1991) Dimethylsulfoniopropionate (DMSP) in ice-algae and its possible biological role. *Marine Chemistry*, **35**, 381–388.

Knowlton, N. (1993) Sibling species in the sea. *Annual Review of Ecology and Systematics*, **24**, 189–216.

Kooistra, W. H. C. F. and Medlin, L. K. (1996) Evolution of the diatoms (Bacillariophyta). IV. A reconstruction of their age from small subunit rRNA coding regions and the fossil record. *Molecular Phylogenetics and Evolution*, **6**, 391–407.

Kornmann, P. (1955) Beobachtungen an *Phaeocystis*-Kulturen. *Helgoländer Wissenschaftliche Meeresuntersuchungen*, 5, 218–233.

Lancelot, C., Billen, G., Sournia, A., Weisse, T., Colijn, F., Veldhuis, M. J. W., Davies, A. and Wassmann, P. (1987) *Phaeocystis* blooms and nutrient enrichment in the continental coastal zones of the North Sea. *Ambio*, **16**, 38–46.

Lange, M. (1997) *Phylogeny and taxonomy of the genus* Phaeocystis *(Prymnesiophyceae)*. Ph.D. dissertation. University of Bremen.

Larsen, A. and Bryant, S. (1998) Growth rate and toxicity of *Prymnesium parvum* and *Prymnesium patelliferum* (Haptophyta) in response to changes in salinity, light and temperature. *Sarsia*, **83**, 409–418.

Larsen, A. and Edvardsen, B. (1998) A study of relative ploidy levels in *Prymnesium parvum* and *P. patelliferum* (Haptophyta) analysed by flow cytometry. *Phycologia*, **37**, 412–414.

Larsen, A. and Medlin, L. K. (1997) Inter- and intraspecific genetic variation in twelve *Prymnesium* (Haptophyceae) clones. *Journal of Phycology*, **33**, 1007–1015.

Larsen, A., Eikrem, W. and Paasche, E. (1993) Growth and toxicity in *Prymnesium patelliferum* (Prymnesiophyceae) isolated from Norwegian waters. *Canadian Journal of Botany*, **71**, 1357–1362.

Larsen, J. and Moestrup, Ø. (1989) *Guide to Toxic and Potentially Toxic Marine Algae*. Fish Inspection Service, Ministry of Fisheries, Copenhagen, 61 pp.

Liss, P. S., Malin, G., Turner, S. M. and Holligan, P. M. (1994) Dimethyl sulphide and *Phaeocystis*: a review. *Journal of Marine Systems*, **5**, 41–53.

Margulis, L., Corliss, J. O., Melkonian, M. and Chapman, D. J. (1990) *Handbook of Protoctista*. Boston: Jones and Bartlett.

Medlin, L. K., Barker, G. L. A., Campbell, L., Green, J. C., Hayes, P. K., Marie, D., Wrieden, S. and Vaulot, D. (1996) Genetic characterisation of *Emiliania huxleyi* (Haptophyta). *Journal of Marine Systems*, **9**, 13–31.

Medlin, L. K., Elwood, H. J., Stickel, S. and Sogin, M. L. (1991) Morphological and genetic variation within the diatom *Skeletonema costatum* (Bacillariophyta): evidence for a new species *Skeletonema pseudocostatum*. *Journal of Phycology*, **27**, 514–524.

Medlin, L. K., Kooistra, W. H. C. F., Potter, D., Saunders, G. W. and Andersen, R. A. (1997) Phylogenetic relationships of the 'golden algae' (haptophytes, heterokont chromophytes) and their plastids. *Plant Systematics and Evolution* (Suppl.), **11**, 187–219.

Medlin, L. K., Lange, M., Barker, G. L. A. and Hayes, P. K. (1995) Can molecular techniques change our ideas about the species concept? *Nato ASI Series*, **38**, 133–152.

Medlin, L. K., Lange, M. and Baumann, M. E. M. (1994) Genetic differentiation among three colony-forming species of *Phaeocystis*, further evidence for the phylogeny of the Prymnesiophyta. *Phycologia*, **33**, 199–212.

Moestrup, Ø. (1994) Economic aspects, 'blooms', nuisance species, and toxins. In *The Haptophyte Algae* (eds J. C. Green and B. S. C. Leadbeater), Systematics Association Special Volume 51, Oxford: Clarendon Press, pp. 265–285.

Moestrup, Ø. and Thomsen, H. A. (1995) Taxonomy of toxic haptophytes (Prymnesiophytes). In *Manual on Harmful Marine Microalgae* (eds G. M. Hallegraeff, D. M. Anderson and A. D. Cembella), Paris: UNESCO, pp. 319–338.

Nanney, D. L., Park, C. Preparata, R. and Simon, E. (1998) Comparison of sequence differences in a variable 23S rRNA domain among sets of cryptic species of ciliated protozoa. *Journal of Eukaryotic Microbiology*, **45**, 91–100.

Ochman, H. and Wilson, A. C. (1987) Evolution in bacteria, evidence for an universal substitution rate in cellular genomes. *Journal of Molecular Evolution*, **26**, 74–86.

Olsen, J. L., Stam, W. T., Berger, S. and Menzel, D. (1994) 18S rDNA and evolution in the Dasycladales (Chlorophyta), modern living fossils. *Journal of Phycology*, **30**, 729–744.

Paasche, E., Edvardsen, B. and Eikrem, W. (1990) A possible alternate stage in the lifecycle of *Chrysochromulina polylepis* Manton et Parke (Prymnesiophyceae). *Nova Hedwigia Beiheft*, **100**, 91–99.

Padan, E., Ginzburg, D. and Shilo, M. (1967) Growth and colony formation of the phytoflagellate *Prymnesium parvum* Carter on solid media. *Journal of Protozoology*, **14**, 477–480.

Palmer, J. D. (1995) Rubisco rules fall – gene transfer triumphs. *Bioessays*, **17**, 1005–1008.

Parke, M., Manton, I. and Clarke, B. (1955) Studies on marine flagellates. II. Three new species of *Chrysochromulina*. *Journal of the Marine Biological Association of the UK*, **34**, 579–609.

Parke, M., Manton, I. and Clarke, B. (1956) Studies on marine flagellates. III. Three further species of *Chrysochromulina*. *Journal of the Marine Biological Association of the UK*, **35**, 387–414.

Parke, M., Manton, I. and Clarke, B. (1958) Studies on marine flagellates. IV. Morphology and microanatomy of a new species of *Chrysochromulina*. *Journal of the Marine Biological Association of the UK*, **37**, 209–228.

Parke, M., Manton, I. and Clarke, B. (1959) Studies on marine flagellates. V. Morphology and microanatomy of *Chrysochromulina strobilus* sp. nov. *Journal of the Marine Biological Association of the UK*, **38**, 169–188.

Pienaar, R. N. (1980) Observations on the structure and composition of the cyst of *Prymnesium* (Prymnesiophyceae). *Electron Microscopy Society of Southern Africa – Proceedings*, **10**, 73–74.

Pienaar, R. N. and Birkhead, M. (1994) Ultrastructure of *Prymnesium nemamethecum* sp. nov. (Prymnesiophyceae). *Journal of Phycology*, **30**, 291–300.

Rousseau, V., Vaulot, D., Cassette, R., Len, J., Gunk, J. and Baumann, M. (1994) The lifecycle of *Phaeocystis* (Prymnesiophyceae): evidence and hypotheses. In *Ecology of Phaeocystis Dominated Ecosystems* (eds C. Lancelot and P. Wassmann). *Journal of Marine Systems*, **5**, 23–39.

Scholin, C. A., Herzog, M., Sogin, M. and Amderson, D. M. (1994) Identification of group- and strain-specific genetic markers for globally distributed *Alexandrium* (Dinophyceae). II. Sequence analysis of a fragment of the LSU rRNA gene. *Journal of Phycology*, **30**, 999–1011.

Simon, N., Brenner, J., Edvardsen, B. and Medlin, L. K. (1997) The identification of *Chrysochromulina* and *Prymnesium* species (Haptophyta, Prymnesiophyceae) using fluorescent or chemiluminescent oligonucleotide probes: a means for improving studies on toxic algae. *European Journal of Phycology*, **32**, 393–401.

Smith, W. O., Codispoti, L. A., Nelson, D. M., Manley, T., Buskey, E. J., Niebauer, H. J. and

Cota, G. F. (1991) Importance of *Phaeocystis* blooms in high-latitude ocean carbon cycle. *Nature*, 352, 514–516.

Sogin, M. L., Elwood, H. J. and Gunderson, J. H. (1986) Evolutionary diversity of eukaryotic small-subunit rRNA genes. *Proceedings of the National Academy of Sciences USA*, 83, 1383–1387.

Sournia, A. (1988) *Phaeocystis* (Prymnesiophyceae), how many species? *Nova Hedwigia*, 47, 211–217.

Starmach, K. (1968) Chrysophyta I. Chrysophycea złotowiciowce. In *Flora Słodkowodna Polski*, vol. 5 (ed. K. Starmach), Warsaw: PWN.

Taberlet, P., Gielly, L., Pautou, G. and Bouvet, J. (1991) Universal primers for amplification of three non-coding regions of chloroplast DNA. *Plant Molecular Biology*, 17, 1105–1109.

Thomsen, H. A., Buck, K. R. and Chavez, F. P. (1994) Haptophytes as components of marine phytoplankton. In *The Haptophyte Algae* (eds J. C. Green and B. S. C. Leadbeater), Systematics Association Special Volume 51, Oxford: Clarendon Press, pp. 187–208.

Thomsen, H. A., Østergaard, J. B. and Hansen, L. E. (1991) Heteromorphic life histories in Arctic coccolithophorids (Prymnesiophyceae). *Journal of Phycology*, 27, 634 –642.

Throndsen, J. (1969) Flagellates of Norwegian coastal waters. *Norwegian Journal of Botany*, 16, 161–216.

Throndsen, J. (1983) Ultra- and nanoplankton flagellates from coastal waters of southern Honshu and Kyushu, Japan (including some results from the western part of the Kuroshio off Honshi). In *Working Party on Taxonomy in the Akashiwo Mondai Kenkyukai Fishing Ground Preservation Division*. Research Department, Fisheries Agency, Japan.

Tilman, D. (1982) *Resource Competition and Community Structure*. Princeton, N.J.: Princeton University Press.

Urbach, E. (1994) Genetic diversity of *Prochlorococcus marinus*, field populations and culture. Abstract in *Molecular Ecology of Aquatic Microbes*, NATO Advanced Study Institute, vol. 38, p. 58.

Valkanov, A. (1964) Untersuchungen über *Prymnesium parvum* Carter und seine toxische Einwirkung auf die Wasserorganismen. *Kieler Meeresforschungen*, 20, 65–81.

van Oppen, M. J. H., Klerk, H., Olsen, J. L. and Stam, W. (1996) Hidden diversity in marine algae: some examples of genetic variation below the species level. *Journal of the Marine Biological Association of the UK*, 76, 239–242.

Vaulot, D., Birrien, J-L., Marie, D., Casotti, R., Veldhuis, M., Kraay, G. and Chrétiennot-Dinet, M-J. (1994) Morphology, ploidy, pigment composition, and genome size of cultured strains of *Phaeocystis* (Prymnesiophyceae). *Journal of Phycology*, 30, 1022–1035.

Zingone, A., Chrétiennot-Dinet, M-J., Lange, M. and Medlin, L. K. (1999) Morphological and genetic characterisation of *Phaeocystis cordata* and *P. jahnii* (Prymnesiophyceae), two new species from the Mediterranean Sea. Journal of Phycology, 35, 1322–1337.

Chapter 15

Occurrence and loss of organelles

Mark A. Farmer and W. Marshall Darley

ABSTRACT

Eukaryotic cells are distinguished from prokaryotes (Bacteria and Archaea) by the presence of organelles. Organelles are specialized parts of a cell that resemble and function as organs: that is, they are differentiated components of a cell that carry out a specific function. Most organelles are defined by the unique properties of their membranes and the specific proteins associated with these membranes. Such organelles include mitochondria, plastids, endoplasmic reticulum (ER), Golgi dictyosomes, peroxisomes, contractile vacuoles, and glycosomes. Some organelles, such as photoreceptors and ejectosomes, may be membrane-bound but are largely defined by the unique nature of their internal constituents. Other organelles such as the eukaryotic flagellum and nucleus represent a unique combination of specialized internal components (axoneme, genome) and a specialized membrane (flagellar membrane, nuclear envelope). This chapter addresses some of the current theories concerning the origins, and in some cases subsequent loss, of organelles in flagellates.

15.1 Introduction

There are two major theories concerning the origin of eukaryotic organelles. The first suggests that certain organelles arose directly by sequestration of various enzymes or cellular constituents by membranes. This theory, known as the Direct Filiation or Autogenesis Theory, proposed by Cavalier-Smith and others, rests on the fact that one major distinction between prokaryotes and eukaryotes is the ability of eukaryotes to carry out endocytosis and exocytosis. Digestive enzymes, nucleic acids, and other specific cellular constituents can be isolated from the cell cytoplasm by infolding or invaginations of plasma membrane (Figure 15.1) and thus be allowed to specialize while remaining in close proximity to other cellular components. The evolution of a cytoskeleton was thought to have been a critical step in this process since it facilitates endocytosis (Mitchison, 1995).

The second major theory is known as the Serial Endosymbiotic Theory (SET), or sometimes more simply as the endosymbiotic theory. First proposed over a century ago (Schimper, 1883; Wallin, 1927) and championed by Lynn Margulis (Margulis, 1993; Taylor, 1974), the SET suggests that eukaryotic organelles are the direct result of permanent unions between various prokaryotes and a host cell. As viewed from this perspective, eukaryotes are actually derived from a consortia of cooperating

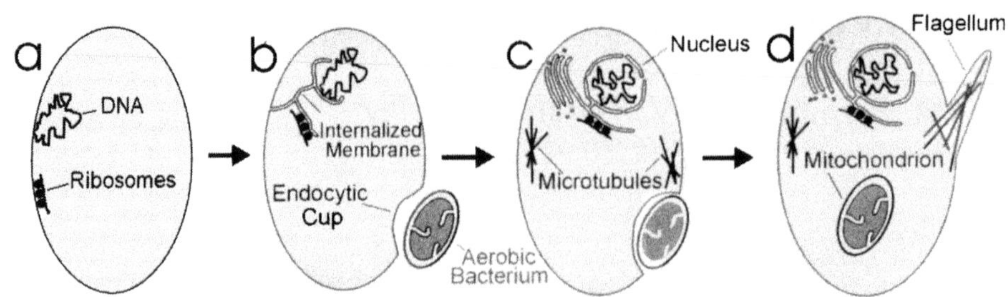

Figure 15.1 Possible pathway for the autogenous evolution of the endomembrane system (nuclear envelope, ER, and Golgi) from invagination of the plasma membrane (a–c), endosymbiotic origin of the mitochondrion (b–d) and autogenous origin of the flagellum from cytoskeletal components (c–d). The second membrane surrounding the mitochondrion is of either host or symbiont origin

cells. The endosymbiotic theory makes certain predictions about the evolution, form and function of an organelle. Initially the symbiont would be an independent organism, with its own complete genome, existing within or on the host cell, and if internalized by phagocytosis, surrounded by a second membrane. The second membrane may represent the remnant of a host ingestion vacuole or it may correspond to the outer of two membranes found in gram negative bacteria (Figures 15.1, 15.2) (Cavalier-Smith, 1982, 1983a). Host and symbiont must coordinate reproduction rates to maintain a stable association. Over time the endosymbiont loses its autonomy to the host cell as some or all of its genes become incorporated into the nucleus of its host. The host must in turn possess a way of regulating the expression of these transferred genes, translating the polypeptides that they encode and transporting the proteins back into the organelle where they are utilized. Thus the mechanism whereby an endosymbiont becomes an organelle is exceedingly complex.

In truth, the origins of organelles are almost certainly a combination of both theories. There is overwhelming evidence that both mitochondria and plastids had their origins as free living prokaryotes which were assimilated into eukaryotic cells. Likewise the idea that specialized membranous organelles such as ER and Golgi dictyosomes arose as a result of internal specialization of existing cellular components is equally compelling. By examining selected organelles separately one can begin to discern the complexity of the cellular constituents of flagellates.

15.2 Nucleus

The presence of a nucleus is the diagnostic feature of eukaryotes that distinguishes them from prokaryotes. The origin of the microtubular cytoskeleton and mitotic spindle allowed eukaryotic chromosomes to separate the site of DNA attachment and segregation from that of DNA replication. This led to an increase in genome size and complexity (Cavalier-Smith, 1993). One possible advantage of the nuclear

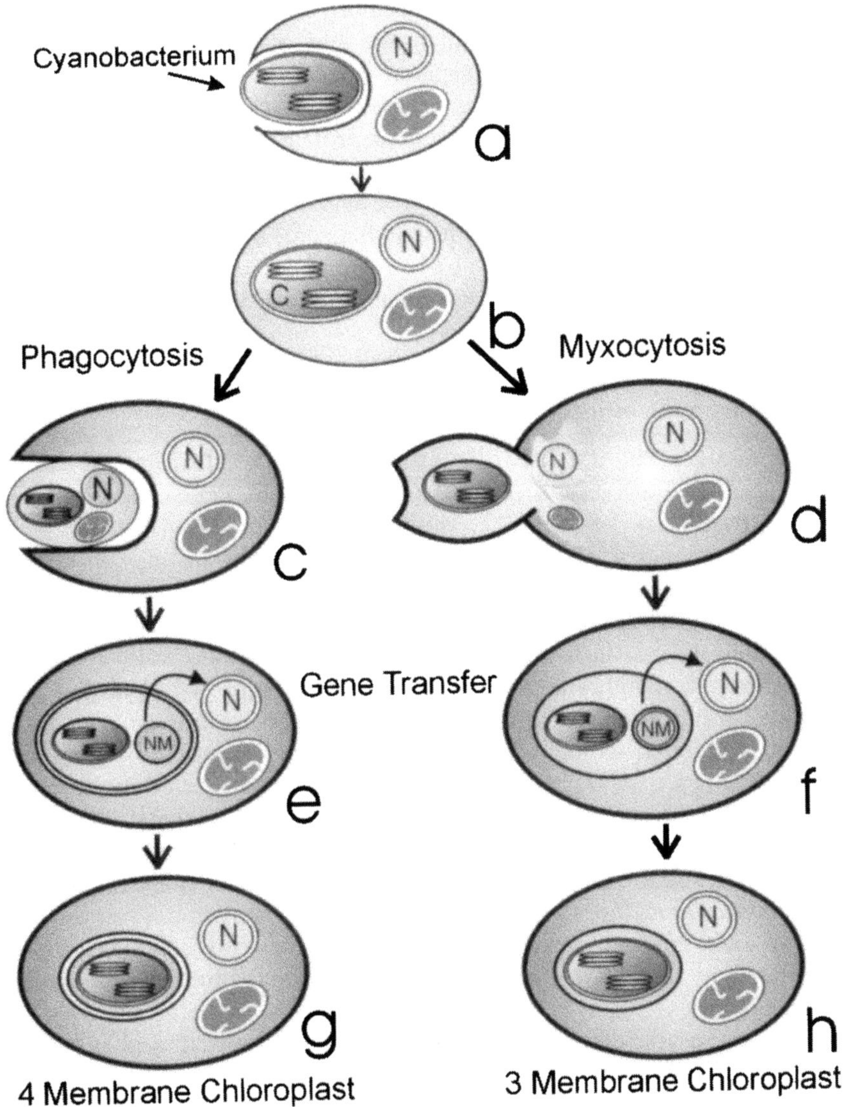

Figure 15.2 Possible pathways for the endosymbiotic origin of plastids. Primary origin from ingestion of a cyanobacterium (a to b) giving rise to green algae, red algae, and glaucocystophytes. Secondary origin through either phagocytosis by a second eukaryote (c) or myxocytosis in which only the cytoplasm of the prey cell is ingested (d). In phagocytosis the end result is plastids with a total of four membranes and a nucleomorph (e) as in chlorachniophytes and cryptomonads, followed by a loss of nucleomorph (g) as in heterokont algae and prymnesiophytes. In myxocytosis the plasma membrane of the prey cell is left behind, with the contents being contained within a food vacuole (f), and following the loss of the nucleus a three-membrane chloroplast remains (h), as in euglenids and dinoflagellates

envelope is the sequestration of the genome from degrading enzymes in the cytoplasm. Another is that by controlling the movement of materials to and away from the chromosomes it provided a spatial separation between transcriptional and translational processes. The nuclear envelope is an elaborate structure consisting of two membranes joined by highly specialized protein complexes known as nuclear pores which regulate movement of materials into and out of the nucleoplasm. The inner portion of the nuclear envelope is lined by a layer of nuclear lamins which are biochemically closely related to intermediate filaments of the cytoskeleton (Minguez *et al.*, 1994). The chromosomes are closely associated with this layer. In many flagellates the nuclear envelope persists during cell division and has been implicated in chromosome segregation (Raikov, 1994).

15.2.1 Autogenous model

There are two aspects of nuclear origin to be considered: origin of the nuclear structure itself and origin(s) of the eukaryotic genome. Prokaryotic genomes are anchored to the plasma membrane, and it has been suggested by Cavalier-Smith (1991) and others that an invagination of this membrane may have led to a sequestering of the early eukaryotic genome within a proto-nucleus (Figure 15.1). A model for such a scenario can be found in the Planctomycetes, a little known group of Bacteria that lack a peptidoglycan reinforced cell wall (Fuerst, 1995). The presence of a membrane-bound nuclear body (Figure 15.3) in this group of prokaryotes is indirect evidence for the independent evolution of an analogous DNA sequestering structure to form the eukaryotic nucleus (Fuerst and Webb, 1991; Lindsay *et al.*, 1997; pers. obser.). The presence of a continuous membrane that extends between the nuclear body and the plasma membrane in Planctomycetes (Fuerst and Webb, 1991) and in the endomembrane system of eukaryotes (nuclear envelope ⇒ ER ⇒ Golgi ⇒ plasma membrane) is consistent with the infolding membrane model. It is perhaps significant to note that both eukaryotes and Planctomycetes lack peptidoglycans as constituents of their outer membranes, and thus have the necessary fluidity to allow for infolding of the plasma membrane.

Another major advantage of the nucleus and its nuclear envelope is the physical separation of RNA transcription and processing from protein synthesis in the cytoplasm. This may have allowed for a proliferation of introns into the protein coding regions of eukaryotes. Introns, particularly splicesomal introns, provide alternative splicing mechanisms, and thereby may afford eukaryotes with certain advantages including increased genomic variation and/or the ability to form different proteins from a single gene (Dibb, 1993). The presence of introns is less likely in prokaryotes, since ribosomes begin translating mRNA almost as soon as it is synthesized, and introns would be translated before they could be edited out (Cavalier-Smith, 1991, 1993). The autogenesis theory of nuclear evolution recognizes that while many flagellates may possess more than one genome (nuclear plus mitochondrial and/or plastid), the earliest eukaryotes had only one. The unique features of the nucleus allowed the eukaryotic genome to evolve in ways profoundly different from those of prokaryotes (Cavalier-Smith, 1993).

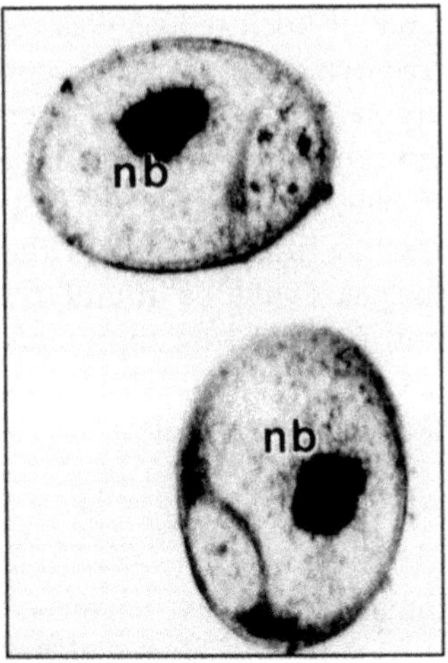

Figure 15.3 Transmission electron micrograph of nuclear bodies (nb) in the planctomycete bacterium *Planctomyces limnophilis*

15.2.2 Symbiotic model

A second, less well accepted, scenario suggests that the nucleus may be the remnant of a symbiosis involving a postulated bacterial ancestor of the eukaryote and an archaebacterium. This could explain why eukaryotes have certain protein sequences (for example, elongation factor, α and β subunits of ATPase) like those of Archaea but rRNA sequences like those in bacteria (Sogin, 1991; Golding and Gupta, 1995). Endosymbiotic acquisition of the nucleus and most of the eukaryotic genome could explain the apparent chimaeric nature of the eukaryotic genome (Gupta, 1995), as well as the origin of the nuclear double membrane. However the biochemistry, structure and continuity of the nuclear envelope with the endomembrane system makes this hypothesis unlikely.

15.2.3 Loss of nuclei

It is very unlikely that a cell could survive the loss of its nucleus. However there are a few examples of nuclear reduction and/or loss in cases involving secondary endosymbiosis. In at least two separate cases a photosynthetic eukaryote has been incorporated into the cytoplasm of a different protist. The nucleus of the symbiont became reduced in size and function to form a remnant nucleus or nucleomorph (Fraunholz *et al.*, 1997; McFadden *et al.*, 1997a). In both cryptomonads and chlorarachniophytes the nucleomorph contains genes organized in three linear chromosomes (chlorachniophytes have spliceosomal introns), that are actively transcribed (Gilson and McFadden, 1996). Most of these genes are devoted to nucleomorph maintenance; only a few play a role in the functioning of the plastid (Gilson and McFadden, 1996, 1997). Evidence suggests that complete loss of the endosymbiont nucleus has occurred in at least four other flagellate groups (dinoflagellates, euglenids, prymnesiophytes, heterokonts) with plastids being the only organelle that was retained from the endosymbiont.

A curious intermediary stage of nuclear loss was reported for the dinoflagellate *Gymnodinium acidotum*, a free-living flagellate that contains within its cytoplasm a complete phototrophic eukaryote, believed to be a cryptomonad (see section on plastids). In one-third of a naturally occurring population of *G. acidotum* all of the internal organelles (nucleus, Golgi, ER, mitochondria, plastids) of the cryptomonad endosymbiont were retained, while in two-thirds of the cells every organelle of the endosymbiont except the nucleus could be found (Farmer and Roberts, 1990). Whether or not the situation in *G. acidotum* represents a true symbiosis is controversial (Schnepf, 1992; Fields and Rhodes, 1991), but the observation that the nucleus of an internalized eukaryote can be selectively lost is intriguing.

15.2.4 Future directions

Detailed analyses of the genomes of archaea, bacteria, and eukaryotes may distinguish between an autogenous or symbiotic origin of the nucleus. The sequencing of the entire genome of an archaebacterium (Bult *et al.*, 1996), and similar projects on flagellates such as *Giardia* (Smith *et al.*, 1998), may reveal patterns not simply in gene sequences, but also in the nature and arrangement of group II introns. The apparent ease with which genes can be transferred to the host genome (Blanchard and Schmidt, 1995; Martin *et al.*, 1998) is certain to complicate any such whole genome analyses.

15.3 Golgi apparatus

The Golgi apparatus typically consists of one or more dictyosomes, each of which is a stack of membranous flattened vesicles (cisternae). The number of cisternae per dictyosome varies from as few as three or four in eustigmatophytes (Santos and Leedale, 1992) to thirty or more in euglenids (Yamaguchi and Anderson, 1994). Golgi carry out post-translational processing of proteins (especially of oligosaccharide side chains) and protein sorting, directing proteins to either lysosomes or the plasma membrane.

The prevailing wisdom is that the Golgi dictyosome arose autogenously by an elaboration and specialization of the endomembrane system (Figure 15.1). When the dictyosome first appeared, and whether it has been secondarily lost or never possessed by some flagellate lineages, remains controversial. Cavalier-Smith (1983b) erected the Archezoa for all protists then believed to have been primitively amitochondriate and lacking peroxisomes. The concept of the Archezoa has undergone many modifications. The Archezoa is now comprised of only the metamonads (vegetatively without dictyosomes) and parabasalids (with prominent dictyosomes), both of which are probably secondarily amitochondriate.

The idea that a Golgi dictyosome was absent from metamonads was challenged by the discovery of a dictyosome in *Giardia* (Reiner *et al.*, 1990). One difficulty with determining whether a true Golgi exists is that a recognizable dictyosome may only be present for a brief portion of the cell cycle. In *Giardia* this only occurs during encystment (Gillin *et al.*, 1996). Despite the lack of stacked cisternae in *Giardia* trophozoites it has been shown that Brefeldin A can inhibit protein secretion, suggesting that the secretory machinery is similar to that of higher eukaryotes (Luján *et al.*, 1995). It is likely that the endomembrane system (nuclear envelope ⇒ ER ⇒ Golgi ⇒ plasma membrane) is a very ancient and highly conserved eukaryote feature that arose by specialization of internal membranes.

15.4 Flagella

The addition of the eukaryotic flagellum must have been extremely important to the previously non-motile eukaryotic lineage, and its origin may have preceded the origin of mitochondria, plastids, peroxisomes, and perhaps even the nucleus itself (Margulis, 1996). This conclusion is based on the observation that the most divergent eukaryotes all possess flagella, and that the ninefold ('9+2') symmetry of eukaryotic flagella is remarkably uniform. Only a few eukaryotes lack a flagellate stage during any portion of the life-cycle (for example microsporidians, red algae, higher fungi, zygnematalean green algae and dictyostelids). Based on small subunit rRNA sequences most of these groups are among the 'crown' of the eukaryotic tree (Barns *et al.*, 1996), and molecular phylogeny has shown that all of these groups have lost their flagellate stages secondarily. What was the origin of this important and complex organelle, containing some 250 different proteins in the axoneme and several hundred more in the kinetosome (basal body)? Two contrasting models have been proposed.

15.4.1 Autogenous model

One hypothesis suggests that the centriole (flagellar base) arose as a light sensory organelle that later gave rise to a locomotory organelle (Albrecht-Buehler, 1992). A more plausible explanation is given by Cavalier-Smith (1991), who proposed that flagella evolved by extension and protrusion of cytoplasmic microtubules (Figure 15.1) whose original function was to attach the centrosome to the cell surface, thereby anchoring the genome within the cytoplasm and preventing misdivision of the cell. His model explains the relationship between the kinetosome/centriole and the flagellar axoneme, as well as the anchoring microtubule rootlets associated with the flagellar apparatus. Other versions of this model suggest that flagella first arose as a mechanisms for food capture (Sleigh, 1986) or as cellular attachment organelles (peduncle) reinforced with microtubules and capable of moving the cell from side to side (Rizzotti, 1995). Restriction of the swinging motion to a single plane (perhaps perpendicular to a current) may have resulted in a nine-fold symmetry, and detachment of the peduncle from the substrate led to the first primitive flagellum. Among extant organisms there are many microtubule reinforced cellular projections: for example in microtubule-containing pseudopods (axopods and reticulopods), and the haptonemata of prymnesiophytes. Microtubules of these structures do not exhibit the ninefold symmetry characteristic of flagella, but do exhibit directed, albeit limited, motility.

Axopods are cytoplasmic extensions characteristic of actinopods (radiolarians, heliozoans and acantharians) which originate from a MTOC and are composed of highly ordered, cross-linked microtubules (Cachon *et al.*, 1990; Febvre, 1990;

Febvre-Chevalier, 1990). Reticulopods of foraminifera are less structured pseudopods with microtubules arranged in discrete, loosely organized bundles. The mechanism of axial movement in reticulopods is believed to be similar to that used in flagella, namely microtubule sliding assisted by the motor protein dynein (Travis and Bowser, 1991; Golz and Hauser, 1994).

A single haptonema is found between the flagella of most prymnesiophytes; haptonematal microtubules are surrounded by three membranes and appear to lack any cytoplasmic structure at their point of origin (Gregson *et al.,* 1993). Longer haptonemata can deliver food particles to the cell posterior for phagocytosis (Kawachi and Inouye, 1995), and also coil rapidly when they encounter an obstacle or when the cell is disturbed. This coiling is responsive to Ca^{2+} concentrations and may involve microtubule sliding (Gregson *et al.,* 1993). While neither pseudopods with microtubules nor haptonemata are believed to represent a transitory step in the evolution of flagella, they serve as non-flagellar models of motile, microtubule-reinforced cellular extensions that are of benefit to the cell (Rizzotti, 1995).

Another example of eukaryotic movement caused by microtubule sliding is the motile axostyle of oxymonads. Although generally it does not extend out from the cell, this axostyle is composed of a great many cross-linked microtubules that apparently move by microtubule sliding, perhaps mediated by dynein (Langford and Inoue, 1979; Woodrum and Linck, 1980). These movements can violently contort the cell, and the oxymonad genus *Saccinobaculus* is native American Choctaw for 'snake in a bag' (Cleveland, 1934).

15.4.2 Symbiotic model

The endosymbiotic theory for the origin of flagella has received attention because of the interesting yet controversial data in support of it. Lynn Margulis has long championed an idea first suggested by B. M. Kozo-Polyansky in 1924 (cited in Margulis, 1993) that flagella (undulipodia) arose from a symbiotic association between a spirochaete bacterium and early eukaryotes (Margulis, 1993, 1996). The most parsimonious version of this scenario posits that flagella preceded cytoplasmic microtubules. Modern 'motility symbioses' in which spirochaetes are attached to, and help move, large protists, provide obvious models; the best studied examples are found as intestinal symbionts in termites (Margulis, 1993). In some associations, the attached end of the spirochaete is specialized and the raised 'docking site' on the surface of the protist has associated cytoplasmic root-fibre systems which are vaguely similar to the root-fibres associated with kinetosomes (Margulis *et al.,* 1979). Synchronous beating by attached spirochaetes renders this model visually appealing, even though these particular associations could not be ancestral since the protist already has functional '9+2' flagella. What are the additional, controversial data to support this model?

According to the spirochaete symbiont hypothesis, spirochaetes should contain microtubules composed of tubulin or tubulin-like proteins. Cytoplasmic tubules have been found in spirochaetes and other bacteria, but molecular data suggest that these tubules are not composed of tubulin (Bermudes *et al.,* 1994). An intriguing bacterial protein called FtsZ may be a possible tubulin ancestor, as it has some sequence homology to tubulin and its published structure is very similar to that of β-tubulin (Lowe and Amos, 1998). FtsZ forms a ring structure at the site of cytokinesis and plays a role in cell division, but does not form tubule-like structures, and its presence in bacteria can be equally well explained by the autogenous theory. It is important to note that motor proteins (dyneins) are also required for microtubule sliding in flagellar function (Cosson, 1996), but neither these proteins nor their genes have ever been reported from prokaryotes (Gibbons, 1995).

By analogy to the reduced genomes in mitochondria and plastids, Margulis predicted that some spirochaete DNA might be retained by the kinetosome/centriole. Hall *et al.* (1989) claimed to show DNA in *Chlamydomonas* kinetosomes using *in situ* hydridization with probes from the *uni* linkage group (ULG) known to contain some of the genes encoding centriolar functions. Subsequent studies using modified *in situ* hybridization indicate that ULG is actually located in the nucleus, not the basal body (Hall and Luck, 1995).

A final difficulty with the spirochaete model involves the prokaryotes themselves. Not only do spirochaetes not move by way of microtubular sliding, but their outer membrane is extremely porous and biochemically quite unlike that of the plasma membrane of eukaryotes. Were the membranes of the spirochaete and the eukaryote to fuse, as posited by the symbiotic theory, it would almost certainly prove lethal to the host cell due to ionic shock (Cavalier-Smith, 1992; Rizzotti, 1995).

15.4.3 Future directions

Neither of these models offers a complete explanation for flagellar origins. Motile, microtubule-containing, cytoplasmic extensions such as axopods and haptonemata are of significant evolutionary advantage, and further research into their microtubule nucleating centres, development, biochemistry and function should provide insights relative to the autogenous model. Recent advances in bacterial genomics and the cultivation of spirochaetes from termites (Leadbetter *et al.*, 1999) will allow us to further address the symbiogenesis model.

15.5 Mitochondria

First called 'mitochondria' (Greek *mito-*, thread + *chondrion*, granule) by Benda in 1897, these organelles were associated with the oxidative respiration of eukaryotes around the turn of the twentieth century. Several years earlier Altmann (1890) had correctly predicted their role in cellular oxidation and described them as 'bioblasts' (Greek *bios-*, life + *blastos*, bud) to signify how mitochondria seem to give rise to other mitochondria. Despite morphological differences in mitochondrial cristae in various flagellate groups, similarities in respiratory enzymes and mitochondrial genomes argue strongly in favour of a monophyletic origin of mitochondria. Likewise, the very presence of a mitochondrial genome, mitochondrial ribosomes, and the mode of mitochondrial biogenesis (binary fission) has led to the universal acceptance that mitochondria are of endosymbiotic origin and represent the union of a cell and an α-proteobacterium (Gray *et al.*, 1999; Sogin, 1997) similar to the obligate parasite *Rickettsia* (Andersson *et al.*, 1998).

Life on Earth is nearly 4 billion years old (Holland, 1997) and for most of that time free oxygen has been an important by-product of photosynthesis (Schopf, 1993). Earth's ability to absorb free oxygen led to the formation of banded iron deposits starting about 3.8 billion years ago (Derry *et al.*, 1992). Approximately 2 billion years ago, the planet's ability to absorb excess oxygen through inorganic chemical reactions was saturated, and molecular oxygen began to accumulate in the atmosphere (Kasting, 1993; Holland, 1995; Nunn, 1998). Free oxygen is a strong oxidizing agent toxic to many organisms. Although aerobic bacteria had evolved and adapted to molecular oxygen, eukaryotes that could not deal with oxygen would have remained in anaerobic or microaerobic environments. Early amitochondriate protists limited to these refugia would have had little opportunity to diversify, as many of their existing habitats became inhospitable. Eventually one fortunate cell established a relationship with a facultatively aerobic α-proteobacterium that allowed it to venture out into oxygenated environments and colonize niches that previously were the domain solely of bacteria. Although not recorded in the fossil record, such an event could have led to the greatest round of speciation in Earth's history; many modifications of the eukaryotic cell plan would have been successful as there were virtually no competitors and a huge variety of unoccupied niches. This putative pre-Precambrian explosion might have been the direct result of the first mitochondrion, and may have been responsible for the radiation of eukaryotic forms that occurred prior to 1.1 billion years ago (Schopf, 1994).

15.5.1 Ancient origins

At one time it was thought that like nuclei, mitochondria were a ubiquitous part of all eukaryotic cells, but today we realize that there are many protists that lack mitochondria. Collectively these protists were called the Archezoa (Cavalier-Smith, 1983b) and originally they included the metamonads (diplomonads, retortamonads and oxymonads), the archamoebae, the microsporidians and the parabasalids. Cavalier-Smith (1987) later removed the parabasalids, arguing that their membrane-bound organelle called the hydrogenosome was actually a modified mitochondrion. Evidence for this has come from examination of the enzymes found in hydrogenosomes (Müller, 1993) as well as from analysis of chaperonin 60 (cpn60) and chaperonin 70 (cpn70) genes in trichomonads, which suggest that these genes are most closely related to those found in α-proteobacteria and mitochondria (Bui *et al.*, 1996; Germot *et al.*, 1996; Horner *et al.*, 1996; Roger *et al.*, 1996). The presence of a mitochondrial-like cpn60 gene in the diplomonad *Giardia* raises the possibility that all extant amitochondriate protists have secondarily lost their mitochondria, and that all eukaryotes are descended from an ancestor that once harboured an α-proteobacterium (Roger *et al.*, 1998), a view that is supported by sequence data for valyl-tRNA synthetase (Hashimoto *et al.*, 1998).

An alternative, and controversial, scenario in which all eukaryotes are derived from a union of two prokaryotic organisms has been proposed by Martin and Müller (1998). In their 'hydrogen hypothesis' the immediate ancestor of eukaryotes was an anaerobic, hydrogen-dependent, archaeal methanogen that formed a union with an aerobic, hydrogen-producing bacterium that later evolved into a mitochondrion.

Arguing largely on the basis of known metabolic pathways in flagellates that, first, lack mitochondria, second, lack mitochondria but possess hydrogenosomes, or third, possess mitochondria, Martin and Müller offer an explanation of how such a union could have become established in the first place and how each partner could have quickly become dependent on the other. Thus the immediate ancestor to the eukaryotes would have been a chimaeric organism (archaebacterium plus α-proteobacterium) that did not even have a nucleus. This hypothesis could explain why cpn60 genes like those of α-proteobacteria have been found in all eukaryotes, even those that now lack mitochondria. Similar hypotheses based on microbial metabolism have been put forward (López-García and Moreira, 1999), and these argue that the increased energy from the union of an archaeal host and an α-purple photosynthetic Gram-negative bacterium allowed for an increase in host genome size and complexity (Vellai *et al.*, 1998).

15.5.2 Eukaryotes first

The alternative, and more conventional, concept of a primitively amitochondriate eukaryote ingesting and retaining an aerobic bacterium which later evolved into the mitochondrion (Figure 15.1) has received support from studies on the flagellate *Reclinomonas*. The SET makes three predictions. First, the ancestral eukaryote was primitively amitochondriate. Second, it was capable of ingesting an α-proteobacterium that gradually became symbiotic, lost autonomy and evolved into the mitochondrion; and third, in the early stages of this process the symbiont would have retained a greater proportion of its genome than would be found in subsequent stages. All three of these predictions seem to have been met in the mitochondrion of *Reclinomonas* (Flavin and Nerad, 1993). The mitochondrial genome of *Reclinomonas* is the largest yet reported for any known eukaryote (Lang *et al.*, 1997; Palmer, 1997). *Reclinomonas* shares a number of ultrastructural features with the retortamonads, a group of amitochondriate flagellates believed to be closely related to *Giardia* and other metamonads (O'Kelly, 1997). Both *Reclinomonas* and the retortamonads have well developed cytostomes for ingesting bacterial prey. The presence of an exceptionally large mitochondrial genome in *Reclinomonas* is consistent with *Reclinomonas* and the retortamonads residing immediately on either side of the mitochondriate–primitively amitochondriate divide.

15.5.3 Loss of mitochondria

While some flagellates may still prove to be primitively amitochondriate, a number of other protists have clearly lost their mitochondria. These include ciliates and flagellates that inhabit the rumen of herbivores (Embley *et al.*, 1995), as well as some free-living flagellates isolated from low oxygen marine environments (Simpson *et al.*, 1997). In most cases, the mitochondria evolved into hydrogenosomes as members of these protistan groups colonized microaerobic environments. In the obligate parasite *Entamoeba*, the mitochondrion has become even further reduced (Mai *et al.*, 1999). Finally, in cases of secondary endosymbiosis (see section on plastids) the mitochondrion of the internalized cell may be retained for a period of time (Farmer and

Roberts, 1990) but is eventually lost due to redundancy of function with the host mitochondria.

15.5.4 Future studies

While organisms may yet be found that will clarify the nature of the first mitochondrion, we must continue to look for examples of how endosymbionts interact with their host cell. If symbiosis with accompanying gene transfer is exceedingly rare, then the presence of mitochondria-like cpn60 in *Giardia* and other basal eukaryotes argues in favour of never finding a primitively amitochondriate eukaryote (Roger *et al.*, 1999). On the other hand, bacterial symbionts can become established in a eukaryote in a few cell generations (Jeon and Lorch, 1967) and may quickly exercise control over expression of the host genome in such a way that the host cell becomes dependent on the symbiont (Choi *et al.*, 1997). It is possible that the mitochondria-like cpn60 genes in basal eukaryotes are remnants of a separate eukaryote/ α-proteobacterium symbiosis that *preceded* the acquisition of the symbiont that gave rise to the mitochondrion. Indeed the previous presence of such a gene might have left the host cell 'primed' for acquisition of a second α-proteobacterium (to become the mitochondrion), as the biochemical machinery necessary for importing and properly folding cytoplasmically translated proteins into the symbiont might already have been in place (see section on peroxisomes).

15.6 Plastids

Overwhelming evidence has led to general acceptance of the idea that eukaryotic plastids had a symbiotic origin (for example Gray, 1992). It is now clear that algae (photoautotrophic protists) are polyphyletic, having evolved several times from different heterotrophic protists, each of which acquired a photoautotrophic organism through endosymbiosis; joint evolution of the association has resulted in modern plastids (Bhattacharya, 1997). Thus the phylogeny of the host cell (the alga and its nucleus) often does not reflect the phylogeny of its plastid (Figure 15.2).

Phylogenetic trees inferred from available small subunit rRNA sequences from plastids show that plastids form a monophyletic lineage rooted among the cyanobacteria. As Turner (1997) points out, it is useful to distinguish between the *ancestry* and *origin* of plastids. *Ancestry* leaves open the possibility that plastids evolved from more than one endosymbiotic event (presumably involving closely related cyanobacteria), whereas *origin* suggests that there was only a single primary endosymbiotic event, and that divergence of plastid lineages occurred subsequently. With the discovery of *Prochloron* (a 'green' cyanobacterium, or prochlorophyte, containing chlorophylls *a* and *b*) (Lewin, 1975), Raven's (1970) earlier hypothesis that green, red, and yellow plastids had evolved independently from endosymbiotic 'green', 'red' and 'yellow' prokaryotes seemed appealing. Since that time, 'red' or 'yellow' prokaryotes have not been discovered. Two additional prochlorophytes have been discovered, but rRNA sequences show that prochlorophytes are polyphyletic, and that none of the known prochlorophytes are close to the root of the plastid lineage (Palenik and Haselkorn, 1992; Urbach *et al.*, 1992; Turner, 1997). Whatever their ancestry or origin, it appears that plastids have been acquired through primary, secondary or even tertiary endosymbioses (Figure 15.2).

15.6.1 Primary symbiosis

Three groups of protists acquired their plastids by primary endosymbiosis: green algae (and their descendants, the land plants), red algae and glaucocystophytes. Primary endosymbiosis involved phagocytosis of a unicellular cyanobacterium followed by joint evolution, for example, genome transfer; see Blanchard and Schmidt, 1995; Thorsness and Weber, 1996) similar to that described above. One notable difference is the retention of a peptidoglycan cell wall remnant between the two membranes of the glaucocystophyte plastid, called a cyanelle (Kies and Kremer, 1990). Gene transfer (endosymbiont to nucleus) and loss has left plastids with approximately 10–20 per cent of the genes coding for plastid proteins (Delwiche and Palmer, 1997). Martin *et al.*, (1998) have compiled and compared 210 protein-coding genes from four algae and five higher plants, and inferred a phylogenetic tree from forty-five genes common to the nine plastids. Their tree supports several others based on rRNA sequences (Bhattacharya and Medlin, 1995; Turner, 1997), suggesting that all three primary plastids are descended from a common ancestor and represent a monophyletic plastid lineage. Martin *et al.* (1998) also documented numerous cases of gene loss from the ancestral cyanobacterial genome during plastid evolution, and found that parallel losses (occurring within more than one plastid lineage) were more common than unique losses within a single plastid lineage (by a margin of four to one).

The genes coding for rubisco, the carbon dioxide-fixing enzyme in photosynthesis, are an interesting case of gene transfer. The genes for the large subunit of rubisco (*rbc*L) are all plastid-encoded, except in dinoflagellates (discussed later). The small subunit genes (*rbc*S) have been transferred to the nucleus in some eukaryotic photoautotrophs (green algae, land plants, euglenids), but not in others where *rbc*S is still plastid-encoded (red algae, glaucocystophytes, heterokont algae, cryptomonads) (Assali *et al.*, 1991). When *rbc*L sequence data are used to infer algal phylogenies, the picture is confusing and incongruent, with phylogenies based on rRNA and other plastid proteins. There are two types (green and red) of the common form of rubisco (form I). One might have expected the 'green' type of form I, which is found in cyanobacteria and in many proteobacteria, to be present in all plastids. Surprisingly the 'green' form is found only in plastids of green algae, glaucocystophytes and chlorarachniophytes; the 'red' type of form I is found in other proteobacteria and in the plastids of red algae, heterokont algae, cryptomonads and haptophytes (Delwiche and Palmer, 1996). The incongruity has been explained as a case of horizontal gene transfer from a bacterium to an ancestral red algal plastid (and its brown plastid descendants, discussed later), although duplication and differential loss have not been ruled out. The other, less common form of rubisco (form II) is found in dinoflagellates (discussed later).

It is not yet clear if the common ancestor in the trees above was a group of closely related cyanobacteria (monophyletic ancestry) or a plastid (monophyletic origin). If plastids have a monophyletic origin, one would expect phylogenies inferred from mitochondrial and nuclear genes of red algae, green algae and glaucocystophytes to be congruent with the plastid phylogenies. Delwiche and Palmer review the evidence and state, 'the total weight of evidence clearly supports a single primary, cyanobacterial origin of plastids, but equally clearly, substantially more data are needed to achieve a robust, unequivocal, and complete picture' (Delwiche and Palmer, 1997). They also discuss several problems with a new hypothesis for the polyphyletic origin of plastids (Stiller and Hall, 1997).

If plastid monophyly turns out to be true, it will require rethinking about the gain and loss of photosynthetic pigments. One possibility is that the ancestral cyanobacterium contained chlorophylls *a, b, c* and phycobilins, and that present pigment distribution was achieved by differential loss. Another possibility is that the ancestral cyanobacterium contained chlorophyll *a* and perhaps chlorophyll *b* in addition to phycobilins, and that chlorophylls *c* and *d* were acquired later (Bhattacharya and Medlin, 1995; Delwiche and Palmer, 1996).

The unusual diversity of carotenoid pigments within the heterokonts shows that pigmentation does change within a monophyletic lineage; heterokont phylogenies based on pigmentation did not match phylogenies based on molecular sequence data (Daugbjerg and Anderson, 1997). The most striking case is in the raphidophytes (Heterokonta) in which marine forms are brown (like chrysophytes) and freshwater forms are green (like xanthophytes) (Potter *et al.*, 1997), even though the raphidophyte plastid appears to have a monophyletic origin (Daugbjerg and Anderson, 1997). Also intriguing is the discovery of a cyanobacterium-type, oxygen-evolving photoautotroph that lacks phycobilins but contains chlorophyll *d* (hitherto only known from red algae) in amounts ten to thirty times higher than chlorophyll *a* (Miyashita *et al.*, 1996).

15.6.2 Secondary symbiosis

In contrast to mitochondria, plastids have also been acquired through secondary endosymbiosis, an idea first proposed by Taylor (1974). Secondary endosymbiosis involves ingestion of a eukaryotic alga, followed by reduction of the alga to a greater or lesser extent, but always (apparently) leaving one or two additional membranes surrounding the plastid. In contrast to the situation with primary endosymbiosis, it is clear that several different lineages of host eukaryotes are involved (Bhattacharya and Medlin, 1995; Delwiche and Palmer, 1997). Of the three primary plastid endosymbioses, only the red and green plastids are known to be involved in surviving secondary endosymbioses.

The nucleomorph (a vestigal nucleus bounded by two membranes with pores, three chromosomes, and that divides amitotically) in cryptomonads and chlorarachniophytes is the best evidence for secondary endosymbiosis. Phylogenies constructed from plastid molecular sequences confirm what pigment similarities suggest, namely that these plastids are derived from a green alga in the case of chlorarachniophytes and a red alga in the case of cryptomonads (Fraunholz *et al.*, 1997; Ishida *et al.*, 1999; McFadden *et al.*, 1997b). In both cases the plastid is surrounded by two additional membranes. The inner of these two membranes is thought to be derived from the plasma membrane of the algal endosymbiont, while the outer membrane may be derived from the food vacuole of the host. In both cases a periplastid space between the plastid and the inner membrane contains a small volume of cytoplasm, the nucleomorph and some ribosomes.

The existence of independently evolved nucleomorphs in two distantly related host organisms provides an opportunity to study the establishment and evolution of secondary endosymbioses (Fraunholz *et al.*, 1997; McFadden *et al.*, 1997b). Comparisons between the genome of the nucleomorph and those of the original green or red alga will be especially interesting. What happened to genes lost from the nucleomorph? Have they been transferred to the nucleus? Which genes have been retained? Have the same genes been retained in both nucleomorphs and are they expressed?

In the other two groups of algae whose plastids are bounded by additional membranes (heterokont algae and prymnesiophytes), the putative endosymbiotic alga has been reduced to nothing more than its plastid and plasma membrane, with no sign of a nucleomorph. The plastid molecular phylogeny data, however, clearly show that the plastids in these two groups belong to the plastid lineage containing the red algal plastid. By analogy with cryptophytes, the simplest explanation is secondary endosymbiosis. The bigger questions with these algae concern the possible monophyly of the host and/or of the plastid. Although prymnesiophytes were once classified as chrysophytes (Heterokonta), rRNA sequence data suggest that the host organisms of these two groups are not closely related (Bhattacharya and Medlin, 1995; Medlin *et al.*, 1997) and despite similarities in plastid structure and pigmentation the plastids may also not be monophyletic. The heterokonts, despite their extraordinary diversity (for example several lineages of unicellular phytoflagellates, diatoms, kelps and the fungus-like oomycetes), appear to be monophyletic in terms of both the host cells (Leipe *et al.*, 1994; Van de Peer *et al.*, 1996) and their plastids (Daugbjerg and Anderson, 1997).

The remaining plastids thought to be derived from secondary endosymbiosis are those with one extra plastid membrane (three membranes total): the euglenids, dinoflagellates and perhaps apicomplexans (discussed later). There is no evidence beyond the number of plastid-bounding membranes to link the euglenids and dinoflagellates, although dinoflagellates and apicomplexans appear to be related (Silberman *et al.*, 1996). Small subunit rRNA trees suggest that the euglenozoan lineage diverged relatively early, and diversified to a considerable extent before acquiring plastids (Linton *et al.*, 1999).

Dinoflagellates, about half of which are heterotrophic, belong to the 'alveolates', a monophyletic lineage that also includes the ciliates and apicomplexans (Van de Peer *et al.*, 1996). The most common (stereotypical) dinoflagellate plastid is surrounded by three membranes, and contains the xanthophyll peridinin as the major accessory photosynthetic pigment. By analogy with the three bounding membranes of euglenids, it is likely that the dinoflagellate plastid is of secondary origin. With the glaring exception of peridinin, it resembles heterokont and prymnesiophyte plastids. Many dinoflagellates engage in phagotrophy (even some of the autotrophic species) and have even established tertiary endosymbioses. In the absence of a putative ancestral peridinin-containing plastid to examine, we must rely on molecular sequence data which is not yet available on dinoflagellate plastids (Delwiche and

Palmer, 1997). Making the search more difficult is the presence of form II of rubisco, found in certain anaerobic proteobacteria, but not in any other alga (Palmer, 1995). Furthermore, *rbc*L is nuclear-encoded in dinoflagellates, whereas it is plastid-encoded in all other algae. Form II does not have a small subunit. This isolated occurrence of form II among eukaryotes has been explained as another case of horizontal gene transfer (Palmer, 1995).

Available evidence suggests that at least six groups of protists acquired their plastids through secondary endosymbiotic events: euglenids, chlorarachniophytes and perhaps apicomplexans acquired their plastids from a green alga, whereas cryptomonads, heterokont algae and prymnesiophytes acquired their plastids from a red alga. These evolutionary lineages are consistent with the location of the *rbc*S gene (plastid-encoded in the 'red-brown' lineage and nuclear-encoded in the 'green' lineage), but the origin of the dinoflagellate plastid remains unclear.

The occurrence of multiple independent secondary endosymbioses requires the transfer of hundreds of genes, first from the original cyanobacterial endosymbiont to the host eukaryotic nucleus, and then again from a 'nucleomorph' to the second host nucleus. What are the selective pressures at work here? Why has the nucleomorph been retained in two cases and not in the others? What is the significance of the difference between three and four bounding membranes? Did the heterokont and prymnesiophyte plastids evolve from separate endosymbiotic events or not?

15.6.3 Tertiary symbiosis

In addition to those cases in which a eukaryote acquired its plastids through secondary symbiosis, there are rare cases in which the process has been repeated yet again, resulting in a single cell that has a minimum of five separate genomes (three eukaryotic and two prokaryotic (mitochondrion and plastid)). The best documented tertiary symbioses are among the dinoflagellates. In *Dinophysis* the plastid pigments suggest that they are of cryptomonad origin (Schnepf and Elbrächter, 1988). Likewise the dinoflagellate *Lepidodinium viride* has plastids containing chlorophylls *a* and *b* that are thought be derived from a prasinophyte green alga (Watanabe *et al.*, 1990). In two *Peridinium* species the evidence for tertiary symbiosis is even more compelling. The plastids of *P. balticum* and *P. foliaceum* actually belong to the remnants of a pennate diatom contained within the dinoflagellate (Chesnick *et al.*, 1997). Nuclear division of the host and its phototrophic 'guest' is coordinated, such that each division cycle results in a cell that contains both a dinoflagellate nucleus (dinokaryon) and a diatom nucleus (Tippit and Pickett-Heaps, 1976).

Finally there is the fascinating case of *Gymnodinium acidotum*. First described ultrastructurally by Wilcox and Wedemayer (1985), this blue-green dinoflagellate is strikingly different from other dinoflagellates, not only in coloration but also by virtue of what appears to be a cryptomonad symbiont within its cytoplasm. Unlike cryptomonad prey found within food vacuoles of other dinoflagellates (Larsen, 1988), the plastids of the photoautotroph in *G. acidotum* are still retained within the plasma membrane of the endosymbiont, but are distributed along the periphery of the host. Some authors claim that the situation in *G. acidotum* is similar to that of another dinoflagellate, *G. aeruginosum*, and represents a transitory phase that does not result in a permanent relationship between host and symbiont. The term 'klepto-plastid' has been used to refer to the temporary, captured nature of the plastid (Schnepf, 1992). While this may be the case, the internalized cryptomonad of *G. acidotum* is retained long enough for both the host and the symbiont to undergo several cycles of cell division and retain all of the cell constituents (plastids, mitochondria, Golgi and so on), but gradually and selectively lose the cryptomonad nucleus (Farmer and Roberts, 1990).

15.6.4 Loss of plastids

A number of algal lineages contain species that are colourless heterotrophs, but it is not always clear if the absence of plastids is primitive, or involves secondary loss from an autotrophic ancestor. The presence of degenerate plastids in *Polytoma* (Chlorophyta) (Gaffal and Schneider, 1980) is highly suggestive of their being derived from an autotrophic ancestor. Likewise the euglenid *Astasia longa* is evolved from a phototrophic *Euglena*-like ancestor (Linton *et al.*, 1999) and has retained a circular 73kb portion of its plastid genome (Gockel *et al.*, 1994); and rRNA sequence data suggest that plastids have been completely lost in at least two groups of algal heterokonts (Cavalier-Smith *et al.*, 1995).

Perhaps the most surprising example of loss of photosynthesis is the remnant plastid of apicomplexans, a group of obligately parasitic protists best known as the causative agents of human diseases such as malaria and toxoplasmosis (Feagin *et al.*, 1991; Wilson *et al.*, 1991). Most apicomplexans possess a 35kb circular genome that contains a number of ORFs and coding regions, including an inverted repeat of large and small subunit rRNA genes that are characteristic for known chloroplast genomes. This genome is present as a single copy per cell, and is localized within a membrane-bound (probably four membranes) organelle near the nucleus (McFadden *et al.*, 1996). While the ancestry of the 'apicoplast' remains uncertain, it has been suggested to be of red (Williamson *et al.*, 1994) or green algal origin (Egea and Lang-Unnasch, 1995; Köhler *et al.*, 1997). Structural and biochemical features of apicomplexans suggest that they are closely related to dinoflagellates (Silberman *et al.*, 1996), and they are most likely descended from either a phototrophic dinoflagellate or a phototrophic flagellate that was a common ancestor to both groups. Studies of the mechanism whereby proteins are targeted to the apicoplast may aid in our understanding of its ancestry (Waller *et al.*, 1998).

15.6.5 Future directions

The most controversial remaining question is whether primary plastids were acquired only once in evolutionary history, or multiple times involving closely related cyanobacterial ancestors. Examination of independent (non-plastid) genes or genomes (mitochondrial) from protists that harbour primary photoautotrophic symbionts, as well as more detailed examination of plastid genes from these protists, should shed light on this question. The search for the heterotrophic ancestor of red, green, and glaucocystophyte algae remains an area of research. Finally, the discovery of form II rubisco in dinoflagellates raises the possibility that this represents an ancient photoautotroph symbiosis in dinoflagellates that predates their having acquired plastids via secondary or tertiary symbiosis. If dinoflagellates, or their ancestors, were photosynthetic before they acquired plastids from other eukaryotes, they may have been able to utilize an existing rubisco for carbon dioxide fixation at the time they entered into a relationship with a photoautotrophic eukaryote. This would have lessened the need for gene transfer from the symbiont to the dinoflagellate host, and may partially explain why secondary, tertiary, and even 'klepto-' plastids are common in this group of flagellates. It would be fascinating to search for either the remnants of a form II rubisco in the apicomplexans or even a 35kb genome like that of the 'apicoplast' in dinoflagellates.

15.7 Peroxisomes

Peroxisomes are membrane-bound (single membrane) organelles approximately 0.5 to 1.0 μm in diameter, which contain catalase and other enzymes involved in the metabolism of H_2O_2. Peroxisomes were first named by DeDuve (1965) as a distinct class of microbodies. Another microbody-like organelle, the glycosome, is found in kinetoplastid flagellates (Vickerman, 1994). Glycosomes contain enzymes involved in glycolysis, and have a number of other enzymes that are also found in peroxisomes and glyoxysomes (Opperdoes and Michels, 1993). It has been suggested that glycosomes are derived from an endosymbiont (Cavalier-Smith, 1987, 1990; Michels and Hannaert, 1994), an idea supported by the contention that the gene responsible for glyceraldehyde-3-phosphate dehydrogenase in kinetoplastid flagellates was acquired by way of horizontal gene transfer (Wiemer *et al.*, 1995). It is interesting to note that in at least one kinetoplastid, a bacterial symbiont has become established and apparently is dependent on the flagellate's glycosomes as a source of ATP (Motta *et al.*, 1997).

The origin of peroxisomes is enigmatic. They were originally thought to be derived from the cell's endomembrane system (DeDuve, 1969) but it has subsequently been suggested that, like mitochondria and plastids, they too are of endosymbiotic origin (DeDuve, 1982). Despite the fact that peroxisomes lack DNA (Kamiryo *et al.*, 1982), several factors suggest an endosymbiotic origin. Like mitochondria and plastids, peroxisomes are formed by division of pre-existing peroxisomes (Lazarow and Fujiki, 1985). Peroxisomal enzymes are synthesized on free ribosomes in the host cytoplasm and then transported into the peroxisome (Small, 1993). As is the case for mitochondria, translocation of cytoplasmically translated proteins into the peroxisome is mediated by chaperone proteins, but unlike mitochondria, protein import into the peroxisome does not require the unfolding of targeted protein and utilizes a dissimilar targeting mechanism (Crookes and Olsen, 1999).

One interesting hypothesis is that peroxisomes, or their bacterial endosymbiotic ancestors, predate the origin of mitochondria (DeDuve, 1996). As such, peroxisomes would have provided their host with a biochemical means of dealing with free oxygen (O_2) that was accumulating in the early Earth's atmosphere. By first converting oxygen into H_2O_2, then combining H_2 with H_2O_2 to form two molecules of water (H_2O) (Masters and Crane, 1995), the peroxisome would have been of significant benefit to an anaerobic host in an increasingly hostile environment of oxygen (DeDuve, 1996). The early eukaryotes would thus have been aerotolerant rather than strictly anaerobic. They would have the ability to deal with rising levels of oxygen, and be able to venture into oxygenated environments where they could encounter an α-proteobacterium. It is possible that the ability to translocate cytoplasmically synthesized proteins across the peroxisome membrane was acquired at this time. Both peroxisomes and mitochondria utilize Hsp70 and Hsp60-like chaperones to aid in protein translocation (Crookes and Olsen, 1999), and this may have made it easier for an endosymbiosis with an α-proteobacterium to become established. The continued presence of peroxisomes in flagellates that also possess mitochondria may be due to the host cell's reliance on peroxisomes for activities such as the metabolism of fatty acids (DeDuve, 1996).

15.7.1 Future directions

The endosymbiotic origin of peroxisomes remains a controversial topic. The absence of a distinct genome makes their prokaryotic ancestry less tenable, but the lack of DNA does not preclude that organelle being of endosymbiotic origin (for example hydrogenosome). This will make it more difficult to find a free-living candidate as has been done for mitochondria (α-proteobacterium) and plastids (cyanobacterium). By careful analysis of those genes that code for peroxisomal enzymes, and by gaining a better understanding of the proteins involved in translocation across the peroxisomal membrane, it may be possible to identify a bacterium that may have given rise to the peroxisomes in flagellates.

15.8 Summary

The vast majority of flagellates, indeed all eukaryotes, are complex organisms formed by the union of two or more cells. When first fully articulated by Wallin (1927) and later championed by Margulis (Sagan, 1967; Margulis, 1970, 1993), the symbiotic theory was heretical. While it is generally accepted as being correct for the origins of mitochondria and plastids, it remains controversial as an explanation for the origins of other organelles. Clearly neither the autogenesis nor the symbiotic theory alone can fully explain the complexity that is the flagellate cell; the answer lies in some combination of the two. The ability of prokaryotes to form simple (planctomycete nuclear bodies) to complex (cyanobacterial thylakoids) internal membranous structures argues in favor of the idea that ER and Golgi are of autogenous origin. Likewise the presence of separate and reduced genomes in plastids and mitochondria, that have a phylogenetic relationship to free living bacteria, clearly supports the idea that these organelles are of endosymbiotic origin. The origins of flagella, peroxisomes and the nucleus itself remain controversial. As has been shown with hydrogenosomes, the lack of a unique genome that segregates with that organelle cannot be taken as evidence that the organelle arose through the process of autogenesis. A careful and complete examination of the unique proteins associated with an organelle, and the origin of the genes that code for those proteins, is needed to achieve a more complete understanding of the history of the eukaryotic cell.

REFERENCES

Albrecht-Buehler, G. (1992) Function and formation of centrioles and basal bodies. In *The Centrosome* (ed. V. I. Kalinis), San Diego: Academic Press, pp. 70–102.

Altmann, R (1890) *Die Elementarorganismen und ihre Beziehungen zu den Zellen*. Leipzig: Veit. u. Comp.

Andersson, S. G. E., Zomorodipour, A., Andersson, J. O., Sicheritz-Ponten, T., Alsmark, U. C. M., Podowski, R. M., Näslund, A. K., Eriksson, A-S., Winkler, H. H. and Kurland, C. G. (1998) The genome sequence of *Rickettsia prowazekii* and the origin of mitochondria. *Nature*, **396**, 133–140.

Assali, N-E., Martin, W. F., Sommerville, C. C. and Loiseaux-de Goër, S. (1991) Evolution of the rubisco operon from prokaryotes to algae: structure and analysis of the *rbc*S gene of the brown alga *Pylailella littoralis*. *Plant Molecular Biology*, **17**, 853–863.

Barns, S. M., Delwiche, C. F., Palmer, J. D., and Pace, N. R. (1996) Perspectives on archaeal diversity, thermophily and monophyly from environmental ribosomal-RNA sequences. *Proceedings of the National Academy of Sciences US*, **93**, 9188–9193.

Bermudes, D., Hinkle, G. and Margulis, L. (1994) Do prokaryotes contain microtubules? *Microbiological Reviews*, 58, 387–400.

Bhattacharya, D. (1997) *Origins of Algae and their Plastids*. Vienna/New York: Springer-Verlag.

Bhattacharya, D. and Medlin, L. (1995) The phylogeny of plastids: a review based on comparisons of small-subunit ribosomal RNA coding regions. *Journal of Phycology*, **31**,489–498.

Blanchard, J. L. and Schmidt, G. W. (1995) Pervasive migration of organellar DNA to the nucleus in plants. *Journal of Molecular Evolution*, **41**, 397–406.

Bui, E. T. N., Bradley, P. J. and Johnson, P. J. (1996) A common evolutionary origin for mitochondria and hydrogenosomes. *Proceedings of the National Academy of Sciences US*, **93**, 9651–9656.

Bult, C. J., White, O., Olsen, G. J., Zhou, L., Fleischmann, R. D., Sutton, G. G., Blake, J. A., Fitzgerald, L. M., Clayton, R. A., Gocayne, J. D., Kerlavage, A. R., Dougherty, B. A., Tomb, J-F., Adams, M. D., Reich, C. I., Overbeek, R., Kirkness, E. F., Weinstock, K. G., Merrick, J. M., Glodek, A., Scott, J. L., Geoghagen, N. S. M., Weidman, J. F., Fuhrmann, J. L., Nguyen, D., Utterback, T. R., Kelley, J. M., Peterson, J. D., Sadow, P. W., Hann, M. C., Cotton, M. D., Roberts, K. M., Hurst, M. A., Kaine, B. P., Borodovsky, M., Klenk, H–P., Fraser, C. M., Smith, H. O., Woese, C. R. and Venter, J. C. (1996) Complete genome sequence of the methanogenic archaeon, *Methanococcus jannaschii*. *Science*, **273**, 1058–1073.

Cachon, J., Cachon, M. and Estep, K. W. (1990) Phylum Actinopoda, classes Polycystina (= Radiolaria) and Phaeodaria. In *Handbook of Protoctista* (eds. L. Margulis, J. O. Corliss, M. Melkonian and D. J. Chapman), Boston: Jones and Bartlett, pp. 334–346.

Cavalier-Smith, T. (1982) The origins of plastids. *Biological Journal of the Linnaean Society*, **17**, 289–306.

Cavalier-Smith, T. (1983a) Endosymbiotic origin of the mitochondrial envelope. In *Endocytobiology* II (eds W. Schwemmler and H. E. A. Schenk), Berlin: de Gruyter, pp. 265–279.

Cavalier-Smith, T. (1983b) A 6–kingdom classification and a unified phylogeny. In *Endocytobiology* II (eds W. Schwemmler and H. E. A. Schenk), Berlin: de Gruyter, pp. 1027–1034.

Cavalier-Smith, T. (1987) The origin of cells: a symbiosis between genes, catalysts, and membranes. *Cold Spring Harbor Symposia on Quantitative Biology*, **52**, 805–824.

Cavalier-Smith, T. (1990) Symbiotic origin of peroxisomes. In *Endocytobiology* IV (eds P. Nardon, V. Gianinazzi-Pearson, A. M. Grenier, L. Margulis and D. C. Smith), Paris: Institut National de la Recherche Agronomique, pp. 515–521.

Cavalier-Smith, T. (1991) The evolution of prokaryotic and eukaryotic cells. In *Fundamentals of Medical Cell Biology, vol. 1*, JAI Press, pp. 217–272.

Cavalier-Smith, T. (1992) Origin of the cytoskeleton. In *The Origin and Evolution of the Cell* (eds H. Hartman and K. Matsuno), Singapore: World Scientific, pp. 79–106.

Cavalier-Smith, T. (1993) Evolution of the eukaryotic genome. In *The Eukaryotic Genome: Organization and Regulation* (eds. P. Orda, S. G. Oliver and P. F. G. Sims), Cambridge: Cambridge University Press, pp. 333–385.

Cavalier-Smith, T., Chao, E. E. and Allsopp, M. T. E. P. (1995) Ribosomal-RNA evidence for chloroplast loss within Heterokonta – pedinellid relationships and a revised classification of ochristan algae. *Archiv für Protistenkunde*, **145**, 209–220.

Chesnick, J. M., Kooistra, W. H. C. F., Wellbrock, U. and Medlin, L. K. (1997) Ribosomal RNA analysis indicates a benthic pennate diatom ancestry for the endosymbionts of the dinoflagellates *Peridinium foliaceum* and *Peridinium balticum* (Pyrrhophyta). *Journal of Eukaryotic Microbiology*, **44**, 314–320.

Choi, J. Y., Lee, T. W., Jeon, K. W. and Tae, L. A. (1997) Evidence for symbiont-induced alteration of a host's gene expression: irreversible loss of SAM synthetase from *Amoeba proteus*. *Journal of Eukaryotic Microbiology*, **44**, 412–419.

Cleveland, L. R. (1934) *The Wood-Eating Roach* Cryptocercus, *its Protozoa, and the Symbiosis Between Protozoa and Roach*. Menasha, Wisconsin: George Banta.

Cosson, J. (1996) A moving image of flagella: news and views on the mechanisms involved in axonemal beating. *Cell Biology International*, **20**, 83–94.

Crookes, W. J. and Olsen, L. J. (1999) Peroxin puzzles and folded freight: peroxisomal protein import in review. *Naturwissenschaften*, **86**, 51–61.

Daugbjerg, N. and Anderson, R. A. (1997) A molecular phylogeny of the heterokont algae based on analysis of chloroplast-encoded *rbc*L sequence data. *Journal of Phycology*, **33**, 1031–1041.

DeDuve, C. (1965) Functions of microbodies (peroxisomes). *Journal of Cell Biology*, **27**, 25A–26A.

DeDuve, C. (1969) Evolution of the peroxisome. *Annals of the New York Academy of Sciences*, **168**, 369–381.

DeDuve, C. (1982) Peroxisomes and related particles in historical perspective. *Annals of the New York Academy of Sciences*, **386**, 1–4.

DeDuve, C. (1996) The birth of complex cells. *Scientific American*, **274**, 50–57.

Delwiche, C. F. and Palmer, J. D. (1996) Rampant horizontal transfer and duplication of rubisco genes in eubacteria and plastids. *Molecular Biology and Evolution*, **13**, 873–882.

Delwiche, C. F. and Palmer, J. D. (1997) The origin of plastids and their spread via secondary symbiosis. *Plant Systematics and Evolution* (Suppl.), **11**, 53–86.

Derry, L. A., Kaufman, A. J., and Jacobsen, S. B. (1992) Sedimentary cycling and environmental change in the late Proterozoic: evidence from stable and radiogenic isotopes. *Geochimica and Cosmochimica Acta*, **56**, 1317–1329.

Dibb, N. J. (1993) Why do genes have introns? *Federation of European Biochemical Societies*, **325**, 135–139.

Egea, N. and Lang-Unnasch, N. (1995) Phylogeny of the large extrachromosomal DNA of organisms in the phylum Apicomplexa. *Journal of Eukaryotic Microbiology*, **42**, 679–684.

Embley, T. M., Finlay, B. J., Dyal, P. L., Hirt, R. P., Wilkinson, M. and Williams, A. G. (1995) Multiple origins of anaerobic ciliates with hydrogenosomes within the radiation of aerobic ciliates. *Proceedings of the Royal Society of London, Series B*, **262**, 87–93.

Farmer, M. A. and Roberts, K. R. (1990) Organelle loss in the endosymbiont of *Gymnodinium acidotum* (Dinophyceae). *Protoplasma*, **153**, 178–185.

Feagin, J. E., Gardner, M. J., Williamson, D. H. and Wilson, R. J. M. (1991) The putative mitochondrial genome of *Plasmodium falciparum*. *Journal of Protozoology*, **38**, 243–245.

Febvre, J. (1990) Phylum Actinopoda, Class Acantharia. In *Handbook of Protoctista* (eds L. Margulis, J. O. Corliss, M. Melkonian and D. J. Chapman), Boston: Jones and Bartlett, pp. 363–379.

Febvre–Chevalier, C. (1990) Phylum Actinopoda, Class Heliozoa. In *Handbook of Protoctista* (eds L. Margulis, J. O. Corliss, M. Melkonian and D. J. Chapman), Boston: Jones and Bartlett, pp. 347–362.

Fields, S. D. and Rhodes, R. G. (1991) Ingestion and retention of *Chroomonas* spp. (Cryptophyceae) by *Gymnodinium acidotum* (Dinophyceae). *Journal of Phycology*, **27**, 525–529.

Flavin, M. and Nerad, T. A. (1993) *Reclinomonas americana* n.g., n. sp., a new freshwater heterotrophic flagellate. *Journal of Eukaryotic Microbiology*, **40**, 172–179.

Fraunholz, M. J., Wastl, J., Zauner, S., Rensing, S. A., Scherzinger, M. M. and Maier, U-G. (1997) The evolution of Cryptophytes. *Plant Systematics and Evolution* (Suppl.), **11**, 163–174.

Fuerst, J. A. (1995) The Planctomycetes: emerging models for microbial ecology, evolution and cell biology. *Microbiology*, **141**, 1493–1506.

Fuerst, J. A. and Webb, R. I. (1991) Membrane-bounded nucleoid in the eubacterium *Gemmata obscuriglobus*. *Proceedings of the National Academy of Sciences US*, **88**, 8184–8188.

Gaffal, K. P. and Schneider, G. J. (1980) Morphogenesis of the plastidome and flagellar apparatus during the vegetative life cycle of the colorless phytoflagellate *Polytoma papillatum*. *Cytobios*, **105**, 43–61.

Germot, A., Philippe, H. and Leguyader, H. (1996) Presence of a mitochondrial-type 70-KDA heat-shock protein in *Trichomonas vaginalis* suggests a very early mitochondrial endosymbiosis in eukaryotes. *Proceedings of the National Academy of Sciences US*, **93**, 14614–14617.

Gibbons, I. R. (1995) Dynein family of motor proteins: present status and future questions. *Cell Motility and the Cytoskeleton*, **2**, 136–144.

Gillin, F. D., Reiner, D. S. and McCaffery, J. M. (1996) Cell biology of the primitive eukaryote *Giardia lamblia*. *Annual Reviews in Microbiology*, **50**, 679–705.

Gilson, P. R. and McFadden, G. I. (1996) The miniaturized nuclear genome of a eukaryotic endosymbiont contains genes that overlap, genes that are co-transcribed, and the smallest known spliceosomal introns. *Proceedings of the National Academy of Sciences US*, **93**, 7737–7742.

Gilson, P. R. and McFadden, G. I. (1997) Good things in small packages – the tiny genomes of chlorarachniophyte endosymbionts. *BioEssays*, **19**, 167–173.

Gockel, G., Hachtel, W., Baier, S., Fliss, C. and Henke, M. (1994) Genes for components of the chloroplast translational apparatus are conserved in the reduced 73-kb plastid DNA of the nonphotosynthetic euglenoid flagellate *Astasia longa*. *Current Genetics*, **26**, 256–262.

Golding, G. B. and Gupta, R. S. (1995) Protein-based phylogenies support a chimeric origin for the eukaryotic genome. *Molecular Biology and Evolution*, **12**, 1–6.

Golz, R. and Hauser, M. (1994) Spatially separated classes of microtubule bridges in the reticulopodial network of *Allogromia*: evidence for a dynein-like ATPase in the filopodial cytoskeleton. *European Journal of Protistology*, **30**, 221–226.

Gray, M. W. (1992) The endosymbiont hypothesis revisited. *International Review of Cytology*, **141**, 233–357.

Gray, M. W., Burger, G., and Lang, B. F. (1999) Mitochondrial evolution. *Science*, **283**, 1476–1481.

Gregson, A. J., Green, J. C. and Leadbeater, B. S. C. (1993) Structure and physiology of the haptonema in *Chrysochromulina* (Prymnesiophyceae). II. Mechanisms of haptonematal coiling and the regeneration process. *Journal of Phycology*, **29**, 686–700.

Gupta, R. S. (1995) Evolution of the chaperonin families (HSP6O, HSP1O and TCP-1) of proteins and the origin of eukaryotic cells. *Molecular Microbiology*, **15**, 1–11.

Hall, J. L., Ramanis, Z. and Luck, D. L. (1989) Basal body/centriolar DNA: molecular genetic studies in *Chlamydomonas*. *Cell*, **59**, 121–132.

Hall, J. L. and Luck, D. L. (1995) Basal body-associated DNA: *in situ* studies in *Chlamydomonas reinhardtii*. *Proceedings of the National Academy of Sciences US*, **92**, 5129–5133.

Hashimoto, T., Sanchez, L. B., Shirakura, T., Müller, M. and Hasegawa, M. (1998) Secondary absence of mitochondria in *Giardia lamblia* and *Trichomonas vaginalis* revealed by valyl-tRNA synthetase phylogeny. *Proceedings of the National Academy of Sciences US*, **95**, 6860–6865.

Holland, H. D. (1995) Atmospheric oxygen and the biosphere. In *Linking Species and Ecosystems* (eds C. G. Jones and J. H. Lawton), New York: Chapman and Hall, pp. 127–140.

Holland, H. D. (1997) Evidence for life on Earth more than 3850 million years ago. *Science*, **275**, 38–39.

Horner, D. S., Hirt, R. P., Kilvington, S., Lloyd, D. and Embley, T. M. (1996) Molecular-data suggest an early acquisition of the mitochondrion endosymbiont. *Proceedings of the Royal Society of London, Series B*, **263**, 1053–1059.

Ishida, K., Green, B. R., and Cavalier-Smith, T. (1999) Diversification of a chimaeric algal group, the chlorarachniophytes: phylogeny of nuclear and nucleomorph small-subunit rRNA genes. *Molecular Biology and Evolution*, **16**, 321–331.

Jeon, K. W. and Lorch, I. J. (1967) Unusual intra-cellular bacterial infection in large, free living amoebae. *Experimental Cell Research*, **48**, 236–240.

Kamiryo, K., Abe, M., Okazaki, K., Kato, S. and Shimamoto, N. (1982) Absence of DNA in peroxisomes of *Candida tropicalis*. *Journal of Bacteriology*, **152**, 269–274.

Kasting, J. F (1993) Earth's early atmosphere. *Science*, **259**, 920–926.

Kawachi, M. and Inouye, I. (1995) Functional roles of the haptonema and the spine scales in the feeding process of *Chrysochromulina spinifera* (Fournier) Pienaar et Norris (Haptophyta = Prymnesiophyta). *Phycologia*, **34**, 193–200.

Kies, L. and Kremer, B. P. (1990) Phylum Glaucocystophyta. In *Handbook of Protoctista* (eds. L. Margulis, J. O. Corliss, M. Melkonian and D. J. Chapman), Boston: Jones and Bartlett, pp. 152–166.

Köhler, S., Delwiche, C. F., Denny, P. W., Tilney, L. G., Webster, P., Wilson, R. J. M., Palmer, J. D. and Roos, D. S. (1997) A plastid of probable green algal origin in apicomplexan parasites. *Science*, **275**, 1485–1487.

Lang, B. F., Burger, G., O'Kelly, C. J., Cedergren, R., Golding, G. B., Lemleux, C., Sankoff, D., Turmel, M. and Gray, M. W. (1997) An ancestral mitochondrial DNA resembling a eubacterial genome in miniature. *Nature*, **387**, 493–497.

Langford, G. M. and Inoue, S. (1979) Motility of the microtubular axostyle in *Pyrsonympha*. *Journal of Cell Biology*, **80**, 521–538.

Larsen, J. (1988) An ultrastructural study of *Amphidinium poecilochroum* (Dinophyceae), a phagotrophic dinoflagellate feeding on small species of cryptophytes. *Phycologia*, **27**, 366–377.

Lazarow, P. B. and Fujiki, Y. (1985) Biogenesis of peroxisomes. *Annual Review of Cell Biology*, **1**, 489–530.

Leadbetter, J . R., Schmidt, T. M., Graber, J. R. and Breznak, J. A. (1999) Acetogenesis from H_2 plus CO_2 by spirochetes from termite guts. *Science*, **283**, 686–689.

Leipe, D. D., Wainright, P. O., Gunderson, J. H., Porter, D., Patterson, D. J., Valois, F., Himmerich, S. and Sogin, M. L. (1994) The stramenopiles from a molecular perspective: 16S-like rRNA sequences from *Labyrinthuloides minuta* and *Cafeteria roenbergensis*. *Phycologia*, **33**, 369–377.

Lewin, R. A. (1975) A marine *Synechocystis* (Cyanophyta, Chroococcales) epizoic on ascidians. *Phycologia*, **14**, 153–160.

Lindsay, M. R., Webb, R. I. and Fuerst, J. A. (1997) Pirellulosomes: a new type of membrane-bounded cell compartment in planctomycete bacteria of the genus *Pirellula*. *Microbiology*, **143**, 739–748.

Linton, E. W., Hittner, D., Lewandowski, C., Auld, T. and Triemer, R. E. (1999) A molecular study of euglenoid phylogeny using small subunit rDNA. *Journal of Eukaryotic Microbiology*, **46**, 217–223.

López-Garcia, P. and Moreira, D. (1999) Metabolic symbiosis at the origin of eukaryotes. *TIBS*, **24**, 88–93.

Lowe, J. and Amos, L. A. (1998) Crystal structure of the bacterial cell-division protein FtsZ. *Nature*, **391**, 203–206.

Luján, H. D., Marotta, A., Mowatt, M. R., Sciaky, N., Lippincott-Schwartz, J. and Nash, T. E. (1995) Developmental induction of Golgi structure and function in the primitive eukaryote *Giardia lamblia*. *Journal of Biological Chemistry*, **270**, 4612–4618.

Mai, Z. M, Ghosh, S., Frisardi, M., Rosenthal, B., Rogers, R. and Samuelson, J. (1999) Hsp60 is targeted to a cryptic mitochondrion-derived organelle ('crypton') in the microaerophilic protozoan parasite *Entamoeba histolytica*. *Molecular Cell Biology*, **19**, 2198–2205.

Margulis, L. (1970) *Origin of Eukaryotic Cells*. New Haven: Yale University Press.

Margulis, L. (1993) *Symbiosis in Cell Evolution: Microbial Communities in the Archean and Proterozoic Eons*, 2nd edn. New York: W. H. Freeman.

Margulis, L. (1996) Archaeal-eubacterial mergers in the origin of Eukarya: phylogenetic classification of life. *Proceedings of the National Academy of Sciences US*, **93**, 1071–1076.

Margulis, L., Chase, D., and To, L. (1979) Possible evolutionary significance of spirochaetes. *Proceedings of the Royal Society of London, Series B*, **204**, 189–198.

Martin, W. and Müller, M. (1998) The hydrogen hypothesis for the first eukaryote. *Nature*, **392**, 37–41.

Martin, W., Stoebe, B., Goremykin, V., Hansmann, S., Hasegawa, M. and Kowallik, K. V. (1998) Gene transfer to the nucleus and the evolution of chloroplasts. *Nature*, **393**, 162–165.

Masters, C. and Crane, D. (1995) *The Peroxisome: A Vital Organelle*. Cambridge: Cambridge University Press.

McFadden, G. I., Reith, M. E., Munholland, J. and Lang-Unnasch, N. (1996) Plastid in human parasites. *Nature*, **381**, 482.

McFadden, G. I., Gilson, P. R., Douglas, S. E., Cavalier-Smith, T., Hofmann, C. J. B. and Maier, U. G. (1997a) Bonsai genomics: sequencing the smallest eukaryotic genomes. *Trends in Genetics*, **13**, 46–49.

McFadden, G. I., Gilson, P. R. and Hofmann, C. J. B. (1997b) Division *Chlorarachniophyta*. *Plant Systematics and Evolution* (Suppl.), **11**, 175–185.

Medlin, L. K., Wiebe, H. C. F., Potter, D., Saunders, G. W. and Anderson, R. A. (1997) Phylogenetic relationships of the 'golden algae' (haptophytes, heterokont chromophytes) and their plastids. *Plant Systematics and Evolution* (Suppl.), **11**, 187–219.

Michels, P. A. M. and Hannaert, V. (1994) The evolution of kinetoplastid glycosomes. *Journal of Bioenergetics and Biomembranes*, **26**, 213–219.

Minguez, A., Franca, S. and Diaz de la Espina, S. M. (1994) Dinoflagellates have a eukaryotic nuclear matrix with lamin-like proteins and topoisomerase II. *Journal of Cell Science*, **107**, 2861–2873.

Mitchison, T. J. (1995) Evolution of a dynamic cytoskeleton. *Philosophical Transactions of the Royal Society of London*, **349**, 299–304.

Miyashita, H., Ikemoto, H., Kurano, N., Adechi, K., Chihara, M. and Miyachi, S. (1996) Chlorophyll *d* as a major pigment. *Nature*, **383**, 402.

Motta, M. C. M., Soares, M. J., Attias, M., Morgado, J., Lemos, A. D. P., Saad Nehme, J.,

Meyer Fernandes, J. R. and DeSouza, W. (1997) Ultrastructural and biochemical analysis of the relationship of *Crithidia deanei* with its endosymbiont. *European Journal of Cell Biology*, **72**, 370–377.

Müller, M. (1993) The hydrogenosome. *Journal of General Microbiology*, **139**, 2879–2889.

Nunn, J. F. (1998) Evolution of the atmosphere. *Proceedings of the Geologists Association*, **109**, 1–13.

O'Kelly, C. J. (1997) Ultrastructure of trophozoites, zoospores and cysts of *Reclinomonas americana* Flavin & Nerad, 1993 (Protista *incertae sedis*: Histionidae). *European Journal of Protistology*, **33**, 337–348.

Opperdoes, F. R. and Michels, P. A. M. (1993) The glycosomes of the kinetoplastida. *Biochimie*, **75**, 231–234.

Palenik, B. and Haselkorn, R. (1992) Multiple evolutionary origins of prochlorophytes, the chlorophyll *b*-containing prokaryotes. *Nature*, **355**, 265–267.

Palmer, J. D. (1995) Rubisco rules fall; gene transfer triumphs. *BioEssays*, **17**, 1005–1008.

Palmer, J. D. (1997) The mitochondrion that time forgot. *Nature,* **387**, 454–455.

Potter, D., Saunders, G. W. and Anderson, R. A. (1997) Phylogenetic relationships of the *Raphidophyceae* and *Xanthophyceae* as inferred from nucleotide sequences of the 18S ribosomal RNA gene. *American Journal of Botany*, **84**, 966–972.

Raikov, I. B. (1994) The diversity of forms of mitosis in protozoa: a comparative review. *European Journal of Protistology*, **30**, 253–269.

Raven, P. H. (1970) A multiple origin for plastids and mitochondria. *Science*, **169**, 641–646.

Reiner, D. S., McCaffery, M. and Gillin, F. D. (1990) Sorting of cyst wall proteins to a regulated secretory pathway during differentiation of the primitive eukaryote, *Giardia lamblia. European Journal of Cell Biology*, **53**, 142–153.

Rizzotti, M. (1995) Cilium: origin and 9-fold symmetry. *Acta Biotheoretica*, **43**, 227–240.

Roger, A. J., Clark, C. G. and Doolittle, W. F. (1996) A possible mitochondrial gene in the early-branching amitochondriate protist *Trichomonas vaginalis. Proceedings of the National Academy of Sciences US*, **93**, 14618–14622.

Roger, A. J., Svard, S. G., Tovar, J., Clark, C. G., Smith, M. W., Gillin, F. D. and Sogin, M. L. (1998) A mitochondrial-like chaperonin 60 gene in *Giardia lamblia*: evidence that diplomonads once harbored an endosymbiont related to the progenitor of mitochondria. *Proceedings of the National Academy of Sciences US*, **95**, 229–234.

Roger, A. J., Morrison, H. G. and Sogin, M. L. (1999) Primary structure and phylogenetic relationships of a malate dehydrogenase gene from *Giardia lamblia. Journal of Molecular Evolution,* **48**, 750–755.

Sagan, L. (1967) On the origin of mitosing cells. *Journal of Theoretical Biology*, **14**, 225–274.

Santos, L. M. A. and Leedale, G. F. (1992) First report of a Golgi body in a uniflagellate eustigmatophycean zoospore. *Phycologia*, **31**, 119–124.

Schimper, A. F. W. (1883) Über die Entwicklung der Chlorophyllkorner und Farbkörper. *Botanische Zeitung*, **41**, 105.

Schnepf, E. (1992) From prey via endosymbiont to plastid: comparative studies in dinoflagellates. In *Origins of Plastids* (ed. R. A. Lewin), New York and London: Chapman and Hall, pp. 53–76.

Schnepf, E. and Elbrächter, M. (1988) Cryptophycean-like double membrane-bound chloroplast in the dinoflagellate, *Dinophysis* Ehrenb: evolutionary, phylogenetic and toxicological implications. *Botanica Acta*, **101**, 196–203.

Schopf, J. W. (1993) Microfossils of the early Archean apex chert: new evidence of the antiquity of life. *Science*, **260**, 640–646.

Schopf, J. W. (1994) The early evolution of life: solution to Darwin's dilemma. *Trends in Ecology and Evolution*, **9**, 375–377.

Silberman, J. D., Sogin, M. L., Leipe, D. D. and Clark, C. G. (1996) Human parasite finds a taxonomic home. *Nature*, **380**, 398.

Simpson, A. G. B., Van der Hoff, J., Bernard, C., Burton, H. R. and Patterson, D. J. (1997) The ultrastructure and systematic position of the euglenozoon *Postgaardi mariagerensis*, Fenchel *et al. Archiv für Protistenkunde*, **147**, 213–225.

Sleigh, M. A. (1986) The origin of flagella: autogenous or symbiontic? *Cell Motility and the Cytoskeleton*, **6**, 96–98.

Small, G. M. (1993) Peroxisome biogenesis. In *Peroxisomes: Biology and Importance in Toxicology and Medicine* (eds G. Gibson and B. Lake), Washington, D.C.: Taylor and Francis, pp. 1–17.

Smith, M. W., Aley, S. B., Sogin, M., Gillin, F. D. and Evans, G. A. (1998) Sequence survey of the *Giardia lamblia* genome. *Molecular and Biochemical Parasitology*, **95**, 267–280.

Sogin, M. L. (1991) Early evolution and the origin of eukaryotes. *Current Opinion in Genetics and Development*, **1**, 457–463.

Sogin, M. L. (1997) History assignment: when was the mitochondrion found? *Current Opinion in Genetics and Development*, **7**, 792–799.

Stiller, J. W. and Hall, B. D. (1997) The origin of red algae: implications for plastid evolution. *Proceedings of the National Academy of Sciences US*, **88**, 8184–8188.

Taylor, F. J. R. (1974) Implications and extensions of the serial endosymbiosis theory of the origin of eukaryotes. *Taxon*, **23**, 229–258.

Thorsness, P. E. and Weber, E. R. (1996) Escape and migration of nucleic acids between chloroplasts, mitochondria and the nucleus. *International Review of Cytology*, **165**, 207–234.

Tippit, D. H. and Pickett-Heaps, J. D. (1976) Apparent amitosis in the binucleate dinoflagellate *Peridinium balticum*. *Journal of Cell Science*, **21**, 273–289.

Travis, J. L. and Bowser, S. S. (1991) The motility of foraminifera. In *Biology of Foraminifera* (eds J. J. Lee and O. R. Anderson), San Diego: Academic Press, pp. 91–155.

Turner, S. (1997) Molecular systematics of oxygenic photosynthetic bacteria. *Plant Systematics and Evolution* (Suppl.), **11**, 13–52.

Urbach, E., Robertson, D. L. and Chisholm, S. W. (1992) Multiple evolutionary origins of prochlorophytes within the cyanobacterial radiation. *Nature*, **355**, 267–270.

Van de Peer, Y., Van der Auwera, G. and De Wachter, R. (1996) The evolution of stramenopiles and alveolates as derived by 'substitution rate calibration' of small ribosomal subunit RNA. *Journal of Molecular Evolution*, **42**, 201–210.

Vellai, T., Takács, K. and Vida, G. (1998) A new aspect to the origin and evolution of eukaryotes. *Journal of Molecular Evolution*, **46**, 499–507.

Vickerman, K. (1994) The evolutionary expansion of the trypanosomatid flagellates. *International Journal for Parasitology*, **24**, 1317–1331.

Waller, R. F., Keeling, P. J., Donald, R. G. K., Striepen, B., Handman, E., Lang-Unnasch, N., Cowman, A. F., Besra, G. S., Roos, D. S. and McFadden, G. I. (1998) Nuclear-encoded proteins target to the plastid in *Toxoplasma gondii* and *Plasmodium falciparum*. *Proceedings of the National Academy of Sciences US*, **95**, 12352–12357.

Wallin, I. E. (1927) *Symbionticism and the Origin of Species*. Baltimore: Williams and Wilkins.

Watanabe, M. M., Suda, S., Inouye, I., Sawaguchi, T. and Chihara, M. (1990) *Lepidodinium viride* gen. et sp. nov. (Gymnodiniales, Dinophyta), a green dinoflagellate with a chlorophyll *a*- and *b*-containing endosymbiont. *Journal of Phycology*, **26**, 741–751.

Wiemer, E. A. C., Hannaert, V., Vandenijssel, P. R. L. A., Vanroy, J. O., Opperdoes, F. R. and Michels, P. A. M. (1995) Molecular analysis of glyceraldehyde-3-phosphate dehydrogenase in *Trypanoplasma borelli*: an evolutionary scenario of subcellular compartmentation in kinetoplastida. *Journal of Molecular Evolution*, **40**, 443–454.

Wilcox, L. W., and Wedemayer, G. J. (1985) Dinoflagellate with blue-green chloroplasts derived from an endosymbiotic eukaryote. *Science*, **227**, 192–194.

Williamson, D. H., Gardner, M. J., Preiser, P., Moore, D. J., Rangachari, K. and Wilson, R. J. M. (1994) The evolutionary origin of the 35 KB circular DNA of *Plasmodium falciparum*: new evidence supports a possible rhodophyte ancestry. *Molecular and General Genetics*, **243**, 249–252.

Wilson, R. J. M., Gardner, M. J., Feagin; J. E. and Williamson, D. H. (1991) Have malaria parasites three genomes? *Parasitology Today*, 7, 134–136.

Woodrum, D. T. and Linck, R. W. (1980) Structural basis of motility in the microtubular axostyle: implication for cytoplasmic microtubule structure and function. *Journal of Cell Biology*, 87, 404–414.

Yamaguchi, T. and Anderson, O. R. (1994) Fine structure of laboratory cultured *Distigma proteus* and cytochemical localization of acid phosphatase. *Journal of Morphology*, **219**, 89–99.

Chapter 16

Flagellate phylogeny

An ultrastructural approach

Serguei A. Karpov

ABSTRACT

Traditionally, morphological characteristics have been almost the sole source of information in determining the possible evolutionary relationships of flagellates. With the advent of electron microscopy in 1960s, much more detailed and reliable information became available. A combination of diverse and conservative structures have been found. The latter have been used to construct *evolutionary rows* of homologous elements, which probably reflect the evolution of organelles and other cell structures.

One of the most informative structures is the flagellar apparatus, which is very complex and seems to have evolved only once. The most conservative parts of flagellar apparatus are the basal body and axoneme, which do not show enough variability to be used for morphological rows. Other flagellar structures are more variable, and are more useful for taxonomic and phylogenetic reconstructions. The distribution of flagellar appendages shows that polysaccharide tubular mastigonemes and scales cover entire flagella, whereas glycoproteinous tubular hairs cover just the anterior flagellum. This characteristic distinguishes green flagellates from heterokonts. There are at least three unique structures of the flagellar transition zone that mark monophyletic taxa, namely the stellate structure, central filament, and transitional helix. The spiral fibre or concentric rings occur in different unrelated taxa. The flagellar rootlets show much broader variability, and therefore more attention has been paid here to this subject. The homology of rootlets permits evolutionary insights for ciliates, green algae and land plants, chromophytes, myxomycetes and bicosoecids.

A new composition of bicosoecids is proposed. Good morphological rows relate to the complexity of the cell, covering a range including the pellicle-theca of dinophytes, pellicle-cuticle of ciliates, tubulemma of kinetoplastids, and folded tubulemma of proteromonads and opalines. The alvelolate layer in ciliates and dinophytes is probably a result of parallel evolution. The diversity of mitochondrial cristae, chloroplast structure, the presence/absence of dictyosomes, and a distribution of unique characteristics in connection with taxonomy and phylogeny are discussed.

In summary, the ultrastructural approach towards evolutionary studies has revealed great diversity within the protists. Some ultrastructural 'peculiarities' may act as specific taxonomic markers, whereas other can be used in the construction of morphological lines.

16.1 The place of morphological characteristics in the estimation of evolution and phylogeny

The interpretation of the phylogeny of organisms is based on the study of heredity, which is manifest in genomic changes and phenotypic expression. Genomic changes can be traced by molecular methods, whereas phenotypic evolution can best be studied by comparative morphological methods. In particular, comparative morphology plays an important part in determining evolutionary trends in larger eukaryotes such as higher plants and animals. It has given us a good indication of the pattern of evolution in vertebrates, and also of many invertebrate groups (Barnes *et al.*, 1998).

However, for small prokaryotes morphological characteristics are much less helpful, whereas biochemical, physiological and molecular characteristics are more important at this level (Kandler, 1985; Kussakin and Drozdov, 1998). In size, protists are located between small prokaryotes and large eukaryotes. Their main characteristics are determined at the cellular level of organization, and therefore it is essential to study both morphological characteristics, and biochemical and molecular features. Problems are encountered in protistan taxonomy and phylogeny when deciding what weight to apportion to each set of characteristics.

Particular difficulty is experienced in determining which morphological characteristics are of significance to understanding protist evolution. Traditionally, in a classical sense, the significance of morphology to understanding evolution has been based on the so-called 'Haeckel's triad', namely one, comparative morphology of fossil records, two, comparative morphology of living organisms, and three, ontogeny. However, amongst protists, and particularly flagellates, there is a limited and patchy fossil record and little information on ontogeny. So we must rely almost entirely on comparative morphology of living organisms to show the similarity or dissimilarity between protistan groups and to search for homologous elements to construct evolutionary series of cell structures.

16.2 The advent of electron microscopy

Before the 1960s the range of morphological characteristics that could be observed on flagellates was limited. The most common characteristics used prior to the application of electron microscopy included the number of flagella; cell size and shape; the thickness and shape of cell coverings; the presence or absence of plastids, and their peculiarities such as colour, shape, and stigma and pyrenoid location. For a long time, based on these characteristics alone, and the dogma current at the time, the flagellates were considered as a separate taxon, which was subdivided into in phyto- and zooflagellates (Honigberg *et al.*, 1964; see also Leadbeater and McCready, Chapter 1 this volume).

The advent of electron microscopy, with modern methods of fixation (Sabbatini *et al.*, 1963), and epoxy embedding for ultrathin sectioning (Glauert, 1962), revealed a large number of hitherto unknown morphological characteristics, and from the 1960s onwards it has been possible to use comparative morphology in a much meaningful way for the reconstruction of protistan phylogeny. As a result of the consequent expansion in our knowledge, the protozoological taxon Mastigophorea

(Flagellata) has ceased to be used, and in its place many distinctive groups have emerged. With respect to the phytoflagellates there has to some extent been a return to Pascher's phycological taxa (compare the classifications of Honigberg *et al.*, 1964 and Corliss, 1984). The polyphyletic nature of the Sarcomastigophora has also been demonstrated, but systematic resolution of the Rhizopoda has not been as successful or conclusive as for the flagellates, and the formal taxon Rhizopoda still remains, at least for the time being, in some classifications (Page, 1987; Corliss, 1994). One possible temporary solution of this problem, which has been proposed by Hausmann and Hülsmann (1996), is to refer the main portion of rhizopods to the Protista *incertae sedis*. Another classification of amoeboid protists, based upon the shape of mitochondrial cristae, has been proposed recently by Patterson (1999).

16.3 The evolution of organelles and cell structures

Our understanding of the evolution of organelles and cell structures is mainly based on the study of comparative morphology, and can be considered independently of the whole organism. By studying the ultrastructure of individual organelles it is possible to derive certain principles which serve as a basis for understanding evolution at the species level or higher. The variations of the same structure in different species of one monophyletic taxon can be ordered into 'morphological series', or 'lines' which, being composed of homologous elements, are invaluable in phylogenetic reconstructions.[1] In fact, the description of these 'morphological series' is a necessary stage in such investigations.

16.3.1 Flagellar/ciliary apparatus

The most informative characteristic of the flagellate cell is the structure of flagellar apparatus, accounting for about 100 characters (Melkonian, 1984). Since the majority of protists have flagellate cells in at least part of their life-cycle, they can therefore be compared with each other. Thus characters associated with the flagellar/ciliary apparatus currently contribute the most information which can be used in the comparative morphology of protists and the reconstruction of their phylogeny.

Because of the uniform morphological and biochemical complexity of flagella and cilia, and the fact that all flagella-bearing protists have the same general structure of the axoneme (9+2) and basal body (9+0), it must be considered very unlikely that the flagellum could have evolved more than once. This must convince us of the monophyly of all flagellated protists in the broadest sense. However, at the same time, it is not necessary to propose that a flagellum-bearing cell was the most ancient eukaryote. Among flagellates the diversity of the flagellar apparatus is so great (see respective chapters in this volume) that at the next (lower) step of hierarchical evolution we have to accept the real possibility of parallel evolution with respect to at least some of its structures.

The most conservative parts of the flagellar apparatus are the basal body and axoneme structures, as they are similar not only in protists, but in all eukaryotic flagellate cells. The absence of specific flagellar structures in some groups of flagellates can be easily explained by reduction, for example the axonematal and basal body microtubules may be reduced in some species (euglenids, gametes and zoospores of algae, sperms of sporozoans). In contrast to the basal body and axoneme, variations have been found in other flagellar structures.

Flagellar appendages

The first electron microscopical observations of whole protists were obtained in the early 1950s, and these confirmed the presence of flagellar appendages, originally called mastigonemes or flimmer (see Manton, 1965 for review; Leadbeater and McCready, Chapter 1 this volume). Later it was shown that mastigonemes may be of different structure and composition. They can be classified into two main groups: tubular (bipartite or tripartite) hairs, and simple hairs (Bouck, 1971; Bouck *et al.*, 1978; Melkonian *et al.*, 1982; Moestrup, 1982). Flagellar scales of differing composition have also been described for some species (Moestrup, 1982). For example, they are composed of polysaccharides in prasinophytes, and mineralized with silica in chrysophytes and some other heterotrophic flagellates (Table 16.1).

Simple flagellar hairs occur in a number of protistan taxa (e.g. bodonids, euglenids, dinophytes, some chlorophytes and haptophytes) and appear to be glycoproteinaceous in nature (Bouck, 1971; Bouck *et al.*, 1978) though in general there is relatively little information on this group. In chromists the hairs are tubular, sometimes bipartite and are glycoproteinaceous in composition (Andersen *et al.*, 1993). They occur on only one flagellum (the immature flagellum) and they are synthesized in the perinuclear space or the rough endoplasmic reticulum, which is a continuation of the perinuclear space. Cryptophytes also possess tubular, bipartite hairs, though their composition is unknown and they exist alongside other types of hairs which are difficult to classify at present (Kugrens *et al.*, 1987). In prasinophytes, the flagella carry both hairs and scales, which appear to be homologous structures, synthesized in the Golgi cisternae. Indeed, the flagellar hairs in this group are often referred to as hair-scales. Both hair-scales and scales appear to be of a polysaccharide nature (Marin *et al.*, 1993).

General conclusions that can be drawn relating to flagellar appendages are as follows:

1 Mastigonemes are absent from multiflagellate cells, ciliate cells and parasitic forms.
2 Scales and simple flagellar hairs always cover every flagellum of the protistan cell. The polysaccharide tubular mastigonemes, or hair-scales of green algae, cover all flagella and are synthesized in the Golgi apparatus.
3 The glycoproteinaceous mastigonemes of chromists cover just the anterior flagellum, and are synthesized in the perinuclear compartment.

The transition zone is less conservative than the remainder of the axoneme, and shows more structural diversity (Pitelka, 1974; Moestrup, 1982; Grain *et al.*, 1988; Sleigh, 1989; Karpov, 1990b; Andersen *et al.*, 1991; Karpov and Fokin, 1995). Some structures, for example concentric rings or coiled fibres (Table 16.2), seem to have co-evolved with other cell features and/or appeared independently in different taxa.

Table 16.1 Diversity of flagellar mastigonemes and scales in protists

Taxa	*Mastigonemes*			*Scales*		
	Anterior flagellum	*All flagella*	*Composition*	*Anterior flagellum*	*All flagella*	*Composition*
Chlorophyceae	—	s	ps	—	—	—
Prasinophyceae	—	tu2	ps	—	+	ps
Charophyceae	—	tu2	ps	—	+	ps
Chrysophyceae	tu3	—	gp	—	+	si
Xanthophyceae	tu3	—	gp	—	—	—
Pelagophyceae	tu2	—	?	—	—	—
Bicosoecida	tu2–3	—	gp?	—	—	—
Pseudodendro-monadida	—	—	—	—	+	?
Eustigmatophyceae	tu3	—	gp	—	—	—
Phaeophyceae	tu3	—	gp	—	—	—
Bacillariophyceae	tu3	—	gp	—	—	—
Raphidophyceae	tu3	—	gp	—	—	—
Haptophyta	—	s[2]	?	—	+	?
Saprolegnea	tu3	—	gp	—	—	—
Labyrinthulea	tu3	—	gp	—	—	—
Thraustochytridea	tu3	—	gp	—	—	—
Hyphochytridea	tu3	—	gp	—	—	—
Pedinellidea	tu3	—	gp?	—	—	—
Cryptophyceae	—	tu2[1]	gp?	—	+	?
Euglenophyceae	—	s	?	—	—	—
Kinetoplastidea	—	s	?	—	—	—
Dinophyceae	—	s	?	—	+	?
Thaumatomonadida	—	—	—	—	+	si

Notes:

Abbreviations:

gp: glycoproteins; si: siliceous; s: simple hairs; ps: polysaccharides; tu2: tubular bipartite; tu3: tubular tripartite mastigonemes; —: absent; +: present; ?: data absent or doubtful

1 Cryptophytes also have another type of mastigoneme

2 In the Pavlovophyceae

At least three unique structures associated within the transition zone are respectively distinctive of three monophyletic protistan taxa. These are, first, the stellate structure of green algae and land plants (Viridiplantae) (Stewart and Mattox, 1975, 1980; Moestrup, 1978; Melkonian, 1982, 1984; O'Kelly and Floyd, 1984; and others); second, the central filament of choanoflagellates (Hibberd, 1975; Zhukov and Karpov, 1985; Karpov and Leadbeater, 1998); and third, the transitional helix of heterokont taxa (Hibberd, 1976; Moestrup, 1982; Patterson, 1989; Preisig, 1989; Karpov, 1990b; Karpov and Fokin, 1995; Sleigh, 1995).

Coiled fibres or concentric rings occur in different taxa such as ciliates (Andersen *et al.*, 1991; Karpov and Fokin, 1995), the choanocytes of sponges (Karpov and Efremova, 1994), thaumatomonads (Karpov, 1987), bicosoecids (Moestrup and Thomsen, 1976; Karpov *et al.*, 1998), chytridiomycetes (Barr and Hadland-Hartmann, 1978), hemimastigophoreans (Foissner *et al.*, 1988), pelagophytes (Andersen *et al.*, 1993), protostelids (Spiegel, 1981), cercomonads (Karpov and Fokin, 1995),

Table 16.2 Flagellar transition zone structures in protists

Taxa	*Long*	*Short*	*Transitional helix*	*Stellate structure*	*Central filament*	*Concentric rings*
Chytridia	—	+	—	—	—	+
Chlorophyta	—	+	—	+	—	—
Charophyta	—	+	—	+	—	—
Chrysophyta	—	+	1–2	—	—	—
Bicosoecida	—	+	—	—	—	+
Pelagophyceae	—	+	—	—	—	+
Eustigmatophyta	—	+	1	—	—	—
Phaeophyta	—	+	—	—	—	—
Raphidophyta	—	+	—	—	—	—
Haptophyta	+	—	—	—	—	—
Saprolegnea	—	+	1–2	—	—	—
Labyrinthulida	—	+	—	—	—	—
Thraustochytrida	—	+	2	—	—	—
Opalinatea	—	+	2	—	—	—
Proteromonadea	—	+	2	—	—	—
Hyphochytrida	—	+	2	—	—	—
Pedinellidea	—	+	1	—	—	—
Cryptophyta	+	—	—	—	—	—
Euglenoidea	+	—	—	—	—	—
Kinetoplastidea	+	—	—	—	—	—
Choanomonada	+	—	—	—	+	—
Polymastigota	—	+	—	—	—	—
Dinophyta	—	+	—	—	—	—
Ciliophora	—	+	—	—	—	+
Plasmodiophora	—	+	—	—	—	+
Protostelia	—	+	—	—	—	+
Cercomonadea	—	+	—	—	—	+
Heliozoa(*Dimorpha*, *Tetradimorpha*)	—	+	—	—	—	+
Hemimastigophorea	—	+	—	—	—	+
Thaumatomonadida	—	+	—	—	—	+

Notes:
Abbreviations: —: absent; +: present; 1: ordinary helix; 2: double helix

plasmodiophorids (Karpov, 1990b), and *Dimorpha* and *Tetradimorpha* (Brugerolle and Mignot, 1984). They are, therefore, of little value as evolutionary markers.

Flagellar rootlets show much greater variability and merit further attention. For example, one can recognize the parallel evolution of multilayered structures (MLS) in different taxa of algae and heterotrophic protists (Karpov, 1986b, 1988), and the independent appearance of other microtubular and also fibrillar rootlets which seem not to be homologous in all flagellate protists.

Rootlet system homology in ciliates

The rootlet systems of ciliates appear to be homologous. By the 1970s, the structure of the ectofibrillar system had been investigated by many authors and sufficient data were available to demonstrate the homology of the main elements (Grain, 1969;

Seravin and Matvejeva, 1972; Gerassimova and Seravin, 1976; Lynn, 1976). It has been shown that the rootlet system associated with cilia is represented by three categories of structure, the transverse ribbon, the postciliary ribbon and the kinetodesmal fibril. These are connected to the basal body in a particular order and are homologous in all ciliates. This set of rootlets is common for ciliates independently of kinetosome number (1–3) in a kinetid. The characteristics of their structure in different taxa of ciliates give us identifiable morphological sequences reflecting the apparent evolutionary trends that can be traced in this group. Seravin, together with Gerassimova, has applied this interpretation for the classification of ciliates in Russia, as has D. Lynn independently in Canada (Seravin and Gerassimova, 1978: Lynn, 1981).

Rootlet system homology in green algae

More recently, a rootlet system homology has been shown for the green algae and land plants. The rootlet system of the ancestral green flagellate is composed of a broad and a narrow microtubular band, originating from one flagellar basal body. Further evolution has led to the duplication of the flagellar apparatus and the appearance of the cruciate microtubular rootlet system (Sluiman, 1983). Both rootlets are homologous to each other in all chlorophyceans, though the number and disposition of microtubules varies from genus to genus (Stewart and Mattox, 1975, 1980; Moestrup, 1978; Melkonian, 1982, 1984; O'Kelly and Floyd, 1984; and others). As a result of these findings, a complete revision of chlorophyte taxonomy has been carried out during the last ten to fifteen years (see *Handbook of Protoctista*: Margulis *et al.*, 1990). The evolutionary sequence of the chlorophytes leads to the vascular plants, the sperm of which in all classes also have the homologous broad MLS. At present, there is no doubt that green algae and land plants represent one monophyletic branch in eukaryote evolution, and this discovery has largely been based on a series of studies using electron microscopy (Stewart and Mattox, 1980, 1984; Melkonian, 1982, 1984; O'Kelly and Floyd, 1984; Sluiman, 1985; McFadden *et al.*, 1986).

Rootlet system homology in chromists

There has also been a revision of our understanding of the taxonomy of the chromophytes (that is, chromists without cryptophytes and dinophytes). The microtubular rootlet homology of the anterior flagellum which bears the mastigonemes (R1 and R2) and of the posterior smooth flagellum (R3 and R4) in the majority of chromists (Chrysophyceae, Xanthophyceae, Eustigmatophyceae, Phaeophyceae, Oomycetes and Hyphochytridiomycetes) has been established by Andersen (1989, 1991). Root R1 normally has conspicuous secondary microtubules, while R3 and R4 pass towards each other and support the ridges of the eyespot cavity on the body surface. However, although there is a good understanding of the broad picture, there are minor details which still have to be worked out in particular groups of Chromista.

Rootlet system homology in Myxomycetes

Recently the homology of the flagellar rootlets in the Myxomycetes and Cercomonads has been demonstrated (Karpov, 1997). The most complete flagellar apparatus of the zoospores in protostelids and myxogastromycetes, which comprise a monophyletic group (Spiegel *et al.*, 1995), has five rootlets numbered 1 to 5 by Spiegel (1981). The most stable rootlets in protostelids and myxogastrids are connected with the basal body of the anterior flagellum (R1, R2 and R3), as well as R4 and R5 that are associated with the posterior basal body but which may be totally absent in some protostelids (Spiegel, 1981, 1991; Spiegel *et al.*, 1995). Investigations on the zoospore cytoskeleton in the myxogastrids *Symphytocarpus confluens, Arcyria cinerea* and *Lycogala epidendrum* have confirmed the conservatism of flagellar rootlets in this taxon also (Karpov *et al.*, 1998).

The flagellar apparatus of the cercomonads *Heteromita* and *Cercomonas* reveals homologous elements corresponding to R1–R3, and in some species to R4 and R5 of the myxomycete flagellar apparatus (Karpov, 1997). *Hyperamoeba*, which is an uncertain myxomycete, possesses characteristics that are transitional between cercomonads and eumycetozoans, and has the same rootlets as the protostelids (Karpov and Mylnikov, 1997).

Rootlet system homology in bicosoecids and pseudodendromonads

Other investigations have revealed rootlet homology in two small groups of colourless flagellates, the bicosoecids and pseudodendromonads, which are very important in understanding the phylogeny and taxonomy of protists. The main rootlet of bicosoecids is a broad microtubular band passing from basal body of the anterior flagellum towards the cytostome region, which may be presented by a lip or a true cytostome with cytopharynx (Moestrup and Thomsen, 1976; O'Kelly and Patterson, 1996; Karpov *et al.*, 1998; O'Kelly and Nerad, 1998; Teal *et al.*, 1998). The distal part of this rootlet encircles the cytostome region and seems to take part in the ingestion of food particles. In its proximal part, fibrillar material is normally present, and in cross section this root has a characteristic L-shape. A mitochondrion with vesicular or tubular cristae is usually associated with the middle part of this rootlet. Another rootlet (dorsal) originates from the same basal body and gives rise to the secondary microtubules, and there are also one or two more lateral microtubular rootlets connected to both basal bodies, for example in *Siluania monomastiga* (Karpov *et al.*, 1998) and *Pseudobodo tremulans* (Karpov, 2000). Surprisingly, the rootlet system of the pseudodendromonad *Adriamonas* (Verhagen *et al.*, 1994) is almost identical to that of *Pseudobodo* (Karpov, 2000). The latter, nevertherless, is a typical bicosoecid (Patterson and Zolffel, 1991; Moestrup, 1995).

Other pseudodendromonads have been less studied, but they also reveal the same main elements of the rootlet system, including an association of the mitochondria with the broad rootlet (Mignot, 1974; Belcher, 1975; Hibberd, 1985; Strüder-Kypke and Hausmann, 1998). Following the description of the bicosoecids with a permanent cytostome, and the expansion of the circumscription of this order (Karpov *et al.*, 1998), we can safely link pseudodendromonads with bicosoecids in a single group. On this basis a new composition of the order Bicosoecida (that is, Bicosoecales) is proposed with the order Pseudodendro-monadida Hibberd, 1985 as a family within it.[2]

Rootlet system homology is not limited to these five groups of protists. It should be broader, of course. Further development of the idea of rootlet homology is connected with the numbering of basal bodies of cilia and flagella, which is becoming clear after study of their duplication during cell division (see Moestrup, Chapter10 this volume).

16.3.2 Cell covering: homologous structures

There is a wide diversity of cell coverings in protists, and there are over 100 terms in use, many of which are synonymous (Preisig *et al.*, 1994; Bouck and Ngö, 1996). All cell coverings have been classified in either five (Dodge, 1973; South and Whittick, 1987), or three classes (Karpov, 1986a, 1990b). Traditional classifications include, first, 'naked forms', covered only by a plasmalemma; second, coverings with additional outer layers (located above the plasmalemma); and third, coverings with additional inner layers (located under the plasmalemma). The second class includes the glycostyles of some amoeboid forms, body scales in algae and colourless flagellates, the perilemma in ciliates, the lorica in some algae and other protists, and the cell wall in algae and zoosporic fungi (Table 16.3).

The third class includes the tubulemma of colourless flagellates, the periplast of cryptophytes, the cuticle of euglenids, the pellicle in ciliates and a few other protists, the derivations of a pellicle (for instance, the frustule of diatoms, and theca, or amphiesma of dinophytes), and the second membrane underlying a plasmalemma in apusomonads.

Good morphological sequences with increasing complexity in the cell covering have been shown for the pellicle-theca of dinophytes (Dodge, 1973), pellicle-cuticle of ciliates (Seravin and Gerassimova, 1978), the tubulemma of kinetoplastids (Frolov and Karpov, 1995) and the folded tubulemma of proteromonads and opalines (Karpov, 1990b). From these morphological data, the origin of the alveolate layer in ciliates and dinophytes may be easily deduced: that is, it is independent in the two groups, showing parallel evolution of the pellicle, leading to the theca in dinophytes and to cuticle in ciliates.

16.3.3 Shape of the mitochondrial cristae

It has been noted by at least two authors independently that mitochondrial cristae can be distinguished by their shape either as flat or tubular, and are uniform in many groups of protists (Taylor, 1978; Seravin, 1980). On the basis of this characteristic, two phylogenetic lines in protist phylogeny have been established (Taylor, 1978), and all flagellates have been divided into two subphyla: Tubulacristata and Lamelli-cristata (Seravin, 1980). More recently, Starobogatov (1986) has used these characteristics as the basis for grouping eukaryotic kingdoms into two empires.

Stephanopogon has been distinguished from the ciliates predominantly because of its flat mitochondrial cristae (Lipscomb and Corliss, 1982). One of the main reasons Hibberd separated *Phalansterium* from choanoflagellates (Hibberd, 1983) was the discovery of vesicular cristae in mitochondria of *Phalansterium*, in contrast to the flat cristae in choanoflagellates.

A more detailed classification of the cristal shape has been carried out by Dodge

(1979), who has also distinguished discoid and sac-like cristae. Patterson (1988) suggested that mitochondria with discoid cristae were ancestral to those with tubular and flat cristae. The most detailed classification of cristal morphology has been that carried out by Seravin (1993), in which each of the three main types (lamellar, vesicular and tubular) is divided into several forms.

At present, the dogma concerning cristal morphology is still employed, though new evidence suggests that cristal form is no longer a reliable indicator of evolutionary trends. For example, first, the Heterolobosea have all transitions from tubular to flat cristae (Page and Blanton, 1985); and second, the variation of this characteristic in kinetoplastids covers all ranges from flat cristae in bodonids to tubular cristae in trypanosomatids: even in the mitochondrion of one cell we can find these two kinds of cristae (Frolov and Karpov, 1995). The distribution of cristal form among protists in connection with other sources of energy is presented in Table 16.4.

16.3.4 Chloroplast structure

The variation in chloroplast ultrastructure is a very important characteristic used in interpreting algal phylogeny. Important characteristics include how many membranes comprise the covering; the number of thylakoids in a lamella; the presence and structure of phycobilin in cyanobacteria and cryptophytes. Some of these characteristics are important for the discussion of flagellate phylogeny.

The number of covering membranes

The resolution of membranes in the chloroplast covering became an important research topic in the period from the 1960s to the early 1970s (see references in Dodge, 1973). The results showed the differences between chloroplasts of green and red algae (two membranes), euglenids and dinophytes (three membranes), and all others (four membranes), and also led to the resurrection of the hypothesis of serial symbiogenesis (Taylor, 1974) (See Table 16.5).

In phylogenetic systematics of autotrophic organisms, the number of covering membranes is a useful characteristic, in spite of the presence of two or three membranes in one chloroplast of some dinophytes (Dodge, 1979), and from two to four membranes in one chloroplast of *Chlorarachnion* (Hibberd and Norris, 1984).

The nucleomorph

The description of a nucleomorph in the periplastid space of cryptophytes (Greenwood, 1974; Gillot and Gibbs, 1980), and more recent evidence of this structure as a rudimentary nucleus (Morrall and Greenwood, 1982; Hausmann *et al.*, 1986; Eschbach *et al.*, 1991) convinced many of the possible origin of the cryptophycean chloroplast in secondary symbiosis with a rhodophycean-like organism. Further discovery of a nucleomorph in the chlorarachniophytes (Hibberd and Norris, 1984; Sitte, 1993) showed that secondary symbiosis as a means of acquiring organelles could be widespread in the evolution of the eukaryotes (Farmer and Darley, Chapter 15 this volume).

Table 16.3 Diversity of coverings in protists

Taxa	*Glycostyles*	*Scales*	*Cell wall*	*Lorica*
Microspora	—	—	+	—
Chytridia	—	—	+	—
Myxospora	—	—	+	—
Chlorophyta	—	+	+	—
Charophyta	—	+	+	—
Rhodophyta	—	—	+	—
Glaucophyta	—	—	+	—
Chrysophyta	—	+	+	+
Bicosoecida	—	—	—	+
Pseudodendromonada	—	+	—	—
Eustigmatophyta	—	—	+	—
Phaeophyta	—	—	+	—
Bacillariophyta	—	—	—	+
Haptophyta	—	+	+	—
Saprolegnea	—	—	+	—
Labyrinthomorpha	—	+	—	—
Hyphochytrida	—	—	+	—
Opalinata	—	—	—	—
Hyphochytrida	—	—	+	—
Cryptophyta	—	+	—	—
Euglenophyceae	—	—	—	+
Kinetoplastidea	—	—	—	—
Choanomastigota	—	—	—	+
Retortamonadida	—	—	—	—
Dinophyta	—	+	+	—
Apicomplexa	—	—	—	—
Ciliophora	—	—	—	+
Plasmodiophora	—	—	+	—
Myxomycetozoa	—	—	+	—
Rhizopoda	+	+	—	+
Foraminifera	—	—	—	+
Heliozoa	—	+	—	—
Radiolaria	—	—	—	+
Apusomonadida	—	—	—	—
Hemimastigophorea	—	—	—	—
Thaumatomonadida	—	+	—	—

Note:
Abbreviations: —: absent; +: present; ft: folded tubulemma

16.3.5 The presence/absence of dictyosomes

The Golgi apparatus in protists may exist in two forms: either as single flat vesicles or as dictyosomes (stacks of cisternae). Typical dictyosomes have not been found in several groups of protists, which have been referred to as primitive organisms (Cavalier-Smith, 1993; Sleigh, 1995). At the same time, there are many protists living in anaerobic conditions, and parasitic protists, which are also without dictyosomes. A technical complication in this context concerns the stability of the dictyosome during chemical preparation for electron microscopy

Table 16.3 continued

Tubulemma	*Pellicle and its derivations*	*Cuticle*	*Periplast*	*Double membrane*
—	—	—	—	—
—	—	—	—	—
—	—	—	—	—
—	—	—	—	—
—	—	—	—	—
—	—	—	—	—
—	+	—	—	—
—	—	—	—	—
—	—	—	—	—
—	—	—	—	—
—	—	—	—	—
—	—	—	—	—
—	+	—	—	—
—	—	—	—	—
—	—	—	—	—
—	—	—	—	—
—	—	—	—	—
ft	—	—	—	—
—	—	—	—	—
—	—	—	+	—
—	—	+	—	—
+	—	—	—	—
—	—	—	—	—
+	—	—	—	—
—	+	—	—	—
—	+	—	—	—
—	+	+	—	—
—	—	—	—	—
—	—	—	—	—
—	—	—	—	—
—	—	—	—	—
—	—	—	—	—
—	—	—	—	—
—	—	—	—	+
—	—	+	—	—
—	—	—	—	—

(Seravin, 1992). On the other hand, there are no dictyosomes in the majority of true fungi (Vasiliev, 1985), which are neither anaerobic nor primitive. A discussion of this matter by Seravin (1992) shows that many acrasids do not have dictyosomes in their cytoplasm regardless of their environment, and similarly some anaerobic ciliates and chytrids from the cow rumen also lack this organelle, though it may be present in others. Many of these protists are not considered to be primitive, and the presence or absence of typical dictyosomes cannot be used as a characteristic showing the primitive status of the organism, nor are they necessarily connected with aerobic/anaerobic habitats.

Table 16.4 The structure of mitochondria and of other metabolic centres in protists

Taxa	*Tubular cristae*	*Vesicular cristae*	*Flat cristae*	*Branched cristae*	*Cristae of indefinite form*	*Other peculiarities*	*Other metabolic centres*
Microsporidia	—	—	—	—	—	No mitochondria	Using host ATP
Chytridiomycetes	—	—	+	—	—	No mitochondria in anaerobes	Hydrogenosomes
Myxosporidia	—	+	+	—	—	—	—
Chlorophyta	—	—	+	—	—	—	—
Charophyta	—	—	+	—	—	—	—
Rhodophyta	—	—	+	—	—	—	—
Glaucophyta	—	—	+	—	—	—	—
Chrysophyta	+	+	—	—	—	—	—
Xanthophyta	+	+	—	—	—	—	—
Sporozoa	+	—	—	—	—	—	—
Ciliophora	+	—	—	—	—	No mitochondria in anaerobes	Hydrogenosomes
Dinophyta	+	—	—	—	—	—	—
Colponema	+	—	—	—	—	—	—
Opalinata	+	—	—	—	—	—	—
Proteromonada	+	—	—	—	—	—	—
Labyrinthula	+	—	—	—	—	Filaments in the cristae	—
Thraustochytridia	+	—	—	—	—	Filaments in the cristae	—
Choanomonada	—	—	+	—	—	Rarely discoidal	—
Kinetoplastida	+	—	+	—	—	Kinetoplast, discoidal, tubular	Glycosomes
Euglenophyceae	+	—	+	—	—	—	—
Stephanopogon	—	—	+	—	—	discoidal	—
Cercomonada	—	+	—	—	—	—	—
Spongomonada	—	+	—	—	—	—	Chemotrophic bacteria
Pseudodendromonada	—	+	—	—	—	—	—
Apusomonada	—	+	—	—	—	—	—
Thaumatomonada	—	+	—	—	—	—	—
Retortamonada	—	—	—	—	—	No mitochondria	Glycolysis in cytoplasm
Oxymonada	—	—	—	—	—	No mitochondria	Glycolysis in cytoplasm

Table 16.4 (continued)

Taxa	*Tubular cristae*	*Vesicular cristae*	*Flat cristae*	*Branched cristae*	*Cristae of indefinite form*	*Other peculiarities*	*Other metabolic centres*
Diplomonada	—	—	—	—	—	No mitochondria	Glycolysis in cytoplasm
Parabasalia	—	—	—	—	—	No mitochondria	Hydrogenosomes
Pedinellomorpha	—	+	—	—	—	—	—
Heliozoa	—	—	+	—	—	—	—
Acantharia	—	+	—	—	—	—	—
Polycystinea	—	+	—	—	—	—	—
Phaeodaria	—	+	—	—	—	—	—
Lobosea	—	+	—	+	—	—	—
Filosea	—	+	—	+	—	—	—
Chlorarachniophyceae	—	+	—	—	—	—	—
Foraminifera	—	+	—	—	—	—	—
Schizopyrenida	—	—	+	+	+	—	—
Acrasida	—	—	+	+	+	—	—
Cryptophyceae	—	—	+	—	—	—	—
Bacillariophyta	+	—	—	—	—	—	—
Raphidophyta	+	—	—	—	—	Filaments in the cristae	—
Phaeophyta	+	—	—	—	—	—	—
Oomycetes	+	—	—	—	—	—	—
Hyphochytridiomycetes	—	+	—	—	—	—	—
Plasmodiophorea	+	—	—	—	—	—	—
Pelomyxa	—	—	—	—	—	No mitochondria	Hydrogenosomes
Entamoeba	—	—	—	—	—	No mitochondria	Glycolysis in cytoplasm
Mastigina	—	—	—	—	—	No mitochondria	Glycolysis in cytoplasm

Note: Abbreviations: —: absent; +: present

Table 16.5 Pigment composition and chloroplast structure in algae and land plants

Taxa	*Chlorophyll composition*	*Membrane number in chloroplast coverings*	*Thylakoid number in lamellae*	*Girdle lamella*	*Phycobilins*	*Nucleomorph*	*Location of stored polysaccharides*
Cyanobacteria	a	1	1	—	+	—	Cytoplasm
Glaucophyta (cyanelles)	a	1	1	—	+	—	Cytoplasm
Rhodophyta	a	2	1	—	+	—	Cytoplasm
Chlorophyta	$a+b$	2	Many	—	—	—	Chloroplast
Charophyta	$a+b$	2	Many	—	—	—	Chloroplast
Plantae	$a+b$	2	Many	—	—	—	Chloroplast
Chlorarachniophyceae	$a+b$	2–4	3	—	—	+	Cytoplasm
Chrysophyta	$a+c_1c_2$	4	3	+	—	—	Cytoplasm
Synurophyceae	$a+c_1$	4	3	+	—	—	Cytoplasm
Xanthophyta	$a+c_1c_2$	4	3	+	—	—	Cytoplasm
Phaeophyta	$a+c_1c_2$	4	3	+	—	—	Cytoplasm
Bacillariophyta	$a+c_1c_2$	4	3	+	—	—	Cytoplasm
Raphidophyta	$a+c_1c_2$	4	3	+	—	—	Cytoplasm
Eustigmatophyta	a	4	3	—	—	—	Cytoplasm
Haptophyta	$a+c_1c_2$	4	3	—	—	—	Cytoplasm
Cryptophyta	$a+c_2$	4	2	—	+	+	Cytoplasm and periplastid space
Euglenophyta	$a+b$	2–3	3	—	—	—	Cytoplasm
Dinophyta	$a+c_2$	2–3	3	—	—	—	Cytoplasm

Notes: Abbreviations: —: absent; +: present

16.3.6 Unique organelles and structures

Ultrastructural investigations have been the means of discovering a number of unusual organelles and structures in some protistan taxa. Many of these structures have proved to be invaluable in the interpretation of the phylogeny and evolution of some groups of organisms (Table 16.6).

The cryptophytes probably have the greatest number of distinctive structures, and there are also a number of characteristics unique to the Viridiplantae, Dinophyta, Choanomonada and Labyrinthulea. Even this feature may be estimated as being synapomorphic for each taxon, which is obviously monophyletic. It demonstrates the evolutionary diversity of flagellate protists. However, unique structures may circumscribe taxa at different levels. For example the Euglenozoa and Alveolata have unique characteristics at the phylum level, while the kinetoplastids (as well as the euglenids, dinophytes and so on) have unique characteristics at the class level. In other groups this tendency is not so obvious, but we should not equate unique characteristics with a particular taxonomic rank. Furthermore, unique characteristics often combine with others in determining the ultrastructural identity of the group (Patterson, 1994).

16.4 Conclusions

In general, the results of ultrastructural investigations, in terms of indicating the possible evolution of cell organelles and structures, may be summarized as follows:

- The electron microscope has revealed the great diversity in the ultrastructure of protists. Some of the ultrastructural peculiarities such as the presence or absence of organelles, flagellar hair morphology, transition zone structures, cristal shape within mitochondria, chloroplast structure, mitotic figures and unique structures, represent distinctive characteristics that may serve as taxonomic markers. However, these may not necessarily be helpful for evolutionary estimations. We can only discuss homogenity/heterogenity of taxa on this basis, and define an *ultrastructural identity* for a taxon. We can also say that the absence of a structure or organelle means that it may have been lost during evolution, but there are no rules to confirm the appearance of a new structure in a group (as an apomorphic characteristic).
- Other peculiarities (flagellar rootlets, coverings) may be considered to be homologous structures, and really can be used to suggest evolutionary trends. Unfortunately, we have found only a few examples of *homologous rows* in flagellates and they are not common for all protists. Thus their significance for evolutionary studies of protists is very restricted.
- The inevitable conclusion is that we cannot use morphological characteristics alone for the estimation of protist evolution. Even in the cases of *monophyletic* taxa we can only describe the homogenous group in a taxonomic sense.

Thus the ultrastructural approach has, on the one hand, confirmed the great diversity of protists, and on the other hand, permitted us to outline '*monophyletic groups*', which are, in fact, merely *homogenous groups*. The real phylogenetic interpretation for this morphological information must be sought from molecular data.

Table 16.6 Distribution of unique characteristics in protists

Taxa	*Unique characteristics*
Microsporidia	Prokaryotic rRNA
Cryptophyta	Periplast, phycobilines inside paired thylakoids, ejectosomes, flagellar mastigonemes
Viridiplantae	Stellate structure in flagellar transition zone, starch inside chloroplast, grana-like thylakoids
Choanomonada	Collar composed by microvilli, central filament in flagellar transition zone
Biliphyta	Plastids (cyanelles) with single thylakoids covered by phycobilisomes
Ramicristates	Branched tubular mitochondrial cristae
Rhodophyta	Cellulose cell wall with mucopolysaccharides penetrated by pores partially blocked by pit connections
Stramenopiles	Tubular cristae with tripartite tubular hairs
Haplosporids	Lidded spore, persistent spindle (Kernstab) in non-dividing nucleus
Polymastigota	Karyomastigont of four basal bodies plus nucleus
Oxymonads	Axostyle constituted of multiple sheets of microtubules
Parabasalia	Parabasal apparatus
Dimorphids	Axopodial axonemes nucleate on an amorphous site linked to kinetosomes
Centroheliozoa	Multilamellate microtubule organizing centre faced with hemispherical structures giving rise to axonemes of microtubules arranged in hexagons and triangles
Pedinellomorpha	MTOC of axopodia scattered on the nuclear surface
Desmothoracids	Heliozoan stage is located in a fenestrated organic lorica
Phaeodarea	Central capsule with three openings, an apical astropyle and two parapyles
Polycystinea	Capsule with fusules separating endoplasm and ectoplasm
Acantharea	Strontium sulphate skeleton, symmetry based on twenty radial elements
Foraminifera	Foramens
Plasmodiophorids	Cruciate mitotic profiles in dividing vegetative cells
Labyrinthulea (incl. Thraustochytridiales)	Ectofibrillar net, sagenogenetosome
Apusomonadida	Double membrane in coverings
Chytridia	Rhumposome
Euglenozoa	Discoid cristae, heteromorphic paraxonemal rods (dorsal flagellum with tubular rod, ventral with lattice structure)
Euglenophyceae	Euglenoid cuticle, paramylon
Kinetoplastida	Kinetoplast
Alveolata	Alveoles
Apicomplexa	Apical complex
Ciliata	Conjugation plus heterokaryosis

Table 16.6 (continued)

Taxa	*Unique characteristics*
Dinophyta	Dinokaryon (loss of histones), theca (amphiesma), pusule
Eustigmatophytes	Stigma association with hairy flagellum
Opalinata (incl. Proteromonadida)	Folded tubulemma
Thaumatomonadida	Body scale formation on mitochondrial surface
Haptophyta	Haptonema

Philippe and Adoutte (1996, 1998) noted that molecular phylogeny did not show the sequence of taxa appearing on the tree, and only confirmed the monophyly of some eukaryotic groups, which was suspected from ultrastructural researches. Molecular data and the morphological information are, therefore, complementary. To be more precise we can add that the molecular data cannot confirm the monophyly of some groups, which could not be shown with morphological data. The molecular approach has really shown for the first time the monophyly of some eukaryotic groups, which is a great advantage of this method.

From this point of view, one can try to estimate the taxonomic relationships between groups. Some connections are more or less clear, and using the information obtained from molecular trees, we can accept some major principles, based on the analysis of *ribosomal* and *protein* trees (Cavalier-Smith, 1995, 1998; Cavalier-Smith and Chao, 1996; Van de Peer and De Wachter, 1997; Philippe and Adoutte, 1998).

At present there is almost no doubt about the monophyly of *chlorobionts, biliphyta, fungi, euglenozoa, alveolates, metazoa and chromists* (Cavalier-Smith, 1993, 1998; Corliss, 1994; Van de Peer and De Wachter, 1997; Philippe and Adoutte, 1998). I would like to discuss here two large groups of eukaryotes, the Opisthokonta and Chromista, which are receiving much consideration at present.

The monophyly of the *Opisthokonta* has been shown by molecular methods (Cavalier-Smith and Chao, 1995; Van de Peer and De Wachter, 1997; Cavalier-Smith, 1998) and the taxon Opisthokonta has been recently established (Patterson, 1999). It includes Metazoa (with Myxozoa), Fungi (with Chytridiomycetes and Microsporidia), choanoflagellates and perhaps pelobionts. The chytridiomycetes share with true fungi the following common characteristics: the presence of chitin in the cell wall, flat cristae in mitochondria, and the lysine synthesis pathway involving diaminopimelic acid. In spite of the presence of flagella, which are absent in true fungi, the chytridiomycetes appear quite isolated from other flagellate protists. Their flagellar apparatus has some common characteristics with ciliates: a pit at the flagellar base, and the spiral fibre (or concentric rings) in the transition zone (Barr and Hadland-Hartmann, 1978; Barr, 1992), but the same characteristics have been found in the choanocytes of the freshwater sponge *Ephydatia fluviatilis* (Karpov and Efremova, 1994). At the same time, choanoflagellates have a similar arrangement of the microtubular rootlet system, which reflects a radial symmetry of the cell (Karpov, 1990b; Karpov and Leadbeater, 1998). This arrangement of the rootlet system occurs also in spongomonads (including *Phalansterium*) (Hibberd, 1983; Karpov, 1990a).

In summary, there are just two common morphological characteristics in the cell structure of choanoflagellates, lower metazoans and fungi: flat cristae in the mitochondria, and one smooth flagellum (which may have 'simple mastigonemes' in some choanoflagellates (Hibberd, 1975)). A radial rootlet system composed of microtubules, which reflects a radial symmetry (likely to appear independently in evolution), occurs just in choanocytes of sponges and in zoospores of some chytrids, and never occurs in sponge larval cells (Wollacot and Pinto, 1995). This symmetry also appears in other epithelial cells of lower metazoans like the Placozoa (Ruthmann *et al.*, 1986; Grell and Ruthmann, 1991) and Turbellaria (Rieger *et al.*, 1991). In spite of this poor morphological support, the molecular and selected biochemical data usually show fungi, animals, and choanoflagellates as one cluster, which is supported at a high bootstrap level (Van de Peer and De Wachter, 1997).

The microsporidia are morphologically very different from metazoa and choanoflagellates, but demonstrate common characteristics with true fungi (Karpov, 1990b; Canning, 1998).

The arrangement of the kingdom Chromista for chromobionts and their relatives is now obvious, but the number of taxa included varies according to different authors. The Chromista includes the protists with heterokont flagellate cells in their life-cycles (with the anterior flagellum covered with tubular glycoproteinaceous mastigonemes), tubular or vesicular cristae in their mitochondria, the presence of a helix in the flagellar transition zone, a similar structure to their chloroplasts, and similar pigments in the autotrophic groups. According to these characteristics the zoosporic fungi (oomycetes, Thraustochytridiae and Hyphochytridiae), opalines and proteromonads (phylum Opalinata) are included in the Chromista (Karpov, 1990b). The Cryptophyta should not be included in the Chromista as they have different pigments, flagellar apparatus structure, mitochondrial cristae, coverings, and extrusomes. The cryptophytes are as unique and separate as the choanoflagellates.

The Dictyochophyceae Silva now includes the pedinellids, *Rhizochromulina* and *Dictyocha* (Moestrup and Thomsen, 1990; O'Kelly and Wujek, 1995). All of them have axopodia originating from the nuclear envelope. On the basis of this characteristic they have been put in the phylum Pedinellomorpha Karpov, which also unites the actinophriid heliozoans, as they have axopodial MTOCs on the surface of the nucleus and the tubular cristae within their mitochondria (Karpov, 1990b). Other heliozoans have flagellate cells of different structure, and predominantly flat cristae in their mitochondria. Their axopodia are produced by either an axoplast or a centroplast. In this respect the phylum Heliozoa can easily be divided in two classes: Axoplasthelidea Febvre-Chevalier et Febvre and Centroplasthelidea Febvre-Chevalie et Febvre.

Unfortunately, many other small groups of protists still remain in unclear positions, which makes the Protozoa a very heterogeneous taxon. Together with hundreds of protists of uncertain affinities (see: Patterson, 1994, 1999) this does not permit us to produce a comprehensive general classification of eukaryotes at present.

ACKNOWLEDGEMENTS

I thank B. S. C. Leadbeater and J. C. Green for useful discussion and corrections to the manuscript.

NOTES

1 In this chapter it is considered that homologous ultrastructural elements demonstrate similarity in morphology and location, and have a similar biochemical composition.
2 According to the IZCN the order Bicosoecida (Grasse) Karpov 1998 includes four families: Bicosoecidae Stein, 1878; lorica present, cytopharynx absent (*Bicosoeca* Stein, 1878); Siluaniidae Karpov 1998; cytopharynx present, lorica absent (*Siluania* Karpov 1998); *Adriamonas* Verhagel *et al.*, 1994; *Caecitellus* Patterson *et al.*, 1992); Cafeteriidae Moestrup, 1995; lorica and cytopharynx absent (*Cafeteria* Fenchel et Patterson, 1998; *Pseudobodo* Griessmann, 1913; *Acronema* Teal *et al.*, 1998; *Discoselis* Vørs, 1988); Pseudodendromonadidae Karpov fam. nov.: lorica absent, body scales and cytopharynx present (*Pseudodendromonas* Bourrelly, 1953; *Cyathobodo* Petersen et Hansen, 1961).

REFERENCES

Andersen, R. A. (1989) The Synurophyceae and their relationship to other golden algae. *Nova Hedwigia*, 95, 1–26.

Andersen, R. A. (1991) The cytoskeleton of chromophyte algae. *Protoplasma*, **164**, 143–159.

Andersen, R. A., Barr, D. J. S., Lynn, D. N., Melkonian, M., Moestrup, Ø. and Sleigh, M. A. (1991) Terminology and nomenclature of the cytoskeletal elements associated with the flagellar/ciliary apparatus in protists. *Protoplasma*, **164**, 1–8.

Andersen, R. A., Saunders, G. W., Paskind, M. P. and Sexton, J. P. (1993) Ultrastructure and 18S rRNA gene sequence for *Pelagomonas calceolata* gen. et sp. nov. and description of a new algal class, the Pelagophyceae *classis nov. Journal of Phycology*, **29**, 701–715.

Barnes, R. S. K., Calow, P., Olive, P. J. W and Golding, D. W. (1998) *The Invertebrates: A New Synthesis*. Oxford: Blackwell Scientific.

Barr, D. J. S. (1992) Evolution and kingdoms of organisms from the perspective of a mycologist. *Mycologia*, **84**, 1–11.

Barr, D. J. S. and Hadland-Hartmann, V. E. (1978) The flagellar apparatus of the Chytridiales. *Canadian Journal of Botany*, **56**, 887–900.

Belcher, J. H. (1975) The fine structure of the loricate colourless flagellate *Bicoeca planktonica* Kisselew. *Archiv für Protistenkunde*, **117**, 78–84.

Bouck, G. B. (1971) The structure, origin, isolation and composition of the tubular mastigonemes of the *Ochromonas* flagellum. *Journal of Cell Biology*, **50**, 362–384.

Bouck, G. B., Rogalski, A. and Valaitis, A. (1978) Surface organization and composition of *Euglena*. II. Flagellar mastigonemes. *Journal of Cell Biology*, 77, 805–826.

Bouck, B., and Ngö, H. (1996) Cortical structure and function of euglenoids with reference to trypanosomes, ciliates, and dinoflagellates. *International Review of Cytology*, **169**, 267–318.

Brugerolle, G. and Mignot, J-P. (1984) The cell characteristics of two helioflagellates related to the Centrohelidian lineage *Dimorpha* and *Tetradimorpha. Origin of Life*, **13**, 305–314.

Canning, E. U. (1998) Evolutionary relationships of Microsporidia. In *Evolutionary Relationships Among Protozoa* (eds G. H. Coombs, K. Vickerman, M. A. Sleigh and A. Warren), Dordrecht: Kluwer, pp. 77–90.

Cavalier-Smith, T. (1993) Kingdom Protozoa and its 18 phyla. *Microbiological Reviews*, **57**, 953–994.

Cavalier-Smith, T. (1995) Zooflagellate phylogeny and classification. *Cytology*, **37**, 1010–1029.

Cavalier-Smith, T. (1998) Neomonada and the origin of animals and fungi. In *Evolutionary Relationships Among Protozoa* (eds G. H. Coombs, K. Vickerman, M. A. Sleigh and A. Warren), Dordrecht: Kluwer, pp. 375–408.

Cavalier-Smith, T. and Chao, E. E. (1995) The opalozoan *Apusomonas* is related to the common ancestor of animals, fungi and choanoflagellates. *Proceedings of the Royal Society of London, Series B,* **261**, 1–6.

Cavalier-Smith, T. and Chao, E. E. (1996) Molecular phylogeny of the free-living archezoan *Trepomonas agilis* and the nature of the first eukaryote. *Journal of Molecular Evolution*, **43**, 551–562.

Corliss, J. O. (1984) The kingdom Protista and its 45 phyla. *Biosystems*, **17**, 87–126.

Corliss, J. O. (1994) An interim utilitarian ('user-friendly') hierarchical classification and characterization of the protists. *Acta Protozoologica*, **33**, 1–51.

Dodge, J. D. (1973) *The Fine Structure of Algal Cells*. London/New York: Academic Press.

Dodge, J. D. (1979) The phytoflagellates: fine structure and phylogeny. In *Biochemistry and Physiology of Protozoa, vol. I*, New York/London: Academic Press, pp. 7–57.

Eschbach, S., Hoffmann, C. J. B., Maier, U-G., Sitte, P. and Hansmann, P. (1991) A eukaryotic genome of 660 kb: electrophoretic karyotype of nucleomorph and cell nucleus of the cryptomonad alga, *Pyrenomonas salina. Nucleic Acid Research*, **19**, 1779–1781.

Foissner, W., Blatterer, H. and Foissner, I. (1988) The Hemimastigophora (*Hemimastix amphikineta* nov. gen., nov. sp.), a new protistan phylum from Gondwanian soils. *European Journal of Protistology*, **23**, 361–383.

Frolov, A. O. and Karpov, S. A. (1995) Comparative morphology of kinetoplastids. *Cytology*, **37**, 1072–1096.

Gerassimova, Z. P. and Seravin, L. N. (1976) Ectoplasmic fibrillar system of Infusoria and its role for the understanding of their phylogeny. *Zoologicheskii Zhurnal*, **55**, 645–656 (in Russian).

Gillot, M. A. and Gibbs, S. P. (1980) The cryptomonad nucleomorph: its ultrastructure and evolutionary significance. *Journal of Phycology*, **16**, 558–568.

Glauert, A. M. (1962) A survey of embedding media for electron microscopy. *Journal of the Royal Microscopical Society*, **53**, 269–277.

Grain, J. (1969) Le cinétosome et ses dérivés chez les ciliés. *Année Biologique*, Ser. **4**, 53–97.

Grain, J., Mignot, J-P. and Puytorac, P. (1988) Ultrastructure and evolutionary modalities of flagellar and ciliary systems in protists. *Biology of the Cell*, **63**, 219–237.

Greenwood, A. D. (1974) The Cryptophyta in relation to phylogeny and photosynthesis. *Proceedings of the 8th International Congress on Electron Microscopy, Canberra*, **2**, 566–567.

Grell, K. G. and Ruthmann, A. (1991) Placozoa. *Microscopic Anatomy of Invertebrates*, **2**, 13–27.

Hansmann, P. (1988) Ultrastructural localization of RNA in Cryptomonads. *Protoplasma*, **146**, 81–88.

Hausmann, K. and Hülsmann, N. (1996) *Protozoology*. Stuttgart/New York: Georg Thieme Verlag.

Hibberd, D. J. (1975) Observations on the ultrastructure of the choanoflagellate *Codosiga botrytis* (Ehr.) Saville Kent with special reference to the flagellar apparatus. *Journal of Cell Science*, **17**, 191–213.

Hibberd, D. J. (1976) The ultrastructure and taxonomy of the Chrysophyceae and Prymnesiophyceae (Haptophyceae): a survey with some new observations on the ultrastructure of the Chrysophyceae. *Botanical Journal of the Linnaean Society*, **72**, 55–80.

Hibberd, D. J. (1983) Ultrastructure of the colonial colourless zooflagellates *Phalansterium digitatum* Stein (Phalansteriida ord. nov.) and *Spongomonas uvella* Stein (Spongomonadida ord. nov.). *Protistologica*, **19**, 523–535.

Hibberd, D. J. (1985) Observations on the ultrastructure of the species of *Pseudodendromonas* Bourrelly *(P. operculifera* and *P. insignis*) and *Cyathobodo* Petersen et Hansen (*C. peltatus* and *C. gemmatus*), Pseudodendromonadida ord. nov. *Archiv für Protistenkunde*, **129**, 3–11.

Hibberd, D. J. and Norris, R. E. (1984) Cytology and ultrastructure of *Chlorarachnion reptans* (Chlorarachniophyta divisio nova, Chlorarachniophyceae classis nova). *Journal of Phycology*, **20**, 310–330.

Honigberg, B. M., Balamuth, W., Bovee, E. C., Corliss, J. O., Gojdics, M., Hall, R. P., Kudo, R. R., Levine, N. D., Loeblich, A. R., Weiser, J. and Wenrich, D. H. (1964) A revised classification of the phylum Protozoa. *Journal of Protozoology*, **2**, 7–20.

Kandler, O. (1985) Evolution and the systematics of bacteria. In *Evolution of Prokaryotes*, London: Academic Press, pp. 335–361.

Karpov, S. A. (1986a) The structure of coverings in flagellates. *Tsitologyia*, **28**, 139–150 (in Russian).

Karpov, S. A. (1986b) Examples of paraphyly in evolution of flagellar rootlets in protists. *Tsitologyia*, **28**, 1135 (in Russian).

Karpov, S. A. (1987) Flagellar apparatus structure in colourless flagellate *Thaumatomonas lauterborni* and estimation of the evolutionary conservatism concept. *Tsitologyia*, **29**, 1349–1354 (in Russian).

Karpov, S. A. (1988) The flagellar rootlets structure of motile cells of algae, zoosporic fungi and colourless flagellates. *Tsitologyia*, **30**, 371–389 (in Russian).

Karpov, S. A. (1990a) Analysis of the orders Phalansteriida, Spongomonadida and Thaumatomonadida. *Zoologicheski Zhurnal*, **69**, 5–12 (in Russian).

Karpov, S. A. (1990b) *System of Protista*. Omsk: Mezhvusovskaya tip. OmPI (in Russian).

Karpov, S. A. (1997) Cercomonads and their relationship to the myxomycetes. *Archiv für Protistenkunde*, **148**, 297–307.

Karpov, S. A. (2000) Ultrastructure of bicosoecid *Pseudobodo tremulans*. *Protistology*, **1**, in press.

Karpov, S. A. and Efremova, S. M. (1994) Ultrathin structure of flagellar apparatus in choanocytes of the sponge *Ephydatia fluviatilis*. *Tsitologiya*, **36**, 403–408 (in Russian).

Karpov, S. A. and Fokin, S. I. (1995) The structural diversity of flagellar transitional zone in heterotrophic flagellates and other protists. *Cytology*, **37**, 1038–1052.

Karpov, S. A. and Leadbeater, B. S. C. (1998) The cytoskeleton structure and composition in choanoflagellates. *Journal of Eukaryotic Microbiology*, **45**, 361–367.

Karpov, S. A. and Mylnikov, A. P. (1997) The ultrathin structure of colourless flagellate *Hyperamoeba flagellata* with special reference to the flagellar apparatus. *European Journal of Protistology*, **33**, 349–355.

Karpov, S. A., Kersanach, R. and Williams, D. M. (1998) Ultrastructure and 18S rRNA gene sequence of a small heterotrophic flagellate *Siluania monomastiga* gen. et sp. nov. (Bicosoecida). *European Journal of Protistology*, **34**, 415–425.

Karpov, S. A., Novozhilov, Y. K. and Chistiakova, L. E. (1998) The comparative study of zoospore cytoskeleton in *Symphytocarpus confluens, Arcyria cinerea* and *Lycogala epidendrum* (Eumycetozoa). In *The Problems of Botany at the Border of XX–XXI c.* Abstracts of the II (X) Congress of Russian Botanical Society, St Petersburg: BIN RAS.

Kugrens, P., Lee, R. E. and Andersen, R. A. (1987) Ultrastructural variations in cryptomonad flagella. *Journal of Phycology*, **23**, 511– 518.

Kussakin, O. G. and Drozdov, A. L. (1998) *Phylema of the Living Beings, Part 2*. St Petersburg: Nauka (in Russian).

Lipscomb, D. L. and Corliss, J. O. (1982) *Stephanopogon*, a phylogenetically important 'ciliate', shown by ultrastructural studies to be a flagellate. *Science*, **215**, 303–304.

Lynn, D. H. (1976) Comparative ultrastructure and systematics of the Colpodida. Structural conservatism hypothesis and a description of *Colpoda steinii* Maupas. *Journal of Protozoology*, **23**, 302–314.

Lynn, D. H. (1981) The organization and evolution of microtubular organelles in ciliated protozoa. *Biological Review*, **56**, 243–292.

Manton, I. (1965) Some phyletic implications of flagellar structure in plants. *Advances in Botanical Research*, **2**, 1–34.

Margulis, L., Corliss, J. O., Melkonian, M. and Chapman, D. J. (eds) (1990) *Handbook of Protoctista*. Boston: Jones and Bartlett.

Marin, B., Matzke, C. and Melkonian, M. (1993) Flagellar hairs of *Tetraselmis* (Prasinophyceae): ultrastructural types and intragenic variation. *Phycologia*, **32**, 213–222.

McFadden, G. I., Preisig, H. R. and Melkonian, M. (1986) Golgi apparatus activity and membrane flow during scale biogenesis in the green flagellate *Scherffelia dubia* (Prasinophyceae). II. Cell wall secretion and assembly. *Protoplasma*, **131**, 174–184.

Melkonian, M. (1982) Structural and evolutionary aspects of the flagellar apparatus in green algae and land plants. *Taxon*, **31**, 255–265.

Melkonian, M. (1984) Flagellar apparatus ultrastructure in relation to green algal classification. In *Systematics of the Green Algae* (eds D. E. G. Irvine and D. M. John), London: Academic Press, pp. 73–120.

Melkonian, M., Robenek, H. and Rassat, J. (1982) Flagellar membrane specializations and their relationships to mastigonemes and microtubules in *Euglena gracilis*. *Journal of Cell Science*, **55**, 115–135.

Mignot, J-P. (1974) Etude ultrastructurale des *Bicoeca*, protistes flagellés. *Protistologica*, **10**, 543–565.

Moestrup, Ø. (1978) On the phylogenetic validity of the flagellar apparatus in green algae and other chlorophyll *a* and *b* containing plants. *BioSystems*, **10**, 117–144.

Moestrup, Ø (1982) Flagellar structure in algae: a review, with new observations particularly on the Chrysophyceae, Phaeophyceae (Fucophyceae), Euglenophyceae, and *Reckertia*. *Phycologia*, **21**, 427–528.

Moestrup, Ø. (1995) Current status of chrysophyte 'splinter groups': synurophytes, pedinellids, silicoflagellates. In *Chrysophyte Algae: Ecology, Phylogeny and Development* (eds C. D. Sandgren, J. P. Smol and J. Kristiansen), Cambridge: Cambridge University Press, pp. 75–91.

Moestrup, Ø. and Thomsen, H. A. (1976) Fine structural studies on the flagellate genus *Bicoeca*. I. *Bicoeca maris* with particular emphasis on the flagellar apparatus. *Protistologica*, **12**, 101–120.

Moestrup, Ø. and Thomsen, H. A. (1990) *Dictyocha speculum* (Silicoflagellata, Dictyochophyceae), studies of armored and unarmored stages. *Det Kongelige Danske Videnskabernes Selskab, Biologiske Skrifter*, **37**, 1–57.

Morrell, S. and Greenwood, A. D. (1982) Ultrastructure of nucleomorph division in species of the Cryptophyceae and its evolutionary implications. *Journal of Cell Science*, **54**, 311–318.

O'Kelly, C. J. and Floyd, G. L. (1984) Correlations among patterns of sporangial structure and development, life histories, and ultrastructural features in the Ulvophyceae. In *Systematics of the Green Algae* (eds D. E. G. Irvine and D. M. John), London: Academic Press, pp. 121–156.

O'Kelly, C. J. and Nerad, T. (1998) Kinetid architecture and bicosoecid affinities of the marine heterotrophic nanoflagellate *Caecitellus parvulus* (Griessmann, 1913) Patterson *et al.*, 1992. *European Journal of Protistology*, **34**, 369–375.

O'Kelly, C. J. and Patterson, D. J. (1996) The flagellar apparatus of *Cafeteria roenbergensis* Fenchel & Patterson, 1988 (Bicosoecales=Bicosoecida). *European Journal of Protistology*, **32**, 216–226.

O'Kelly, C. J. and Wujek, D. E. (1995) Status of the Chrysamoebales (Chrysophyceae): observations on *Chrysamoeba pyrenoidifera, Rhizochromulina marina* and *Laginion delicatulum*. In *Chrysophyte Algae: Ecology, Phylogeny and Development* (eds C. D. Sandgren, J. P. Smol and J. Kristiansen), Cambridge: Cambridge University Press, pp. 361–372.

Page, F. C. (1987) The classification of 'naked' amoebae (Phylum Rhizopoda). *Archiv für*

Protistenkunde, **133**, 199–217.
Page, F. C. and Blanton, R. L. (1985) The Heterolobosea (Sarcodina: Rhizopoda), a new class uniting the Schizopyrenida and the Acrasidae (Acrasida). *Protistologica*, **21**, 121–132.
Patterson, D. J. (1988). The evolution of Protozoa. *Memorias Instituto Oswaldo Cruz*, Rio de Janeiro, **83** (suppl. 1), 580–600.
Patterson, D. J. (1989) Stramenopiles: chromophytes from a protistan perspective. In *The Chromophyte Algae: Problems and Perspectives* (eds J. C. Green, B. S. C. Leadbeater and W. L. Diver), pp. 357–379, Oxford: Clarendon Press.
Patterson, D. J. (1994) Protozoa: evolution and systematics. In *Progress in Protozoology* (eds K. Hausmann and N. Hülsmann), Stuttgart: Gustav Fisher Verlag, pp. 1–14.
Patterson, D. J. (1999). The diversity of eukaryotes. *American Naturalist*, **154**, Sup., 96–124.
Patterson, D. J. and Zölffel, M. (1991) Heterotrophic flagellates of uncertain taxonomic position. In *The Biology of Free-Living Heterotrophic Flagellates* (eds D. J. Patterson and J. Larsen), Oxford: Clarendon Press, pp. 427–475.
Philippe, H. and Adouttte, A. (1996) How far can we trust the molecular phylogeny of protists? *Verhandlungen der Deutschen Zoologischen Gesellschaft*, **89**, 49–62.
Philippe, H. and Adoutte, A. (1998) The molecular phylogeny of Eukaryota: solid facts and uncertainties. In *Evolutionary Relationships Among Protozoa*, (eds G. H. Coombs, K. Vickerman, M. A. Sleigh and A. Warren), Dordrecht: Kluwer, pp. 25–56.
Pitelka, D. R. (1974) Basal bodies and root structures. In *Cilia and Flagella* (ed. M A. Sleigh), London and New York: Academic Press, pp. 437–469.
Preisig, H. R. (1989) The flagellar base ultrastructure and phylogeny of chromophytes. In *The Chromophyte Algae: Problems and Perspectives* (eds J. C. Green, B. S. C. Leadbeater and W. L. Diver), Oxford: Clarendon Press, pp. 167–187.
Preisig, H. R., Anderson, O. R., Corliss J. O. *et al.* (1994) Terminology and nomenclature of protist cell surface structures. *Protoplasma*, **181**, 1–28.
Rieger, R. M., Tyler, S., Smith, J. P. S. and Rieger, G. E. (1991) Platyhelminthes, Turbellaria. In *Microscopic Anatomy of Invertabrates, vol. 3: Platyhelminthes and Nemertinea* (eds F. W. Harrison and B. J. Bogitsh), New York: Wiley-Liss, pp. 7–140.
Ruthmann, A., Behrendt, G. and Wahl, R. (1986) The ventral epithelium of *Trichoplax adhaerens* (Placozoa): cytoskeletal structures, cell contacts and endocytosis. *Zoomorphology*, **106**, 115–122.
Sabbatini, D. D., Bensch, K. and Barnett, R. J. (1963) The preservation of cellular ultrastructure and enzymatic activity by aldehyde fixation. *Journal of Cell Biology*, **17**, 19–58.
Seravin, L. N. (1980) The macrosystem of flagellates. In *Principles of Construction of the Macrosystem of Unicellular Animals*, vol. 94 (eds M. V. Krylov and Y. I. Starobogatov), Leningrad: Zoological Institute RAS, pp. 4–22.
Seravin, L. N. (1992) Eukaryotes lost the main cellular organelles (flagella, Golgi apparatus, mitochondria) and the main task of organellology. *Tsitologiya*, **34**, 3–33 (in Russian).
Seravin, L. N. (1993) Main types and forms of mitochondrial cristae structure: their evolutionary conservatism (ability for morphological transformation). *Tsitologiya*, **35**, 3–12 (in Russian).
Seravin, L. N. and Gerassimova, Z. P. (1978) A new macrosystem of Ciliophora. *Acta Protozoologica*, **17**, 399–418.
Seravin, L. N. and Matvejeva, Z. P. (1972) Ultrastructure of the cortical fibrillar systems of the marine ciliate *Helicoprorodon gigas* Kahl, 1933. *Acta Protozoologica*, **9**, 263–274.
Sitte, P. (1993) Symbiogenetic evolution of complex plastids. *European Journal of Protistology*, **29**, 131–143.
Sleigh, M. A. (1989) *Protozoa and Other Protists*. Cambridge: Cambridge University Press.
Sleigh, M. A. (1995) Progress in understanding the phylogeny of flagellates. *Cytology*, **37**, 985–1009.
Sluiman, H. J. (1983) The flagellar apparatus of the zoospore of the filamentous green alga

Coleochaete pulvinata: absolute configuration and phylogenetic significance. *Protoplasma*, **115**, 160–175.

Sluiman, H. J. (1985) A cladistic evaluation of the lower and higher green plants (Viridiplantae). *Plant Systematics and Evolution*, **149**, 217–232.

South, G. R. and Whittick, A. (1987) *Introduction to Phycology*. Oxford: Blackwell Scientific.

Spiegel, F. M. (1981) Phylogenetic significance of the flagellar apparatus in protostelids (Eumycetozoa). *BioSystems*, **14**, 491–499.

Spiegel, F. M. (1991) A proposed phylogeny of the flagellated protostelids. *BioSystems*, **25**, 113–120.

Spiegel, F. M., Lee, S. B. and Rusk, S. A. (1995) Eumycetozoans and molecular systematics. *Canadian Journal of Botany*, **73** (Suppl.), S738–S746.

Starobogatov, Ya. I. (1986) On the number of kingdoms of eukaryotic organisms. *Trudy Zoologicheskogo Instituta*, **144**, 4–25 (in Russian).

Stewart, K. D. and Mattox, K. R. (1975) Comparative cytology evolution of the green algae with some consideration of the origin of other organisms with chlorophylls *A* and *B*. *Botanical Review*, **41**, 104–135.

Stewart, K. D. and Mattox, K. R. (1980) Phylogeny of phytoflagellates. In *Developments in Marine Biology, vol. 2: Phytoflagellates* (ed. E. Cox), New York: Elsevier North Holland, pp. 433–462.

Strüder-Kypke, M. C. and Hausmann, K. (1998) Ultrastructure of the heterotrophic flagellates *Cyathobodo* sp., *Rhipidodendron huxleyi* Kent, 1880, *Spongomonas sacculus* Kent, 1880, and *Spongomonas* sp. *European Journal of Protistology*, **34**, 376–390.

Taylor, F. J. R. (1974) Implications and extensions of the serial endosymbiosis theory of the origin of eukaryotes. *Taxon*, **23**, 229–258.

Taylor, F. J. R. (1978) Problems in the development of an explicit hypothetical phylogeny of the lower eukaryotes. *BioSystems*, **10**, 67–89.

Teal, T. H., Guillemete, T., Chapman, M. and Margulis, L. (1988) *Acronema sippewissettensis* gen. nov., sp. nov., microbial mat bicosoecid (Bicosoecales = Bicosoecida). *European Journal of Protistology*, **34**, 402–414.

Van de Peer, Y. and De Wachter, R. (1997) Evolutionary relationships among the eukaryotic crown taxa taking into account site-to-site rate variation in 18S rRNA. *Journal of Molecular Evolution*, **45**, 619–630.

Vasiliev, A. E. (1985) About the primitiveness of the fungal cell and the origin of eukaryotes. *Botanical Journal*, **70**, 1145–1156 (in Russian).

Verhagen, F. J. M., Zolffel, M., Brugerolle, G. and Patterson, D. (1994) *Adriamonas peritocrescens* gen. nov., sp. nov., a new free-living soil flagellate (Protista, Pseudodendromonadida *incertae sedis*). *European Journal of Protistology*, **30**, 295–308.

Wollacott, R. M. and Pinto, R. L. (1995) Flagellar basal apparatus and its utility in phylogenetic analysis of the Porifera. *Journal of Morphology*, **226**, 247–265.

Zhukov, B. F. and Karpov, S. A. (1985) *Fresh-Water Choanoflagellates*. Leningrad: Nauka.

Chapter 17

Flagellate megaevolution

The basis for eukaryote diversification

T. Cavalier-Smith

ABSTRACT

Flagellates are a vastly disparate spectrum of related life forms, grouped here in 111 orders and sixty classes in sixteen phyla in three kingdoms: Protozoa, Plantae, Chromista. Many (especially chromists and dinoflagellates) are photophagotrophic exceptions to classical animal/plant dichotomies; zooflagellates and phytoflagellates are both polyphyletic. Multiple evolutionary losses of mitochondria, chloroplasts and flagella confused earlier classifications. Molecular phylogeny is clarifying flagellate evolution, but conflicts between different molecular trees show widespread and misleading biases, leaving many major questions unresolved. Probably all extant eukaryotes are either flagellates or evolved directly or indirectly from them. I discuss major events in flagellate diversification and, briefly, how I think they gave rise to major non-flagellate groups. Archezoan flagellates (Metamonada; Parabasalia), which evolved by losing mitochondria or converting them to hydrogenosomes, are basal eukaryotes on rRNA trees, but protein trees and their flagellar tetrakonty suggest that they evolved from aerobic biflagellates and may be derived from other protozoa, not truly basal. Aerobic zooflagellates are very diverse: their phylogeny and classification (including some new taxa) are discussed in detail. From them all other heterotrophic groups evolved, as did the ancestral phytoflagellate, by a single symbiogenetic origin of chloroplasts from cyanobacteria; glaucophytes, green plants and red algae are derived from this ancestor. Chromistan phytoflagellates arose by symbiogenetic uptake of a red alga and may be sisters of Alveolata; in my view dinoflagellates and sporozoa got their plastids from the same symbiotic red alga. Plastids have been lost many times by chromists and alveolates. Euglenoids and chlorarachneans are chimaeras of a flagellate host and a green algal symbiont, and may have had a photosynthetic common ancestry.

17.1 Introduction

17.1.1 What are flagellates?

Flagellates are protists (that is, unicellular, plasmodial or colonial eukaryotes) characterized by having one or more cilia or flagella in the trophic phase and no macronuclei. As there is no valid distinction between cilia and eukaryotic flagella, it is illogical and confusing to maintain a different name for them. I therefore prefer to use the older and more general term 'cilium' for the vibratory locomotory organelle of both flagellates and ciliates, and reserve 'flagellum' for the non-homologous bacterial flagella, as I have advocated earlier (Cavalier-Smith, 1986a). Although speaking of flagellates with cilia may seem odd, it emphasizes that the fundamental distinction between ciliates and multiciliate flagellates (such as hypermastigotes, opalinids, hemimastigids, *Stephanopogon, Multicilia*) is the presence of macronuclei in ciliates alone; their cilia, by contrast, have no consistent differences from those of flagellates. However, at the editors' request I use the more common term 'flagella' in the rest of this chapter.

Ancestrally phagotrophic and heterotrophic, flagellates diversified nutritionally to form phytoflagellates by the symbiogenetic origin of chloroplasts and by the lateral transfer of chloroplasts in secondary symbiogeneses. Some phytoflagellates have secondarily given rise to phagotrophic zooflagellates by chloroplast loss, while phytoflagellates and zooflagellates have both given rise to saprotrophic flagellates by losing photosynthesis or phagotrophy. Phytoflagellates and zooflagellates are both polyphyletic; although invalid as taxa, these terms often remain convenient in ecological and evolutionary discussions. But some phytoflagellates are photophagotrophs (mainly among dinoflagellates, chromists and chlorarachneans), so the simple division between phytoflagellates and zooflagellates is nutritionally oversimplified.

The body plans of flagellates, defined by their organelle structure and arrangement, are tremendously varied. Because many (notably excepting dinoflagellates, euglenoids, pelobionts and multiciliate forms) are very small, this became apparent only after extensive study by electron microscopy. These fundamental ultrastructural differences, supplemented by chemical and DNA sequence data, have led to a complete overhaul of flagellate classification over the past two decades. I now group them in sixty classes in sixteen phyla (eight consisting predominantly of flagellates) in three separate kingdoms (Protozoa, Plantae, Chromista: see Table 17.1), that is, in half the kingdoms of life (Cavalier-Smith, 1998a). For phytoflagellates (those with plastids) this systematic revolution is probably largely complete, although the recent discovery of Bolidophyceae (Guillou *et al.*, 1999), here grouped with diatoms in a new ochrophyte subphylum Khakista, emphasizes that a few equally distinctive groups may still elude recognition. Understanding of the woefully neglected smaller zooflagellates (Patterson and Zölfell, 1991) remains very inadequate; a consensus view of what classes and phyla deserve recognition should develop over the next decade. The present system is my current view: several new major taxa and substantial rearrangements can be expected as knowledge advances.

The systematic appendix lists all flagellate classes, but I cannot discuss every one. I overview flagellate phylogeny, which constitutes the basic framework for the eukaryote part of the tree of life, and highlight the importance of flagellates for eukaryote cell megaevolution.

17.1.2 What is megaevolution?

The distinction between microevolution, macroevolution and megaevolution is somewhat arbitrary but nonetheless useful (Simpson, 1953). Megaevolution refers to

Table 17.1 Classification of the sixteen phyla with flagellates in their three kingdoms

	Taxa with flagellates		*Flagella/ kinetid*	*Mitochondria*	*Plastids*
	Classes	*Orders*			
Kingdom Protozoa (13 phyla in all: 40 flagellate classes)					
Subkingdom Eozoa					
Infrakingdom Archezoa					
Metamonada	2	4	4	—	—
Parabasalia*	2	3	4	-/H	—
Infrakingdom Loukozoa					
Loukozoa	2	2	2/4	+/H	—
Subkingdom Neozoa (Retaria, Heliozoa and Ciliophora have no flagellates)					
Infrakingdom Sarcomastigota					
Neomonada*	5	11	2(1)	+	—
Amoebozoa** (incl. Mycetozoa**)	3	3	1(2)	-/+	—
Cercozoa*	7	12	2(1)	+	-/+[a,e]
Infrakingdom Discicristata					
Percolozoa** (incl. Heterolobosea)	5	7	2,4	+/-/H	—
Euglenozoa	4	10	2	+	-/+[b, e]
Infrakingdom Alveolata					
Miozoa** (Dinozoa*[b,f] and Sporozoa#[h])	8	18	2	+	-/+
Kingdom Plantae (5 phyla in all: 4 flagellate classes)					
Subkingdom Biliphyta (Rhodophyta have no flagellates)					
Glaucophyta*	1	1	2	+	+[c,g]
Subkingdom Viridaeplantae (Bryophyta and Tracheophyta have no flagellates)					
Chlorophyta**	3	8	2(4,1)	+	+[c,e]
Kingdom Chromista (5 phyla in all: 16 flagellate classes)					
Subkingdom Cryptista					
Cryptophyta	2	3	2	+	+[d,f,g]/-
Subkingdom Chromobiota					
Infrakingdom Heterokonta					
Ochrophyta**	7	12	2(1)	+	+[d,f]/-
Bigyra** (incl. Oomycetes#, Opalinata)	3	4	2(1)	+	—
Sagenista* (Labyrinthulea; Bicoecea)	2	4	2	+	—
Infrakingdom Haptophyta					
Haptophyta*	2	4	2	+	+[d,f]

Notes:

* these phyla also include a usually small minority of non-flagellates (amoeboid, fungoid, coccoid or saprotrophic nets)

** these phyla have a majority of non-flagellates (amoebae, sporozoa or non-flagellate algae)

\# non-flagellates

H hydrogenosomes: double enveloped respiratory organelles that evolved from (and so replace) mitochondria in some anaerobes

a with two-membraned envelope and chlorophyll b;located within periplastid membrane and a fourth, smooth membrane

b with three-membraned envelope; located in cytosol

c with two-membraned envelope; located in cytosol

d with two-membraned envelope; located within periplastid membrane inside rough endoplasmic reticulum lumen

e with chlorophyll b if photosynthetic

f with chlorophyll c (except in eustigmatophytes) if photosynthetic

g with phycobilins if photosynthetic

h plastid envelope with four (or three) membranes

the origin of the characters that distinguish taxa of higher categories (Simpson, 1944); originally it referred to origins of taxa of family or higher rank. However, I have restricted it to the highest categories (taxa of class rank or higher), since these typically involve major qualitative innovations in body plan (Cavalier-Smith, 1990a), whereas evolution at ordinal and familial levels predominantly involves quantitative changes (for example in size, relative proportions or numbers of pre-existing structures or relatively easy qualitative changes in them) similar to those common in generic and specific differentiation. Such macroevolutionary changes are usually adaptive modifications of pre-existing body plans; convergent and parallel evolution is more common than in the much rarer megaevolutionary origin of new body plans.

In protists, megaevolutionary changes often involve the gain of substantially new cell organelles, either by symbiogenesis (for example mitochondria, chloroplasts, and possibly peroxisomes) or by autogenous modifications of pre-existing structures (for example flagella, axopodia, cortical alveoli from endomembranes, cell walls, multicellularity, macronuclei, mitochondria into hydrogenosomes). But they may also involve organelle multiplications (for example kinetids, nuclei) or losses (for example mitochondria, flagella, plastids, phagotrophy). Though some have apparently been unique events (such as the origin of flagella, mitochondria and plastids), others have occurred more often (such as losses of these organelles; origins of hydrogenosomes, axopodia and pseudopods; acquisition of chloroplasts by secondary symbioses) making it more difficult than for multicellular organisms to establish protist phyla and classes on sound phylogenetic lines. Syntheses of ultrastructural and molecular sequence data can disentangle the confusing parallelisms and convergences among protists, and thus improve protist megasystematics (Cavalier-Smith, 1998a, 1998b).

17.2 Are flagellates the basal group from which all other eukaryotes evolved?

Haeckel (1866) thought that the first cell was an anucleate amoeba, that nuclei evolved before flagella, and that flagellates are more advanced than amoebae. Many later nineteenth-century protozoologists thought that eukaryotic flagella and cilia are related to bacterial flagella, and therefore argued that flagellates must be ancestral, and amoebae secondarily derived by loss of flagella. Although we now know that eukaryotic flagella are not derived from bacterial flagella (Cavalier-Smith, 1992a), it is possible that flagella and nuclei evolved simultaneously, and that the first flagellate and the first eukaryote originated in a single megaevolutionary event (Cavalier-Smith, 1987a): I argued that loss of the bacterial cell wall, essential for DNA segregation and cell division, put a premium on early pre-eukaryotes evolving a rigid cell surface to facilitate reliable chromosome segregation and division, and that Haeckel's idea that the first eukaryote was a simple amoeba is incompatible with this early requirement for surface rigidity (Cavalier-Smith, 1987a, 1981a). One must distinguish, however, between the first or ancestral eukaryote (the first cell with a nucleus), and the latest common ancestor (sometimes called the cenancestor) of all extant eukaryotes, since stem eukaryotes (lineages now extinct but diverging from the ancestral eukaryote before the cenancestor) might have had very different phenotypes from crown eukaryotes (properly defined as all those derived from the cenancestor: Jefferies, 1979). In principle the cenancestor could have been a flagellate even if the ancestral eukaryote was an amoeba. The dichotomy itself may be oversimplified, since either might have been an amoeboflagellate (a protozoan like *Naegleria* alternating between flagellate and amoeba) or a rhizoflagellate/mastigamoebid (simultaneously a flagellate and amoeba) (Bütschli, 1888; Cavalier-Smith, 1991a).

Eventually molecular phylogeny should be able to decide between these possibilities. But it cannot yet, because molecular data are missing for many important flagellates and amoebae, and trees for known sequences conflict for different genes. These conflicts do not make molecular phylogeny useless; it simply needs to be evaluated critically and with insight. Figure 17.1 is an 18S rRNA tree for a greater variety of flagellate taxa than previously published. It emphasizes that most non-flagellate taxa are derived from flagellates, and illustrates some problems with molecular trees, discussed more fully, but somewhat over-pessimistically in my view, by Philippe and Adoutte (1998). rRNA genes are technically easier and quicker to amplify and sequence than other genes, making the total database much larger. They are therefore an excellent initial pointer to relationships of ill-characterized taxa. Clusters thus identified often correlate well with ultrastructure. However, like all genes, rRNA evolves at dissimilar rates in different taxa; when changing especially fast, it gives very much longer branches on trees. Different parts of the molecule also evolve at vastly different rates: the regions which are fast-evolving or invariant change idiosyncratically in different lineages, as does base composition. All these systematic biases violate assumptions behind standard phylogenetic algorithms, so almost any molecular tree is historically misleading in places.

Fortunately the problems often affect separate genes differently, so data from several contrasting genes can show which parts of trees are congruent and phylogenetically informative, and which conflict and are potentially misleading. Long branches, especially if unbroken by sidebranches, are particularly problematic, since algorithms often cluster them artefactually or place them too low in the tree, near or at its root, thus partially concealing their excessive length. Locating the root is most difficult of all (Philippe and Forterre, 1999). Some gene trees are rooted by sequences from known outgroups (for example eukaryote rRNA or EF1α trees, rooted using archaebacteria, which are either sisters (Cavalier-Smith, 1987a) or paraphyletic ancestors (Baldauf *et al.*, 1996) of eukaryotes (but which?)). Others where close homologues are absent in the outgroup (such as tubulins, actin) have been rooted using paralogous sister genes that diverged close to the group's cenancestor. But both methods are often confounded by long branch effects: paralogues typically have different functions and long branches. Despite earlier optimism, we do not know where the root is (Philippe and Adoutte, 1998; Embley and Hirt, 1998) and until we do we cannot be sure that the eukaryote cenancestor was a flagellate, as I consider most likely.

Microsporidia and Mycetozoa, both very long branches on rRNA trees, exemplify these problems and how they are conclusively soluble by studying sufficiently many different genes. Microsporidia branch very close to the base of rRNA trees. Doubts that they were as primitive as this suggested (Cavalier-Smith, 1993a) were amply confirmed by protein trees (Keeling and Doolittle, 1996; Müller, 1997). Several (tubulins, chaperones, RNA polymerase) conclusively show that microsporidia are not early diverging protozoa but fungi (Hirt *et al.*, 1999), in which

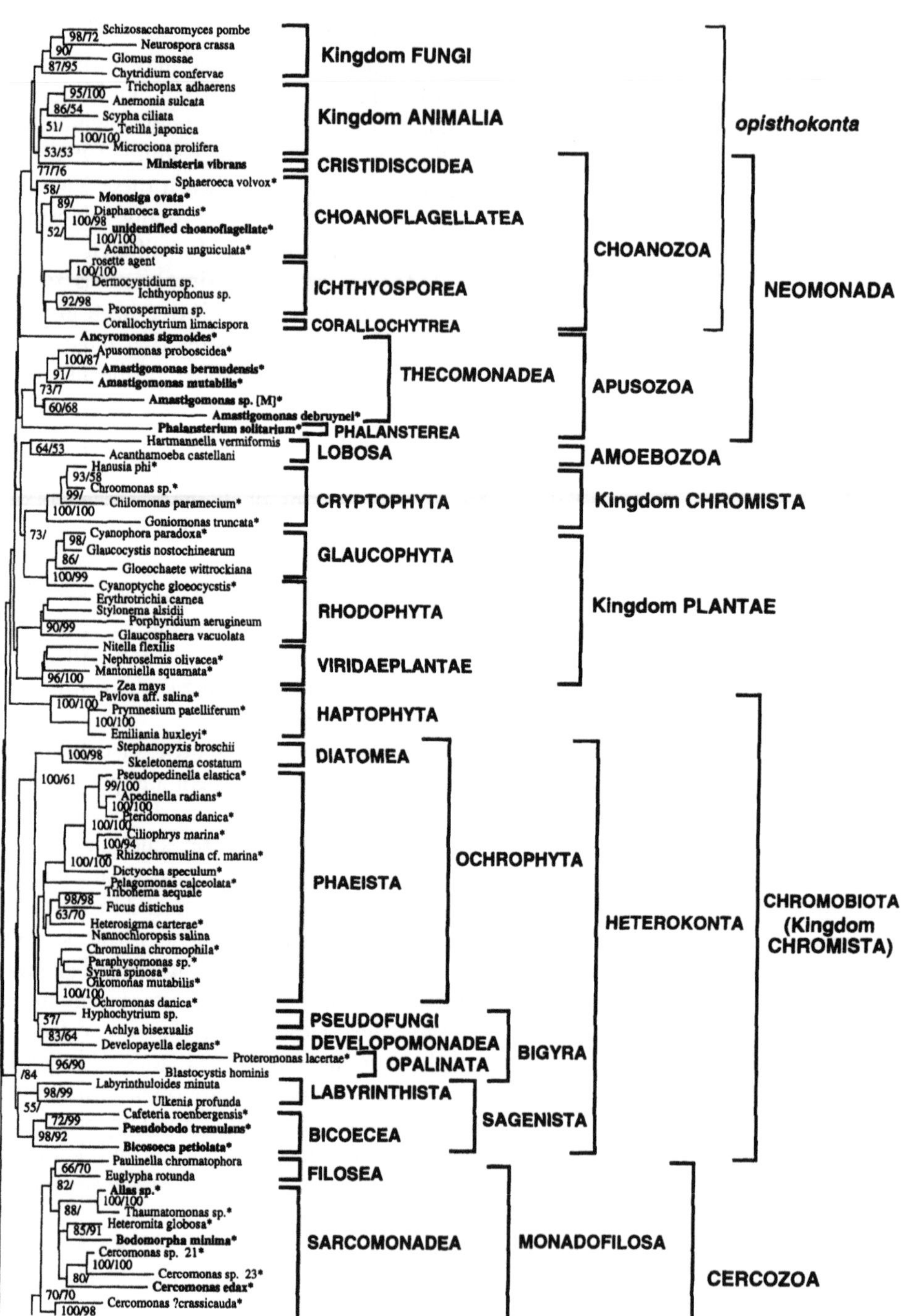

Figure 17.1 Eukaryote 18S rRNA phylogeny emphasising flagellates (marked *). This neighbour joining (NJ) tree (using the maximum likelihood distance model of PHYLIP v. 3.5 for 1515 alignment positions) was rooted assuming jakobids are the outgroup (see text). Numerals at nodes are bootstrap percentages (100

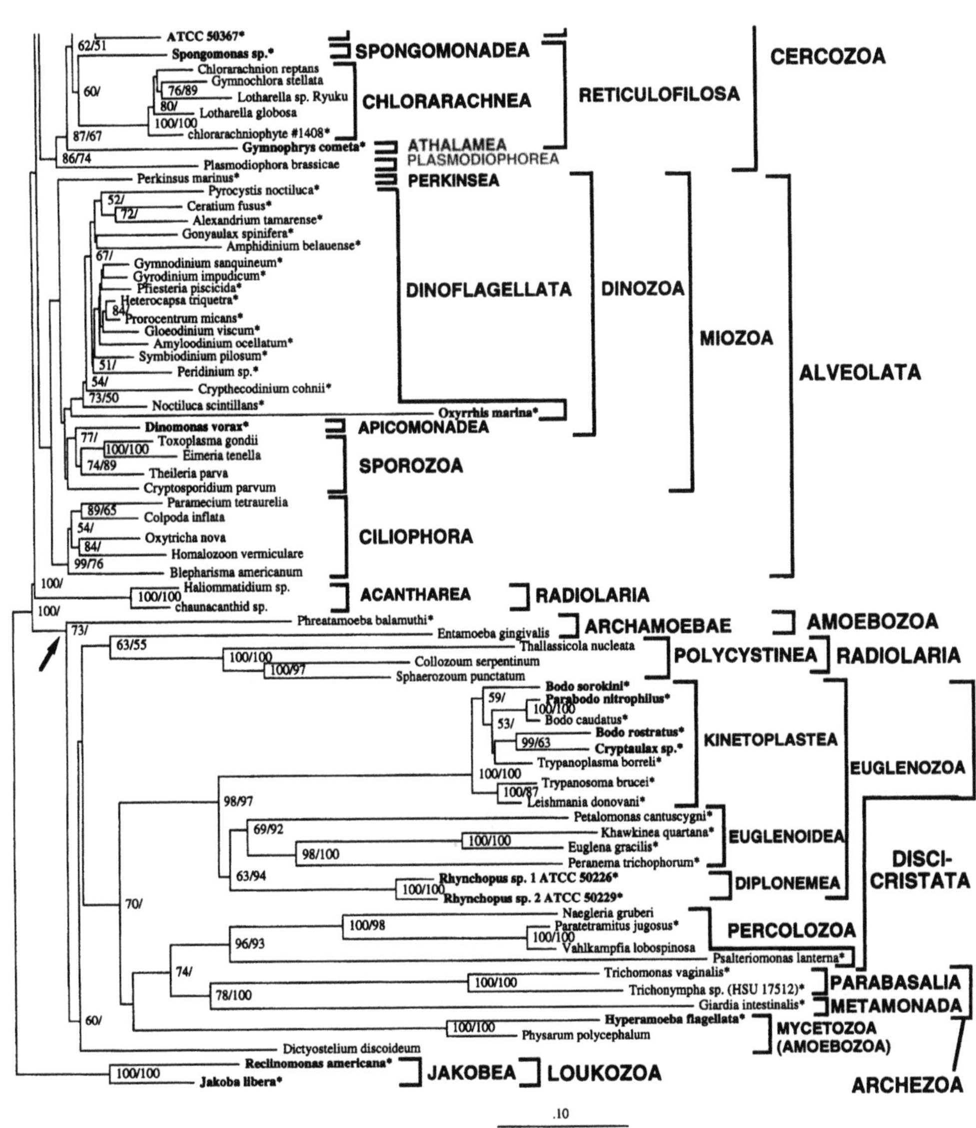

resamplings) over 50% (NJ on left; parsimony with global rearrangements on right); taxa were added randomly. The 27 bold taxa are new sequences (Cavalier-Smith and Chao, unpublished); the rest are from Genbank. The arrow marks the probably artefactual cluster of long-branch taxa kingdom I have formally put them (Cavalier-Smith, 1998a). Mycetozoa are typically in the upper part of the cluster of long branch taxa near the base of rRNA trees (for example Figure 17.1), but trees for many proteins show that this is incorrect and that Mycetozoa are more closely related to animals and fungi, and that Mycetozoa *sensu* Cavalier-Smith (1993a) are monophyletic (Baldauf and Doolittle, 1997) even though many (not all) rRNA trees place the longer-branched *Physarum* substantially below *Dictyostelium*, not as its sister.

Philippe and Adoutte (1998) suggest that all 'early' diverging lineages on rRNA trees (the cluster on Figure 17.1 comprising Archezoa, discicristates and Mycetozoa) are much too low in the tree through long branch artefacts, and that all major eukaryote groups diverged in a single unresolvable radiation. Even without taking such an extreme view (which might be correct), we must acknowledge that these long rRNA branches greatly distort the relative timing of the origin of mitochondria and plastids compared with protein trees, and that the tree topology is fundamentally and grossly wrong for at least Microsporidia, and usually, but not invariably, wrong for Mycetozoa. I consider rRNA distance trees (including Figure 17.1) equally misleading for the amitochondrial Archamoebae *Entamoeba* and *Phreatamoeba* (treated as *Mastigamoeba* by Simpson *et al.*, 1997): some maximum likelihood trees group them together and as sisters to Lobosa (Cavalier-Smith and Chao, 1996). Bootstrap support in Figure 17.1 for the relative branching order of all major eukaryotic clusters is very low; it will be harder than often thought to establish the true historical pattern. The difficulty of resolving the relative branching order of major eukaryotic lineages on molecular trees implies that they all diverged as a 'big bang' over a relatively short timespan (Cavalier-Smith, 1978; Philippe and Adoutte 1998), I suggest within the period 500–600 million years ago. The origin of the first eukaryote and zooflagellate was probably as recent as 600 million years ago: much later than previous estimates (Cavalier-Smith, 1990).

17.3 Flagellar transformation and flagellate diversification

Intuitively one might think that the origin of flagella initially created just one flagellum per kinetid, and that kinetids with two to eight flagella evolved later (Cavalier-Smith, 1992a). Most eukaryote phyla, however, have two centrioles and usually also two flagella per kinetid. All such biflagellate eukaryotes undergo flagellar transformation (Beech *et al.*, 1991); the two flagella are chemically and usually structurally, and often functionally, differentiated. Often the two flagella are anisokont: that is, one points anteriorly and the other posteriorly, the anterior flagellum typically being younger and the posterior one at least a generation older. In every cell cycle, the two centrioles/flagella segregate into separate daughter cells, the former anterior flagellum transforms into a posterior one, and each cell assembles a new anterior flagellum. Thus the centriolar life-cycle and morphogenesis extends over two successive cell cycles. Several uniflagellate groups clearly arose independently from such biflagellate ancestors by losing the posterior flagellum, though its centriole often remains.

But are all uniflagellate groups thus derived? Or is one or more ancestrally uniflagellate, as I proposed for pelobiont Archamoebae (Cavalier-Smith, 1991a, 1992a), because of their single centriole per kinetid with a simple cone of microtubules as root, and the possibility that they ancestrally lack mitochondria? RNA polymerase trees suggest that the rhizoflagellate archamoeba *Mastigamoeba* might have diverged before other studied eukaryotes (Stiller *et al.*, 1998), but its sequence is so divergent that this could be another long branch artefact. Interestingly the parsimony tree for the same data as Figure 17.1 grouped Archamoebae with *Dictyostelium* (but not the very long branch *Physarum*) and Lobosa as sisters to *Phalansterium*, an aerobe with a single flagellum and centriole (Hibberd, 1983), another candidate for being ancestrally uniflagellate. Since *Entamoeba* is secondarily amitochondrial (Clark and Roger, 1995), but usually groups with *Phreatamoeba* on maximum likelihood rRNA trees (Cavalier-Smith and Chao, 1996), Archamoebae as a whole are probably secondarily amitochondrial. The possibility that the eukaryotic root lies close to *Phalansterium* or another uncharacterized uniflagellate (there are several: Patterson and Zölfell, 1991) cannot be excluded and is attractive from the point of view of flagellar evolution.

Recently I grouped Archamoebae with Mycetozoa as subphylum Conosa (named after the cone of flagellar root microtubules subtending their nuclei, also found in *Phalansterium*) of phylum Amoebozoa (Cavalier-Smith, 1998a), which also contains the amoeboid Lobosa (including, because of its glycostyles, the flagellate *Multicilia*: Mikrjukov and Mylnikov, 1996). A separate phylum for *Multicilia* (Mikrjukov and Mylnikov, 1998) is unneccessary: its short centrioles and aberrant ciliary (I strongly resist renaming the cilia of *Multicilia* 'flagella') axonemes with many singlet outer microtubules suggest an affinity with pelobiont Archamoebae; its interkinetid microtubular connective might be a homologue of the archamoebal lateral microtubule band. Even its truncated open pericentriolar microtubular cone, although inverted compared with the larger more open cone of Conosa, might be modified from such a cone as a result of ciliary multiplication.

Golgi images (Mikrjukov and Mylnikov, 1998: figs 4, 5) suggest that *Multicilia* makes its own glycostyles, so they are probably not simply derived from those of its vanellid lobosan food. Therefore the combination of lobosan and conosan characters in this very odd multiciliate amoebozoan flagellate emphasizes the phylogenetic links between the two subphyla of Amoebozoa; molecular studies are needed to see if *Multicilia* belongs in the Conosa with Archamoebae, not in the Lobosa. Mitochondrial gene sequences, notably several uniquely shared features (such as fused cox I and II genes) of the lobosan *Acanthamoeba* and the secondarily non-flagellate conosan *Dictyostelium* (Gray *et al.*, 1998), also strongly support the monophyly of Amoebozoa, despite the often contradictory 18S rRNA trees. Although many mycetozoan swarmers are uniflagellate some are biflagellate; as their posterior flagellum is younger (Wright *et al.*, 1980), not older as in all other studied flagellates, their biflagellate condition perhaps evolved secondarily, and the ancestral amoebozoan might have been an aerobic uniflagellate amoeboflagellate similar to *Hyperamoeba*. The closeness of *Hyperamoeba* to *Physarum* in Figure 17.1 suggests that it lacks fruiting bodies secondarily (Cavalier-Smith and Chao, 1999), not primarily like the postulated amoebozoan ancestor, which could have evolved from a unicentriolar *Phalansterium*-like zooflagellate.

The finding that the flagellum-bearing centriole of choanoflagellates is the younger one (Karpov and Leadbeater, 1997) suggests that this may be the ancestral condition for all neomonads. If this were so, one might expect the anterior flagellum to be younger in thecomonads as in slime moulds. If this were confirmed there would be a remarkable congruence between the distribution of the two types of flagellar transformation and the basic split on the eukaryote tree between scotokaryotes (defined in section 17.6) and photokaryotes shown on Figure 17.2. All flagellates so far shown to have a younger anterior flagellum are photokaryotes, and all those with an older one are scotokaryotes. From the point of view of flagellar evolution, an attractive place for the eukaryotic root would be between scotokaryotes and photokaryotes, which is where it lies on reciprocally-rooted tubulin trees.

The earliest diverging protozoan phyla on rRNA trees are Metamonada and Parabasalia, comprising a revised infrakingdom Archezoa (Cavalier-Smith, 1998a). Both have kinetids with four centrioles as the ancestral (tetrakont) state. Oxymonads, with two bicentriolar kinetids linked by a paracrystalline preaxostylar ribbon, are here transferred to Percolozoa (for molecular evidence that oxymonads are not close to diplomonads see Moriya *et al.*, 1998; Dacks and Roger, 1998); Metamonada now comprise only Trepomonadea and Retortamonadea. Flagellar transformation is likely in Archezoa, but yet unstudied; one might expect their flagellar differentiation to be spread over three successive cell cycles as in certain prasinophytes. The complex tetrakont kinetids of Archezoa are probably secondarily derived from biflagellate ancestors, not an early flagellate phenotype. In agreement with this morphogenetic argument, trees for β-tubulin, EF-1α, cpn60 and valyl-tRNA synthetase (Hirt *et al.*, 1999) show Metamonada and Parabasalia as sister groups, making Archezoa *(sensu* Cavalier-Smith, 1998a) holophyletic, implying that rRNA trees with archaebacterial outgroups (Leipe *et al.*, 1993; Cavalier-Smith and Chao, 1996) showing Archezoa as paraphyletic are incorrectly rooted. The evidence that diplomonads are secondarily amitochondrial (Roger *et al.*, 1998; Embley and Hirt, 1998; Hashimoto *et al.*, 1998) and rRNA evidence that retortamonads are related (Sogin *et al.*, pers. comm.) suggest that all archezoa are secondarily amitochondrial.

Flagellar transformation probably originated during the transition from an ancestral uniflagellate to the first biflagellate. The basic asymmetry of most biflagellate zooflagellates is likely to be an early adaptation to phagotrophy (Stewart and Mattox, 1980; Cavalier-Smith, 1992a); differentiation between the anterior locomotory flagellum and posterior flagellum associated with a feeding groove, as in the recently established phylum Loukozoa (jakobids and *Trimastix*: Cavalier-Smith, 1999), may have been the primary reason why flagellar transformation and anisokont kinetids evolved. Isokont flagella, characteristic of chlorophytine green algae and many haptophytes, are likely secondary adaptations to loss of phagotrophy by phytoplankton.

17.4 Mitochondria and early flagellate evolution

Further clues to the nature of the ancestral biflagellate zooflagellate come from the gene-rich, exceptionally proteobacterium-like, mitochondrial genome of the jakobid zooflagellate *Reclinomonas* (Lang *et al.*, 1997). It alone of studied mitochondrial genomes retains bacterial RNA polymerase genes; all others have replaced this by a nuclear-encoded RNA polymerase with viral affinities. If this happened only once, then jakobids are the outgroup for all other mitochondrial eukaryotes. I used this criterion to root the tree in Figure 17.1, not an archaebacterial outgroup. The hypothetical eukaryote phylogeny of Figure 17.2 assumes that jakobids and other Loukozoa are the earliest diverging eukaryotes and that there are no known primitively uniflagellate eukaryotes; if this rooting is correct, *Phalansterium* and Amoebozoa must be secondarily uniflagellate. But as the ancestral flagellate might have had both proteobacterial and viral mitochondrial RNA polymerases, enabling multiple losses of the former, an alternative rooting near *Phalansterium* instead needs careful testing.

The lateral flanges on the flagella and similarities in flagellar root structure of jakobids and retortamonads led O'Kelly (1993) to suggest that jakobids are the most basal eukaryote group with mitochondria and arose from retortamonads. Because of the diversity in mitochondrial cristae of jakobids, he suggested they are paraphyletic and gave rise independently to other eukaryotes with flat or tubular cristae. Figure 17.1 does not support this, since *Jakoba* with flat cristae and *Reclinomonas* with tubular cristae appear more closely mutually related than either is to other eukaryotes with similar cristal form. If Archezoa are secondarily amitochondrial, retortamonads cannot be ancestral to jakobids; more likely the vanes and other features shared between Loukozoa and retortamonads are common ancestral states. On this basis I grouped Archezoa and Loukozoa as a revised subkingdom Eozoa (Cavalier-Smith, 1999).

Heat shock protein (Hsp) sequences make it unlikely that any primitively amitochondrial eukaryotes exist. Cytosolic, ER and mitochondrial Hsp70 proteins are more similar to each other and to those of Negibacteria (the bacterial subkingdom with two bounding membranes: Cavalier-Smith, 1998a) than to homologues from Unibacteria (the bacterial subkingdom (Cavalier-Smith, 1998a) with one bounding membrane, comprising Archaebacteria and Posibacteria) from which the nucleocytoplasmic component of the eukaryote cell probably derived. Thus cytosolic and ER paralogues probably evolved from negibacterial genes, as did those of mitochondria. Gupta (1998) interprets this as evidence for a separate symbiogenetic fusion of a negibacterium and unibacterium before the uptake of an α-proteobacterium to form mitochondria. A much more parsimonious interpretation assumes only a single symbiogenetic merger with all three Hsp70 paralogues coming from the α-proteobacterial ancestor of mitochondria; if so, the differentiation of ER and cytosolic Hsp70s (presumably during the origin of the RER and the eukaryotic cell itself) did not precede the origin of mitochondria, and there are no primitively amitochondrial eukaryotes, whether flagellates or not.

Metamonada and Parabasalia are the only eukaryote phyla where a reasonable number of protein-coding genes have been sequenced without finding any spliceosomal introns (Logsdon, 1998). I postulated that spliceosomal introns originated from group II introns that entered eukaryotes within the genome of the α-proteobacteria that became mitochondria (Cavalier-Smith, 1991b). If they arose after the initial divergence of eukaryotes then some may primitively lack spliceosomal introns altogether. Total absence of spliceosomal introns in a group might be another criterion helpful in rooting the eukaryotic tree; but are they totally absent in

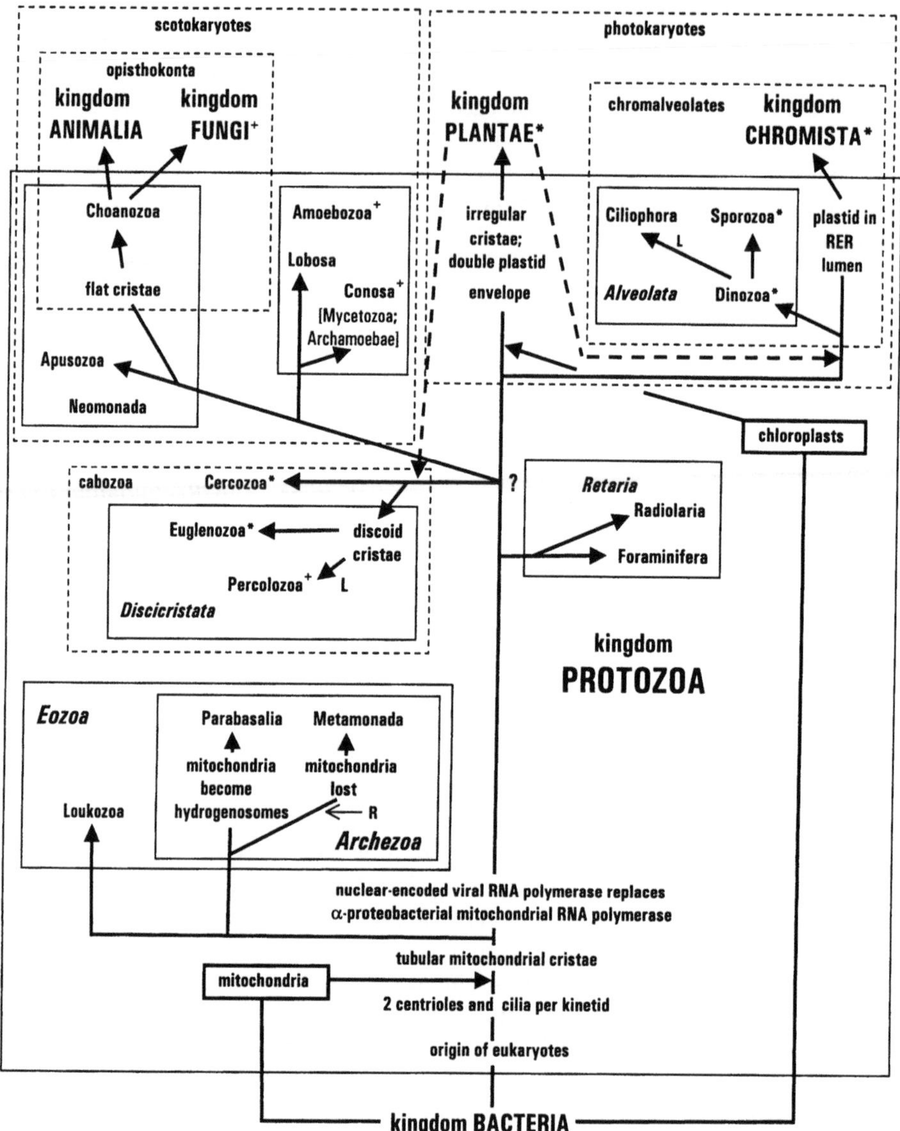

Figure 17.2 Postulated relationships of the major organismal groups, emphasising the centrality of flagellates. Taxa with some or many species with plastids are marked *. The single symbiogenetic origins of mitochondria and chloroplasts from eubacteria are shown in boxes. The two secondary symbiogenetic events (i.e. cabozoan and chromalveolate origins) are shown by dashed arrows; different biflagellate protozoa phagocytozed early eukaryote algae: a red alga to form chromalveolates and a green alga making the cabozoan algae. Continuous boxes delimit higher order taxa and dashed boxes clades not recognized as taxa. Two postulated major plastid losses are marked L; others occurred within Chromista, Dinozoa and Euglenozoa, and are postulated within Cercozoa. Mitochondria were lost or converted to hydrogenosomes in Archezoa and within the three taxa marked +. The position of the

Archezoa? If they are merely very rare, as suggested by their possession of a conserved spliceosomal protein, PRP8 (Fast and Doolittle, 1999), little reason other than the somewhat suspect rRNA and elongation factor trees would remain for considering them early diverging eukaryotes.

17.5 The origin of eukaryotic flagella and zooflagellates

Although the first eukaryote may have been a zooflagellate, eukaryote flagella are not evolutionarily related to bacterial flagella: they differ fundamentally chemically, structurally and functionally (Cavalier-Smith, 1978, 1992a). The idea that eukaryotic flagella evolved symbiogenetically from ectosymbiotic spirochaetes (Margulis, 1970) has been firmly refuted by complete spirochaete genome sequences, which reveal no closer similarities to eukaryotic flagellar proteins than in any other bacteria (Fraser *et al.*, 1998). Tubulin, the major protein of eukaryotic flagella and cilia, apparently originated instead from FtsZ, a key constituent of bacterial (and chloroplast) cell division mechanisms (Faguy and Doolittle, 1998). Dynein (the motor responsible for ciliary and eukaryotic flagellar motion and mitotic chromosome movements), kinesins (also microtubule motors), myosins (the actin motors) and G-proteins though much more distant in sequence, all have striking three-dimensional and functional similarities to tubulins and FtsZ (Erickson, 1998), suggesting their origin also from FtsZ.

Ever since ciliary and flagellar basal bodies and centrioles were equated (Henneguy, 1898; Lenhossek, 1898: I use 'centriole' for all three), it has been accepted that eukaryotic flagella, cilia and the mitotic apparatus have co-evolved. The fact that ciliary and flagellar axonemes contain not only microtubules and dynein but also actin and intermediate-filament-like tektins (Norrander and Linck, 1994) shows that they are essentially specializations of the eukaryotic cytoskeleton. Despite fundamental differences between bacterial and eukaryotic cells – the greatest discontinuity in the living world – there may have been an essential molecular continuity between the mechanism of bacterial DNA segregation and cell division, and eukaryotic mitosis; eukaryotic flagella are autogenous complexifications of this mechanism that secondarily became involved in cell locomotion (Cavalier-Smith, 1978, 1987a, 1992a).

Because of their complexity (around a thousand genes are needed for ciliary/eukaryotic flagellar structure) we may never reconstruct how this happened in detail. But by elucidating mechanisms of bacterial DNA segregation (Hirano, 1999) and division, as well as those of early zooflagellates, we may come to understand some major steps. We need better knowledge of the proteins linking centrioles to zoo-flagellate nuclei and axonemes to the flagellar membrane, and of the morphogenesis of centrioles and triplet and doublet microtubules: axoneme assembly and

eukaryote root (between Eozoa and the rest) is very uncertain: possibly instead (for pros and cons see text): (a) Metamonada (position R) and/or Parabasalia, (b) Loukozoa alone, or rather less likely (c) Percolozoa, or even (d) *Phalansterium*/Amoebozoa, diverged earliest from other eukaryotes. Other possible positions lack a specific rationale, but cannot yet be ruled out cytoskeleleton/membrane coevolution are keys to flagellar origins. Many proteins may have evolved from each other by gene duplication, chimaerization and divergence; few will have direct bacterial homologues. Rather than flagella springing virtually preformed from spirochaetes (Margulis, 1970), immense macromolecular innovation was needed.

The other key steps in making the first zooflagellate (endomembranes and nuclear pore complexes) were also major innovations not present in bacteria (Cavalier-Smith, 1987a). The probable simultaneous origin of the cytoskeleleton, endomembrane system, nucleus and flagella are most plausibly explained as co-evolutionary consequences of the origin of phagocytosis (De Duve and Wattiaux, 1966; Stanier, 1970) triggered by the loss of the bacterial cell wall (Cavalier-Smith, 1975, 1987a) and the consequent replacement of an exoskeleton by an endoskeleton (Cavalier-Smith, 1992a). Wall loss and the origin of at least a primitive form of phagotrophy were also prerequisites for the symbiogenetic origin of mitochondria and chloroplasts (Cavalier-Smith, 1975, 1982, 1983) and possibly also peroxisomes (Cavalier-Smith, 1990b). Mitochondria may have originated during eukaryogenesis as soon as the origin of primitive phagocytosis enabled the uptake of bacteria to become endosymbionts. On this view the first eukaryote was a zooflagellate, probably a facultative aerobe/anaerobe like the engulfed α-proteobacterium.

17.6 Phylum Neomonada: basal group to scotokaryotes?

Tubulin trees (Keeling and Doolittle, 1996; Keeling *et al.*, 1998) suggest a bifurcation relatively early in eukaryote evolution between three exclusively heterotrophic groups (animals, fungi, Amoebozoa) and three predominantly photosynthetic groups (kingdoms Plantae, Chromista and alveolate Protozoa, that I collectively called photokaryotes: Cavalier-Smith, 1999). Ribosomal RNA trees (Cavalier-Smith and Chao, 1995; Cavalier-Smith, 1998b) show that the recently established flagellate phylum Neomonada (Cavalier-Smith, 1997a, 1998b) also belongs with the heterotrophic groups; I here name this assemblage of four exclusively heterotrophic groups 'scotokaryotes' (Greek *scoto*: dark). This divergence between scotokaryotes and photokaryotes seems rather basic to flagellate evolution.

Neomonads comprise two subphyla: the basically uniflagellate Choanozoa with predominantly flat cristae, and the typically biflagellate Apusozoa mostly with tubular cristae (flat in *Ancyromonas* only). The non-flagellate Cristidiscoidea are now assigned to Choanozoa (Cavalier-Smith, 1998b), as rRNA phylogeny indicates that *Ministeria* is related to choanoflagellates and animals (Figure 17.1). There have been multiple origins of non-rhizopod non-flagellate heterotrophs in Choanozoa, probably from choanoflagellate ancestors: *Corallochytrium*, Ichthyosporea, *Ministeria*. Hemimastigida are now placed within Thecomonadea; the similarity of their two plastron-like cortical plates to the single cortical plates of Apusomonadida suggests that they evolved from an *Amastigomonas*-like ancestor by doubling the cortical plates and flagella in two-fold symmetry, and further multiplying the flagella in two rows; these novelties justify separate ordinal but not phylum status (Foissner *et al.*, 1988). Figure 17.1 weakly supports a relationship between *Phalansterium* and thecomonads, but no other

Apusozoa have been sequenced; several may really belong elsewhere, for example, the groove of *Diphylleia* is reminiscent of that of Loukozoa.

17.6.1 Choanoflagellate origins of kingdoms Animalia and Fungi

Animals almost certainly evolved from a choanoflagellate, while fungi evolved either from a choanoflagellate (most likely) or an unknown choanozoan flagellate (Cavalier-Smith, 1987b, 1998a,b, 2000). The plasma membrane of such a zooflagellate would already have been differentiated into an apical or flagellar domain, with integral membrane proteins different from those on the general surface (basal domain). Choanoflagellates (and even zooflagellates in general) might already have evolved the differential apical/basolateral targeting machinery found in epithelial cells, and if so would be good model systems for elucidating its mechanisms. The zooflagellate cell may be regarded as the prototype polarized eukaryotic cell with distinct apical (flagellar associated) and basal plasma membrane domains, as in animal epithelia. The validity of the clade opisthokonta (*sensu* Cavalier-Smith, 1987b: animals, fungi and Choanozoa) is one of the few supraphyletic features of the eukaryotic tree supported by virtually all types of evidence.

17.7 The primary diversification of phytoflagellates

The serial endosymbiosis theory (Taylor, 1974) assumes that chloroplasts evolved substantially after the symbiogenetic origin of mitochondria, and that most aplastidic protozoa are primitively so. Though this is widely accepted, the alternative view that the symbionts ancestral to mitochondria and chloroplasts became established simultaneously (Cavalier-Smith, 1987c) cannot be totally dismissed. The relatively high position on rRNA trees of algae compared with heterotrophic protozoa (Figure 17.1) is not as strong support for the serial symbiosis theory as is widely assumed; it is strongly contradicted by many protein trees where the basal eukaryotic radiation scarcely, if at all, precedes the major algal radiations (such as Keeling *et al.*, 1998). If the simultaneous symbiogenesis theory were true, one might expect to find relict cyanobacterial genes in all heterotrophic eukaryotes: conceivably the resemblance of integrin domains to a cyanobacterial protein is an example, not a case of lateral gene transfer from eukaryotes to cyanobacteria (May and Ponting, 1999).

17.7.1 Symbiogenetic origin of chloroplasts and the kingdom Plantae

Early proponents of the symbiogenetic theory suggested that chloroplasts originated several times independently from differently pigmented prokaryotes (Mereschkowsky, 1910; Margulis, 1970; Raven, 1970; Stanier, 1970). But I argued (Cavalier-Smith, 1982) that separate origins were phylogenetically unnecessary, as all chloroplasts could have evolved from cyanobacteria by differential loss and gain of pigments, and separate origins were highly improbable, as they would have required multiple origins of chloroplast-specific protein-import machinery (Cavalier-Smith, 1992b,

1993b; 1995). Recent evidence supports pigment diversification from a single ancestor (Wolfe *et al.*, 1994) and a basic similarity in protein-import mechanism of plastids of green plants, red algae and glaucophytes (Apt *et al.*, 1993; Schwartzbach *et al.*, 1998); together with shared derived features of chloroplast genomes, and the monophyly of all plastids on 16S and 23S rRNA and protein trees, this has led to widespread acceptance of the monophyletic origin of all plastids from cyanobacteria (Palmer and Delwiche, 1998; Martin *et al.*, 1998; Douglas, 1998).

Kingdom Plantae *sensu* Cavalier-Smith (1981b: Viridaeplantae, Rhodophyta, Glaucophyta; see Cavalier-Smith, 1998a for its higher level classification) is therefore probably monophyletic (well supported also by mitochondrial gene trees; Gray *et al.*, 1999), not polyphyletic. We cannot yet determine whether Plantae are holophyletic, as I think likely, or paraphyletic. Plantae would be paraphyletic if the host in the origin of the Chromista (discussed later) was a eukaryotic alga (Häuber *et al.*, 1994) rather than a heterotrophic protozoan (Whatley *et al.*, 1979; Cavalier-Smith, 1982, 1986b). Although one chromist group (Cryptophyta) commonly branches just within the Plantae, and another (Haptophyta) sometimes does, this is very weak evidence for the theory of Häuber *et al.* (1994) or plant paraphyly; rRNA trees have insufficient resolution to establish the branching order of the three major plant and chromist taxa (Cavalier-Smith, 1998a). The zooflagellate host for the symbiotic origin of plastids was almost certainly biflagellate (Cavalier-Smith, 1982), but molecular phylogeny has not yet revealed one specifically related to Plantae. Correlating molecular trees with the fossil record suggests that the first phytoflagellate arose only slightly before animals (Cavalier-Smith, 1990); a reasonable date is about 550 million years ago (the end of the great Varangian glaciation), later than my previous estimate of 700 million years (Cavalier-Smith, 1990) because of the recently lowered estimates of animal origins (Erwin *et al.*, 1997).

Descended from the ancestral phytoflagellate, the ancestral red alga lost flagella and replaced its cyanobacterial carbon-fixing enzyme rubisco by one from a proteobacterium (Palmer and Delwiche, 1998). This lateral transfer of a single gene is sharply distinguished from the lateral transfer of whole genomes and associated membranes in symbiogenetic organelle origins. Within green plants, coccoid or filamentous algae evolved many times from phytoflagellates; sometimes flagellate stages were retained (for example in the ancestors of land plants and ulvophytes) and sometimes flagella were totally lost (for example Conjugophyceae).

17.8 The predominantly flagellate protozoan phylum Cercozoa

Ribosomal RNA genes have now been sequenced from many ill-studied zooflagellates of unclear taxonomic position (Cavalier-Smith and Chao, 1995, 1996, 1997; and the over two dozen unpublished sequences on Figure 17.1). An unsuspected major clade thus revealed is phylum Cercozoa (Cavalier-Smith, 1998a, 1998b), which includes *Cercomonas* and other sarcomonads, euglyphid testate amoebae, chlorarachnean algae, plasmodiophorids, zooflagellates of previously uncertain affinities such as *Spongomonas* and the reticulose amoeboflagellate *Gymnophrys*. The composition of this novel assemblage (temporarily called Rhizopoda: Cavalier-

Smith, 1997a; for reasons for the name change see Cavalier-Smith, 1998a, 1998b) is summarized in the appendix. Many other poorly characterized zooflagellates probably belong in Cercozoa, so its classification will eventually be substantially modified. Many cercozoan zooflagellates have marked pseudopodial tendencies; multiple losses or phase-specific suppressions of flagella have yielded filose (euglyphids), reticulose (*Gymnophrys*) or filose/reticulose (*Chlorarachnion*) rhizopods. So cercozoan cells are morphologically much more diverse than in an almost entirely flagellate phylum like Euglenozoa.

Ancestrally Cercozoa were probably biflagellate aerobes, but one class (Chlorarachnea) is exclusively uniflagellate. The relationship of chlorarachneans and sarcomonads, and thus the monophyly of Cercozoa is supported by α-tubulin (Keeling *et al.*, 1998) as well as rRNA trees. Cercozoa branch close to the bifurcation between scotokaryotes and photokaryotes: in some trees just below, and in others just above, on the photokaryote side. It is important for eukaryote megaphylogeny to determine which is correct.

17.8.1 Origin of chlorarachnean algae by secondary symbiogenesis

Chlorarachnea are the only cercozoan algae, typically and ancestrally possessing amoeboid, coccoid and flagellate stages, but one or two of them have been lost in different lineages; thus in chlorarachneans simple phytoflagellates are derived (Ishida *et al.*, 1999), not ancestral as in other algae. Chlorarachnea originated by secondary symbiogenesis (the permanent merger of an endosymbiotic eukaryote alga with its host; McFadden *et al.*, 1994a): a possibly ulvophycean green alga (Ishida *et al.*, 1998; 1999) merged with a heterotrophic flagellate host (Cavalier-Smith and Chao, 1997). Their closest identified relative is *Spongomonas* (Figure 17.1). As all other cercozoans are heterotrophic, it has been assumed that their common ancestor was also. But I recently suggested that euglenoids obtained their green plastids in the very same symbiogenetic event as chlorarachneans (Cavalier-Smith, 1999).

17.8.2 Are Cercozoa and Discicristata (that is, Euglenozoa and Percolozoa) sister groups?

I postulated that Cercozoa and Discicristata form a clade designated cabozoa because their plastids have chlorophyll *a* and *b* and cab proteins (Cavalier-Smith, 1999). This cabozoan theory requires more plastid losses than the conventional assumption of two independent origins (twice or thrice within Cercozoa; in the progenitors of Percolozoa and Kinetoplastea; and several within euglenoids). But if there was a common origin for protein-targeting to euglenoid and chlorarachnean plastids via Golgi vesicle fusion (Cavalier-Smith, 1999), then gene transfers from the green algal nucleus to the host nucleus (and the acquisition by every transferred chloroplast protein gene of signal sequences to ensure RER/Golgi targeting) had to occur only once. By reducing the number of necessary independent mutations by over a thousand, this theory seems more parsimonious. In euglenoids the nucleomorph was lost and the triple plastid envelope probably evolved by losing the periplastid membrane after the euglenoids and chlorarachneans diverged (Cavalier-Smith, 1999). Plastid starch was lost and its function replaced by cytosolic β-glucan (paramylum), probably once in the ancestral biflagellate cabozoan. This theory assumes that the low position of discicristates on rRNA trees (as in Figure 17.1) is a long-branch artefact and their high position relatively close to Cercozoa on tubulin trees (Keeling *et al.*, 1998) is correct. The monophyly of discicristates is weakly supported by tubulin trees, but like the cabozoan theory this needs stringent testing. An early divergence of Percolozoa, suggested by their possibly primitively unstacked Golgi (Cavalier-Smith, 1993c), is supported by rRNA trees (Cavalier-Smith, 1993a; Cavalier-Smith and Chao, 1996) but contradicted by tubulin trees (Keeling *et al.*, 1998); more protein trees are needed to clarify their position.

17.9 Origin of Chromista by incorporating a red algal endosymbiont

Chromista (Cryptophyta, Heterokonta and Haptophyta) arose by the symbiogenetic incorporation of a red alga into a biflagellate host, as shown by the fact that all chromistan algae have the red algal type of proteobacterial rubisco. The clustering of all three chromist groups as a clade within red algal rubisco sequences (Chesnick *et al.*, 1996) supports the view that Chromista are monophyletic (Cavalier-Smith, 1982, 1986b, 1995, 1998a) and there was only a single symbiogenetic event. But this evidence is disputable, as other arrangements may occur (also with low bootstrap support: Daugbjerg and Andersen, 1997). As the three taxa do not consistently cluster on nuclear 18S rRNA or chloroplast 16S rRNA trees, many espouse the earlier view of three separate symbioses (Whatley *et al.*, 1979; Palmer and Delwiche, 1998; Douglas, 1998). This lack of phylogenetic resolution implies a 'big bang' rapid divergence of the seven major photokaryote lineages, with an origin of chromists about 530 million years ago (the same time as the Cambrian explosion of animals). In all three groups the red algal plasma membrane remains as the periplastid membrane within the RER lumen. The improbability of all three independently evolving a protein-import mechanism into chloroplasts across these extra membranes argues for a single origin for chromistan phytoflagellates; a claim to have largely removed this difficulty by assuming a borrowing of the Golgi dependent mechanism of dinoflagellate and euglenoid plastids (Bodyl, 1997) is unsound, as the symbiont was a red alga with two envelope membranes, not a dinoflagellate with three, and there is no evidence for import via Golgi in red algae or chromists. 18S RNA sequences of cryptomonad nucleomorphs, the relict red algal nucleus lost by other chromists (Cavalier-Smith *et al.*, 1996b), group within red algae; the cryptomonad chloroplast genome is also red algal in character (Douglas, 1998; Douglas and Penny, 1999), supporting a red alga as the chromistan symbiont.

Nucleomorph genomes have a very high gene density; most genes are nuclear, ribosomal or proteosomal housekeeping genes (McFadden *et al.*, 1997; Zauner *et al.*, 2000) or structural genes like histones or tubulins; essentially all red algal metabolic genes were lost. Several nucleomorph-encoded proteins are probably imported into the plastid, explaining nucleomorph retention. Import of nuclear-encoded proteins to chromist chloroplasts involves presequences with initial co-translational import into the RER lumen. The most direct tests of chromist monophyly are to elucidate the unknown mechanisms of transfer across the periplastid membrane, which should be the same in all three groups (Cavalier-Smith, 1999), and to determine whether heterokont and cryptomonad flagellar hair proteins are homologous, as predicted (Cavalier-Smith, 1986b).

Differential losses and gains of pigments; losses of the posterior flagellum or phagotrophy; losses of flagellar hairs associated with the origin of the haptonema (Cavalier-Smith, 1994); and changes in mitochondrial morphology and cortical structures diversified the three phyla of chromistan phytoflagellates (Cavalier-Smith, 1986b). Although most chromistan flagellates are phytoflagellates, a few are zooflagellates: goniomonads, related to cryptomonads but lacking plastids and nucleomorphs; and four heterokont groups: pedinellids, oikomonads, bicoeceans and bigyromonads (Cavalier-Smith, 1997b). Oikomonads and heterotrophic pedinellids clearly evolved by plastid loss (Cavalier-Smith *et al.*, 1995, 1996a). The others must have done so also if chromistan algae arose by one symbiogenesis, but this remains controversial: goniomonads and early diverging heterokont zooflagellates are sometimes postulated to lack plastids ancestrally, and regarded as host relatives for independent symbiotic acquisitions of red algae by cryptophytes and ochrophytes (McFadden *et al.*, 1994b; Leipe *et al.*, 1996). The bigyran zooflagellate *Developayella* and Pseudofungi branch marginally within ochrophyte algae in Figure 17.1, but with some taxon samples go just below them: the failure of *Proteromonas/Blastocystis* to group with other Bigyra is, I suggest, a long branch artefact.

17.10 Alveolate evolution

The monophyly of Alveolata is widely accepted. Morphologically the basal group is the dinozoan Protalveolata, postulated to be ancestral to Dinoflagellata, Sporozoa and Ciliophora (Cavalier-Smith, 1991c). Their multiploid macronuclei enabled ciliate cell size to increase vastly compared with protalveolates, yet maintain quite rapid multiplication (Cavalier-Smith, 1985); with ciliary multiplication and more complex cortical differentiation, they established a novel adaptive zone.

17.10.1 Origin and diversification of dinozoan flagellates: Miozoa now a phylum

Molecular trees confirm that Dinoflagellata and Sporozoa are much more closely related than either is to Ciliophora. It has been problematic whether to group the flagellates *Perkinsus* and *Colpodella* with Sporozoa (Cavalier-Smith, 1993a) or Dinozoa (Cavalier-Smith, 1998a,b). The arbitrariness of the boundary between these two phyla is emphasized by the close grouping of *Dinomonas vorax* (Figure 17.1), with discrete cortical alveoli like Dinozoa (Mylnikov, 1991), with Sporozoa, and by the discovery of plastids in Sporozoa (McFadden *et al.*, 1995). I have therefore decreased Dinozoa and Sporozoa in rank to subphylum, and Miozoa (the taxon embracing both: Cavalier-Smith, 1987a) from superphylum to phylum (Cavalier-Smith, 1999).

As *Perkinsus* branches more deeply than other Miozoa, the original taxon Apicomplexa grouping *Perkinsus* with Sporozoa (Levine, 1970) is probably polyphyletic. The ancestor of Sporozoa was probably a predatory protalveolate flagellate like *Colpodella*, the cell cortex of which is more similar to Sporozoa than either *Perkinsus* or *Dinomonas*.

17.10.2 The chromalveolate theory of miozoan plastid origins

Evidence for a secondary symbiogenetic origin of dinoflagellate and sporozoan plastids is reviewed elsewhere (Cavalier-Smith, 1999); this includes indications that nuclear-encoded plastid proteins are targeted to the chloroplast via the Golgi in both groups, and that coccidiomorph Sporozoa (either sisters of, or more likely derived from, Dinozoa) have plastids with four envelope membranes (Köhler *et al.*, 1997).

It is commonly assumed that dinoflagellates and sporozoans obtained their plastids by separate symbiogenetic events, both distinct from the one that created Chromista (Palmer and Delwiche, 1998). However, peridinin-containing dinoflagellate chloroplast rRNA is distinctly similar to that of Sporozoa, as are their chloroplast-encoded proteins with those of red algae and chromists (Zhang *et al.*, 1999). I recently argued that chromophyte algae (those with chlorophyll *c*) are all directly related, and that Alveolata and Chromista are sister groups and form a clade designated the chromalveolates (Cavalier-Smith, 1999). I argue that the ancestral chromalveolate was photosynthetic, obtained its plastid from a single red alga, and evolved chlorophyll c_2 and chlorophyll *c*-binding proteins (which cluster on trees: Durnford *et al.*, 1999) prior to the divergence of chromists and alveolates, which must have taken place before the perialgal vacuole fused with the nuclear envelope to create Chromista by placing the plastid and periplastid membrane within the RER lumen. On this theory, miozoans retained the ancestral chromalveolate protein-targeting via the Golgi, the evolution of which was the key step in the chromalveolate symbiogenesis. The red algal nucleus was lost independently by miozoans and chromobiotes, and dinoflagellates lost the periplastid membrane, unlike chromists and sporozoa. This chromalveolate theory greatly reduces the number of secondary symbioses, each involving thousands of mutations, that need be postulated, and is thus very parsimonious. It relates specifically to typical peridinin-containing dinoflagellate plastids, not the few with fucoxanthin obtained in independent symbioses from diatoms (Chesnick *et al.*, 1996) or haptophytes (Delwiche, personal communication) or the diverse temporarily harboured kleptochloroplasts.

Even if peridinin-containing dinoflagellates arose instead by an independent secondary symbiosis (Gibbs, 1981), one would still have to postulate several plastid losses within Dinoflagellata given present rRNA trees (Saunders *et al.*, 1997); if as argued (Cavalier-Smith, 1999) Alveolata are ancestrally photosynthetic, one must postulate a few more losses (such as in an ancestor of ciliates). Since acceptance of the symbiogenetic theories of the origins of mitochondria and plastids, there has been a strong prejudice against postulating losses of these organelles. But one lesson of molecular phylogeny is that losses of mitochondria have been many times more common (probably about thirteen times) than their symbiotic origin (only once).

There is no a priori reason to think that plastid loss is rarer. We already have phylogenetic evidence for at least three losses within ochrophyte algae (Cavalier-Smith *et al.*, 1996a), one or two within euglenoids (Linton *et al.*, 1999), and several within dinoflagellates (Saunders *et al.*, 1997, and Figure 17.1): the eventual number may be over twenty (Cavalier-Smith, 1992b). I expect at least thirty chloroplast losses by phytoflagellates to be eventually demonstrated.

APPENDIX

Classification of the flagellate-containing kingdoms: Protozoa, Plantae and Chromista

All sixty classes with flagellates are listed; related taxa lacking flagellates are marked +. The number of flagellate orders is shown within square brackets. Probably paraphyletic taxa are marked with an asterisk (two if certainly paraphyletic).

Kingdom 1 Protozoa** Owen 1858 [75]
  Subkingdom 1 Eozoa Cavalier-Smith 1997
    Infrakingdom 1 Archezoa Cavalier-Smith 1983 em. 1998 stat. nov. 1999
      Phylum 1 Metamonada Grassé 1952 stat. nov. Cavalier-Smith 1981 em. [4]
        Class 1 Trepomonadea Cavalier-Smith 1993 (Diplomonads e.g. *Giardia, Hexamita*; Enteromonads e.g. *Enteromonas*) [3]
        Class 2 Retortamonadea Cavalier-Smith 1993 (*Retortamonas, Chilomastix*) [1]
      Phylum 2 Parabasalia Honigberg 1973 stat. nov. Cavalier-Smith 1981 [3]
        Class 1 Trichomonadea Kirby 1947 stat. nov. 1974 (e.g. *Trichomonas, Tritrichomonas, Dientamoeba*+) [1]
        Class 2 Hypermastigea Grassi et Foà 1911 stat. nov. Margulis 1974 (e.g. *Trichonympha, Calonympha*) [2]
    Infrakingdom 2 Loukozoa Cavalier-Smith 1999a [2]
      Phylum Loukozoa Cavalier-Smith 1999a
        Class 1 Anaeromonadea Cavalier-Smith 1997 (*Trimastix*) [1]
        Class 2 Jakobea Cavalier-Smith 1997 em. (order Histionida: *Jakoba, Histiona, Reclinomonas*) [1]
  Subkingdom 2 Neozoa** Cavalier-Smith 1993 stat. nov. 1997 em. 1999
    Infrakingdom 1 Sarcomastigota** Cavalier-Smith 1983 stat. nov. em. 1999
      Phylum 1 Cercozoa Cavalier-Smith 1998 [12]
        Subphylum 1 Phytomyxa Cavalier-Smith 1997 stat. nov.
          Class Phytomyxea Engler et Prantl 1897 orthogr. em. Cavalier-Smith 1993 (e.g. *Phagomyxa, Plasmodiophora*) [2]
        Subphylum 2 Reticulofilosa Cavalier-Smith 1997
          Class 1 Chlorarachnea Hibberd et Norris 1984 orthogr. em. Cavalier-Smith 1993 (e.g. *Chlorarachnion, Cryptochlora, Gymnochlora, Lotharella*) [1]
          Class 2 Proteomyxidea Lankester 1885 em. Cavalier-Smith 1993 (e.g. *Pseudospora, Leucodictyon*+) [2]
          Class 3 Spongomonadea cl. nov. (diagnosis as for order Spongomonadida Hibberd 1983 p. 534) (*Spongomonas, Rhipidodendron*) [1]
          Class 4 Athalamea Haeckel 1862 stat. nov. Lee 1990 (e.g. *Biomyxa, Gymnophrys*) [1]
        Subphylum 3 Monadofilosa Cavalier-Smith 1997
          Class 1 Sarcomonadea Cavalier-Smith 1993 stat. nov. 1995 em. 1997 (e.g. *Cercomonas, Massisteria, Bodomorpha, Heteromita, Metopion, Thaumatomonas, Thaumatomastix, Allas, Cryothecomonas*) [3]
          Class 2 Filosea+ Leidy 1879 em. Cavalier-Smith 1997 (e.g. *Vampyrella, Euglypha, Paulinella, Gromia*)

Class 3 Ramicristea Cavalier-Smith 1997 stat. nov. (*Dimorpha, Tetradimorpha, Commation*) [2]
Phylum 2 Neomonada* Cavalier-Smith 1997 [11]
Subphylum 1 Apusozoa Cavalier-Smith 1997 em.
Class 1 Thecomonadea Cavalier-Smith 1993 stat. nov. 1995 em. (*Apusomonas, Amastigomonas, Ancyromonas, Spironema, Hemimastix*) [3]
Class 2 Phalansterea cl. nov. (diagnosis: as for order Phalansteriida: Hibberd 1983 p. 534) (*Phalansterium*) [1]
Class 3 Anisomonadea Cavalier-Smith 1993 em. (*Diphyllea, Collodictyon, Phagodinium*) [3]
Subphylum 2 Choanozoa* Cavalier-Smith 1981 stat. nov. 1997 em. 1998
Class 1 Choanoflagellatea Kent 1880 stat. nov. Cavalier-Smith 1998 (e.g. *Monosiga, Diaphanoeca*) [2]
Class 2 Ichthyosporea Cavalier-Smith 1998 (*Dermocystidium, Ichthyophonus, Psorospermium*) [2]
Class 3 Corallochytrea+ Cavalier-Smith 1995 (*Corallochytrium*)
Class 4 Cristidiscoidea+ Page 1987 stat. nov. Cavalier-Smith 1997 (e.g. *Nuclearia, Fonticula, Ministeria*)
Phylum 3 Amoebozoa Lühe 1913 stat. nov. em. Cavalier-Smith 1998 [3]
Subphylum 1 Lobosa Carpenter 1861 stat. nov. Cavalier-Smith 1997 em.
Class 1 Amoebaea+ Ehrenberg 1830 stat nov. Cavalier-Smith 1997 (e.g. *Amoeba, Acanthamoeba*)
Class 2 Testacealobosea+ De Saedeleer stat. nov. Cavalier-Smith 1997 (e.g. *Arcella, Difflugia*)
Class 3 Holomastigea Cavalier-Smith 1997 (*Multicilia*) (possibly belongs instead in Conosa) [1]
Subphylum 2 Conosa Cavalier-Smith 1998
Infraphylum 1 Archamoebae Cavalier-Smith 1983 stat. nov. 1998
Class 1 Pelobiontea Page 1976 stat. nov. 1991 (e.g. *Pelomyxa, Mastigamoeba*) [1]
Class 2 Entamoebea+ Cavalier-Smith 1991 (e.g. *Entamoeba+*)
Infraphylum 2 Mycetozoa De Bary 1859 stat. nov. Cavalier-Smith 1998
Class 1 Protostelea+ Olive et Stoianovitch 1966
Class 2 Myxogastrea Fries 1829 stat. nov. Cavalier-Smith 1993 (e.g. *Physarum+, Hyperamoeba*) [1]
Class 3 Dictyostelea+ Lister 1909 stat. nov. Cavalier-Smith 1993
Phylum 4 Heliozoa+ Haeckel 1886 stat. nov. Margulis 1974
Sarcomastigota incertae sedis: Discocelida Cavalier-Smith 1997 (*Discocelis*) [1]
Infrakingdom 2 Retaria+ Cavalier-Smith 1999
Phylum 1 Foraminifera+ Eichwald 1830 stat. nov. Margulis 1974 (incl. *Reticulomyxa*)
Phylum 2 Radiolaria+ Müller 1858 stat. nov. Cavalier-Smith 1999
Infrakingdom 3 Discicristata Cavalier-Smith 1998
Phylum 1 Percolozoa Cavalier-Smith 1991 [7]
Subphylum 1 Tetramitia Cavalier-Smith 1993
Class 1 Percolomonadea Cavalier-Smith 1993 (*Percolomonas*) [1]
Class 2 Heterolobosea Page et Blanton 1985 (e.g. *Naegleria*) [2]
Class 3 Lyromonadea Cavalier-Smith 1993 (*Psalteriomonas, Lyromonas*) [1]
Subphylum 2 Axostylaria Grassé 1952 stat. nov. em. Cavalier-Smith 1993
Class Oxymonadea Grassé 1952 stat. nov. Margulis 1974 (e.g. *Oxymonas, Pyrsonympha*) [2]
Subphylum 3 Pseudociliata Cavalier-Smith 1993
Class Pseudociliatea Cavalier-Smith 1981 (*Stephanopogon*) [1]
Phylum 2 Euglenozoa Cavalier-Smith 1981 [10]
Subphylum 1 Plicostoma Cavalier-Smith 1998
Class 1 Diplonemea Cavalier-Smith 1993 (*Diplonema, Rhynchopus*) [1]
Class 2. Euglenoidea Bütschli 1884 stat. nov. Pascher 1931 (e.g. *Euglena, Petalomonas, Peranema*; three classes of Cavalier-Smith (1993a) now subclasses) [6]

Subphylum 2 Saccostoma Cavalier-Smith 1998
Class 1 Kinetoplastea Honigberg 1963 stat. nov. Margulis 1974 (e.g. *Bodo, Parabodo, Cryptaulax, Trypanosoma, Leishmania*) [2]
Class 2 Postgaardea Cavalier-Smith 1998 (*Postgaardi, Calkinsia*) [1]
Infrakingdom 4 Alveolata Cavalier-Smith 1991 em. 1998
Phylum 1 Miozoa* Cavalier-Smith 1987 stat. nov. 1999 [18]
Subphylum 1 Dinozoa* Cavalier-Smith 1981 em. stat. nov. 1999
Infraphylum 1 Protalveolata* Cavalier-Smith 1991 em. stat. nov. 1999
Class 1 Colponemea Cavalier-Smith 1993 (e.g. *Colponema*) [1]
Class 2 Perkinsea Levine 1970 (*Perkinsus*) [1]
Class 3 Apicomonadea Cavalier-Smith 1993 (*Colpodella, Dinomonas*) [1]
Class 4 Ellobiopsea+ Loeblich III 1970 (e.g. *Ellobiopsis*)
Infraphylum 2 Dinoflagellata Bütschli 1885 stat. nov. Cavalier-Smith 1999
Class 1 Syndinea Chatton 1920 (e.g. *Amoebophrya, Hematodinium*) [1]
Class 2 Oxyrrhea Cavalier-Smith 1987 (*Oxyrrhis*) [1]
Class 3 Noctilucea Haeckel 1866 (e.g. *Noctiluca*) [1]
Class 4 Haplozooidea Poche 1911 (e.g. *Blastodinium, Amyloodinium*) [1]
Class 5 Peridinea Ehrenberg 1831 stat. nov. Wettstein 1901 (syn. Dinophyceae Pascher 1914; e.g. *Crypthecodinium, Amphidinium, Dinophysis, Heterocapsa, Gonyaulax*) [11]
Subphylum 2 Sporozoa+ Leuckart 1879 stat. nov. Cavalier-Smith 1999
Infraphylum 1 Gregarinae+ Haeckel 1866 stat. nov. Cavalier-Smith 1999
Infraphylum 2 Coccidiomorpha+ Doflein 1901 stat. nov. Cavalier-Smith 1999
Infraphylum 3 Manubrispora+ Cavalier-Smith 1998 stat. nov. 1999 (metchnikovellids)
Phylum 2 Ciliophora+ Doflein 1901 stat. nov. Copeland 1956 em. auct.
Neozoa incertae sedis:
Class 1 Telonemea Cavalier-Smith 1993 (*Telonema, Nephromyces, Kathablepharis, Leucocryptos*) [3]
Class 2 Ebridea Lemmerman 1901 em. Deflandre 1936 stat. nov. Loeblich III 1970 (e.g. *Ebria, Hermesinum*) [1]
Kingdom 2 Plantae Haeckel 1866 [9]
Subkingdom 1 Biliphyta* Cavalier-Smith 1981
Infrakingdom 1 Glaucophyta Cavalier-Smith 1998
Phylum Glaucophyta Skuja 1954 (unnecessary syn. Glaucocystophyta Kies et Kremer 1986) [1]
Class Glaucophyceae Bohlin 1901 (e.g. *Cyanophora, Cyanoptyche*) [1]
Infrakingdom 2 Rhodophyta+ Cavalier-Smith 1998
Phylum Rhodophyta+ Wettstein 1922 (red algae)
Subkingdom 2 Viridiplantae Cavalier-Smith 1981 (green plants)
Infrakingdom 1 Chlorophyta Cavalier-Smith 1993
Phylum Chlorophyta auct. [8]
Subphylum 1 Chlorophytina Cavalier-Smith 1998
Infraphylum 1 Prasinophytae* Cavalier-Smith 1998
Class 1 Micromonadophyceae* Mattox et Stewart 1984 em. (e.g. *Mamiella, Pyramimonas, Mesostigma*) [3]
Class 2 Nephrophyceae Cavalier-Smith 1993 (e.g. *Nephroselmis, Pseudoscourfieldia*)[1]
Infraphylum 2 Tetraphytae Cavalier-Smith 1998
Class 1 Chlorophyceae (e.g. *Chlamydomonas, Tetraselmis*) [4]
Class 2 Trebouxiophyceae+ (e.g. *Chlorella*)
Class 3 Ulvophyceae+ (e.g. *Ulva, Bryopsis, Codium, Acetabularia*)
Subphylum 2 Phragmophytina*+ Cavalier-Smith 1998 (*Charophyceae; Chaetophyceae*)
Infrakingdom 2 Cormophyta+ Endlicher 1836 (syn. Embryophyta auct.) stat. nov. Cavalier-Smith 1998

Phylum 1 Bryophyta+ Braun 1864 em. Eichler 1883 (hornworts, liverworts, mosses)
Phylum 2 Tracheophyta+ (vascular plants: pteridophytes, seed plants)
Kingdom 3 Chromista Cavalier-Smith 1981 [2]
Subkingdom 1 Cryptista Cavalier-Smith 1989
Phylum Cryptophyta Cavalier-Smith 1986 [3]
Class 1 Cryptomonadea Stein 1878 syn. Cryptophyceae Pascher 1914 (e.g. *Cryptomonas, Guillardia, Chilomonas*) [2]
Class 2 Goniomonadea Cavalier-Smith 1993 (*Goniomonas*) [1]
Subkingdom 2 Chromobiota Cavalier-Smith 1991
Infrakingdom 1 Heterokonta Cavalier-Smith 1986 stat. nov. 1995 em.
Superphylum 1 Sagenista Cavalier-Smith 1998
Phylum Sagenista Cavalier-Smith 1995 [4]
Subphylum 1 Bicoecia Cavalier-Smith 1989
Class Bicoecea Cavalier-Smith 1986 (e.g. *Bicosoeca, Cafeteria, Pirsonia, Caecitellus*) [3]
Subphylum 2 Labyrinthista+ Cavalier-Smith 1986 stat. nov. 1989 (e.g. *Labyrinthula+, Thraustochytrium+*)
Sagenista incertae sedis: Class Cyathobodonea Cavalier-Smith 1993 em. sole order Pseudodendromonadida Hibberd 1985 (*Cyathobodo, Pseudodendromonas, Adriamonas*) [1]
Superphylum 2 Gyrista Cavalier-Smith 1998
Phylum 1 Ochrophyta* Cavalier-Smith 1986 stat. nov. 1995 [12]
Subphylum 1 Phaeista Cavalier-Smith 1995
Infraphylum 1 Hypogyrista Cavalier-Smith 1995
Class 1 Pelagophyceae Andersen et Saunders 1993 (e.g. *Pelagomonas, Sarcinochrysis+*) [1]
Class 2 Actinochrysea Cavalier-Smith 1995 (e.g. *Pedinella, Ciliophrys, Dictyocha*) [3]
Infraphylum 2 Chrysista Cavalier-Smith 1986 stat. nov. 1995
Class 1 Raphidophyceae Chadefaud ex Silva 1980 (e.g. *Heterosigma*) [1]
Class 2 Eustigmatophyceae+ Hibberd et Leedale 1971 (e.g. *Vischeria*)
Class 3 Chrysophyceae Pascher ex Hibberd 1976 (e.g. *Ochromonas*) [4]
Class 4 Oikomonadea Cavalier-Smith 1996 (*Oikomonas*) [1]
Class 5 Chrysomerophyceae+ Cavalier-Smith 1995 (*Chrysomeris*)
Class 6 Phaeothamniophyceae+ Andersen et Bailey 1998 (e.g. *Stichogloea, Phaeothamnion*)
Class 7 Xanthophyceae Allorge ex Fritsch 1935 (e.g. *Chloromeson*) [1]
Class 8 Phaeophyceae+ Kjellman 1891
Subphylum 2 Khakista subphyl. nov. (diagnosis: Chloroplastum cum lamella cingenti et annulus e desoxyribo-nuclei-acidis. Sine radicibus ciliorum; pars transitoria ciliorum sine helice. With girdle lamella and annular chloroplast DNA but no transition helix or flagellar roots)
Class 1 Bolidophyceae Guillou et Chretiennot-Dinet 1999 [1]
Class 2 Diatomeae+ Dumortier 1821 stat. nov.
Phylum 2 Bigyra Cavalier-Smith 1998 [4]
Subphylum 1 Bigyromonada Cavalier-Smith 1998
Class Bigyromonadea Cavalier-Smith 1998 (*Developayella* Tong 1995) [1]
Subphylum 2 Pseudofungi+ Cavalier-Smith 1986 em. 1989 (oomycetes, hyphochytrids)
Subphylum 3 Opalinata Wenyon 1926 stat. nov. em. Cavalier-Smith 1993, 1997
Class 1 Proteromonadea Cavalier-Smith 1997 (*Proteromonas*) [1]
Class 2 Blastocystea+ Cavalier-Smith 1997 (*Blastocystis+*)
Class 3 Opalinea Wenyon 1926 (e.g. *Karotomorpha, Opalina*) [2]
Infrakingdom 2 Haptophyta Cavalier-Smith 1995
Phylum Haptophyta Cavalier-Smith 1986 [4]
Class 1 Pavlovophyceae Cavalier-Smith 1993 (e.g. *Pavlova*) [2]
Class 2 Prymnesiophyceae Hibberd 1976 (e.g. *Prymnesium*) [2]

ACKNOWLEDGEMENTS

I thank NSERC for a research grant and the Canadian Institute for Advanced Research for Fellowship support.

REFERENCES

Apt, K. E., Hofmann, N. E. and Grossman, A. R. (1993) The γ subunit of R-phycoerythrin and its possible mode of transport into the plastid of red algae. *Journal of Biological Chemistry*, **268**, 16208-16215.

Baldauf, S. L. and Doolittle, W. F. (1997) Origin and evolution of the slime moulds. *Proceedings of the National Academy of Sciences US*, **94**, 12007-12012.

Baldauf, S. L., Palmer, J. D. and Doolittle, W. F. (1996) The root of the universal tree and the origin of eukaryotes based upon elongation factor phylogeny. *Proceedings of the National Academy of Sciences US*, **93**, 7749–7754.

Beech, P. L., Heimann, K. and Melkonian, M. (1991) Development of the flagellar apparatus during the cell cycle in unicellular algae. *Protoplasma*, **164**, 23–37.

Bodyl, A. (1997) Mechanism of protein targeting to the chlorarachniophyte plastids and the evolution of complex plastids with four membranes – a hypothesis. *Botanica Acta*, **110**, 395–400.

Bütschli, O. (1880–1889) Protozoa. In *Klassen und Ordnung des Thier-Reichs, 1* (ed. H. G. Bronn) Leipzig: Winter, 1–2035.

Cavalier-Smith, T. (1975) The origin of nuclei and of eukaryote cells. *Nature*, **256**, 463–468.

Cavalier-Smith, T. (1978) The evolutionary origin and phylogeny of microtubules, mitotic spindles and eukaryote flagella. *BioSystems*, **10**, 93–113.

Cavalier-Smith, T. (1981a) The origin and early evolution of the eukaryotic cell. In *Molecular and Cellular Aspects of Microbial Evolution* (eds M. J. Carlile, J. F. Collins and B. E. B. Moseley), Cambridge: Cambridge University Press, pp. 33–84.

Cavalier-Smith, T. (1981b) Eukaryote kingdoms, seven or nine? *BioSystems*, **14**, 461–481.

Cavalier-Smith, T. (1982) The origins of plastids. *Biological Journal of the Linnaean Society*, **17**, 289–306.

Cavalier-Smith, T. (1983) Endosymbiotic origin of the mitochondrial envelope. In *Endocytobiology II* (eds W. Schwemmler and H. E. A. Schenk), Berlin: de Gruyter, pp. 265–279.

Cavalier-Smith, T. (1985) Cell volume and the evolution of eukaryotic genome size. In *The Evolution of Genome Size* (ed. T. Cavalier-Smith), Chichester: Wiley, pp. 105–184.

Cavalier-Smith, T. (1986a) Cilia versus undulipodia. *BioScience*, **36**, 293–294.

Cavalier-Smith, T. (1986b) The kingdom Chromista, origin and systematics. In *Progress in Phycological Research, vol. 4* (eds F. E. Round and D. J. Chapman), Bristol: Biopress, pp. 309–347.

Cavalier-Smith, T. (1987a) The origin of eukaryote and archaebacterial cells. *Annals of the New York Academy of Sciences*, **503**, 17–54.

Cavalier-Smith, T. (1987b) The origin of Fungi and pseudofungi. In *Evolutionary Biology of the Fungi* (eds A. D. M. Rayner, C. M. Brasier and D. M. Moore), Cambridge: Cambridge University Press, pp. 339–353.

Cavalier-Smith, T. (1987c). The simultaneous symbiotic origin of mitochondria, chloroplasts, and microbodies. *Annals of the New York Academy of Sciences*, **503**, 55–71.

Cavalier-Smith, T. (1990a) Microorganism megaevolution: integrating the living and fossil evidence. *Revue de Micropaléontologie*, **33**, 145–154.

Cavalier-Smith, T. (1990b) Symbiotic origin of peroxisomes. In *Endocytobiology IV* (eds P. Nardon, V. Gianinazzi-Pearson, A. M. Grenier, L. Margulis and D. C. Smith), Paris: Institut National de la Recherche Agronomique, pp. 515–521.

Cavalier-Smith, T. (1991a) Archamoebae: the ancestral eukaryotes? *BioSystems*, **25**, 25–38.

Cavalier-Smith, T. (1991b) Intron phylogeny: a new hypothesis. *Trends in Genetics*, **7**, 145–148.

Cavalier-Smith, T. (1991c) Cell diversification in heterotrophic flagellates. In *The Biology of Free-Living Heterotrophic Flagellates* (eds. D. J. Patterson and J. Larsen), Oxford: Clarendon Press, pp.113–131.

Cavalier-Smith, T. (1992a) Origin of the cytoskeleton. In *The Origin and Evolution of the Cell* (eds H. Hartman and K. Matsuno), Singapore: World Scientific, pp. 79–106.

Cavalier-Smith, T. (1992b) The number of symbiotic origins of organelles. *BioSystems*, **28**, 91–106.

Cavalier-Smith, T. (1993a) Kingdom Protozoa and its 18 phyla. *Microbiological Reviews*, **57**, 953–994.

Cavalier-Smith, T. (1993b) The origin, losses and gains of chloroplasts. In *Origin of Plastids: Symbiogenesis, Prochlorophytes and the Origins of Chloroplasts* (ed. R. A. Lewin), New York: Chapman and Hall, pp. 291–348.

Cavalier-Smith, T. (1993c) Percolozoa and the symbiotic origin of the metakaryote cell. In *Endocytobiology V* (eds H. Ishikawa, M. Ishida and S. Sato), Tübingen: Tübingen University Press, pp. 399–406.

Cavalier-Smith, T. (1994) Origin and relationships of Haptophyta. In *The Haptophyte Algae* (eds J. C. Green and B. S. C. Leadbeater), Oxford: Clarendon Press, pp. 413–435.

Cavalier-Smith, T. (1995) Membrane heredity, symbiogenesis, and the multiple origins of algae. In *Biodiversity and Evolution* (eds R. Arai, M. Kato and Y. Doi), Tokyo: National Science Museum Foundation, pp. 75–114.

Cavalier-Smith, T. (1997a) Amoeboflagellates and mitochondrial cristae in eukaryote evolution: megasystematics of the new protozoan subkingdoms Eozoa and Neozoa. *Archiv für Protistenkunde*, **147**, 237–258.

Cavalier-Smith, T. (1997b) Sagenista and Bigyra, two phyla of heterotrophic heterokont chromists. *Archiv für Protistenkunde*, **148**, 253–267.

Cavalier-Smith, T. (1998a) A revised six-kingdom system of life. *Biological Reviews*, **73**, 203–266.

Cavalier-Smith, T. (1998b) Neomonada and the origin of animals and fungi. In *Evolutionary Relationships Among Protozoa* (eds G. H. Coombs, K. Vickerman, M. A. Sleigh and A. Warren), London: Kluwer, pp. 375–407.

Cavalier-Smith, T. (1999) Principles of protein and lipid targeting in secondary symbiogenesis: euglenoid, dinoflagellate, and sporozoan chloroplast origins and the eukaryote family tree. *Journal of Eukaryotic Microbiology*, **46**, 347–366.

Cavalier-Smith, T. (2000) What are Fungi? In *The Mycota, vol. VII* (ed. D. J. McLaughlin), Springer-Verlag, in press.

Cavalier-Smith, T. and Chao, E. E. (1995) The opalozoan *Apusomonas* is related to the common ancestor of animals, fungi, and choanoflagellates. *Proceedings of the Royal Society of London Series B*, **261**, 1–6.

Cavalier-Smith, T. and Chao, E. E. (1996) Molecular phylogeny of the free-living archezoan *Trepomonas agilis* and the nature of the first eukaryote. *Journal of Molecular Evolution*, **43**, 551–562.

Cavalier-Smith, T. and Chao, E. E. (1997) Sarcomonad ribosomal RNA sequences, rhizopod phylogeny, and the origin of euglyphid amoebae. *Archiv für Protistenkunde*, **147**, 227–236.

Cavalier-Smith, T. and Chao, E. E. (1999) *Hyperamoeba* rRNA phylogeny and the classification of the phylum Amoebozoa. *Journal of Eukaryotic Microbiology*, **46**, Suppl. 5A.

Cavalier-Smith, T. Chao, E. E. and Allsopp, M. T. E. P. (1995) Ribosomal RNA evidence for chloroplast loss within Heterokonta, pedinellid relationships and a revised classification of ochristan algae. *Archiv für Protistenkunde*, **145**, 209–220.

Cavalier-Smith, T., Chao, E. E., Thompson, C. and Hourihane, S. (1996a) *Oikomonas*, a distinctive zooflagellate related to chrysomonads. *Archiv für Protistenkunde*, **146**, 273–279.

Cavalier-Smith, T., Couch, J. A., Thorsteinsen, K. E., Gilson, P., Deane, J., Hill, D. A. and McFadden, G. I. (1996b) Cryptomonad nuclear and nucleomorph 18S rRNA phylogeny. *European Journal of Phycology*, **31**, 315–328.

Clark, G. and Roger, A. (1995) Direct evidence for secondary loss of mitochondria in *Entamoeba histolytica. Proceedings of the National Academy of Sciences US*, **92**, 6518–6521.

Chesnick, J. M., Kooistra, W. H., Wellbrock, U. and Medlin, L. K. (1997) Ribosomal RNA analysis indicates a benthic pennate diatom ancestry for the endosymbionts of the dinoflagellates *Peridinium foliaceum* and *Peridinium balticum* (Pyrrhophyta). *Journal of Eukaryotic Microbiology*, **44**, 314–320.

Dacks, J. and Roger, A. J. (1999) The first sexual lineage and the relevance of facultative sex. *Journal of Molecular Evolution*, **16**, 779–783.

Daugbjerg, N. and Andersen, R. A. (1997) Phylogenetic analyses of the rbcL sequences from haptophytes and heterokont algae suggest their chloroplasts are unrelated. *Molecular Biology and Evolution*, **14**, 1242–1251.

De Duve, C. and Wattiaux, R. (1966) Functions of lysosomes. *Annual Review of Physiology*, **28**, 435–492.

Douglas, S. (1998) Plastid evolution: origins, diversity, trees. *Current Opinion in Genetics and Development*, **8**, 655–661.

Douglas, S. E. and Penny, S. L. (1999) The plastid genome of the cryptomonad alga, *Guillardia theta*: complete sequence and conserved synteny groups confirm its common ancestry with red algae. *Journal of Molecular Evolution*, **48**, 236–244.

Durnford, D. G., Deane, J., Tan, S., McFadden, G. I., Gantt, E. and Green, B. R. (1999) A phylogenetic assessment of the eukaryotic light-harvesting antenna proteins. *Journal of Molecular Evolution*, **48**, 59–68.

Embley, T. M. and Hirt, R. P. (1998) Early branching eukaryotes? *Current Opinion in Genetics and Development*, **8**, 624–629.

Erickson, H. P. (1998) Atomic structure of tubulin and FtsZ. *Trends in Cell Biology*, **8**, 133–137.

Erwin, D., Valentine, J. and Jablonski, D. (1997) The origin of animal body plans. *American Scientist*, **85**, 126–137.

Faguy, D. M. and Doolittle, W. F. (1998) Cytoskeletal proteins: the evolution of cell division. *Current Biology*, **8**, R338–R341.

Fast, N. M. and Doolittle, W. F. (1999) *Trichomonas vaginalis* possesses a gene encoding the essential spliceosomal component, PRP8. *Molecular and Biochemical Parasitology*, **99**, 275–278.

Foissner, W., Blatterer, H. and Foissner, I. (1988) The Hemimastigophora (*Hemimastix amphikineta* nov. gen., nov. spec.), a new protistan phylum from Gondwanian soils. *European Journal of Protistology*, **23**, 361–383.

Fraser, C. M. and 32 others (1998) Complete genome sequence of *Treponema pallidum*, the syphilis spirochaete. *Science,* **281**, 375–382.

Gibbs, S. P. (1981) The chloroplasts of some algal groups may have evolved from endosymbiotic eukaryotic algae. *Annals of the New York Academy of Sciences*, **361**, 193–207.

Gray, M. W., Lang, B. F., Cedergren, R., Golding, G. B., Lemieux, C., Sankoff, D., Turmel, M., Brossard, N., Delage, E., Littlejohn, T. G., Plante, I., Rioux, P., Saint-Louis, D., Zhu, Y. and Burger, G. (1998) Genome structure and gene content in protist mitochondrial DNAs. *Nucleic Acids Research*, **26**, 865–878.

Gray, M. W., Burger, G. and Lang, B. F. (1999) Mitochondrial evolution. *Science*, **283**, 1476–1481.

Guillou, L., Chrétiennot-Dinet, M-J., Medlin, L. K., Claustre, H., Loiseaux-de Goër, S. and Vaulot, D. (1999) *Bolidomonas*: a new genus with two species belonging to a new algal class, the Bolidophyceae (Heterokonta). *Journal of Phycology*, **35**, 368–381.

Gupta, R. S. (1998) Protein phylogenies and signature sequences: a reappraisal of evolutionary relationships among Archaebacteria, Eubacteria, and eukaryotes. *Microbiology and Molecular Biology Reviews*, **62**, 1435–1491.

Haeckel, E. (1866) *Generelle Morphologie der Organismen*. Berlin: Reimer.

Häuber, M. M., Müller, S. B., Speth, V. and Maier, U-G. (1994) How to evolve a complex plastid? A hypothesis. *Botanica Acta*, **107**, 383–386.

Hashimoto, T., Sánchez, L. B., Shirakura, T., Müller, M. and Hasegawa, M. (1998). Secondary absence of mitochondria in *Giardia lamblia* and *Trichomonas vaginalis* revealed by valyl-tRNA synthetase phylogeny. *Proceedings of the National Academy of Sciences US*, **95**, 6860–6865.

Henneguy, L-F. (1898) Sur les rapports des cils vibratils avec les centrosomes. *Archives d'Anatomie, Microscopie et Morphologie Expérimentale*, **1**, 481–496.

Hibberd, D. J. (1983) Ultrastructure of the colourless zooflagellates *Phalansterium digitatum* Stein (Phalansteriida ord. nov.) and *Spongomonas uvella* Stein (Spongomonadida ord. nov.). *Protistologica*, **19**, 523–535.

Hirano, T. (1999) SMC-mediated chromosome mechanics: a conserved scheme from bacteria to vertebrates? *Genes and Development*, **13**, 11–19.

Hirt, R. P., Logsdon, J. M., Healy, B., Dorey, M. W., Doolittle, W. F. and Embley, T. M. (1999) Microsporidia are related to Fungi: evidence from the largest subunit of RNA polymerase II and other proteins. *Proceedings of the National Academy of Sciences US*, **96**, 580–585.

Ishida, K., Cao, Y., Hasegawa, M., Okada, N. and Hara, Y. (1997) The origin of chlorarachniophyte plastids, as inferred from phylogenetic comparisons of amino acid sequences of EF-Tu. *Journal of Molecular Evolution*, **45**, 682–687.

Ishida, K., Green, B. R. and Cavalier-Smith, T. (1999) Diversification of a chimaeric algal group, the chlorarachniophytes: phylogeny of nuclear and nucleomorph small subunit rRNA genes. *Molecular Biology and Evolution*, **16**, 321–331.

Jefferies, R. S. (1979) The origin of chordates: a methodological essay. In *The Origin of Major Invertebrate Groups* (ed. M. R. House), London: Academic Press, pp. 443–477.

Karpov, S. A. and Leadbeater, B. S. C. (1997) Cell and nuclear division in a freshwater choanoflagellate, *Monosiga ovata* Kent. *European Journal of Protistology*, **33**, 323–334.

Keeling, P. and Doolittle, W. F. (1996) α-tubulins from early diverging eukaryotic lineages: divergence and evolution of the tubulin family. *Molecular Biology and Evolution*, **13**, 1297–1305.

Keeling, P., Deane, J. A. and McFadden, G. I. (1998) The phylogenetic position of alpha- and beta- tubulins from the *Chlorarachnion* host and *Cercomonas* (Cercozoa). *Journal of Eukaryotic Microbiology*, **45**, 561–570.

Köhler, S., Delwiche, C. F., Denny, P. W., Tilney, L. G., Webster, P., Wilson, R. J. M., Palmer, J. D. and Roos, D. S. (1997) A plastid of probable green algal origin in apicomplexan parasites. *Science*, **275**, 336–342.

Lang, B. F., Burger, G., O'Kelly, C. J., Cedergren, R., Golding, G. B., Lemieux, C., Sankoff, D., Turmel, M. and Gray, M. W. (1997) An ancestral mitochondrial DNA resembling a eubacterial genome in miniature. *Nature*, **387**, 493–497.

Leipe, D. L., Gunderson, J. H., Nerad, T. A. and Sogin, M. L. (1993) Small subunit ribosomal RNA of *Hexamita inflata* and the quest for the first branch in the eukaryotic tree. *Molecular and Biochemical Parasitology*, **59**, 41–48.

Leipe, D. D., Tong, S. M., Goggin, C. L., Siemenda, S. B., Pieniazek, N. J. and Sogin, M. L. (1996). 16S-like rRNA sequences from *Developayella elegans*, *Labyrinthuloides haliotidis*, and *Proteromonas lacertae* confirm that the stramenopiles are a primarily heterotrophic group. *European Journal of Protistology*, **32**, 449–458.

Lenhossék, M. (1898) Ueber Flimmmerzellen. *Verhandlung der Anatomische Gesellschaft, Kiel*, **12**, 106–128.

Levine, N. D. (1970) Taxonomy of the Sporozoa. *Journal of Parasitology*, **56**, 208–209.

Linton, E. W., Hittner, D., Lewandowski, C., Auld, T. and Triemer, R. E. (1999) A molecular study of euglenoid phylogeny using small subunit rDNA. *Journal of Eukaryotic Microbiology*, **46**, 217–223.

Logsdon, J. M. (1998) The recent origins of spliceosomal introns revisited. *Current Opinion in Genetics and Development*, **8**, 637–648.

Margulis, L. (1970) *Origin of Eukaryotic Cells*. New Haven: Yale University Press.

Martin, W., Stoebe, B., Goremykin, V., Hansmann, S., Hasegawa, M. and Kowallik, K. V. (1998) Gene transfer to the nucleus and the evolution of chloroplasts. *Nature*, **393**, 162–165.

May, A. P. and Ponting, C. P. (1999) Integrin α and β 4-subunit domain homologues in cyanobacterial proteins. *Trends in Biochemical Sciences*, **24**, 12–13.

McFadden, G. I., Gilson, P. R., Hoffman, C. J. B., Adcock, G. J. and Maier, U-G. (1994a) Evidence that an amoeba acquired a chloroplast by retaining part of an engulfed eukaryotic alga. *Proceedings of the National Academy of Sciences US*, **91**, 3690–3694.

McFadden G. I., Gilson, P. and Hill, D. R. A. (1994b) *Goniomonas*: rRNA sequences indicate that this phagotrophic flagellate is a close relative of the host component of cryptomonads. *European Journal of Phycology*, **29**, 29–32.

McFadden, G. I., Reith, M. E., Munholland, J. and Lang-Unnasch, N. (1995) Plastid in human parasites. *Nature*, **381**, 482.

McFadden, G. I., Gilson, P. R., Douglas, S. E., Cavalier-Smith, T., Hofmann, C. J. B. and Maier, U-G. (1997) Bonsai genomics: sequencing the smallest eukaryotic genomes. *Trends in Genetics*, **13**, 46–49.

Mereschkowsky, C. (1910) Theorie der zwei Plasmaarten als Grundlage der Symbiogenesis, einer neuen Lehre von der Entstehung der Organismen. *Biologisches Centralblatt*, **30**, 278–303, 321–47, 353–67.

Mikrjukov, K. A. and Mylnikov, A. P. (1996) Protist *Multicilia marina* Cienk.: flagellate or heliozoon? *Doklady Akademii Nauk*, **346**, 136–139 (in Russian).

Mikrjukov, K. A. and Mylnikov, A. P. (1998) The fine structure of a carnivorous multiflagellar protist, *Multicilia marina* Cienkowski, 1881 (Flagellata incertae sedis). *European Journal of Protistology*, **34**, 391–401.

Moriya, S., Ohkuma, M. and Kudo, T. (1998) Phylogenetic position of symbiotic protist *Dinenympha exilis* in the hindgut of the termite *Reticulotermes speratus* inferred from the protein phylogeny of elongation factor 1 alpha. *Gene*, **210**, 221–227.

Müller, M. (1997) What are microsporidia? *Parasitology Today*, **13**, 455–456.

Mylnikov, A. P. (1991) The ultrastructure and biology of some representatives of order *Spiromonadida* (protozoa). *Zoologicheskii Zhurnal*, **70**, 5–15.

Norrander, J. M. and Linck, R. W. (1994) Tektins. In *Microtubules* (eds J. Hyams and C. W. Lloyd), New York: Wiley-Liss, pp. 201–220.

O'Kelly, C. J. (1993) The jakobid flagellates: structural features of *Jakoba, Reclinomonas* and *Histiona* and implications for the early diversification of eukaryotes. *Journal of Eukaryotic Microbiology*, **40**, 627–636.

Palmer, J. D. and Delwiche, C. F. (1998) The origin and evolution of plastids and their genomes. In *Molecular Systematics of Plants II. DNA Sequencing* (eds D. E. Soltis, P. S. Soltis and J. J. Doyle), Norwall, Mass.: Kluwer, pp. 375–409.

Patterson, D. J. and Zölffel, M. (1991) Heterotrophic flagellates of uncertain taxonomic position. In *The Biology of Free-Living Heterotrophic Flagellates* (eds D. J. Patterson and J. Larsen) , Oxford: Clarendon Press, pp. 427–476.

Philippe, H. and Adoutte, A. (1998) The molecular phylogeny of Eukaryota: solid facts and uncertainties. In *Evolutionary Relationships Among Protozoa* (eds G. H. Coombs, K. Vickerman, M. A. Sleigh and A. Warren), London: Kluwer, pp. 25–56.

Philippe, H. and Forterre, P. (1999) The rooting of the universal tree of life is not reliable. *Journal of Molecular Evolution*, **4**, 509–523.

Raven, P. H. (1970) A multiple origin for plastids and mitochondria. *Science*, **169**, 641–646.

Roger, A. J., Svärd, S. G., Tovar, J., Clark, C. G., Smith, M. W., Gillin, F. D. and Sogin, M. L. (1998) A mitochondrial-like chaperonin 60 gene in *Giardia lamblia*: evidence that diplomonads once harboured an endosymbiont related to the progenitor of mitochondria. *Proceedings of the National Academy of Sciences US*, **95**, 229–234.

Saunders, G. W., Hill, D. R. A., Sexton, J. P. and Andersen, R. A. (1997) Small-subunit ribosomal RNA sequences from selected dinoflagellates: testing classical evolutionary hypotheses with molecular systematic methods. *Plant Systematics and Evolution* (Suppl.), **11**, 237–259.

Schwartzbach, S. D., Osafune, T. and Löffelhardt, W. (1998) Protein import into cyanelles and complex chloroplasts. *Plant Molecular Biology*, **38**, 247–263.

Simpson, A., Bernard, C., Fenchel, T. and Patterson, D. J. (1997) The organisation of *Mastigamoeba schizophrenia* n. sp.: more evidence of ultrastructural idiosyncrasy and simplicity in pelobiont protists. *European Journal of Protistology*, **33**, 87–98.

Simpson, G. G. (1944) *Tempo and Mode in Evolution*. New York: Columbia University Press.

Simpson, G. G. (1953) *The Major Features of Evolution*. New York: Columbia University Press.

Stanier, R. (1970) Some aspects of the biology of cells and their possible evolutionary significance. *Symposia of the Society for General Microbiology*, **20**, 1–38.

Stewart, K. D. and Mattox, K. (1980) Phylogeny of phytoflagellates. In *Phytoflagellates* (ed. E. Cox), New York: Elsevier/North Holland, pp. 433–462.

Stiller, J. W., Duffield, E. C. S. and Hall, B. D. (1998) Amitochondriate amoebae and the evolution of DNA-dependent RNA polymerase II. *Proceedings of the National Academy of Sciences US*, **95**, 11769–11774.

Taylor, F. J. R. (1974) Implications and extensions of the serial endosymbiosis theory of the origin of eukaryotes. *Taxon*, **23**, 229–258.

Whatley, J. M., John, P. and Whatley, F. R. (1979) From extracellular to intracellular: the establishment of mitochondria and chloroplasts. *Proceedings of the Royal Society of London Series B*, **204**, 165–187.

Wolfe, G. R., Cunningham, F. X., Durnford, D., Green, B. R. and Gantt, E. (1994) Evidence for a common origin for all plastids and their differently pigmented light-harvesting complexes. *Nature*, **367**, 566–568.

Wright, M., Moisand, A. and Mir, L. (1980) Centriole maturation in the amoebae of *Physarum polycephalum*. *Protoplasma*, **105**, 119–160.

Zauner, S., Fraunholz, M., Wastl, J., Penny, S., Beaton, M., Cavalier-Smith, T., Maier, U-G. and Douglas, S. (2000) Chloroplast protein and centrosomal genes, a tRNA intron, and odd telomeres in an unusually compact eukaryotic genome: the cryptomonad nucleomorph. *Proceedings of the National Academy of Sciences US*, **97**, 200–205.

Zhang, Z., Green, B. R. and Cavalier-Smith, T. (1999) Single gene circles in dinoflagellate chloroplast genomes. *Nature*, **400**, 155–159.

Index

Note: Several groups of organisms are indexed under various terms, formal and informal (e.g. dinoflagellates, Dinoflagellida, Dinophyta). This reflects both the current fluid state of protistan taxonomy and the usage by the authors of the various chapters.

Systematics Association Publications

1. Bibliography of key works for the identification of the British fauna and flora, 3rd edition (1967)†
Edited by G. J. Kerrich, R. D. Meikie and N. Tebble
2. Function and taxonomic importance (1959)†
Edited by A. J. Cain
3. The species concept in palaeontology (1956)†
Edited by P. C. Sylvester-Bradley
4. Taxonomy and geography (1962)†
Edited by D. Nichols
5. Speciation in the sea (1963)†
Edited by J. P. Harding and N. Tebble
6. Phenetic and phylogenetic classification (1964)†
Edited by V. H. Heywood and J. McNeill
7. Aspects of Tethyan biogeography (1967)†
Edited by C. G. Adams and D. V. Ager
8. The soil ecosystem (1969)†
Edited by H. Sheals
9. Organisms and continents through time (1973)†
Edited by N. F. Hughes
10. Cladistics: a practical course in systematics (1992)*
P. L. Forey, C. J. Humphries, I. J. Kitching, R. W. Scotland, D. J. Siebert and D. M. Williams
11. Cladistics: the theory and practice of parsimony analysis (2nd edition) (1998)*
I. J. Kitching, P. L. Forey, C. J. Humphries and D. M. Williams

* Published by Oxford University Press for the Systematics Association
† Published by the Association (out of print)

Systematics Association Special Volumes

1. The new systematics (1940)
Edited by J. S. Huxley (reprinted 1971)
2. Chemotaxonomy and serotaxonomy (1968)*
Edited by J. C. Hawkes
3. Data processing in biology and geology (1971)*
Edited by J. L. Cutbill
4. Scanning electron microscopy (1971)*
Edited by V. H. Heywood
5. Taxonomy and ecology (1973)*
Edited by V. H. Heywood
6. The changing flora and fauna of Britain (1974)*
Edited by D. L. Hawksworth

7 . Biological identification with computers (1975)*
Edited by R. J. Pankhurst
8. Lichenology: progress and problems (1976)*
Edited by D. H. Brown, D. L. Hawksworth and R. H. Bailey
9. Key works to the fauna and flora of the British Isles and northwestern Europe, 4th edition (1978)*
Edited by G. J. Kerrich, D. L. Hawksworth and R. W. Sims
10. Modern approaches to the taxonomy of red and brown algae (1978)
Edited by D. E. G. Irvine and J. H. Price
11. Biology and systematics of colonial organisms (1979)*
Edited by C. Larwood and B. R. Rosen
12. The origin of major invertebrate groups (1979)*
Edited by M. R. House
13. Advances in bryozoology (1979)*
Edited by G. P. Larwood and M. B. Abbott
14. Bryophyte systematics (1979)*
Edited by G. C. S. Clarke and J. G. Duckett
15. The terrestrial environment and the origin of land vertebrates (1980)
Edited by A. L. Pachen
16. Chemosystematics: principles and practice (1980)*
Edited by F. A. Bisby, J. G. Vaughan and C. A. Wright
17. The shore environment: methods and ecosystems (2 volumes) (1980)*
Edited by J. H. Price, D. E. C. Irvine and W. F. Farnham
18. The Ammonoidea (1981)*
Edited by M. R. House and J. R. Senior
19. Biosystematics of social insects (1981)*
Edited by P. E. House and J-L. Clement
20. Genome evolution (1982)*
Edited by G. A. Dover and R. B. Flavell
21. Problems of phylogenetic reconstruction (1982)*
Edited by K. A. Joysey and A. E. Friday
22. Concepts in nematode systematics (1983)*
Edited by A. R. Stone, H. M. Platt and L. F. Khalil
23. Evolution, time and space: the emergence of the biosphere (1983)*
Edited by R. W. Sims, J. H. Price and P. E. S. Whalley
24. Protein polymorphism: adaptive and taxonomic significance (1983)*
Edited by G. S. Oxford and D. Rollinson
25. Current concepts in plant taxonomy (1983)*
Edited by V. H. Heywood and D. M. Moore
26. Databases in systematics (1984)*
Edited by R. Allkin and F. A. Bisby
27. Systematics of the green algae (1984)*
Edited by D. E. G. Irvine and D. M. John
28. The origins and relationships of lower invertebrates (1985)‡
Edited by S. Conway Morris, J. D. George, R. Gibson and H. M. Platt
29. Infraspecific classification of wild and cultivated plants (1986)‡
Edited by B. T. Styles

30. Biomineralization in lower plants and animals (1986)‡
Edited by B. S. C. Leadbeater and R. Riding
31. Systematic and taxonomic approaches in palaeobotany (1986)‡
Edited by R. A. Spicer and B. A. Thomas
32. Coevolution and systematics (1986)‡
Edited by A. R. Stone and D. L. Hawksworth
33. Key works to the fauna and flora of the British Isles and northwestern Europe, 5th edition (1988)‡
Edited by R. W. Sims, P. Freeman and D. L. Hawksworth
34. Extinction and survival in the fossil record (1988)‡
Edited by G. P. Larwood
35. The phylogeny and classification of the tetrapods (2 volumes) (1988)‡
Edited by M. J. Benton
36. Prospects in systematics (1988)‡
Edited by J. L. Hawksworth
37. Biosystematics of haematophagous insects (1988)‡
Edited by M. W. Service
38. The chromophyte algae: problems and perspective (1989)‡
Edited by J. C. Green, B. S. C. Leadbeater and W. L. Diver
39. Electrophoretic studies on agricultural pests (1989)‡
Edited by H. D. Loxdale and J. den Hollander
40. Evolution, systematics, and fossil history of the Hamamelidae (2 volumes) (1989)‡
Edited by P. R. Crane and S. Blackmore
41. Scanning electron microscopy in taxonomy and functional morphology (1990)‡
Edited by D. Claugher
42. Major evolutionary radiations (1990)‡
Edited by P. D. Taylor and G. P. Larwood
43. Tropical lichens: their systematics, conservation and ecology (1991)‡
Edited by G. J. Galloway
44. Pollen and spores: patterns of diversification (1991)‡
Edited by S. Blackmore and S. H. Barnes
45. The biology of free-living heterotrophic flagellates (1991)‡
Edited by D. J. Patterson and J. Larsen
46. Plant-animal interactions in the marine benthos (1992)‡
Edited by D. M. John, S. J. Hawkins and J. H. Price
47. The Ammonoidea: environment, ecology and evolutionary change (1993)‡
Edited by M.R. House
48. Designs for a global plant species information system (1993)‡
Edited by F. A. Bisby, G. F. Russell and R. J. Pankhurst
49. Plant galls: organisms, interactions, populations (1994)‡
Edited by M. A. J. Williams
50. Systematics and conservation evaluation (1994)‡
Edited by P. L. Forey, C. J. Humphries and R. I. Vane-Wright
51. The Haptophyte algae (1994)‡
Edited by J. C. Green and B. S. C. Leadbeater

52. Models in phylogeny reconstruction (1994)‡
Edited by R. Scotland, D. I. Siebert and D. M. Williams
53. The ecology of agricultural pests: biochemical approaches (1996)**
Edited by W. O. C. Symondson and J. E. Liddell
54. Species: the units of diversity (1997)**
Edited by M. F. Claridge, H. A. Dawah and M. R. Wilson
55. Arthropod relationships (1998)**
Edited by R. A. Fortey and R. H. Thomas
56. Evolutionary relationships among Protozoa (1998)**
Edited by G. H. Coombs, K. Vickerman, M. A. Sleigh and A. Warren
57. Molecular systematics and plant evolution (1999)
Edited by P. M. Hollingsworth, R. M. Bateman and R. J. Gornall
58. Homology and Systematics (2000)
Edited by R. Scotland and R. T. Pennington
59. The flagellates: unity, diversity and evolution (2000)
Edited by B. S. C. Leadbeater and J. C. Green

* Published by Academic Press for the Systematics Association
† Published by the Palaeontological Association in conjunction with Systematics Association
‡ Published by the Oxford University Press for the Systematics Association
** Published by Chapman & Hall for the Systematics Association

Milton Keynes UK
Ingram Content Group UK Ltd.
UKHW052022071024
449327UK00027B/2376